Laser-Plasma Interactions

Scottish Graduate Series

Laser-Plasma Interactions

D. A. Jaroszynski
University of Strathclyde, Scotland, UK

R. Bingham
STFC, Rutherford Appleton Laboratory
England, UK

R. A. Cairns
University of St Andrews, Scotland, UK

CRC Press
Taylor & Francis Group
Boca Raton London New York

CRC Press is an imprint of the
Taylor & Francis Group, an **informa** business

Cover Picture: Simulation of laser plasma wakefield acceleration using OSIRIS. We gratefully acknowledge the OSIRIS Consortium, consisting of UCLA, IST (Lisbon, Portugal), and USC, for the use of OSIRIS, and IST for providing access to the OSIRIS 2.0 framework.

CRC Press
Taylor & Francis Group
6000 Broken Sound Parkway NW, Suite 300
Boca Raton, FL 33487-2742

First issued in paperback 2017

© 2009 by The Scottish Universities Summer School in Physics
CRC Press is an imprint of Taylor & Francis Group, an Informa business

No claim to original U.S. Government works

ISBN-13: 978-1-58488-778-2 (hbk)
ISBN-13: 978-1-138-11436-4 (pbk)

Library of Congress Cataloging-in-Publication Data

Scottish Universities Summer School in Physics (60th : 2005 : St Andrews, Scotland)
　　Laser-plasma interactions / editors, Dino A. Jaroszynski, R. Bingham, R.A. Cairns.
　　　　p. cm. -- (SUSSP proceedings ; 60)
　　Includes bibliographical references and index.
　　ISBN 978-1-58488-778-2 (hardcover : alk. paper)
　　1. Laser Plasmas--Congresses. 2. Laser-plasma
　　interactions--Congresses. I. Jaroszynski, Dino A. II. Bingham, R. III. Cairns, R. A.
　　IV. Title. V. Series.

QC718.5.L3S38 2005
539.7'3--dc22　　　　　　　　　　　　　　　　　　　　　　　　　　2008054114

Visit the Taylor & Francis Web site at
http://www.taylorandfrancis.com

and the CRC Press Web site at
http://www.crcpress.com

SUSSP Schools

/continued

SUSSP Schools (continued)

Lecturers

Robert Bingham	Rutherford Appleton Laboratory & University of Strathclyde, UK
Chris Clayton	University of California, Los Angeles, USA
Mikhail Fedorov	A.M. Prokhorov General Physics Institute, Moscow, Russia
Ricardo Fonseca	Instituto Superior Técnico, Lisboa, Portugal
Valeri Goncharov	Laboratory for Laser Energetics, Rochester, NY, USA
Simon Hooker	Clarendon Laboratory, Oxford University, Oxford, UK
Steven Jamison	University of Abertay Dundee & Daresbury Laboratories, UK
Dino Jaroszynski	University of Strathclyde, SUPA, Glasgow, UK
Karl Krushelnick	Imperial College, London, UK & University of Michigan, USA
William L. Kruer	University of California, Davis, USA
Tito Mendona	Instituto Superior Técnico, Lisboa, Portugal
Nikolay Narozhnyi	Moscow Engineering Physics Institute, Moscow, Russia
Peter Norreys	Rutherford Appleton Laboratory, STFC, UK
Francesco Pegoraro	Università di Pisa, Pisa, Italy
Mordecai Rosen	Lawrence Livermore National Laboratory, Livermore, USA
Steven Rose	Imperial College, London, United Kingdom
Damian Swift	Los Alamos National Laboratory, Los Alamos, USA

Organising Committee

Dino Jaroszynski	University of Strathclyde	*Co-Director*
Mikhail Federov	A.M. Pokhorov General Physics Institute	*Co-Director*
Alan Cairns	St Andrews University	*Treasurer*
Robert Bingham	University of Strathclyde & RAL	*Scientific Secretary*
Allan Gillespie	University of Dundee	*Bursar*
David Jones	University of Strathclyde	*Social Organiser*
Niki Legge	Rutherford Appleton Laboratory	*SUSSP60 Secretary*
Gisle Mackenzie Smith	University of Strathclyde	*Social Secretary*
Sarah Nicholls	Rutherford Appleton Laboratory	*Web Design*

International Advisory Committee

Alan Hauer	Los Alamos National Laboratory
Robert McCrory	Laboratory for Laser Energetics, Rochester
Bill Kruer	Lawrence Livermore National Laboratory
Tito Mendonca	Institute Superior Tecnico, Lisboa
Robert Bingham	University of Strathclyde & Rutherford Appleton Laboratory

Contents

14 Fundamentals of ICF Hohlraums
Mordecai ("Mordy") D. Rosen
325

15 Dense Plasma Physics – the Micro Physics of Laser-Produced Plasmas
S. J. Rose
353

Preface

D.A. Jaroszynski, R. Bingham and R.A. Cairns

The 60th Scottish Universities Summer School in Physics, which was also a NATO Advanced Study Institute, was the 6th to be devoted to the topic of laser plasma interactions. In the ten years since the last school in 1995 exciting new developments have taken place in the area of laser plasma accelerators and numerical simulation techniques. This has been brought about by advances in laser technology. In particular the dream beam results that realized the potential for laser wakefields to produce mono-energetic electron beams in the 100's MeV to GeV energy range has opened up the field to a far greater audience. Ion acceleration to 10's and 100's of MeV energies is also possible with the new breed of terawatt to petawatt lasers opening up potential applications such as radiation light sources, nuclear physics medicine (oncology) and laboratory astrophysics. The intervening years has seen exciting new developments in laser fusion with the fast ignition concept where short pulse high intensity lasers are used to ignite a directly driven fusion pellet. A great deal of progress has also been made in numerical simulations with full 3-D particle in cell codes being able to model exact experimental parameters. The school also for the first time has an element of practical work involving the running of computational models based on the particle in cell (PIC) method. All students had access to a computing network and PIC codes. The school was held at John Burnet Hall, University of St Andrews, and attracted 49 Students from 19 Countries, up by 30% on previous years.

The organizing committee for the school was formed form a team led by two directors, Professor D.A. Jaroszynski from the University of Strathclyde and Prof. Mikhail Federov from the A.M. Pokhorov General Physics Institute, Russian Academy of Sciences, and Prof. Allan Gillespie (University of Dundee) as Bursar. To advise on the programme an International Advisory Committee was set up consisting of Prof. Alan Hauer (Los Alamos National Laboratory), Prof. Robert Mccrory (Laboratory for Laser Energetics - Rochester), Prof. Bill Kruer (Lawrence Livermore National Laboratory), Prof. Tito Mendonca (Institute Superior Tecnico, Lisboa, Portugal), Prof. R. Bingham (Rutherford Appleton Laboratory and the University of Strathclyde). Without the heroic efforts of the Local Organizing Committee, with Prof. R. Bingham as Scientific Secretary, Prof. D.A. Jaroszynski as Director, Prof. R.A. Cairns as Treasurer, Dr. David Jones (University of Strathclyde) as Social Organiser, Niki Legge (Rutherford Appleton Laboratory) as Secretary, Sarah Nicholls (Rutherford Appleton Laboratory) as Web site designer, and, last but not least, the most memorable events organized by Gisèle Mackenzie Smith (Strathclyde) as Social Secretary.

The organisers wish to acknowledge the assistance provided by Conference Services at St Andrews and especially Jackie Matthews and the rest of the staff at John Burnet Hall for their friendly and efficient service. Other staff at St Andrews to whom thanks are due for their help are Peter Lindsay, who set up computer access for the participants, and Tricia Watson who helped the Treasurer keep track of the accounts. We would also like to thank Dr. Enrico Brunetti for his efforts in compiling and assistance with editing the manuscript. The financial contributions of NATO, EPSRC, SUSSP, IOP, AWE, LLE Rochester are greatly appreciated.

The editors, Prof. D.A. Jaroszynski, Prof. R. Bingham and Prof. R.A. Cairns
University of Strathclyde, Rutherford Appleton Laboratory and St Andrews University, 1995.

1

Laser-Driven Plasmas: a Gateway to High Energy Density Physics and a Test-Bed for Nonlinear Science

William L. Kruer

University of California, Davis

Abstract

Laser-driven plasmas provide a key gateway to the study of high energy density physics. A brief introduction is given to this exciting field, which is rich both in physics and in applications. Laser plasma interactions are a key component of high energy density physics. These interactions are a fascinating and challenging test bed for nonlinear physics, as will be illustrated by several examples of particular interest to inertial fusion.

1 Introduction

High power lasers are proving to be remarkable tools for the creation and study of high energy density phenomena in the laboratory. Applications include inertial confinement fusion, laboratory astrophysics, compact particle accelerators and radiation sources, and many fundamental studies of dense, hot matter. A brief introduction will be given to this exciting field of high energy density physics, which is described in more detail in a recent National Academy report (Davidson et al., 2004).

Optimal use of high power lasers for the numerous applications motivate an ongoing strong effort to improve the understanding of laser matter interactions. These interactions are remarkably rich (Kruer, 1998), ranging from classical inverse bremmsstrahlung absorption and many laser-driven plasma instabilities, to strongly relativistic plasma phenomena. The interaction processes are coupled with energy transport phenomena, which are very complex for intense heat fluxes. Laser plasma interactions are proving to be a fascinating and challenging test bed for nonlinear physics, as will be illustrated by several examples of particular interest to ignition-scale hohlraum targets.

2   High energy density physics

The study of high energy density phenomena in the laboratory encompasses many different areas of science, including plasma physics, materials science and condensed matter physics, atomic and nuclear physics, fluid dynamics, magnetohydrodynamics, and astrophysics. The research has spawned many exciting applications, which are intellectual challenges due to the many complex and coupled nonlinear processes which occur in high energy density regimes. This is also a very opportune time for advances in high energy density physics for many reasons. A new generation of large experimental facilities which include lasers and other drivers, such as Z-pinches, are coming on line with unprecedented capabilities. A wide range of sophisticated diagnostics have been developed in the inertial fusion programme, and both space-based and ground-based instruments are measuring astrophysical phenomena with remarkable accuracy and detail. Lastly, the ongoing development of sophisticated computer models and the development of terascale (and beyond) computing provides a unique capability to model high energy density phenomena.

2.1   Definition and examples of high energy density physics

The criterion for high energy density is somewhat arbitrary and has been taken (Davidson et al, 2004) as an energy density exceeding 10^{11} J/m^3 (or 10^5 J/cm^3). Such an energy density corresponds to a pressure exceeding 1 Mbar. A bar is the pressure of the atmosphere under standard conditions and can be readily shown to correspond to .1 J/cm^3.

It is instructive to give several concrete examples of high energy density systems with a pressure of 1 Mbar. A plasma with an electron density of 6×10^{20} cm^{-3} and a temperature of 1 keV has a pressure of about 1 Mbar; i.e.,

$$p = nk_BT_e \tag{1}$$

$$p = 6 \times 10^{20} \frac{1}{\text{cm}^3} 1\text{keV} \frac{1.6 \times 10^{-16}\text{J}}{\text{keV}} \simeq 10^5 \frac{\text{J}}{\text{cm}^3}. \tag{2}$$

The light pressure of a laser beam with an intensity $I = 3 \times 10^{15}$ W/cm^2 is 1 Mbar; i.e.,

$$p = \frac{I}{c} = 3 \times 10^{15} \frac{\text{W}}{\text{cm}^2} \frac{1}{3 \times 10^{10}} \frac{\text{sec}}{\text{cm}} = 10^5 \frac{\text{J}}{\text{cm}^3}. \tag{3}$$

The pressure of a magnetic field $B = 5 \times 10^6$ Gauss is also about 1 Mbar; i.e.,

$$p = \frac{B^2}{8\pi} = \frac{\left(5 \times 10^6\right)^2}{8\pi} \frac{\text{ergs}}{\text{cm}^3} \frac{\text{Joule}}{10^7\text{ergs}} \simeq 10^5 \frac{\text{J}}{\text{cm}^3}. \tag{4}$$

Many additional examples are shown in Table 1.

Matter at high energy density is routinely investigated in inertial fusion experiments with high power lasers. In many of these experiments, plasma with an electron density of about 10^{21} cm^{-3} is heated to a temperature of many keV. With a temperature of 3 keV, such a plasma has a pressure of about 5 Mbar. In the simplest form of inertial fusion (so-called direct drive), intense laser light heats the outer layers of a capsule filled with DT fuel (Nuckolls et al., 1972). The capsule is compressed and heated by the ablation pressure (p_{ABL}) due to the blowing-off of heated matter. This pressure is roughly I/v_s, where I is the absorbed intensity and v_s is the sound velocity in the matter which is blowing off. For 0.35 μm laser light with an intensity $I = 10^{15}$ W/cm^2, the ablation pressure is about 100 Mbar. In the mainline approach to inertial fusion (Lindl, 1998), laser light is directed into a hohlraum and converted to x-rays, which are then used to ablate and compress the capsule. The ablation pressure is greater for x-rays, since

Energy Density Parameter Corresponding to $\sim 10^{11}$ J/m^3	Value
Pressure	1 Mbar
Electromagnetic radiation	
Electromagnetic wave (laser) intensity (l) $(p \sim l)$	3×10^{15} W/cm^2
Blackbody radiation temperature (T_{rad}) $(p \sim T_{rad}^{1/4})$	4×10^2 eV
Electric field strength (E) $(p \sim E^2)$	1.5×10^{11} V/m
Magnetic field strength (B) $(p \sim B^2)$	5×10^2 T
Plasma Pressure	
Plasma density (n) for a thermal temperature (T)	
of 1 keV $(p \sim nT)$	6×10^{26} m^{-3}
Plasma density (n) for an energy per particle temperature (T)	
of 1 GeV $(p \sim nT)$	6×10^{20} m^{-3}
Ablation Pressure	
Laser intensity (l) at 1 μm wavelength (λ) $(p \sim (l/\lambda)^{2/3})$	4×10^{12} W/cm^2
Blackbody radiation temperature (T_{rad}) $(p \sim T_{rad}^{3.5})$	75 eV

Table 1. *Examples of physical systems with a pressure of 1 Mbar (from Davison et al., 2004). The scaling of the pressure with the appropriate physical quantity is shown parenthetically in the first column.*

they penetrate to higher density. The ablated matter is then cooler, and thus the effective sound speed is lower. The ablation pressure for blackbody radiation with a temperature of 300 eV is about 100 Mbar. When a hot spot finally forms and ignites in the compressed fuel, its pressure is of order 10^{12} bars!

ICF physics is extremely rich and challenging. To achieve high gain, it is necessary to compress the DT fuel to a density of order $10^3 - 10^4$ times that of liquid DT. This requires an implosion where the radius of the capsule is reduced by a factor of about 20-30. The implosion is hydrodynamically unstable, since in effect a light fluid (the hot, ablating plasma) accelerates a heavier fluid. A great deal of research has been focused on understanding and controlling this Rayleigh-Taylor instability and related effects. Achieving high compression requires a quite smooth capsule to begin with, excellent symmetry in the ablation pressure driving the capsule, as well as careful pulse shaping and control of preheating of the fuel by high energy x-rays and electrons. Clearly, ICF research is rich in physics challenges on many topics, which include intense heat transport, radiation generation and transport, equation of state physics, nonlinear hydrodynamics, and a wide range of laser plasma physics.

As shown in Figure 1, matter at high energy density also exists in planetary interiors and in stars. Note that the pressure in planetary cores exceeds a Mbar. The pressure in stellar interiors exceeds 10^3 Mbars. There is still a great to be learned about our universe. For example, accurate models for planetary interiors require a better understanding of the equation of state of matter at high energy density.

Many astrophysical phenomena can be investigated in laboratory experiments with intense lasers (Remington, et al., 2000). These include turbulent hydrodynamics in supernovae, the equation of state of dense hydrogen in Jupiter, relativistic plasma behavior associated with gamma-ray bursts, and radiative hydrodynamic jets. As an example, let's consider supernovae explosions, which are the fate of stars with a sufficient mass. Stars first burn their hydrogen, then

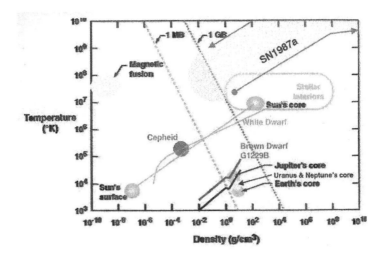

Figure 1. *A temperature versus density plot illustrating the conditions in planetary interiors and in stars.*

contract to become hotter in order to burn progressively heavier elements. When the fusion fuel is finally exhausted, the core of the star collapses and violently explodes, sending a very strong shock wave through the surrounding multi-layered structure. The dynamics of the explosion is unstable to the Rayleigh-Taylor instability, since in effect lighter fluids in the outer layers are trying to slow down heavier fluid blasting into them. Strongly turbulent hydrodynamic mixing results, as is apparent from the images of supernova remnants. Core ejection depends on this turbulent hydrodynamics, which needs to be better understood.

Supernova phenomena can be scaled (Ryutov et al., 1999) to laboratory experiments, at least in an ideal limit in which dissipation and heat transport are neglected. The dynamics are then described by Euler's equations, which represent the conservation of mass, momentum, and energy:

$$\frac{\partial \rho}{\partial t} + \nabla \cdot (\rho u) = 0 \tag{5}$$

$$\frac{\partial u}{\partial t} + u \cdot \nabla u = -\frac{\nabla p}{\rho} \tag{6}$$

$$\frac{\partial p}{\partial t} - \gamma_a \frac{p}{\rho} \frac{\partial \rho}{\partial t} + u \cdot \nabla p - \gamma_a \frac{p}{\rho} u \cdot \nabla \rho = 0 \tag{7}$$

$$p = \rho^{\gamma_a} \tag{8}$$

Here ρ is the mass density, u the velocity, and p the pressure, which obeys an adiabatic equation of state. It is straightforward to show that these equations are invariant under a scale transformation in which $\rho = a_1\rho$, $p = a_2p$, $h = a_3h$, and $t = a_3(a_1/a_2)^{1/2}t$. The key point is that one can scale from supernova explosions (for example) with a characteristic length scale of 10^{12} cm and time scale of 10^3 sec to laboratory experiments with a length scale of 100 μm and a time scale of 10 ns.

Such scaled experiments (Kane et al., 1999; Miles et al., 2004) have been carried out. In experiments on the Omega laser at the University of Rochester (Miles et al., 2004), a small disc was accelerated to a high velocity by a strong shock and then slowed down by passing it through a low density foam. The turbulent structure due to the Rayleigh-Taylor instability was diagnosed using x-ray backlighting and compared using the appropriate scaling factors with structure observed in SN1987A. Good agreement was also found with the structure predicted in detailed computer simulations.

As a final example of high energy density physics, let's consider some applications of short pulse, ultrahigh power lasers. These lasers are enabling numerous laboratory studies of relativistic plasmas, ultrahigh field physics, and the behavior of very intense electron and ion fluxes. Such lasers have been focused to an intensity of about 10^{21} W/cm^2. Then the light pressure is about 330 Gigabar, and the electron energy of oscillation in the transverse field of the laser is about 10 MeV! Computer simulations using so-called particle-in-cell codes are ideal for exploring such strongly relativistic and nonlinear behavior. In fact, some of the early simulations (Wilks et al., 1992) were carried out in order to motivate building the first Petawatt laser using a beam line of the Nova laser at the Lawrence Livermore National Laboratory. These ideal two dimensional simulations sent a focused 1.06 μm light beam with a peak intensity of 10^{19} W/cm^2 onto an overdense plasma slab. The simulations showed sizeable absorption (efficiency about 33%) into electrons with an effective temperature T_h of about 1 MeV (T_h scaled as approximately the square root of intensity), strong hole boring, high energy ion generation both directly by the ponderomotive force and indirectly by acceleration in the sheath at the plasma-vacuum interface, as well as very large self-generated magnetic fields (with a magnitude about 3×10^8 Gauss at $I = 10^{19}$ W/cm^2 and about 10^9 Gauss at $I = 10^{21}$ W/cm^2). All these effects have now been observed in experiments and are being studied in detail.

With ultra-intense lasers, a new approach to inertial fusion may be possible. In this so-called fast ignition (Tabak et al., 1994), compression of the fuel and creation of a hot spot are separated. The compression then requires a less sophisticated implosion. The fuel is lit by creating a hot spot in the compressed fuel using, say, 1 MeV electrons generated by a short pulse of ultra-intense laser light. The ideal hot spot radius R is simply the range of an alpha particle ($\rho R \sim .4$ g/cm^2, where ρ is the fuel density), which is also approximately the classical range of an electron with an energy of 1 MeV. As an example, take a compressed fuel density of 300 g/cm^3. R is then about 13 μm, and the energy required to raise the hot spot to 10 keV is about 3×10^3 J. This energy must be delivered in a disassembly time, which is $\sim R/v_s$, where v_s is the sound velocity ($\sim 10^8$ cm/sec). The required laser power is then > 300 terawatts, and the laser intensity is $> 6 \times 10^{19}$ W/cm^2.

Higher target gains are possible with fast ignition, since less energy has to be invested in compressing the fuel. However, a vital issue is the transport of the energetic electrons from the laser absorption region to the hot spot in the fuel. A flux of 10^{20} W/cm^2 of 1 MeV electrons corresponds to a current density of $\sim 10^{14}$ amps/cm^2, which can only propagate if highly neutralized by return currents in a dense plasma. The physics associated with the propagation of such intense fluxes in a plasma is quite complex and continues to be actively investigated. Various schemes to move the energy deposition region closer to the compressed fuel are being explored, including hole boring by a sacrificial laser beam and the use of cone targets (Norreys et al., 2000). The possibility of using proton beams is also being considered.

2.2 Enabling technology: high power lasers and other drivers

The impressive development of high power lasers over the last 30 years has given great impetus to the study of high energy density physics in the laboratory. On a bright sunny day, the power of the sun light striking the surface of the earth is about 2 Watts. With a magnifying glass, one can readily focus this sunlight to an intensity of about 10^2 W/cm^2, sufficient to start a leaf burning. In comparison, the National Ignition Facility (projected for completion in 2008) will deliver over 10^{14} Watts of laser power, which will be focused to intensities of about 10^{15} W/cm^2. According to calculations, this facility (NIF) can ignite a fusion burn and produce a miniature star in the laboratory (Lindl, 1998).

Figure 2 illustrates the remarkable development in high power lasers over the last few decades. The energy of high power lasers has increased from less than a joule to over a megajoule for the NIF laser now under construction. Laser power has progressed from megawatts to many hundreds of terawatts, a rapid increase analogous to that for computer power. In addition, short pulse, ultrahigh power lasers have been developed using so-called chirped pulse amplification (Perry and Mourou, 1994). These lasers have achieved an output power of up to a petawatt and have been focused to intensities of order 10^{21} W/cm^2. They are enabling studies of high field physics, relativistic plasmas, and a new approach to fusion, called fast ignition, briefly discussed above.

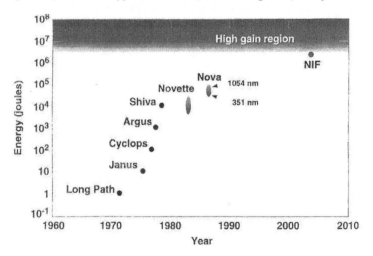

Figure 2. *The energy versus year of high power lasers developed in the inertial confinement fusion programme at Lawrence Livermore National Laboratory.*

Other drivers have also been developed to generate matter at high energy density. These include high current Z-pinches (Liberman et al., 1999), such as the 20 MA facility at Sandia National Laboratory. This facility has produced over 1 MJ of soft x-rays and has enabled a number of fundamental studies of matter at high energy density. Likewise, particle accelerators in high energy physics readily generate high energy density conditions. For example, a beam of 30 GeV electrons with a current density of 100 kA/cm^2 corresponds to an energy density of 10^5 J/cm^3 (or a pressure of 1 Mbar).

2.3 Enabling technology: diagnostics and integrated modeling

An impressive array of diagnostic tools are now available for investigating high energy density systems. Indeed, space-based telescopes continue to routinely reveal new marvels in the heavens. These instruments are now rivaled by ground-based telescopes using adaptive optics. In addition, several decades of research with lasers have led to the development of an impressive array of diagnostics for laboratory experiments. For example, scattered light, x-ray emission, and fusion burn products are routinely measured in detail. In addition, x-ray backlighting and Thomson scattering are commonly employed. Many features are diagnosed with a spatial resolution of 10's of microns and a temporal resolution of 10's of picoseconds. Figure 3 exhibits an excellent example of the impressive and detailed diagnostics now available. Shown here is a space and time-resolved measurement (Labaune et al., 1998) via Thomson scattering of a probe beam of the density fluctuations associated with both electron plasma waves and ion acoustic waves in a laser-irradiated plasma. Such measurements enable quite detailed comparisons with theory and simulations.

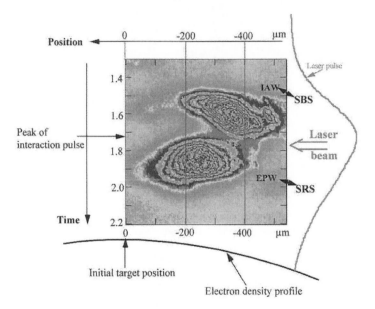

Figure 3. *A space and time-resolved measurement using Thomson scattering of a probe beam of the amplitudes of the electron plasma wave associated with stimulated Raman scattering and of the ion acoustic wave associated with stimulated Brillouin scattering in a laser-irradiated plasma (Labaune et al., 1998).*

Finally, integrated modeling is a very important part of high energy density physics. Integrated modeling (Zimmerman and Kruer, 1975) refers to the use and ongoing development of large computer codes for the modeling of the entire range of physical process taking place. These processes include hydrodynamics, heat transport, radiation production and transport, energy deposition, equation of state, and thermonuclear burn. Each of these processes is an important ongoing area of study in its own right. The codes are invaluable tools for the interpretation and design of experiments. They are a test bed for understanding the many physical processes occurring and their important interplay with one another. The codes are also very important for driving the ongoing development of improved models for the key processes and for testing models with experiments.

Integrated modeling continues to greatly benefit from the remarkable growth in computer power. This growth over the last 50 years is illustrated in Figure 4, which shows the peak speed (in floating point operations per second) of computers at the Lawrence Livermore National Laboratory versus year. Note the evolution from kiloflops to many hundreds of teraflops. The

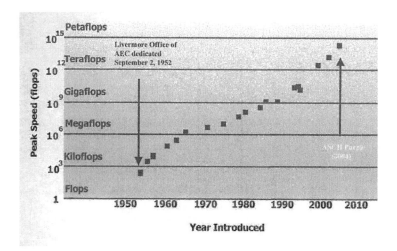

Figure 4. *The peak speed versus year introduced of computers at Lawrence Livermore National Laboratory.*

growth has been particularly impressive over the last decade. The peak speed is now approaching 10^{15} flops! This increased speed and associated increases in memory and storage capabilities are enabling more detailed modeling, including three-dimensional calculations. 3-D particle-in-cell codes, 3-D radiation-hydrodynamic codes, and 3-D wave propagation codes are now commonly used.

3   Laser plasma physics

There have been many developments in laser plasma physics since the last Scottish Universities' Summer School on Laser Plasmas in 1994. At that time, a key challenge in inertial fusion was to understand and control laser plasma coupling in Nova-scale hohlraum targets. This was especially important after the severe difficulties with laser plasma instabilities encountered in earlier experiments at Lawrence Livermore National Laboratory with the Shiva laser. Significant progress had been made in both characterizing and controlling the instabilities (Kruer, 1995). Key control mechanisms included use of shorter wavelength (i.e. 0.35 μm) laser light, laser beam smoothing, and plasma composition. Indeed, laser plasma coupling turned out to be quite good in Nova-scale hohlraum targets (Glenzer, et al., 1998), a success which helped to motivate proceeding to the construction of a much larger ($\sim$2 MJ) laser for ignition studies. The initial phase of this National Ignition Facility (NIF) has now successfully operated, and initial ignition experiments with the complete laser are currently scheduled for 2010.

Attention is now focused on understanding how the laser plasma coupling scales to the much larger regions of plasma in ignition-scale hohlraum targets. For NIF, the irradiated plasma will have a scale length of $\sim$ 4 − 5 mm rather than the scale length of $\sim$ 1 mm in Nova targets, and the plasma can be significantly hotter (electron temperature $\sim$ 5 keV rather than $\sim$ 3 keV). Additional interaction physics issues also arise due to the requirement for excellent implosion symmetry for an ignition capsule. Some representative examples of this more challenging inter-action physics will be discussed. The first two examples illustrate interaction processes which can impact the implosion symmetry in NIF hohlraums. The third example will illustrate the rich and complex nonlinear behavior of laser-driven instabilities and the importance of kinetic effects.

3.1 Energy transfer between crossing laser beams

Exchange of energy between laser beams as they cross one another is a potential issue (Kruer et al., 1995) for x-ray driven inertial fusion as well as an excellent test bed for understanding laser plasma interactions. An example of the irradiation pattern of a hohlraum target for the National Ignition Facility is shown in Figure 5. The 192 beams are arrayed in inner and outer cones directed to different locations within the hohlraum. The time-dependent symmetry of the drive on the fuel capsule is controlled by adjusting the relative power in the inner and outer cones. Note that the laser beams cross near the laser entrance holes. Clearly, energy exchange among the crossing laser beams can modify the symmetry.

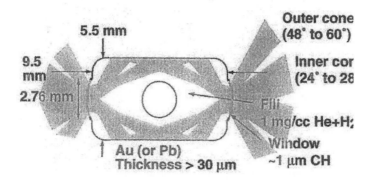

Figure 5. *An example of an ignition scale hohlraum being irradiated by inner and outer cones of beams.*

Energy transfer from one beam to another can occur through stimulated Brillouin scattering (SBS). Indeed, studies of this effect were initially motivated by the proposed use of a so-called four-colour scheme for laser beam smoothing. In this scheme, which was an attempt to enhance beam smoothing in order to reduce filamentation and other instabilities, each beam is composed of four distinct frequencies with a total frequency separation $\delta\omega$ of about $10^{-3}\omega_0$, where ω_0 is the average frequency. It is straightforward to see that induced Brillouin scattering can occur. The wave vector k and the frequency ω of the ion wave associated with scattering of one light wave into another (downshifted in frequency) light wave propagating at an angle θ_s are $k = 2k_0 \sin(\theta_s/2)$ and $\omega = kc_s$, where c_s is the ion sound speed and k_0 is the wave number of the pump wave. As shown in Figure 5, there is a range of angles of interest. For example, if we take $\theta_s = 50°$, then $k = .85k_0$. For a CH plasma with an electron density of $0.1n_{cr}$ (the critical density), an electron temperature of 4 keV, and an ion-electron temperature ratio of $0.2\omega/\omega_0 \approx 1.3 \times 10^{-3}$, which is approximately the proposed frequency separation between the highest and lowest frequency lines. For $\theta_s = 25°$, $k = 0.43k_0$, and $\omega/\omega_0 \approx 6.5 \times 10^{-4}$, which is also approximately resonant (with different lines).

It is clearly important then to better understand the interaction between crossing laser beams. More generally, experiments with crossing laser beams can allow well-controlled studies of stimulated scattering, which requires improved understanding for optimal target design in inertial fusion applications. Via independent variation of the frequency separation, polarization, and relative beam intensities, the instability can be studied in both linear and nonlinear regimes.

Let's here consider the simplest problem of two crossing laser beams with the lower frequency one with a much lower amplitude. To address the magnitude of the energy transfer, we compute the gain exponent Q using the coupled mode equations in the limit in which the plasma is uniform and the ion wave is heavily damped, giving $Q = Q_{max}I_G$. Here I_G is the factor which takes into account frequency mismatch, and Q_{max} is the maximum gain in a homogeneous plasma at perfect resonance

$$Q_{max} = \frac{\pi}{2} \frac{n}{n_{cr}} \left(\frac{v_{os}}{v_e}\right)^2 \frac{kc_s}{\nu} \frac{L}{\lambda_0} \frac{1}{(1 + 3T_i/ZT_e)\cos\theta_s}. \tag{9}$$

Here v_{os} (v_e) is the electron oscillation (thermal) velocity, T_i (T_e) is the ion (electron) temperature, Z is the ion charge state, n (n_{cr}) is the electron plasma (critical) density, λ_0 (k_0) is the laser wavelength (wave number), and θ_s is the angle between the laser beams. The wave number of the ion wave associated with the stimulated scattering is k, its energy damping rate is ν, and c_s is the ion sound speed. In the absence of plasma flow, I_G is the familiar Lorentzian:

$$I_G = \frac{1}{1 + 4(\Delta\omega/\nu)^2}, \tag{10}$$

where $\Delta\omega$ is the frequency mismatch.

For $\Delta\omega = 0$, the spatial gain rate is the same as that for backscatter, except for a geometrical factor. In this limit, the gain rate is simply

$$Q = \frac{\gamma_0^2}{2\nu v_{gz}}, \tag{11}$$

where γ_0 is the homogeneous growth rate in an undamped plasma and v_{gz} is the component of the group velocity of the scattered light wave along the direction of propagation of the pump wave. Since the square of the growth rate and the damping both depend on the wave number of the ion wave, there is no explicit dependence on k. The gain over a distance L is $\exp(QL)$. Corrections due to pump depletion are readily obtained by returning to the original coupled mode equations.

Given the discrete frequencies in, say, the four colour scheme, the induced scattering from one beam into another is not likely to be perfectly resonant. Detuning can significantly reduce the gain, as shown in Equation (10). The resonance width is approximately the amplitude damping rate $\nu/2$. For $\Delta\omega = \nu = 0.5kc_s$, the gain is reduced by a factor of about 5. What makes this induced scattering particularly dangerous is that a relatively small gain could lead to a significant transfer of energy from one beam to another, since the crossing beams have comparable amplitudes.

Energy transfer has been observed in experiments with crossing laser beams (Kirkwood et al., 1996). In these experiments using gas bag plasmas, a low intensity interaction beam was crossed at an angle of 53° with an intense laser beam, and the power transfer was monitored as a function of the frequency separation. As shown in Figure 6, the expected resonance was observed when the frequency separation was near the acoustic frequency. However, the maximum observed energy transfer corresponded to a gain of about 1, which was much less than the gain estimated for an ideal homogeneous plasma. This reduction in gain has been attributed to long-wavelength velocity modulations in the plasma, which act to detune the resonance. Many processes could produce significant velocity modulations in the plasma, such as the blast wave associated with the explosion of the skin of the gas bag. These modulations are apparent in hydrodynamic calculations. Other effects include ion waves launched by filamentation of the beams which ionize and heat the gas bag or even frequency changes associated with modifications of the velocity distributions of the ions and electrons.

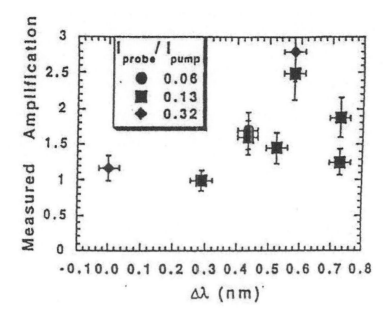

Figure 6. *The measured amplification of a probe beam as a function of the frequency separation (Kirkwood et al., 1996).*

Even though the four-colour scheme has been abandoned, energy transfer between crossing laser beams remains an issue for ignition-scale hohlraums. Induced scattering can take place even when the crossing beams have the same frequency, provided the beat fluctuation is Doppler-shifted by the plasma flow into resonance. This can in fact happen in the expanding plasma near the laser entrance holes. Such energy transfer between laser beams with the same frequency in a flowing plasma has also been confirmed in experiments (Kirkwood, et al., 2002) and continues to be studied in NIF hohlraum designs (Williams et al., 2004).

3.2 Laser beam deflection in nonlinearly generated plasma flow profiles

An improved understanding of how laser beams are deflected in plasmas is also important for inertial fusion and other applications. For example, in hohlraums filled with low density plasma, laser beams have been observed to have an intensity-dependent deflection in angle (Kauffman et al., 1998). Beam pointing must then be changed in order to recover symmetric implosions. Such a deflection has also been observed in experiments with preformed plasmas.

This beam bending has been attributed to filamentation in a plasma with a flow transverse to the laser beam (Rose, 1996; Hinkel et al., 1996). It is now well known that filamentation induces an angular spread in a laser beam. If the symmetry is broken, say, by a transverse plasma flow, this spreading can become asymmetrical; i.e., a mean deflection can occur. The dispersion relation for the ponderomotive filamentation instability in a flowing plasma becomes (Short et al., 1982)

$$\frac{\kappa^2}{k_\perp^2} = \frac{n}{8n_{cr}} \frac{v_{os}^2}{v_e^2} / \left(1 - \frac{v_\perp^2}{c_s^2} - \frac{k_\perp^2}{4k_0^2} \right). \tag{12}$$

Here κ is the spatial gain rate, $k_\perp$ the transverse wavenumber, $v_\perp$ the transverse flow velocity, c_s the ion sound speed, k_0 the wavenumber of the laser light, v_{os} the oscillation velocity of an electron in the laser electric field, v_e the electron thermal velocity, n the plasma density, and n_{cr} the critical density. It is assumed that $\kappa << k_\perp$. It is apparent that the spatial growth rate increases as $v_\perp$ approaches c_s. For $v_\perp \sim c_s$, $\kappa \sim k_\perp$, and one must solve the full dispersion relation, finding that the most unstable filaments grow at an angle to the direction of propagation of the laser beam.

In general, nonlinear deflection of a laser beam only depends on the existence of induced changes in the plasma density profile. Useful insights can be obtained by considering a simple model of a laser beam propagating transverse to a freely expanding plasma near its sonic point (Kruer and Hammer, 1997). We first estimate the nonlinear steepening of the density profile and then the laser beam deflection in this steepened density profile. Filamentation can modify the profile changes and the beam deflection, although we note that the profile changes can help stabilize filamentation.

As a simple model, let's consider a one-dimensional expanding plasma traversed by a laser beam with its propagation vector orthogonal to the gradient of the plasma density and expansion velocity. The beam radius is assumed to be much less than the plasma scale lengths, which in turn are sufficiently long to allow the neglect of refraction in the unperturbed profiles. The beam traverses a region of plasma that includes the sonic point. We anticipate that the plasma profile will be locally steepened in the neighborhood of the sonic density (n_s, where $u = c_s$) from a density $n_2 > n_s$ to a density $n_1 < n_s$. If we use the two-fluid model in a frame moving with the sonic density,

$$\frac{\partial}{\partial x} nu = 0 \tag{13}$$

$$\frac{\partial}{\partial x} \frac{u^2}{2} = -c_s^2 \frac{\partial}{\partial x} \ln n - \frac{Zm}{4M} \frac{\partial}{\partial x} v_w^2. \tag{14}$$

Here, u is the flow velocity, Z the ion charge state, m (M) the electron (ion) mass, and v_w is the oscillation velocity of an electron in the laser light electric field in the plasma. For $v_{os}/v_e << 1$, we readily find that the density profile is steepened over a density range $\Delta n \sim n_s(v_{os}/v_e)$ in a distance of about $2r_0$, where r_0 is the beam radius, v_{os} is the peak oscillatory velocity in the laser beam, and v_e is the electron thermal velocity. Note that Δn is proportional to the square root of the laser intensity, since the plasma is resonantly perturbed near the sonic point.

The deflection of the laser beam by refraction in the locally steepened density profile is now readily estimated. The equation for a light ray is

$$\frac{\partial^2}{\partial t^2} \xi = \frac{-c^2}{2n_{cr}} \nabla n, \tag{15}$$

where ξ is the ray displacement and c the velocity of light. If we approximate $\nabla n \sim n(v_{os}/v_e)/(2r_0)$, we estimate the deflection angle θ as

$$\theta \simeq \frac{l}{4r_0} \frac{n}{n_{cr}} \frac{v_{os}}{v_e}, \tag{16}$$

where $l = ct$ is approximately the distance traversed by the laser beam in the steepened plasma profile. Let's consider an ideal case in which there's a shear in the transverse flow velocity along the beam path. The flow is then only resonantly perturbed over a distance $l \sim l_v(v_{os}/v_e)$, where l_v is the shear velocity gradient length. For a laser beam (or a hot spot) with a half-width r_0, we then estimate

$$\theta \simeq \frac{l_v}{4r_0} \frac{n}{n_{cr}} \frac{v_{os}^2}{v_e^2}. \tag{17}$$

The basic physics of the nonlinear beam deflection was isolated and characterized in experiments using the Janus laser (Young et al., 1998). In these experiments the plasma was preformed by irradiating a thin disk with 0.53 μm laser light. The temperature, density, and flow profiles were then measured via interferometry and Thomson scattering. Finally, a 1.06 μm interaction beam with a well-characterized intensity profile was passed through the plasma at an angle to the plasma flow. Indeed, these experiments were motivated and initially designed using the predictions of the simple model just discussed. This model emphasizes that enhanced deflection can occur even in the absence of filamentation (which would of course enhance the bending) and provides a prediction for the deflection angle as a function of the plasma parameters.

The beam deflection was directly observed via interferometric measurements of the density channel made by the beam. The beam was focused with a cylindrical lens to facilitate these measurements. An interferogram is shown in Figure 7. The interaction beam, incident along the white arrow, indeed deflects at a relatively low density where the measured transverse flow is nearly sonic. The channel formed by the deflected beam is apparent and denoted by the second arrow. For this case with an interaction beam intensity of 1.5×10^{15} W/cm^2, the observed deflection is about $10°$ in agreement with the early predictions. In addition, the deflection is found to depend on intensity. It is also noteworthy that these experiments show deflection independent of laser beam filamentation, which is found to occur at another, significantly higher density in these experiments. Finally, these well-diagnosed experiments have provided a detailed test bed for a three-dimensional wave propagation code pF3D (Berger et al., 1998). With the measured plasma and laser conditions, calculations using this code were in good agreement with the experimental results.

Figure 7. *An interferogram which shows the channel formed by an interaction beam with a peak intensity of 1.5x 10^{15} W/cm^2 (Young et. al., 1998).*

Finally, note that the nonlinear beam deflection is especially important for laser beams with small scalelength spatial structure. For a hot spot in a 0.35 μm laser beam with a half-width of $20\lambda_0$ (λ_0 is the free-space wavelength) and an intensity of 4×10^{15} W/cm^2 in a plasma with a density of $0.1n_{cr}$, an electron temperature of 3 keV, and $l_v \sim 1500\lambda_0$, $\theta \sim 9°$. Hence it is important to reduce the fine-scale intensity structure with some form of temporal smoothing, such as the use of SSD (smoothing by spectral dispersion). A strong reduction of nonlinear beam deflection through the use of SSD has been confirmed in experiments (Glenzer, et al., 1998).

3.3   Kinetic effects on laser plasma instability evolution

Another exciting current issue is to better understand the role of kinetic effects on the saturation of stimulated Raman and Brillouin scattering in long scalelength plasmas. Instability thresholds and gains commonly depend explicitly on the Landau damping rate, which in turn is significantly modified by the distortion of the distribution functions of the electrons and ions. As an example, let us consider stimulated Brillouin scattering (SBS) on heavily damped ion waves, a regime very relevant to gas-filled hohlraum targets. The instability gain varies inversely with the Landau damping rate of ion acoustic waves. However, the ion Landau damping rate is only valid for a time $t \ll 1/\omega_{bi}$, where ω_{bi} is the bounce frequency of an ion in the potential trough of the ion sound wave. For example, for a wave with an amplitude of $\delta n/n = .01$ and typical plasma parameters appropriate for ignition targets, $t < 1$ ps. On a longer time scale, the slope of the ion distribution function flattens, and the Landau damping vanishes, leading to an enhancement of SBS. This effect has been demonstrated in simulations of SBS (Rambo et al., 1997) and also in simulations of stimulated Raman scattering (Vu et al., 2001), where the enhanced SRS is described as kinetic inflation.

In the simulations of SBS using a particle ion, fluid electron code, it has been shown that sometimes ion-ion collisions can maintain the slope of the ion distribution function and hence the ion Landau damping. Figure 8(a) shows the temporal evolution of the velocity perturbation associated with an initially excited ion sound wave in simulations with and without ion collisions. In this example, the plasma consists of an equal mixture of Au^{+50} and Be^{+4} with an electron temperature of 3 keV, an ion temperature of 1.5 keV, and an electron plasma density of $0.25n_{cr}$ for 0.35 μm light. The ion wave has a wave number twice that of the 0.35 μm light wave. Initially the ion wave Landau damps at the predicted rate. Without collisions, the damping ceases at $t \sim \pi/\omega_{bi}$. However, with collisions, the slope of the distribution function is preserved, and the damping continues until the amplitude of the ion wave is quite small. A snapshot of the distribution function of the ions in Figure 8(b) shows the distribution function distorting in the collisionless case but remaining nearly Maxwellian in the other case.

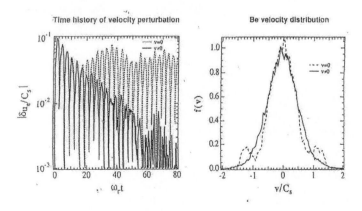

Figure 8. *a) The time history of the velocity perturbation δu_e of an initially excited ion sound wave with and without ion-ion collisions and (b) the velocity distribution function of the Be ions in the nonlinear state (Rambo et al., 1997).*

We next consider a case where the ion waves are self-consistently driven by SBS. The system consists of a plasma slab composed of the same Au/Be plasma just considered. Laser light with a wavelength of 0.35 μm and with an intensity of 4×10^{15} W/cm^2 is incident from the left side of the slab, which is composed of an active 11 μm long plasma region, buffered between fixed ion regions. The fixed ion regions prevent plasma expansion driven by the ambipolar field at the slab edge, which have linear density ramps to minimize reflection at the edges. The density profile shown in Figure 9(a) for the collisionless case illustrates the strong density fluctuations due to SBS. The time-dependent reflectivity is shown in Figure 9(b) for both the collisionless simulation and a simulation including ion-ion collisions. Without collisions, the reflectivity reaches a high value of $\sim 80\%$. Inspection of the ion phase space shows strong trapping of resonant ions in large amplitude ion waves with $\delta n/n \sim 15 - 20\%$. Note that the peak reflectivity is much less in the simulation with collisions. The ion waves saturate at a much lower level of $\delta n/n \sim 2 - 3\%$. Although the ion velocity distribution shows significant heating, the slope of the distribution is observed to remain monotonic.

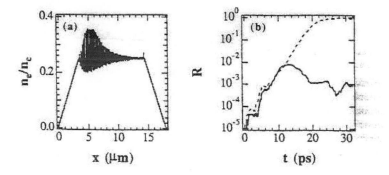

Figure 9. *Simulations of SBS in a Be/Au plasma. (a) Electron density from a collisionless simulation and (b) the time-dependent reflectivity from collisionless (dashed line) and collisional (solid line) simulations (Rambo et al., 1997).*

When the distribution flattens around the phase velocity of the wave, Landau damping vanishes, but a frequency shift arises. This frequency shift due to trapped particles is proportional to the bounce frequency of the particles trapped in the wave and can act to detune the instability driving the wave. For ion waves the frequency shift is due both to trapped ions and to trapped electrons. Indeed, for $ZT_e/T_i >> 1$, the latter frequency shift is dominant, because there are very few trapped ions, unless the wave amplitude becomes very large.

Trapped electron effects become important for ion waves of remarkably small amplitude. The electron bounce frequency in the potential trough of an ion sound wave with amplitude δn is $\omega_{be} = k v_e (\delta n/n)^{1/2}$, where k is the wave number, v_e the electron thermal velocity, and n the plasma density. Note that $\omega_{be} > \omega_{ia}$ (the ion wave frequency) when $\delta n/n > Zm/M$, where Z is the ion charge state, m the electron mass, and M the ion mass. Alternatively, the trapping velocity, $v_{tr} \sim \omega_{be}/k$, exceeds the sound speed under the same conditions. This flattening of the electron distribution function around the phase velocity of the sound wave occurs very rapidly and over a sizeable range of velocities. For example, for an ion wave with an amplitude of $\delta n/n = 5\%$ associated with SBS of a 0.35 μm light wave in a 5 keV plasma, $\omega_{be}^{-1} \sim 5$ fs and $v_{tr} \sim v_e/3!$ Clearly, the oft-quoted electron Landau damping of an ion wave in general only occurs for extremely small amplitude waves.

It is straightforward to estimate the frequency shift associated with electron trapping and

William L. Kruer

the concomitant flattening of the electron distribution function. The dispersion relation for an ion sound wave is $\epsilon(k, \omega) = 0$, where

$$\epsilon(k, \omega) = 1 + \chi_e(k, \omega) + \chi_i(k, \omega). \tag{18}$$

Here χ_e (χ_i) is the electron (ion) susceptibility. In particular,

$$\chi_e(k, \omega) = -\frac{\omega_{pe}^2}{k^2} \int \frac{\partial f_e}{\partial v} dv / (v - \omega/k), \tag{19}$$

where f_e is the electron distribution function. In accordance with the above discussion, we approximate $\partial f_e / \partial v = 0$ for velocities in the range $c_s \pm v_{tr}$. In the limit of $v_{tr} >> v_s$ and $k\lambda_{De} << 1$,

$$\epsilon(k, \omega) = \frac{1 - \text{erf}(\sqrt{\delta n / n})}{k^2 \lambda_{De}^2} - \frac{\omega_{pi}^2}{\omega^2}, \tag{20}$$

where erf is the error function, ω_{pi} is the ion plasma frequency, and λ_{De} is the electron Debye length. The frequency shift $\Delta\omega$ is then

$$\Delta\omega \cong 0.5 k c_s \sqrt{\delta n / n}. \tag{21}$$

This estimated frequency shift for $v_{tr} >> v_s$ is within a factor of two of the trapped electron frequency shift calculated for very low amplitude ion waves (Dewar, R., et al, 1972). Note that this frequency shift is positive, since the principal effect is that the trapped electrons no longer contribute to the shielding of the low frequency charge density fluctuation. In contrast, the frequency shift due to trapped ions is negative, since the trapped ions reduce the restoring force for the oscillation.

Finally, kinetic effects can play a key role in saturation models for laser plasma instabilities. Indeed, saturation of stimulated Raman scattering (SRS) by the frequency shift due to trapped electrons (Vu, et al., 2001) and of stimulated Brillouin scattering by the frequency shift due to trapped ions (Williams, E., et al., 2004) has been proposed. Even saturation models which invoke the excitation of secondary instabilities can be sensitive to kinetic effects. These models include the saturation of SRS by excitation of the Langmuir wave decay instability (Bezzerides et al., 1993) and of SBS by excitation of the two-ion wave decay instability (Cohen et al., 1997). For example, in both cases the threshold of the secondary instability depends on the Landau damping rate, which in turn depends on the slope of the distribution function in the nonlinear state. Kinetic effects can even introduce new modes of oscillation, such as kinetic electron electrostatic nonlinear waves (Afeyan, et al., 2003).

4   Summary

In summary, laser plasmas continue to be an exciting and remarkably fruitful field of research. This field has been given great impetus by the growth of the inertial confinement fusion program, both in the US and the rest of the world. Indeed, this core application is rapidly approaching an exciting time where it can be tested with mega-joule lasers which are being used to explore ignition physics. Many other applications have also blossomed, including compact laser plasma accelerators and radiation sources, laboratory astrophysics, and ultra-intense laser matter interactions. These varied applications and related fundamental studies have focused new attention on laser plasma physics and high energy density science.

Acknowledgements

This paper is based on lectures presented at the 6th Scottish Universities' Summer School on Laser Plasmas in St. Andrews, Scotland. I am indebted to my colleagues on the various published works which are reviewed in part of this paper, including S. Wilks, P. Rambo, P. Young, R. Kirkwood, J. Hammer, Max Tabak, R. Davidson, B. Still, D. Hinkel, R. Berger, B. Langdon, and E. Williams. This work was performed under the auspices of the U.S. Department of Energy by the Lawrence Livermore National Laboratory under Contract No. W-7405-Eng-48.

References

Afeyan, B, *et al.*, 2003, Kinetic electrostatic electron nonlinear (KEEN) waves and their interaction driven by the ponderomotive force of crossing laser beams, in *Inertial Fusion Sciences and Applications 2003* (American Nuclear Society, La Grange Park, Illinois)

Berger, *et al.*, 1993, Theory and three-dimensional simulation of light filamentation in laser-produced plasma, *Phys. Fluids B* 5, 2243

Bezzerides, B., DuBois, D., and Rose, H., 1993, Saturation of stimulated Raman scattering by excitation of strong Langmuir wave turbulence, *Phys. Rev. Lett.* 70, 2569

Cohen, B., *et al.*, 1997, Resonantly excited nonlinear ion waves, *Phys. Plasmas* 4, 956

Davidson, R., *et al.*, 2004, *Frontiers of High Energy Density Physics* (National Academies Press, Washington, D.C.)

Dewar, R., Kruer, W., and Manheimer, W., 1975, Modulational instabilities due to trapped electrons, *Phys. Rev. Lett.* 28, 215

Glenzer, G., *et al.*, 1998, Energetics of inertial confinement fusion hohlraum plasmas, *Phys. Rev. Lett.* 80, 2845

Hinkel, D., Williams, E., and Still, C. H., 1996, Laser beam deflection induced by transverse flow, *Phys. Rev. Lett.* 77, 1298

Kane, J., *et al.*, 1999, Scaling supernova hydrodynamics to the laboratory, *Phys. Plasmas* 6, 2065

Kauffman, R., *et al.*, 1998, Improved gas-filled hohlraum performance on Nova with beam smoothing, *Phys. Plasmas* 5, 1927

Kirkwood, R. *et al.*, 1996, Observation of energy transfer between frequency mismatched laser beams in a large scale plasma, *Phys. Rev. Lett.* 76, 2065

Kirkwood, R., *et al.*, 2002, Observation of saturation of energy transfer between co-propagating beams in a flowing plasma, *Phys. Rev. Lett.* 89, 215003-4

Kruer, W., *et al.*, 1996, Energy transfer between crossing laser beams, *Phys. Plasmas* 3, 382

Kruer, W., 1996, Laser plasma interactions in hohlraums, in *Laser Plasma Interactions V*, editor Hooper, M. B. (SUSSP, Edinburgh)

Kruer, W, and Hammer, J., 1997, Laser beam deflection in nonlinearly-generated flow profiles, *Comments Plasma Phys. Cont. Fusion* 18, 85

Kruer, W., *et al.*, 1998, Interplay between laser plasma instabilities, *Physica Scripta* T75, 7

Kruer, W., 2003, *The Physics of Laser Plasma Interactions* (Westview Press, Boulder, CO)

Labaune, C., *et al.*, 1998, Time-resolved measurements of secondary Langmuir waves produced by the Langmuir wave decay instability in a laser-produced plasma, *Phys. Plasmas* 5, 234

Liberman, M., *et al.*, 1999, *Physics of High-Density, Z-Pinch Plasmas* (Springer-Verlag, New York)

Lindl, J., 1998, *Inertial Confinement Fusion* (Springer-Verlag, New York)

Miles, A., *et al.*, 2004, Numerical simulation of supernova-relevant laser-driven hydro experiments on Omega, *Phys. Plasmas* 11, 3631

Norreys, P., *et al.*, 2000, Experimental studies of the advanced fast ignitor scheme, *Phys. Plasmas* 7, 3721

Nuckolls, J., *et al.*, 1972, Laser compression of matter to super-high densities: thermonuclear applications, *Nature* 239, 139

Perry, M. and Mourou, G., 1994, Terawatt to petawatt subpicosecond lasers, *Science* 264, 917

Rambo, P., Wilks, S., and Kruer. W., Hybrid PIC simulations of stimulated Brillouin scattering including ion-ion collisions, 1997, *Phys. Rev. Lett.* 79, 83

Remington, B., *et al.*, 2000, Review of astrophysics experiments on intense lasers, *Phys. Plasmas* 7, 1641

Rose, H., 1996, Laser beam deflection by flow and nonlinear self-focusing, *Phys. Plasmas* 3, 1709

Ryutov, D., *et al.*, 1999, Similarity criteria for the laboratory simulation of supernova hydrodynamics, *Ap. J.* 518, 821

Short, R., Bingham, R. and Williams, E., 1982, Filamentation of laser light in flowing plasmas, *Phys. Fluids* 25, 2392

Tabak, M., *et al.*, 1994, Ignition and high gain with ultra powerful lasers, *Phys. Plasmas* 1, 1629

Vu, H., *et. al.*, 2001, Transient enhancement and detuning of laser-driven parametric instabilities by particle trapping, *Phys. Rev. Lett.* 86, 4306

Wilks, S., *et al.*, 1992, Absorption of ultra-intense laser pulses, *Phys. Rev. Lett.* 69, 1383

Williams, E., *et al.*, Effects of ion trapping on crossed-laser-beam stimulated Brillouin scattering, 2004, *Phys. Plasmas* 11, 231

Young, P. *et. al.*, 1998, Observations of laser beam bending due to transverse plasma flow, *Phys. Rev. Lett.* 81, 1425

Zimmerman, G. and Kruer, W., 1975, The Lasnex simulation code, *Comments Plasma Phys. Cont. Fusion* 2, 51

2

An Experimentalist's Perspective on Plasma Accelerators

Christopher E. Clayton

University of California at Los Angeles

1   Introduction

Experimental plasma physics has always been a multi-disciplinary field, especially at high electron densities n_e ($\gg 10^{12}$cm^{-3}). At these densities, the apparatus needed to produce, manipulate, and study plasma is usually large and complex. Laboratories performing work in laser-plasma accelerators, such as laser-driven "laser wakefield accelerators" (LWFA) or electron/positron-driven "plasma wakefield accelerators" (PWFA) are no exception, becoming increasingly complex as this field moves further away from "proof-of-principle" experiments. From the recent breakthroughs in the last few years, such as the observation of "monoenergitic" GeV electrons from a LWFA experiment (Leemans et al 2006) and the energy-doubling of electrons initially at $\sim$ 40GeV in a PWFA experiment (Blumenfeld et al 2006), one can draw the happy conclusion that modern laboratories can indeed create the plasma accelerator structures that will have high impact on other areas of science and technology; those requiring light sources or high-energy particles. On the other hand, one could observe that these recent successes were the natural outcome of improvements to the laboratory infrastructure; for example, bigger lasers and better electron beams. As a practical experimentalist, this author leans towards the latter view and that we are very much still in the research stage with "development" still well down the road. Indeed, the growth in the plasma accelerator field has and continues to be built upon the breakthroughs of the past decades—we remain in this evolutionary period and many of the challenges ahead have been challenges from the start.

Consider just the space- and time-scales upon which we wish to create, manipulate and control aspects of the experiments: time scales $> \omega_p^{-1} = 18$fs at $n_e = 10^{18}$cm^{-3} where (in cgs units), $\omega_p = (4\pi n_e e^2/m)^{1/2}$ is the plasma frequency. Similarly, we must control the transverse spacial scales on the order of $\sim \lambda_p \cong 2\pi/k_p = 33\mu$m at $n_e = 10^{18}$cm^{-3}. At the same time, the longitudinal scales vary from /mum to cm to m, depending on the experiment and the details within the that experiment. These short scales are $\sim 10^{-3}$ smaller than those of conventional accelerators. The accelerated electron beams have space- and time-scales commensurate with the plasma accelerator structure and are unprecedented in the history of linear accelerators. Also, while sections of a conventional accelerator can be taken to the workshop for troubleshooting and repair, plasma-accelerator sections clearly cannot. Thus, apart from the issues of the small-scale nature of the plasma accelerator, it is a transient manifestation of what has historically been a difficult (or stubborn!) substance to mold. Clearly, certain aspects of the plasma, plasma wave, and electrons must be highly correlated and may change from shot to shot.

To pull together as much information as possible about the plasma, plasma wave, and electrons is the next stage in the evolution of our field. Even now, as we struggle to do this, the drive beams (the laser pulse or the electron/positron pulse) often drift with time and become another unknown quantity in our shot-to-shot analysis of the physics. Thus, it appears to this author, that the fastest-growing additions to a typical laboratory infrastructure is (or should be) a collection of hardware—distinct from the apparatus used to produce the plasma, plasma wave, and electrons—collectively called our diagnostics. The term "diagnostics" of course includes such essentials as photodiodes, oscilloscopes, energy meters, burn paper, charged-coupled device (CCD) cameras, infrared (IR) viewers, spectrographs, digital volt meters, data-acquisition computer boards, pressure gauges, phosphorescent or fluorescent materials, and so forth. But we need to expand on these essentials for quantitative measurements of the entire chain of events that starts with switching on the apparatus to producing accelerated electrons.

In particular, researchers in our and related fields, pressed with the need to characterize the laser, plasma, plasma wave, and electrons on such small space- and time-scales, are literally inventing new apparatus and/or techniques to make these measurements. Sometimes, however, sophisticated measurements can be done with "simple" diagnostics by employing clever experimental *techniques*. An excellent example of this is in the paper of Hsieh et al (2006) where, in a LWFA experiment that was producing monoenergetic electrons, the authors used a laser pulse coming in from the side and timed such that it reached the gas-jet target much earlier (few ns) than the pump beam. With this heating pulse, they were able to remove, bit by bit, sections of the plasma until they discovered where (along the direction of the pump) the electrons originated and also where they stopped gaining energy (due to dephasing). We mention this to remind the reader that breakthroughs require breakthrough thinking as well as advances in technology. The scope and complexity of the apparatus needed for plasma accelerator experiments, once understood, is an attractive feature for those who enjoy "hands-on" science and problem solving.

In this article, we will explain more fully the central role of diagnostics in plasma accelerator experiments, give some example of such diagnostics and the priniciples upon which they work. These principles can be expressed in terms of laser-, plasma-, and beam-physics and the interplay of the physics with the diagnostics. While we construct the mathematical description of the interplay, we will also attempt to highlight the most general principles and, hopefully, by the end of this chapter, develop a deeper intuition about how the diagnostics work. In this author's experience, this intuition is necessary to interpret the measurements while it also appears to play an integral part of the evolution, and indeed the invention, of diagnostic techniques.

2 Pursuing experimental goals

Figure 1 shows this author's impression of a typical experimental campaign to discover or reveal the unknown. Someone has come up with an Idea and wants to test it in the laboratory. This Idea may have come from a previous Experiment or from some Theory. In Figure 1, the term "Experiment" refers to the *purpose* for showing up in the laboratory; what you want to do that day—what is it that is not known that you are prepared to reveal or discover? Obviously, as indicated by the solid arrows emanating from the Experiment, this requires Apparatus and

Diagnostics. It is very helpful to be motivated by the immediate purpose or goal of today's Experiment, but this goal should *not limit* the choice of Diagnostics! Ideally, as many Diagnostics are active in parallel as possible. It is perhaps obvious, but note that the source of Data can only be the Diagnostics and nothing else (in other words, not intuition, calculations, simulations, etc.).

The dashed arrows pointing back towards Experiment represent a not uncommon mode of inquiry where one just blasts away and observes what happens. Although this is not normally recommended, it can lead to an Idea which can "kick-start" a more focused campaign. Also indicated in Figure 1 is the inevitable fact that the Apparatus will often not work as expected and Diagnostics must be used to sort out those problems. Implicit in Figure 1 is the flow of Ideas representing discussions among colleagues, publications, and interactions at conferences and workshops; just a few of the avenues for Ideas to travel.

Once the potentially overwhelming amount of raw Data is recorded or "logged", it is the experimentalist's duty to perform the Analysis. It is surprising how often this step is put off to the future; i.e., probably never looked at, especially when there is an impression that today's Data is "not good". The key point here is that, if there are several Diagnostics acquiring synchronized Data (where each record from each Diagnostic is correlated to that of another), then the Analysis of even "bad" Data can bring insight, perhaps as to even why the Data appears to be "bad". Finally, the Data-Analysis-Theory loop is where the discovery of the unknown happens. The availability of multiple streams of Data is often key to closing this loop: if Analysis and Theory require this-or-that to be true, one can bring in another recorded Data stream (as indicated by the dashed arrow from Theory to Data) to check whether this-or-that is indeed true and, eventually, *quod erat demonstrandum*, Voilà, Eureka! However, there is another reason as to why the prudent experimentalist will keep as many parallel Diagnostics on-line for a given Experiment. This additional Data may suggest that the Experiment is not tracking the Idea or Theory—a discovery itself!

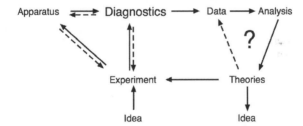

Figure 1. *A diagram of how the unknown (the "?" in the Figure) becomes known. Here, "Experiment" is the purpose for showing up in the laboratory, usually a result of someone's Idea or Theory. The solid arrows show a somewhat ideal path toward the known. However, as indicated by the dashed arrows, sometimes the available Apparatus and/or Diagnostics influence the choice of the Experiment while it is often the case that Diagnostics must be used to understand the Apparatus. Note: the Data-Analysis-Theories loop, if non-infinite, is the source of new Ideas and Theories.*

3 Plasma-accelerator laboratories

A generic laboratory for acceleration of electrons by plasma waves is shown schematically in Figure 2. If this were a PWFA laboratory, where an electron (or positron) bunch, with a

Lorentz factor γ_b, excites the plasma wave, then consider Figure 2 as showing electrons coming in, a "quadrapole triplet" (a lens for charged particles) for focusing the bunch into the plasma, another triplet to capture the electrons and an insertion device (a dipole magnet, for example) to characterize the electrons coming out of the plasma. If Figure 2 represented a LWFA laboratory, where a photon pulse excites the plasma wave, there is a chirped-pulse amplification (CPA) laser system which produces the photon bunch which is focused by an off-axis parabolic (OAP) mirror to produce intensities high enough to drive a plasma wave. This laser is typically optically pumped by another, long-pulse laser. Such high-energy ($\sim$ J level) pump lasers can also be used to produce the plasma itself and to pump other CPA amplifiers to produce multiple probe beams, as indicated in Figure 2. In the LWFA experiments, one may rely on self-trapped electrons to be accelerated or one could use externally injected electrons. In either case, the diagnostics for the accelerated bunch of electrons look similar to the PWFA case. Common to both the PWFA and LWFA schemes are the need to produce long, high-density plasmas and probes/diagnostics to understand what is going on within the plasma.

The maximum energy obtainable in a single stage, un-tapered plasma accelerator is governed by the phase velocity v_{ph} of the plasma wave which must be relativistic (Tajima and Dawson 1979); i.e. we need "relativistic plasma waves". For the LWFA, the laser's group velocity $v_{gr} = c(1 - n_e/n_{cr})^{1/2}$ is relativistic—that is, it has an associated Lorentz factor $\gamma_{gr} \gg 1$—then v_{ph} is also relativistic since it is synchronous with the laser pulse. Thus it will have an associated Lorentz factor equal to γ_{gr}; namely, $\gamma_{ph} = \gamma_{gr} = (1 - v_{gr}^2/c^2)^{-1/2} \approx \omega_o/\omega_p \approx (n_{cr}/n_e)^{1/2}$. Here, n_{cr} is the "critical density" and is defined as the plasma density for which the plasma frequency is equal to the laser frequency ω_o and is give by $\omega_0 = (4\pi n_{cr}e^2/m)^{1/2}$ where we have assumed that $\omega_o \gg \omega_p$. Similarly, for the PWFA, $\gamma_{ph} = \gamma_b$; i.e., the wave is synchronous with the drive beam.

In Section 3.1 and 3.2, we will more fully describe the makeup of a LWFA laboratory and a PWFA laboratory, respectively. Since there has been a revolution in laser technology in the last two decades, the makeup of a LWFA laboratory has evolved substantially. Thus Section 3.1 will have an historical bent. Conventional electron/positron linear accelerators (or "linacs") have not changed greatly over this time, thus very little history is given in Section 3.2. Section 4 lists some common diagnostic tasks while Sections 5, 6, 7, and 8 address some of the underlying physics and/or examples of the diagnostics of lasers, plasmas, plasma waves, and electrons, respectively.

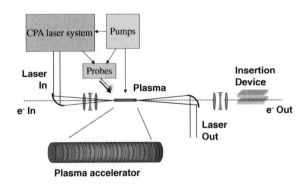

Figure 2. *Cartoon of a generic LWFA or PWFA laboratory. (From Clayton (2004). With permission.)*

3.1 A laser wakefield acceleration experiment

The original concept and theoretical underpinnings of the LWFA were given by Tajima and Dawson (1979) in their seminal paper (the Idea). They noted that efficient conversion of laser energy into plasma wave energy would occur if the laser pulse length τ_L was approximately equal to half of a plasma wavelength; i.e., that there is a "resonant density" where $\lambda_p \approx 2c\tau_L$. This paper came out about 6 years prior to the invention of the chirped-pulse amplification (CPA) scheme (Strickland and Mourou 1985). The sub-picosecond dye lasers available at the time were of quite low energy ($\ll 1$J) and the Q-switched YAG lasers, although having huge energies ($\gg 1$J) suffered from long (ns) pulses such that the resonant density would be so low that the accelerating field (from an engineering formula in GeV/m) $E_z \approx 96\, \epsilon_{epw}\sqrt{n_{e18}}$ would be only about 10MeV/m for a wave amplitude ϵ_{epw} of unity and where n_{e18} is the plasma density in units of $10^{18}\mathrm{cm}^{-3}$ and we have kept $\lambda_p \approx 2c\tau_L$ for these ns pulses. Note that 10MeV/m $< E_z <$ 50MeV/m is typical of current radio-frequency (RF) metallic accelerating structures: plasmas as accelerating structures can in principal sustain E_z fields $\approx 1000\times$ that of RF structures.

Tajima and Dawson (1979) suggested two ways to overcome (at that time) the lack of suitable intense, short-pulse lasers, both of which involved transforming a long pulse (now with $\tau_L \gg \lambda_p/c$) into a train of short pulses which we shall call an "electromagnetic beat wave", or EMBW for short. In one case, the EMBW would be self-generated through the parametric instability called Raman forward scattering (RFS) (Forslund et al 1975, Drake et al. 1974). In the other case, the EMBW would be externally imposed by a two-color laser (Kroll et al. 1964, Rosenbluth and Liu 1972, Cohen et al. 1972).

The Raman forward scattering (RFS) instability can be though of as the decay of an incident laser photon into a "plasmon" and another, frequency down-shifted (Stokes) photon; i.e., one wave into two "daughter waves". As the RFS instability grows via propagation into a plasma, the laser pulse interferes with the growing Stokes (and anti-Stokes) sidebands in space and time producing an EMBW of increasing intensity. The feedback loop causing this growth is the enhancement of the daughter plasma wave via the ponderomotive force of the EMBW. The ω- and k-matching required for the instability ensures that the frequency difference between the main laser and the sidebands is ω_p and the process is resonant. Plasma accelerators in this regime have been coined SM-LWFA for self-modulated LWFA (Andreev et al. 1992, Krall et al. 1993). The other EMBW approach, using a two-color laser, *could* be called the PM-LWFA for pre-modulated LWFA, but has historically been called the plasma beat wave accelerator (PBWA) (Joshi et al. 1984). In this case, the system is not self-resonant and the plasma density must be chosen to satisfy ω-matching; if the two colors have frequencies ω_1 and ω_2, then the plasma density must be chosen such that $\omega_p = \omega_1 - \omega_2$.

Even after the invention of the CPA scheme, single pulses were typically 0.4ps $< \tau_L <$ 1ps due to the narrow gain-bandwidth of even mixed Nd:YAG and Nd:Glass laser chains. These pulses were still much too long for direct LWFA (which need $\lambda_p \approx 2c\tau_L$) as they would need many kJ of energy to drive up a substantial wave. Although PBWA experiments using CO_2 lasers had pulses of comparable durations (0.2–0.6ps), these occurred in a train of pulses where each subsequent pulse would add to the plasma wave amplitude from the proceeding pulses, reducing the laser energy requirements to < 100J, which was readily obtained. Due to the flexibility of existing CO_2 laser systems to operate simultaneously at two wavelengths and the wide availability of glass laser systems, all experiments on plasma accelerators, from 1985 to about 2000 were either PBWA or SM-LWFA) (see the review article by Esarey et al. 1996). The first Experiment conceived to produce and directly measure relativistic plasma waves was done in 1985 (Clayton et al. 1985) using a two-color CO_2 laser using the PWBA scheme. This success led to an Experiment that eventually led to the acceleration of externally injected electrons in 1992 (Clayton et al. 1993). As implied in Section 2 on pursuing experimental goals, even a seven-year gap between building the Experiment and its fruition was such an engine of Ideas that the graduate students filled much of that time producing excellent laser-plasma and wave-wave experimental papers. This is yet another benefit to the implementation of parallel, robust Diagnostics!

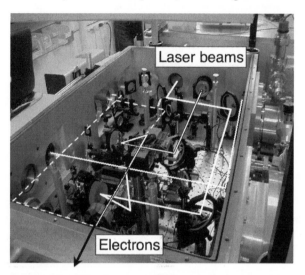

Figure 3. *A photograph of the interior of a LWFA target chamber. The top of this vacuum box has been raises to reveal the complex and highly adaptable optical and mechanical components. The apparatus is setup for studying the colliding- pulse optical injection. (From Leemans, W.P., Lawrence Berkeley National Laboratory. With permission.)*

For the 1 μm Nd:YAG lasers, since the collection of laser apparatus was the largest at Lawrence Livermore National Laboratory, it is not surprising that one of the first 10TW, 0.8ps lasers was built there by 1991 using the CPA scheme (Patterson et al. 1991). Although 10TW was achieved immediately after the compressor gratings, after several meters of air, the beam would break up, limiting the usable power available to the experiment to $\sim$ 3TW. Nevertheless, an Experiment was conceived and built to measure electrons from the RFS instability a underdense ($n_e < n_{cr}$) gas jet target. The first electrons from this SM-LWFA experiment were observed in 1994 (Coverdale et al. 1995). In 1996, Umstadter et al. (1996) suggested that the 100% energy spread of the electron spectra observed (in the now several SM-LWFA experiments reporting results around the world (Esarey et al. 1996)) could be made "monochromatic" if electrons were ponderomotively injected into the wave before wave-breaking occurred. The Idea was to use an auxiliary laser beam entering from the side to optically "inject" background plasma electrons into a narrow accelerating phase of the wave, thus narrowing the spectrum of the accelerated electrons.

In an invited talk at the 1990 Conference on Lasers and Electro-Optics, Professor Wilson Sibbett of St. Andrews University in St. Andrews, Scotland astonished the audience when he announced the first stable, self-mode-locking laser oscillator using Ti-doped sapphire (Ti:sapphire). Although Ti:sapphire crystals were produced in the early 1980's and intracavity dispersion-compensating prisms were routinely used in colliding mode-locked dye lasers, were it not for the invention of the stable Kerr-lens mode-locking technique (Spence et al. 1991) we would never complain about "only 50fs pulses" as we do today.

The first commercial Ti:sapphire CPA system was sold in 1991. Laser laboratories which did not have enormous investments in prior laser technology and, indeed, *new* laser laboratories soon took advantage of this new technology. Now, 15 years later, there are dozens (if not more) large Ti:sapphire CPA systems in dozens of countries. For example, 1991 was also the year when a former PBWA researcher, Wim Leemans, joined Lawrence Berkeley National Laboratory (LBNL) and saw the possibilities of the new laser technology. So perhaps it is no accident that one of the largest (100 TW) systems is located in his laboratory at LBNL.

One of the LWFA experimental areas at LBNL is shown in Figure 3. This figure shows the implementation of an "Experiment", again in the sense of Figure 1, in that the *purpose* of the Experiment is to test out an Idea. In this case, to optically-inject electrons into a LWFA using an auxiliary laser beam that is approximately counter-propagating to the pump (Esarey et al. 1997), which, under certain circumstances, the ponderomotive force of the resulting EMBW of the "colliding pulses" could be more efficient and controllable than the original Idea of Umstadter et al. (1996). There are three beams entering the chamber on the far side of the photograph. The central beam (shown in white with a black border) is the main pump. The injector (colliding) beam (indicated by the thick white line) collides with the pump at their mutual foci within the plume of the helium gas jet. The third beam (indicated by the wavy line) is a frequency-doubled short probe beam for interferometry and shadowgraphy.

The experiment shown in Figure 3 is a good example of how the import and export of Ideas discussed with respect to Figure 1 can cross from one laboratory to another. This naturally occurs through the publication of peer-reviewed journal articles and the personal interactions that occur at conferences and workshops. The results from Experiments from both the transverse ponderomotive injection of Umstadter et al. (1996) and the colliding-pulse ponderomotive injection of Esarey et al. (1997) both showed an enhancement of the amount of accelerated charge. The Idea migrated to other laboratories and, with help of the inevitable rise of infrastructure in any particular laboratory, finely came to fruition where, in the absence of the colliding pulse, no electrons were observed whereas, with the colliding pulse, tunable, monoenergetic electrons could be produced (Faure et al 2006). The two dashed lines in Figure 3 indicate locations where a portion of the pump beam can be sent out of the chamber for characterization, either to a FROG setup, a single-shot autocorrelator, a spectrometer, or to a CCD camera. Some of the laser, plasma, and plasma-wave diagnostics utilized in this and similar experiments will be the discussed in more detail in Sections 5 through 7, respectively. There are electron diagnostics not visible in Figure 3, some of which will be discussed in general in Section 8.

3.2 The plasma wakefield acceleration experiment at SLAC

The connection between the Tajima and Dawson (1979) concept of the laser-driven LWFA and the electron/positron beam driven PWFA concept surprisingly took about 6 years. Linear accelerators are incredibly expensive and, at the time, were devoted as tools for high-energy physics and nuclear physics. Perhaps the very idea that a plasma experiment could be brought into these environments may have delayed the inevitable observation that wakefields from electrons were very similar to those from laser pulses (Chen et al 1985). The first PWFA experiments were performed at Argonne National Laboratory by Rosenzweig et al (1988) which eventually lead to the concept of the so-called "blowout" regime (sometimes referred to as the "bubble" regime). From the success of these experiments and the subsequent ability of simulations to extrapolate the physics to higher energies, a proposal to study PWFA experiments at the end of the 3km-long Stanford Linear Accelerator Center's (SLAC) linac was approved.

Figure 4 shows schematically the layout of the PWFA experiment housed in the final 150m of the SLAC beamline. This section of beamline, which is a few hundred meters from the final accelerator sections of the SLAC linac, was built in the 1990's to (successfully) test new ideas for

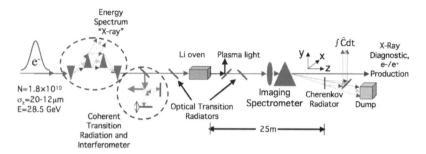

Figure 4. *A schematic of the experimental setup in the FFTB beamline (not to scale). From Muggli, University of Southern California. With permission.*

a relatively-compact, broadband magnetic-focusing system for a linear collider: hence its name, the Final Focus Test Beam (FFTB) facility. For radiation-safety reasons FFTB, although parallel to the linac, was offset with a "dog-leg" section of beamline so as to go around a radiation shield. This turned out to be very fortunate, since bends in beamlines can be quite useful. Just as there is a "group delay" in a grating-pair for compression of chirped optical pulses, where different colors have different times-of-flight through the system, one can similarly think of a bend in a beamline arranged such that different electron energies have different total path lengths through the bend. With the addition of the FFTB, there were now two bends in the beamline (the other in the transport line from the damping ring to the linac). Since these both have a negative "group delay" or a negative "R56", by adding a four-dipole chicane magnet at the $\sim$ 1 km mark of the 3 km long linac with as a well-chosen positive R56, the chirp in the first 1/3 of the linac could be fully removed, compressing the pulse to 50μm RMS. Finally, the new, additional chirp due to the wakefields in the final 2/3 of the linac could be nearly fully removed in the FFTB, allowing for even more pulse compression. The pulses went from a RMS bunch length σ_z of about 600 μm down to as low as 12 μm by the addition of this chicane.

Moving from left to right in Figure 4: the electron beam, while still in a dispersive section of the dog-leg beamline, is bumped up and down by a wiggler to measure its energy spectrum via synchrotron radiation; through a 1 μm-thick titanium (Ti) foil to produce coherent transition radiation (CTR) for bunch-length measurements; through a focusing system (not shown, but like the triplet in Figure 2); focused down into a column of Lithium (Li) vapor contained in a heat-pipe oven at about 1000°C; and finally energy-analyzed by an imaging spectrometer before heading to the beam dump. About 1 m away from each end of the Li oven is another 1 μm-thick foil for optical transition radiation (OTR) measurements of the beam size and position. Another Ti foil just downstream of the Li oven is used to pick off plasma and Cherenkov light for spectral measurements. The recent bunch-length compression allowed by the recently installed chicane had two beneficial side effects for PWFA experiments. First, the "resonant" n_e is now higher to keep the relation $\lambda_p \approx 2\sigma_z$ satisfied. (Note, the RMS bunch length σ_z in this PWFA scheme essentially takes the place of the laser pulse length τ_L of the LWFA scheme). Since the operating density is higher, the maximum accelerating field E_z in a plasma wave of a given amplitude ϵ_{epw} is also higher (recall the engineering formula $E_z \approx 96\,\epsilon_{epw}\sqrt{n_{e18}}$ giving the field in GeV/m). Second and more impressively, the radial electric field of the focused electron beam is now well over that needed to tunnel-ionize the Li (Ammosov et al. 1987), eliminating the need to photo-ionize the plasma (previously done with a large eximer laser).

The electron "detector" in the energy-dispersion-plane of the spectrometer is a Cherenkov emitter which is either a piece of aerogel or an length of ambient air. The Cherenkov light is imaged onto a CCD camera with no time resolution, hence the notation $\int \hat{C} dt$ in the figure. The spectrometer consists of a long dipole magnet to disperse the electrons vertically according to their energies as well as quadrapole magnets to image the exit of the plasma onto the dispersion plane. Thus the finite-size image of the output of the plasma is the equivalent to the slit size in an optical spectrometer. Without this imaging, there is no way to distinguish between exit energy and exit angle unless either the angle is explicitly measured for every shot or the range of possible angles is limited enough so that the projected extent of these angles onto the dispersion plane, which is now the equivalent slit size, is smaller than the desired spectral resolution. Some of these and other electron diagnostics will be discussed in Section 8.

4 Diagnostics in theory and practice

Advances in experimental physics rarely occur without the persistent determination by the experimentalist to overcome occasional "technical problems". Some of the perceived problems may lie on the real-axis, and some may lie on the imaginary-axis, while most are simply complex and thus more Data is needed to resolve the problem at hand. For example, in LWFA and PWFA experiments, there are many re-occuring problems that can bring progress to a halt. When this happens, certain phrases are heard over and over in the laboratory. Some of these are:

1. Something is wrong with the laser.

2. Something is wrong with the plasma.

3. Something is wrong with the plasma wave.

4. Something is wrong with the electron beam.

5. Something is wrong with the vacuum.

6. Something is wrong with the alignment.

7. Something is wrong with the timing.

8. Something is wrong with the data-acquisition or control computers.

These problems are exacerbated by: a lack of or misuse of diagnostics; uncalibrated diagnostics; and perhaps by false expectations i.e., not seeing what you expected to see. The first four issues in the above list (laser, plasma, plasma wave, and electron beam "problems") will be covered to some extent in Sections 5, 6, 7, and 8, respectively, in terms of diagnostics.

Although this article is mainly about diagnosing the aforementioned experimental parameters, simply listing experimental apparatus to perform these diagnostics is insufficient, to say the least. To make judgments on what the diagnostic is revealing requires: (a) an understanding of the physics relating the measured quantities with the interaction of the diagnostic with the plasma (or laser or plasma wave or electron beam, etc.), and (b) an intuition which is based, in part, on the knowledge of the physics (and mathematics!) for these interactions and in part on experience.

5 Laser diagnostics

The quality of the drive laser, and to some extent the probe beams, must be established. Some of the parameters to measure are summarized in the second column of Table 1. We would like to know the frequency content of the pulses as well as the pulse shape. In fact, for ultra-short

Aspect	Incoming beams	Outgoing beams
Spectrum	Bandwidth, spectral and/or temporal phase	Spectral shifts, phase modulation
Wavefront	Amplitude and phase-uniformity, spectral shear	Scattering, guiding
Power	Energy, pulse shape, pre-pulse, ASE	Absorption

Table 1. *A short list of measurements associated with laser beams both prior to interacting with a plasma and after the interaction.*

pulses, it is important to know the actual time-history of the electric field of the pulse, both in amplitude and phase. Of course, the quality of the wavefront and the temporal contrast of the focused intensity are also needed.

In addition, once the drive beam or one of the probe beams has traversed the plasma, there will be features embedded into the transmitted or scattered light that contain information about the plasma. Some of these features are tabulated in the third column of Table 1. Spatial, angular, temporal, and spectral features may be encoded onto the transmitted probe beams. Much of the laser-plasma physics can be discerned from these measurements and thus are crucial to the understanding of experiments.

We will not cover all the parameters in Table 1 (even though the contents of the Table is far from exhaustive). We will, however, say that some of these parameters can be obtained by the appropriate use of common laboratory instrumentation such as spectrometers, joule meters, photodiodes, CCD cameras, etc.. However, the pulse shape and especially the temporal phase require specialize apparatus, one of which we will now describe.

Most of the laser laboratories doing work in plasma accelerators use drive-laser pulses of sub-picosecond duration. A robust diagnostic for short pulses is FROG, which stands for Frequency-Resolved Optical Gating (Kane and Trebino 1993). Suppose the laser pulse is fully described by its electric field $E(t)$ (which is true in our laboratory settings). Equation (1) shows a related mathematical quantity $E_{sig}(t, \tau)$ which, if measured for all possible values of delay τ, could in principle be used to calculate the unknown waveform $E(t)$.

$$E_{sig}(t, \tau) = E(t) \left| E(t - \tau) \right|^2 \tag{1}$$

On the right-hand-side of Equation (1), the unknown waveform is multiplied by (i.e., gated by) the intensity of a replica pulse that is delayed by a time τ. This looks like some kind of "nonlinear autocorrelation".

The measurement of $E_{sig}(t, \tau)$ is illustrated in Figure 5. The "pulse to be measured" ($E(t)$) is sent into the optical system shown. It first hits a beam splitter to produce its own replica. The pulse continues along the bottom path in Figure 5. We shall call this pulse the "first pulse". The replica is delayed with respect to the first pulse by an adjustable delay line. We will refer to the delayed replica as the "second pulse". Following the path of the first pulse in Figure 5, we note that it is sent through two crossed polarizers. Thus in the absence of the second pulse, the magnitude the "polarization-gated" first pulse, defined as the "signal" pulse $E_{sig}(t, \tau)$, seen by the detector — a CCD camera coupled to an imaging spectrometer — will be zero. The function of the relatively intense second pulse, with a polarization rotated by 45° with respect to the first pulse (introduced by a waveplate, not shown), is to very briefly induce a birefringence

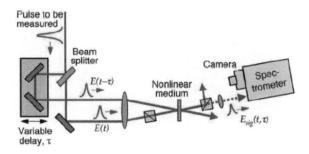

Figure 5. *Experimental setup for the polarization-gated FROG diagnostic. The polarization of the gaiting pulse is at 45° with respect to that of the pulse that enters the spectrometer (From Trebino, R., Georgia Institute of Technology). With permission.*

in the nonlinear (NL) medium via the optical Kerr effect. If the first and second pulses have any temporal overlap at the plane of the NL medium, a "temporal slice" of the first pulse will be rotated in polarization and thus be recorded by the camera. By scanning the delay line over many shots, sufficient different time-slices (different τs) can be recorded to reconstruct the waveform $E(t)$.

In practice, for very short pulses, this can be done on a single shot! By having the two pulses cross at an angle, the gating pulse can overlap the head of the first pulse at one transverse position within the NL medium and overlap the tail of the first pulse at another transverse position providing that the NL medium is thick enough and other geometrical factors are chosen correctly. Now the gating time τ is mapped into transverse spatial position across the NL medium. The final optic before the spectrometer is an imaging lens which images the crossing location (proportional to the gating time) onto the slit of the imaging spectrometer, thus encoding the delay time onto one of the axes of the CCD image. The spectrometer will, of course, decompose $E_{sig}(t, \tau)$ into a frequency spectrum. In particular, if the entire time duration of $E_{sig}(t, \tau)$ is less than the integration time of the spectrometer (typically many ps), then the effect of the spectrometer is to perform a Fourier transform of $E_{sig}(t, \tau)$. Since the CCD is a photon detector (a "square law" detector), the camera records the square of the magnitude of this Fourier transform. Equation (2) describes the resultant 2-dimensional image $I_{FROG}^{PG}(\omega, \tau)$.

$$I_{FROG}^{PG}(\omega, \tau) = \left| \int_{-\infty}^{+\infty} E(t) \left| E(t - \tau) \right|^2 \exp(-i\omega t) dt \right|^2 \tag{2}$$

Without going into details (Kane and Trebino 1993), it turns out that $E(t)$ is sufficiently constrained through Equation (2) so that only one solution for $E(t)$ exists if the FROG image is "well-measured". In this author's intuitive view, the constraint lies in the fact that the measurement, expressed by Equation (2), is a function only of gate time and frequency which, being almost conjugate quantities, are not wholly independent.

Figure 6(a) shows an experimental FROG trace where wavelength is on the vertical axis and the gate-time is on the horizontal axis. One nice feature of the polarization-gated version of FROG is that the raw data is sometimes quite illustrative of what is actually happening in the phase or frequency dynamics within the pulse. Here in Figure 6(a) we see that the frequency increases as a function of the gate time and we may conclude that the pulse has a positive chirp (Siegman 1986) that looks to be fairly linear with time. The FROG data is used to "recover" $E(t)$ by use of a 2-dimensional "phase-retrieval" algorithm. This is a computational-intensive, iterative procedure that generally converges to provide the true waveform.

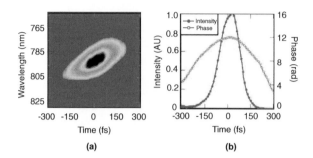

Figure 6. *(a) An experimental FROG trace and (b) the FROG-derived instantaneous intensity and phase of the laser electric field obtained from the data in (a). (From Kohler, B., Ohio State University. With permission.*

The FROG-derived intensity and phase corresponding to the data in Figure 6(a) are shown in Figure 6(b). The pulse is roughly Gaussian in time and the temporal phase is parabolic in time. The definition of instantaneous frequency is the derivative of the temporal phase. Thus, the parabola in Figure 6(b) implies that $\omega(t) \propto t$; i.e., the frequency is changing linearly in time. This is just what we see in the raw data: a linear chirp! Since the original Idea of Kane and Trebino (1993) for FROG, many variations of FROG have been invented and most of these are summarized in very-readable book edited by Trebino (2002).

The Fourier-transform properties of a spectrograph have long been known. Ordinary light from an incandescent bulb, if reflected off, for example, a microscope slide and sent into a spectrometer can show "spectral fringes" if the path-length-difference between the front and rear reflections is less than the coherence length of the light. Similarly, as we will fully explain in Section 7.4, when two short laser pulses are sent into a spectrograph, they too will produce spectral fringes if they originate from the same laser. Recently, there has been a proliferation of work, for example, in chemistry, exploiting this phenomena. Such techniques may be called "spectral phase interferometry" techniques. Out of this work came a new method of measuring the amplitude and phase of $E(t)$ called SPIDER which stands for "**S**pectral **P**hase **I**nterferometry for **D**irect **E**lectric-field **R**econstruction of ultrashort optical pulses" (Iaconis and Walmsley 1998).

Here the beam is split into *three* replicas where one of them is chirped by a grating-pair and subsequently "mixes", via sum-frequency generation in a crystal, with the other two pulses separated by some time τ. These two pulses are nearly identical except that their central frequencies are slightly offset by mixing at different portions of the chirped replica. This relative difference in central frequency Ω is sometimes referred to as a "spectral shear". If $\tau \ll$ the integration time of the spectrometer that these two pulses are sent into, there will be "spectral fringes" and, if one knows Ω, analysis of the spectral fringes can reveal $E(t)$, again, in a single shot! This laser-pulse diagnostic is intimately related to the plasma-wave diagnostic technique that is the subject of Section 7.4.

Obtaining $E(t)$ from FROG or SPIDER is in principle simply a matter of putting the data through a numerical recipe. Generally, FROG, being an iterative algorithm, will take much more computing time than SPIDER which is a purely linear calculation. Some people prefer SPIDER for this reason: that the amplitude and phase of the laser pulse can be calculated and displayed in real time, say, at 10Hz. However, each has its advantages and disadvantages more in the experimental details rather than the computation time. We will not go into these measurement techniques any further. For now, we know that the full $E(t)$ is experimentally available.

6 Plasma diagnostics

Before we characterize the plasma, it is usually a good idea to first characterize the local environment where the plasma is to be created. For example, if the plasma is to be created in a gas jet, one would need to know the density of the gas (or clusters) in the jet. Similarly, if the experiment takes place in a capillary, it would be nice to know the uniformity of the gas within the tube. Since the properties of these "targets" are not step functions in space, one would like to know how their boundaries fall off into the vacuum at either end. Vacuum quality, vibrations, and other environmental quantities can also be important to know.

If the plasma is preformed, one may have the luxury of studying the plasma in the absence of the drive beam. Even if the plasma formation requires the drive/pump beam, one may want to know how the plasma evolves during and after the passage of the pump through the plasma. Some of the plasma properties that are quite useful to know are longitudinal and transverse density profiles, the ionization state, and the hydrodynamic evolution. In addition to active probing, one can observe the emission from the plasma. For example: x-rays, fast ions, neutrons, as well as plasma line- and continuum-radiation.

The remainder of this section will be on using a probe laser beam to study the plasma itself. Typically, one can measure the plasma density, the amplitude of a plasma wave at a particular wavenumber, or the plasma temperature (not covered here). There are three commonly used mathematical formulations to describe the interaction of a probe beam with a plasma; the *microscopic* description, the *fluid* description, and the *macroscopic* description. The microscopic description, in which the plasma is treated as an ensemble of electrons and ions with no assumed "fluid" or "material" properties, is the subject of Section 6.1 below.

Although we will not explicitly discuss the fluid description of the scattering of a probe laser beam by a plasma, it is worth outlining here. The probe beam is an external field in Maxwell's equations, the charge sources are regions where the electron and ion densities are different, and the current sources are due to these charge sources oscillating in the field of the probe beam. More specifically, the incident probe field is $E_i(\vec{r}, t)$, the charge source (treating the heavy ions as uniformly dispersed throughout the plasma) is $\tilde{n}(\vec{r}, t)$, and to first order the current density is $J(\vec{r}, t) = -e\tilde{n}(\vec{r}, t)\vec{v}_{os}(\vec{r}, t)$ where $\vec{v}_{os}(\vec{r}, t) = -\mathrm{i}(eE_i(\vec{r}, t))/(m\omega_i)$ and ω_i is the probe's frequency (Slusher and Surko 1980). Since the radiated or "scattered" field is due to the current $J(\vec{r}, t)$, the scattered field contains information about $\tilde{n}(\vec{r}, t)$.

The third description of a probe-plasma interaction comes from treating the plasma material with a refractive index $\eta_p(\vec{r}, t) \equiv k_i c/\omega_i = (1 - \omega_p^2(\vec{r}, t)/\omega_i^2)^{1/2}$ where, generally, the refractive index is defined as the light-speed in vacuum c divided by the phase velocity of light within the material given by $\omega_i/(k_i(\vec{r}, t))$. This description is quite useful for scattered-light angles near zero (near-forward scattering) if the plasma is relatively stationary during the transit time of the probe beam through the plasma. This will be the subject of Section 6.2.

We should note that these three formalisms can also be applied to laser probing of the relativistic plasma waves: i.e., the plasma accelerators. For example, if the refractive index appears quasi-stationary in the group-velocity frame of the probe beam — as is approximately the case if the probe and plasma wave are co-propagating — then a refractive-index approach which keeps track on the *longitudinal phase* of the probe beam can employed, as will be covered in Sections 7.1 and 7.3.

6.1 Thomson scattering

In the microscopic treatment (Sheffield 1975), the plasma is an ensemble of electrons and ions which oscillate in the incident probe beam's electric field $E_{io} \cos(\vec{k}_i \cdot \vec{r} - \omega_i t)\hat{i}$. Here $\vec{k}_i$ and ω_i

are the wavenumber and frequency of the probe beam, respectively. We assume that the probe pulse has a pulse length τ_i much greater than τ_p so that we can neglect the variation of the temporal envelope of the probe pulse in the following analysis. Here, $\tau_p \equiv 2\pi/\omega_p$ is the plasma period. Also, we assume that the probe beam has a transverse size which is greater than the transverse size of the volume of plasma that contributes to the scattered light. Under these assumptions, we can take the probe beam as a constant amplitude plane wave for the following analysis. Furthermore, we neglect any depletion of the probe beam so that its amplitude is constant throughout the scattering volume. Finally, we neglect any ion motion so that the only particles that can move are electrons.

As the charged particles oscillate, they radiate a "scattered field" $\vec{E}_S(\vec{R}, t)$ as illustrated in Figure 7 in which the scattered wave is viewed in the $\hat{s}$-direction. The total scattered field $E_S^T(\vec{R}, t)$ is simply the sum over all charged particles where we must take care to include the phase of the scattered light (at the detector) for each electron. Typically, one considers two contributions to E_S^T. First, since the number of radiating charges within the scattering volume is finite, there can never be perfect cancellation of the scattered field $E_S^T(\vec{R}, t)$ due to random phases from each electron. (Mathematically, the total scattered field from an infinite number of uncorrelated sources is zero, but this is not the case here.) This is the so-called "incoherent" contribution to the scattered field — the contribution from uncorrelated electrons. We will return to the issue of radiation from uncorrelated electrons in a completely different context in Section 8.3 where, for a beam of electrons that "self-radiate" in the presence of a dielectric or a bending magnet, the correlation depends on the wavelength of observation. This is not the case here since the electrons are not self-radiating, but are essentially re-radiating the probe beam frequency. This will clearly scale as N, the number of electrons in the scattering volume contributing to E_S^T. The second contribution typically included in E_S^T comes from the fact that, in a plasma with a finite temperature, the charges are not generally randomly distributed! There *are* spatial and temporal correlations between electrons. Why is this?

The electrons in a warm plasma will try to shield out a locally-induced potential or voltage. However, due to their finite inertia, this shielding is imperfect. If, for example, a charge Q is approximately stationary in a plasma with electron temperature T_e, the electrons will rearrange themselves to "neutralizing" the potential of the charge Q, but because they have a "thermal velocity" $v_{th} \equiv (2kT_e/m)^{1/2}$, they cannot hold their positions (total neutralization). Rather, they "swarm" around the charge Q, like bees *in flight* swarm around the queen bee. The characteristic size of this swarm is the Debye length $\lambda_{De} = (kT_e)^{1/2}/(4\pi n_e e^2)^{1/2}$. So, at least over a Debye length or so, plasma is good at neutralizing the charge Q. Now, if we set this charge Q into motion with velocity v_Q, not much will change if $v_Q \ll v_{th}$; the plasma electrons can still catch up and shield the slowly moving charge. However, when $v_Q \gg v_{th}$, the (instantaneously) local electrons will still move to try to shield the charge, but by the time they get there, the charge is gone. Further along the trajectory of the charge, the same thing happens: electrons react, but to no avail. Meanwhile, in the act of rearranging themselves to shield the moving charge, the electrons (on average), have moved away from the neutralizing ions. Thus, once the charge Q has disappeared over the λ_{De} horizon, the electrons must move back to shield the ions that they left behind. Of course, due to their finite inertia, they overshoot and oscillate. It seems that by disturbing the background electrons in such a way, the moving charge sets up an electron density fluctuation or a plasma wave or a wakefield! And since this relies on the inability of electrons to shield "over the λ_{De} horizon", the wavelengths of these electron density fluctuations, or "plasmons" in the jargon of weak turbulence theory, must $\gg \lambda_{De}$.

Since we did not specify the magnitude of Q, this appears to be true if the moving charge is simply just one of the plasma electrons! Indeed, this is the case. Electrons moving with a velocity v_Q somewhat greater than v_{th} will emit small-amplitude plasma waves or plasmons, that, by causality, have a phase velocity $v_{ph} = v_Q$. This is the single-particle analog to the PWFA (where an intense bunch of electrons or positrons set up the wake). To summarize, there are electron density fluctuations or collective waves which exist in plasmas with finite electron temperature T_e, even in thermal equilibrium. From the plasma dispersion relation for electron plasma waves (EPW) or plasmons

$$
\begin{aligned}
\omega_{EPW}^2 &= \omega_p^2 + (3/2)v_{th}^2 k_{EPW}^2 \\
&\approx \omega_p^2 \left(1 + 3\lambda_{De}^2 k_{EPW}^2\right) \\
&\approx \omega_p^2 \left(1 + \frac{3}{2}\frac{v_{th}^2}{v_{ph}^2}\right)
\end{aligned}
\tag{3}
$$

we note that $\omega_{EPW} \approx \omega_p$ plus a "thermal correction". Here v_{ph} is the phase velocity of the plasmon. For that electron with $v_Q > v_{th}$, the plasmon will have a frequency approximately equal to ω_p and thus the k_{EPW} of the emitted plasmon must apparently be such that $\omega_{EPW}/k_{EPW} \approx v_Q$. Now, since there are a large number of electrons, each with a different v_Q, there will be a large range of thermal plasmon wavenumbers k_{EPW} with frequencies close to ω_p. We have discussed the extreme cases where $v_Q \ll v_{th}$ and $v_Q \gg v_{th}$. If $v_Q \sim v_{th}$, many of the plasma electrons can still react to the passage of the charge Q and one could say that these plasmons are heavily "beam-loaded" (Landau damped) due to the large number of electrons with average velocities near v_Q. So a characteristic transition in k-space of the thermal plasmons is when the thermal correction terms in Equation (3) becomes $\sim 0.10\omega_p - 0.15\,\omega_p$ or when $k_{EPW}\lambda_{De} \approx 0.3$. We will return to this $k\lambda_{De}$ product below in connection with the Thomson scattered spectrum.

Note that we have described these density fluctuations in ω- and $\vec{k}$-space; i.e., as $\tilde{n}(\vec{k},\omega)$. These are the natural units for Thomson scattering since, the scattered light will differ from the probe beam in both frequency and angle (the angular spectrum of a light wave is given by its $\vec{k}$-spectrum). The more physical $\tilde{n}(\vec{r},t)$ is given by the inverse Fourier transform

$$
n_e(\vec{r},t) = \int \frac{\mathrm{d}\vec{k}}{(2\pi)^2} \int \frac{\mathrm{d}\omega}{2\pi} n_e(\vec{k},\omega)\mathrm{e}^{-\mathrm{i}(\vec{k}\cdot\vec{r}-\omega t)}.
\tag{4}
$$

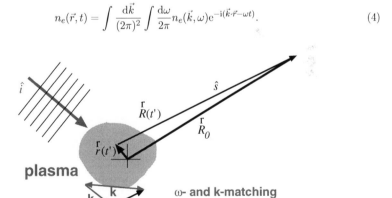

Figure 7. *Geometry for Thomson scattering. The unit vectors $\hat{i}$ and $\hat{s}$ define the directions of the incident and scattered light wavevectors, respectively. The coordinate $\vec{r}(t')$ is the location of an electron that scatters light and $\vec{R}(t')$ is the vector from the electron to the detector (From Clayton (2004). With permission.)*

Figure 7 shows a probe beam incident onto the plasma along the direction $\hat{i}$. We would like to know the field scatted by one electron located at $\vec{r}(t')$ as seen by a detector located at $\vec{R}_o$.

The scattered light from this electron is collected in the $\hat{s}$-direction with the exact electron-to-detector vector given by $\vec{R}(t') = \vec{R}_o - \vec{r}(t')$. Note that the origin of the coordinate system is somewhere within the plasma. Also shown in Figure 7 is a "k-matching" diagram. This will be discussed near the end of this Section (Section 6.1), but we point out now that for a scattered wave frequency of ω_S, $k_S = \omega_S/c$ and the two possible directions of $\vec{k}$ in the vector k-matching diagram allow for the scattered wave to be up-shifted in frequency for the left-pointing $\vec{k}$ or downshifted for the right-pointing $\vec{k}$. Both are valid and the choice is up to the experimentalist. If, as in a typical thermal plasma, the plasmon $\vec{k}$ spectrum is isotropic in $|\vec{k}|$, then the scattered wave will have two peaks, one up-shifted and one down-shifted. Each peak will be shifted by the plasma frequency (if the thermal correction term for our geometry is small).

For a probe beam with an incident field $E_{io}\cos(\vec{k}_i \cdot \vec{r} - \omega_i t)\hat{i}$, the acceleration of an electron is $\dot{\vec{\beta}}(\vec{r},t) = -(e)/(mc)E_{io}\cos(\vec{k}_i \cdot \vec{r} - \omega_i t)\hat{i}$ where $\vec{\beta}$ is the oscillation velocity normalized to c. The resulting scattered field at time t and far away from the plasma is due to the the the electron oscillations at the "retarded time" t' as

$$\vec{E}_S(\vec{R},\vec{r},t) \approx \frac{e}{Rc}\left[\hat{s} \times \left(\hat{s} \times \dot{\vec{\beta}}(t')\right)\right] \tag{5}$$

The retarded time is t minus the time it takes for the radiated light to travel from the electron to the detector

$$t' = t - \frac{\vec{R} - \hat{s} \cdot \vec{r}}{c} \approx t - \frac{R}{c} + \frac{\hat{s} \cdot \vec{r}}{c} \tag{6}$$

where we have assumed that the characteristic size of the plasma is much smaller than $|\vec{R}_o|$ and therefore $|\vec{R}(t')| \simeq |\vec{R}_o|$ and $|\hat{s} \cdot \vec{R}_o| \simeq |\vec{R}_o|$. Note that we have also neglected the electron's velocity in Equation (6) and thus Doppler-shift effects. The total scattered field is the sum of that from each electron which we will approximate by the integral over the plasma volume V

$$\begin{aligned}
\vec{E}_S^T(\vec{R},t) &= \int_V d\vec{r}\, n_e(\vec{r},t')\left(\vec{E}_S + \vec{E}_S^{\,*}\right)/2 \\
&= \int_V d\vec{r}\left\{\int \frac{d\vec{k}}{(2\pi)^2}\int \frac{d\omega}{2\pi}n_e(\vec{r},\omega)e^{-i(\vec{k}\cdot\vec{r}-\omega t')}\right\}\left(\vec{E}_S + \vec{E}_S^{\,*}\right)/2
\end{aligned} \tag{7}$$

where, in the top line of Equation (7), $n_e(\vec{r},t')$ is real and $\vec{E}_S$, also real, is now expressed in complex notation, and we have used Equation (4) to substitute for $n_e(\vec{r},t')$ in the bottom line of Equation (7). Note that it is the oscillating part (in space and time) of $n_e(\vec{r},t')$ that is emphasized in Equation (7). This is related to the amplitude of the plasma density fluctuations $\tilde{n}$ that this Thomson scattering diagnostic is designed to measure.

In an experiment, there would be a collection lens at position $\vec{R}$ in Figure 7 which would sustain a solid angle $d\Omega$. The scattered power per unit solid angle at this (far-field) position is given by $dP_S(R)/d\Omega = R^2\vec{S}$ where

$$\vec{S} = (c/4\pi)\left[\vec{E}_S^T \times (\hat{s} \times \vec{E}_S^T)\right] \approx c(\vec{E}_S^T)^2/(4\pi)$$

is the Poynting vector. (Note that $\vec{S} \sim 1/R^2$ (Equation (5)), and so $dP_S(R)/d\Omega$ is independent of R in the far field.) Recall that the spectral measurement of the FROG signal field of Equation (1) became its Fourier transform (magnitude squared), as was shown in Equation (2). Similarly, since we want the frequency spectrum of the Thomson scattering, we can express the measured spectrum within a bandwidth of $\delta\omega_S$ and a collection solid angle of $\delta\Omega$

$$\frac{dP_s(R,\omega_S)}{d\Omega}\delta\Omega\delta\omega_S = \frac{cR^2}{4\pi}\delta\Omega\delta\omega_S\left|\int_{-\infty}^{+\infty}dt E_S^T(t)e^{-i\omega_S t}\right|^2. \tag{8}$$

It is possible to obtain an estimate of the amplitude of the plasma waves contributing to the scattered power in Equation (8) through knowledge of the geometry of the focused probe beam within the plasma and the subsequent diffraction angle of the scattered light (Clayton 2004). If the collection lens captures all the scattered light (the full range of $d\Omega$) and all the scattered frequencies are within the spectral acceptance of the spectrograph (both are usually obtained) then the scattered power $P_S \propto P_i \times (\lambda_i L \tilde{n})^2$ where P_i is the incident power and L is the interaction length between the probe and plasma as seen by the collection lens and we have assumed that $\lambda_S^2 \simeq \lambda_i^2$. Here, we have also assumed that we have chosen a scattering geometry which yields "collective" scattering as opposed to "incoherent" scattering. This depends on the k of the wave being probed compared to λ_{De} as it pertains to Equation (3). This is discussed next.

6.1.1 ω and k matching

We said earlier that ω and k were the natural units for Thomson scattering yet that is not evident in Equation (8) because we did not do all the algebra. It is simple enough, however, to collect together all the fast (oscillatory) time and space factors. These are summarized in Table 2 in column 2 with column 1 indicating where these factors came from.

Source	arguments of exponential
Eq (8), ω_S transform for spectrometer	$-i\omega_S t$
Eq (7), Fourier transform of density at t'	$-i(\vec{k} \cdot \vec{r} - \omega t')$
Eqs (5) and (7), total scattered field	$i(\vec{k_i} \cdot \vec{r} - \omega_i t') + cc$

Table 2. *Sources of space- and time-dependancies in Equation (8).*

Summing up all the arguments in the second column of Table 2, substituting for t' using Equation (6), and combining the time terms and space terms separately, we find that the integrand of Equation (8) contains $\exp[i(\triangle\vec{k} \cdot \vec{r} - \triangle\omega t)]$ where $\triangle\omega \equiv \omega_S - (\mp\omega_i - \omega)$ and $\triangle\vec{k} \equiv (-1/c)(\mp\omega_i - \omega)\hat{s} + (\mp\vec{k_i} - \vec{k})$. Note that $\triangle\vec{k}$ contains a term which is also in $\triangle\omega$. Note also that the $\mp$ coefficients, which came from combining the complex conjugate terms from the bottom row of Table 2, must be chosen the same for ω_i and $\vec{k_i}$ in order for the probe to be traveling in the specific direction given by $\hat{i}$. Similarly, ω_S can be replaced by $\pm\omega_S$ as long as the resultant $\vec{k_S}$ changes sign in the same way.

After making this replacement, the phase terms that appear in Equation (8) are given by

$$\exp[i(\triangle\vec{k} \cdot \vec{r} - \triangle\omega t)] =$$
$$\exp\left\{i\left[((1/c)(\omega_S - \triangle\omega)\hat{s} + \vec{k_i} \pm \vec{k})) \cdot \vec{r} - (\omega_S - \omega_i \mp \omega))t\right]\right\}$$
$$\exp\left\{i\left[(\vec{k_S} - (\vec{k_i} \pm \vec{k})) \cdot \vec{r} - (\omega_S - (\omega_i \pm \omega))t\right]\right\}, \tag{9}$$

where, in the last line of Equation (9), we have used the self-evident fact that $\vec{k_S} \equiv \frac{\omega_S}{c}\hat{s}$ and the non-selfevident fact that for Equation (8) to be finite, $\triangle\omega$ must be approximately zero. The latter can be instantly seen by putting the phase terms of Equation (9) into Equation (8): typically a probe beam has a duration of 10s – 1000s of plasma periods so that, if $\triangle\omega$ were not zero, the temporal oscillation of this factor will produce a very small integrated scattered power. Similarly, for a spatial extent $\gg \lambda_p$, the $\triangle\vec{k}$ factors in the integration over $d\vec{r}$ of Equation (7) should also add up to $\approx$ zero for obtaining a large scattered power. Indeed, since we must use $\triangle\omega = 0$ to also have a finite integration over *space* (middle line of Equation (9)), ω- and $\vec{k}$-matching must be *simultaneously* obtained. Since these matching conditions correspond to energy conservation ($\triangle\hbar\omega = 0$) and momentum conservation ($\triangle\hbar\vec{k} = 0$), respectively, they should indeed hold simultaneously in any physical system.

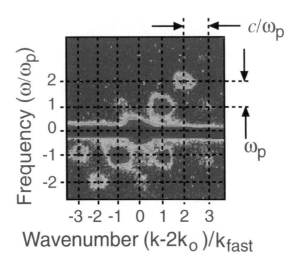

Figure 8. *Thomson scattering from a mixture of a Raman backscatter mode (at -1, -1) and a relativistic plasma wave with wavenumber k_{fast}. Some evidence of the relativistic plasma wave is seen in the fact that the slope connecting the dots is c. (From Everett et al. (1995). With permission.)*

The discussion around Equation (3) suggested that the amplitude of the density fluctuations $n_e(\vec{k}, \omega)$ scales inversely with the thermal correction. A figure-of-merit is the "scattering parameter" $\alpha \equiv 1/(k\lambda_{De})$. For $\alpha > 1$ ("collective" or "coherent" scattering), the Thomson scattering will view weakly-damped plasmons or plasma waves and a typical spectrum would show peaks at $\pm\omega_p$ away from the probe beam, thus giving a good measure of the plasma density. For $\alpha \ll 1$ ("incoherent" or "thermal" scattering), there are no correlations or long-range fields and hence no plasmons to observe. Had we included the electron velocity to the retarded time in Equation (6), we would find that a typical scattered spectrum would show the Doppler shifts from the isotropic electron thermal motion and thus give a measure of T_e. Returning to Figure 7, the $\vec{k}$-matching diagram shows that, by choosing the directions $\hat{i}$ and $\hat{s}$, one can probe a wide range of $\vec{k}$s (and therefore αs) in a plasma — again, a choice made by the experimentalist.

During a series of experiments at UCLA where plasma waves were driven well above the thermal levels by one- and two-color pump beams, it became clear that the frequency spectrum of the Thomson-scattered light alone may be deceptive. Spectral features observed near zero frequency shift and near $2\omega_p$ each could have multiple explanations. So an Experiment was devised measure very fine details in the angular spectrum of the scattered wave, the Idea being that wave-wave coupling may lead to "harmonics" in the frequency domain — including the "harmonic" at zero frequency — without the corresponding harmonics in $\vec{k}$ (Everett et al. 1995).

Figure 8 shows a Thomson scattered spectrum where the horizontal axis is the angular shift in $\vec{k_s}$ and the vertical axis is the frequency shift. The slit of the spectrograph was wide open so the frequency resolution was limited by the spot-size of the probe beam at the slit plane. These were not thermal plasmons but the mixture of two laser-driven waves. One of the waves was a relativistic wave with $\omega_p/k_{fast} \simeq c$ and with $\gamma_{ph} \approx 30$ (the "fast wave") and the other was from Raman back scatter (RBS) with a wavevector $k_{RBS} \approx 60k_{fast}$. In the k-matching diagram, all these plasma waves would be clustered around $k \approx k_{RBS}$ since $k = k_{RBS} \pm nk_{fast} \approx k_{RBS}$. For a frequency-doubled YAG-laser probe beam, the scattering angle $(\hat{i} \cdot \hat{s})$ was around 3.7° so the variation in scattering angle for each resolved dot on the horizontal axis was 3.7°/60! The corresponding ω-matching condition is $\omega = \omega_{RBS} \pm n\omega_{fast} \approx \omega_p(1 \pm n)$. Therefore Figure 8 shows that the validity of Equation (9), where ω- and $\vec{k}$-matching must be simultaneously obtained, is demonstrably true!

6.2 Refractive index description

In the introductory remarks of Section 6, we discussed the three mathematical approaches used to describe probe-plasma interactions. Here we disuss the refractive-index representation. Figure 9 shows schematically a probe beam traversing an essentially static plasma (transit time < plasma evolution time) and a parallel reference beam that misses the plasma.

Figure 9. *A typical geometry for treating the plasma as a refractive medium.*

It is much more convenient to treat this class of geometries using the refractive-index description of the plasma, although the probe is actually "scattered" in the forward direction. What does "refractive index" have to do with "forward scattering"? We can make this connection by finding the analogy between the refractive-index description and the Thomson scattering theory presented in the previous Section. Consider that a plane wave in a homogenous medium of refractive index η_p will satisfy Maxwell's equations if each of the vector components $V_j(\vec{r}, t)$ of the electromagnetic field individually satisfies the scalar wave equation $[\nabla^2 - (1/v^2)\,\partial^2/\partial t^2]\,V_j(\vec{r}, t) = 0$ where $v = c/\eta_p$, as discussed in Chapter 1 of Born and Wolf (1980). By the definition of "plane wave", any field component that is only a function of the phase term $\vec{r} \cdot \hat{s} - vt$ (the only significant space-time dependence in our plane-wave description of the probe beam, where we continue to assume a probe pulse length $\tau_i \gg \tau_p$, the plasma period, and plasma dimensions $\gg \lambda_i$) will indeed satisfy the scalar wave equation. In particular, for a harmonic field component of frequency ω_i, the phase term can be written as

$$\omega_i\left[t - (\vec{L} + \vec{r}) \cdot \hat{s}/v\right] = \omega_i t - \phi(\eta_p) - \vec{k_S} \cdot \vec{r}, \tag{10}$$

where, within the medium, $\vec{k_S} = (\omega_i/v)\hat{s}$ and, for forward scattering, we have assumed that $\vec{L} \| \hat{s}$ so that $\vec{L} \cdot \hat{s} = L$ resulting in the slowly-varying phase-advance term

$$\phi(\eta_p) \equiv \eta_p(t)|\vec{k_i}| = \omega_i\eta_p(z, t)L/c. \tag{11}$$

We recognize the left-hand-side of Equation (10) as ω_i times the retarded time, t', as in Equation (6). In this case, t' is t minus the time it takes a point of constant phase within the probe beam to traverse the plasma of length L (in the forward direction). Note that in this approximation, the frequency is conserved while the wavenumber can change. Thus it appears that the concept of a plane wave moving forward (in the refractive-index description) is identical to the concept of the plane wave being the "forward scattered" incident wave evaluated at the retarded time, just as in the microscopic description! We will make a further connection between Thomson scattering and the refractive-index description in Section 7.2 with respect to a discussion of the similarity seen between laser-beam refraction from an acousto-optic modulator and the scattering of a probe beam from a plasma wave.

If the plasma has dimensions $\gg \lambda_i$ and edges (where the density rises or falls) with transverse and longitudinal "scale lengths" $\Lambda_{\perp,\parallel} \equiv [(1/n_e)(dn_e/d\ell_{\perp,\parallel})]^{-1} \gg \lambda_i$, then the above plane-wave argument for the refractive-index description is accurate. Here, $\ell_\perp$ and $\ell_\parallel$ represent the transverse and longitudinal coordinates of the density, respectively, along the path of the probe beam. These constraints are similar to those used in "geometric optics" which takes $k_i \Lambda \to 0$.

The dispersion relation for light waves in a plasma is $\omega_i^2 = \omega_p^2 + c^2 k_i^2$. If we express this as $k_i c/\omega_i = (1 - \omega_p^2/\omega_i^2)^{1/2}$, we see that, since $k_i c/\omega_i$ is less than unity, the phase velocity $v_{ph,i} = \omega_i/k_i$ is greater than c. This ratio is essentially the definition of the refractive index η_p. Thus

$$\begin{aligned}
\eta_p(\vec{r}, t) &\equiv k_i(\vec{r}, t)c/\omega_i = (1 - \omega_p^2(\vec{r}, t)/\omega_i^2)^{1/2} \approx 1 - \omega_p^2(\vec{r}, t)/(2\omega_i^2) \\
&\approx 1 - \frac{1}{2}\frac{\bar{\omega}_p^2}{\omega_i^2}\left(1 + \frac{\tilde{n}(r, z, t)}{n_e} + \frac{\Delta n(r, z, t)}{n_e}\right).
\end{aligned} \tag{12}$$

In the second form of top line of Equation (12), we have assumed that $\omega_p^2 \ll \omega_i^2$. In the bottom line of Equation (12), the refractive index has been expanded and we have highlighted some physics of interest in LWFA plasmas which tend to be cylindrical in shape. The three terms within the brackets are due to an expansion of $\omega_p^2(r, z, t)$ (where $r = (x^2 + y^2)^{1/2}$) correspond to some background plasma density (the $\bar{\omega}_p^2$ term), a plasma wave of amplitude $\tilde{n}(r, z, t)$, and a radial variation of the plasma density $\Delta n(r, z, t)$, respectively (relativistic and ponderomotive effects have been neglected in this expansion). The term involving our plasma wave on interest, $\tilde{n}(r, z, t)$, will be discussed in Section 7 and so we will neglect it here.

Since $v_{ph,i} > c$, the probe will exit the plasma with a "phase advance" $\int_L dz\, \eta_p k_i$ which will be smaller than that of the reference beam of $\int_L dz\, \eta_{ref} k_i = k_i L$. If our plane-wave probe beam is entering transverse to $\hat{z}$, i.e., transverse to the axis of cylindrical symmetry, such that $\hat{s} \cdot \hat{z} = 0$, then the phase-advance difference $\Delta\phi(x, z)$, due to the different phase velocities in the plasma and the reference-beam environment (or, equivalently, due to the different retarded times), is

$$\Delta\phi(x, z) = \int_0^L (\eta_p(\vec{r}) - \eta_{ref})2\pi/\lambda_i dy, \tag{13}$$

where we have assumed that the probe is propagating in the $\hat{y}$-direction. The first integral on the right-hand-side of Equation (13) says that the probe beam carries the *projected* phase-advance variation (due to the integration in y) onto the detector. A simple CCD camera or photodiode (a photon counter or planar square-law detector) cannot measure an absolute phase advance, but it can measure a *relative* phase advance, and that is the purpose of the reference beam in Figure 9 and Equation (13).

We have not included any time dependence in Equation (13) although in some cases, slow phase changes can be tracked in time as in a homodyne detection system where the reference beam acts as the local oscillator (if the probe beam duration is longer than the plasma duration) and the two beams "mix" on a photodiode producing an output proportional to $\sin(\Delta\phi(t))$. A a 5mW He-Ne laser can thus be used to measure the evolution of the line-averaged density in discharge-produced plasmas if the bandwidth of the photodiode and oscilloscope are sufficient to track the rise or fall time of the plasma. As we will see in Section 7, it is precisely the time-dependence of the plasma density, albeit quasi-static in the group-velocity frame of the probe beam, that underlies some of the diagnostics for relativistic plasma waves by capturing longitudinal phase shifts $\Delta\phi(z - v_{gr}t)$. These methods include: "forward scattering" of long-pulse probe beams ($\tau_{pr} \gg 2\pi/\omega_p$) off of the plasma wave in which the longitudinal (temporal) phase modulation produces sidebands in the scattered wave; and "longitudinal interferometry" using *short* probe beams; i.e., short compared to the integration time of a spectrograph while simultaneously being comparable to the plasma period ($\tau_{pr} \sim 2\pi/\omega_p$).

An "interferogram" is produced by splitting a wide (compared to the plasma diameter) laser beam into a probe and reference beam and then recombining them onto a CCD camera with a slight angle between the two, say $\theta_{i,ref} \ll 1$. Figure 10(a) shows a Mach-Zehnder interferometer (as opposed to a Michelson interferometer). The small angle makes "background fringes" which, in the absence of plasma, are straight lines. With a sufficiently short probe pulse, a snapshot of the projected phase advance can be visualized via "fringe shifts" which are proportional to $\Delta\phi(x, z)$. Fringe "shifts" as low as 0.1 fringe can easily be seen by eye in the resulting interferogram. In effect, the periodic background fringes produce a "carrier wave" onto which the small phase perturbation is applied. The angular combination of the reference and probe beams produces spatial fringes: the carrier wave. An experimental setup to measure such fringe shifts is shown in Figure 10(a).

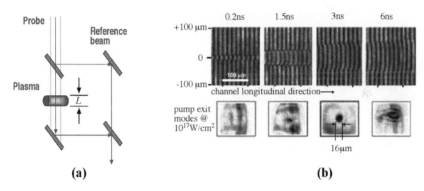

(a)   **(b)**

Figure 10. *(a) Schematic of a typical Mach-Zehnder setup for transverse interferometry. (b) Top: A sequence of images (interferograms) from a setup similar to (a) showing the time evolution of the plasma density due to hydrodynamic expansion. Bottom: Exit images of the spatial quality of a laser beam injected at the entrance to the plasma structures shown above. (From Zgadzaj et al. (2004). With permission.)*

Figure 10(b) shows refractive-index data taken in two ways (Zgadzaj 2004). The purpose of this Experiment was test the performance and check the theory of a *plasma fiber*. The Idea is, for a given input spot size of the longitudinally-propagating beam, say, $w_{o,i}$ there should be a specific radial shape to $\Delta n(r)$ that will guide this incoming beam over the length of the plasma at a fixed spot size of w_M. By measuring two characteristics of the plasma fiber simultaneously: namely the transverse profile and the guiding, the Idea can be experimentally verified. Figure 10(b) shows four interferograms taken at various delay times t_j with respect to when a separate laser produced the plasma (top of Figure 10(b)), and four images of the exit of the plasma column where yet another laser beam was sent along the (z) axis of the cylindrical plasma at the same time as the interferogram was taken (bottom of Figure 10(b)). The interferograms at the top of Figure 10(b) reveal the evolution of the "side view" of $\Delta n(r, z, t_j)$ due to hydrodynamic expansion while the images at the bottom of Figure 10(b) reveal the resultant "end view" of $\Delta n(r, z, t_j)$. Note that the interferograms are recorded as projections onto the $x - y$ plane. For a cylindrical plasma, the line-integral of Equation (13) will vary in x (and/or y), even if the density is constant within the cylinder. The true radial profile must be extracted from this projection by use of the Abel transform, yielding the unique radial variation of the refractive index that would produce the observed projected phase shifts and this was done in this experiment (not shown).

While the images from the end view are not interferograms, they do measure, in a way the, z-projection of the radial refractive index variation. In particular, the third image, with a time delay τ_d of 3ns, indicates that the longitudinal probe beam was "guided" as in a Graded Refrative INdex (GRIN) optical fiber. For this time delay, there must have been a suitable $\eta_p(r)$ to guide the beam. For a optical fiber with $\eta(r)$, the spot size evolution within the fiber is given by $\nabla_\perp^2 E + k_o^2 \eta(r)^2 E = 0$ where $E = E(r,z) \propto \exp(r^2/w(z)^2) \times \exp(ik_o\eta(r))$. A suitable radial variation of the refractive index $\eta_p(r)$ is one that forces the phase fronts of $E(r,z)$ to be independent of r for all z. This condition leads to a constant $|E(r,z)|$ and thus to a constant spot size $w(z)$. If $n_e(r) = n_{eo} + \Delta n_e r^2/a^2$ then, for a gaussian beam of spot size w_M, the laser will be guided at this "matched spot size" if $w_M(z_{in}) = w_{o,i} = (a^2/(\pi r_e \Delta n_e))^{1/4}$ (Durfee and Milchberg 1993) and $R(z_{in}) = \inf$. Here, z_{in} is the entrance to the plasma fiber and $R(z_{in})$ is the radius of curvature of the Gaussian beam at the entrance. Note that this guiding is only a function of the input spot size and the second derivative of $n_e(r)$ and not a function of the on-axis density.

In the Experiment of Figure 10, the second derivative of $n_e(r, \tau_d = 3\text{ns})$ (where $n_e(r)$ was obtained from the Abel inversion of the measured $\Delta\phi(x,y)$ for $\tau_d = 3\text{ns}$) agrees well with that needed to guide the longitudinal probe beam with its particular input spot size $w_{o,i}$. Although the result may seem anticlimactic, this self-affirming measurement also marks a milestone in the evolution of the apparatus and of the laboratory itself. If the longitudinal probe intensity were to be increased to the point where one can no longer neglect, for example, transverse ponderomotive perturbations to the plasma fiber, would an identical measurement still yield such good agreement? This is now not clear since one has to consider the fact that the perturbation is transient and may be time-integrated in the interferogram. Also, a wakefield would likely be exited and that, too, modifies $\Delta\phi(x,y)$. Thus these baseline measurements are more than a sanity check: they pave the way for future progress.

Although the Apparatus for the guiding experiment of Figure 10 was unique to that laboratory, one can get an idea of what is involved by noting that the third beam (indicated by the wavy line) in Figure 3 can be used for the transverse interferometry while the optical path indicated by the forward-going dashed line could be used to image the plasma exit. We will note also that, although the particular Experiment represented in Figure 10 did not have the diagnostics for $\tilde{n}(r, z, t_j)$, such diagnostics are available and some of these will be discussed in the next Section.

7   Plasma wave diagnostics

As mentioned before, the refractive-index description can be applied to diagnostics of the relativistic plasma wave itself if the longitudinal phase terms in the probe beam are considered. In particular, the phase term $\phi(\eta_p) = |\vec{k_S}|L$ in Equations (10) and (11) will still be "slowly varying" if the probe beam is co-propagating with the plasma wave. Thus this term will contain the longitudinal variations in the phase advance. For Sections 7.1 and 7.2, we assume that the pulse length of the probe beam $\tau_{pr} \gg \tau_p \equiv 2\pi/\omega_p$ while in Sections 7.3 and 7.4 the probe beam is short — in particular, $\tau_{pr} \sim \tau_p$ while also being much shorter to the integration time of a typical spectrograph.

7.1   Forward scattering at $0°$ incident angle

Figure 11 shows a cartoon of forward scattering. Here the wave at the top is a snap shot of the probe beam $E_i \cos(k_i z_1 - \omega_i t)$ as would be measured near the entrance to the plasma at $z = z_1$. The wave in the middle is the plasma wave which we are trying to diagnose and we will describe

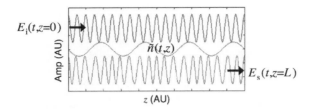

Figure 11. *Cartoon of the incident wave (top), modulating plasma wave (middle) and output or scattered wave (bottom) from the Bessel Identity for phase modulation. (From Clayton (2004). With permission.)*

this wave as $\eta_p(z, t)$ since we are using the refractive-index description. The wave at the bottom is the scattered wave $E_S(z_2, t)$ as would be measured near the exit of the plasma at $z = z_2$ where the length of the plasma is $L = z_2 - z_1$. As discussed in Section 6.2, we can consider $E_S(z_2, t)$ as the (forward) scattered incident wave evaluated at the retarded time. Therefore, from Equation (10), the phase term for E_S is $\omega_i t - \phi(\eta_p) - \eta_p(z, t)\vec{k}_i \cdot \vec{r}$. We will take the origin of the coordinate system to be near $(x, y, z) = (0, 0, z_1)$. Thus the term $\eta_p(z, t)\vec{k}_i \cdot \vec{r} = 0$ and $z_2 = L$ so that $E_S(L, t) \approx E_i \cos(\omega_i t') = E_i \cos(\omega_i t - \phi(\eta_p))$.

Using Equation (11) and the expansion of η_p from the bottom line of Equation (12) (with $\Delta n = 0$), the phase of the scattered wave at the exit of the plasma is

$$
\begin{aligned}
\omega_S(t, L)t &= \omega_i t - \phi(\eta_p) \\
&= \omega_i t - \eta_p(t)|\vec{k}_i| = \omega_i t - \omega_i[1 - \omega_p^2(\vec{r}, t)/(2\omega_i^2)]L/c \\
&\approx \omega_i t - \phi_o(L) + \epsilon_1 \sin(\omega_p t),
\end{aligned} \tag{14}
$$

where $\epsilon_1 \equiv \omega_i \left(0.5\bar{\omega}_p^2/\omega_i^2\right)(\tilde{n}/n_e)(L/c) = \pi(\tilde{n}/n_{cr})(L/\lambda_i)$ and, dropping the mean phase advance $\phi_o(L) \equiv k_i L(1 - 0.5\bar{\omega}_p^2/\omega_i^2)$, the scattered field is

$$
E_s(t, L) \approx \left(\frac{\pi}{2} \frac{\tilde{n}}{n_{cr}} \frac{L}{\lambda_i}\right) E_i \left[(\cos(\omega_i t + \omega_p t) - \cos(\omega_i t - \omega_p t)\right], \tag{15}
$$

where we have used the Bessel identity

$$
\cos(\omega_i t + \epsilon_1 \sin(\omega_p t)) = J_o(\epsilon_1)\cos(\omega_i t) + \sum_{k=\pm 1}^{\pm n} \frac{J_n(\epsilon_1)}{n!(-1)_{n<0}^n} \cos(\omega_i t + n\omega_p t).
$$

In Equation (15) we have suppressed the $n = 0$ (unshifted) term and only carried out the Bessel expansion to $n = \pm 1$. Interestingly, the coefficient for the frequency shifted light ($= \epsilon_1/2$) is the usual collective Thomson scattering efficiency. But this should not be too surprising since this geometry can also be treated with the Thomson scattering theory of Section 6.1 giving the same result.

Figure 12(a) shows schematically the setup of an experiment similar to the cartoon of Figure 11. Here, the Idea was to drive a relativistic plasma wave with a two-color CO_2 laser, but where the difference frequency ($\omega_1 - \omega_2$) was a fraction of ω_p: i.e., drive the wave off-resonance (which will reduce the wave amplitude) but use a higher plasma density (providing a higher accelerating field for a given wave amplitude) to recover the high accelerating gradient. In this way, density non-uniformities and relativistic de-tuning of the resonant density become unimportant (Filip et al. 2004). To check the validity of this Idea — that is, if one can recover a reasonably large E_z off-resonance by increasing the plasma density — the value of $\tilde{n}$ needs to be measured.

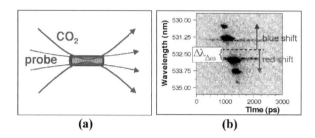

(a)   **(b)**

Figure 12. *(a) Schematic for collinear probing of a relativistic plasma wave. (b) Result of collinear probing showing a series of up- and down-shifted sidebands in scattered frequency. The probe frequency was blocked. The "tilt" of the sidebands with time is due to an uncompensated reflection from a grating — the timing of the sidebands are nearly simultaneous. (From Filip et al. (2004). With permission.)*

Figure 12(a) shows a blown up of the interaction region where the approximately 12 cm diameter CO_2 laser beam was tightly focused in the center of a vacuum chamber back-filled with H_2 gas. The large OAP mirror (not shown) used to focus the beam had a small hole in its center through which a frequency-double Nd:YAG probe beam was focused weakly such that the spot sizes of the two beams would roughly match within the plasma. Since the probe beam had a pulse duration of $\approx 10\times$ longer than the CO_2 beam, most of the forward-scattered light was not modulated in phase. To block this unwanted light, a custom "notch filter" composed of a multi-passed etalon and a grating plus a focusing lens (which amounts to half of a pulse stretcher) with a narrow mask placed at the unwanted frequency ω_i at the focal (dispersion) plane. The remaining scattered light, which passed either side of the narrow mask, was transported to a spectrograph coupled to a streak camera, with the spot at the slit of the spectrograph being a (filtered) image of the exit of the plasma. The time-resolved forward-scattered spectrum is shown in Figure 12(b) where we see both the red- and blue-shifted sidebands due to the phase modulation described by Equation (15) (had we expanded Equation (15) out to $n = 3$ using the Bessel identity). The "chirp" observed in the spectrum is due to the half pulse stretcher used in the notch filter. We can clearly see the terms $n = \pm 1, \pm 2$, and -3. The facts that blue-shifted third harmonic is missing and the red-shifted third harmonic is relatively weak are due to the finite bandwidth that could be transported to the spectrometer.

Equation (15) gives us two ways to estimate the wave amplitude $\widetilde{n}$ (the $n = \pm 1$ terms) in the data of Figure 12(b). We can convert both E_i and E_S to the incident and scattered intensity, respectively (essentially by squaring both sides). By knowing the transverse geometrical overlap of the probe beam with the plasma wave, these intensities can be converted to the measured quantities, incident power P_i and scattered power P_S. Equation (15) then gives $\widetilde{n}L$ as a function of P_S/P_i and, taking L as the longitudinal overlap of the probe beam with the plasma wave we have our measure of $\widetilde{n}$ up to a calibration factor which accounts for any losses in P_i and P_S. All the losses in the transport system could be estimated, but it is much easier to replace some mirrors by uncoated wedges and use calibrated attenuators to reduce P_i and then let this probe beam travel all the way to the P_S detector. By determining in this way the total amount of attenuation needed to replicate the level of the scattered light, one can have a direct measurement of P_S/P_i and hence of $\widetilde{n}$ (Lal et al. 1997).

The second and independent method of estimating $\widetilde{n}$ is to take the ratio of the $n = 2$ or second harmonic scattered power signal to that of the fundamental at $n = 1$ in the data of Figure 12(b). When Equation (15) is expanded out to $n = 2$ we see that P_i and L drop out and $\widetilde{n}$ depends only on the measured $P_S(\omega_p)/P_S(2\omega_p)$. Although this bypasses the need to obtain accurate measurements of the absolute incident and scattered powers as well as L, some corrections are required. This is due to the fact that, in 2- or 3-dimensions, there is a radial-dependence of the phase advance over the area of the probe beam. This will tend to produce a measured $n = 2$ term smaller than expected (compared with the fundamental) due to nonlinearity in going from E_S to P_S.

Another correction to this second method is needed when the plasma wave itself has harmonic content which was neglected here. As a plasma wave becomes large (say, $\tilde{n}/n_e > 0.3$), $E_z(z)$ begins to take on a "saw tooth" shape while the density becomes spiked, eventually accumulating to values greater than n_{eo} (the average plasma density) at the the falling edge of the "teeth" of $E_z(z)$, as described by Poisson's equation. The Fourier transform of such a non-sinusoidal variation of the density perturbation naturally has strong harmonic content. It can be shown that the phase-modulation sidebands beyond $n = 2$ are actually smaller (due mainly to the $1/n!$ term in the Bessel expansion) than the contribution from the harmonic components of plasma wave to the scattered light, although for $n = 2$, the contributions are comparable (Lal et al. 1997).

7.2 Forward scattering near $90°$ incident angle

The forward-scattering theory developed in Sections 6.2 and 7.1 used the refractive index description of the plasma that was made possible by the probe beam seeing a "quasi-static" variation of $\eta(z,t)$, allowing for a simple substitution within $E_i(L,t)$ of t for the retarded time t' (Equations (10)) which led to Equation (15); i.e., $E_S(L,t) \approx E_i(L,t')$.

Apart from the relatively simple analytic treatment, there appears to be an experimental benefit to collinear scattering in that the phase modulation is integrated along the entire length of the interaction region L, making the sidebands (supposedly) larger than they would have been in any other geometry. However, this line of reasoning is flawed. For example, suppose we attempt forward scattering at an incident angle of 0.1rad (about $6°$). Moreover, suppose the width of the plasma wave is $\sim \lambda_p$. Then, since the component of the probe-beam phase velocity parallel to $\vec{k_p}$ is now only about $0.9c$, the the phase slippage between the probe and plasma wave will be about 2π when the probe exits the plasma after an interaction length $L_{eff} \approx 10\lambda_p$. The phase of E_S would now be $\approx \omega_i t - \phi_o(L_{eff})$, independent of the wave amplitude, due to $\epsilon_1 \sin(\omega_p L_{eff}/v_{ph}(t)) = \epsilon_1 \sin(2\pi) = 0$ for all t: i.e., no phase modulation and therefore no sidebands! But this is not what happens. Had we used the microscopic treatment of the plasma (as in Equation (5) and forward) the loss in scattered power would only be down by $(L_{eff}/L)^2$ which is not necessarily small; the loss of *signal* (sideband power) at an incident angle of $6°$ may be factors of two or four whereas the drop in *noise* (due, as we will see below, to the fact that the scattered light is shifted in angle), can increase the overall signal-to-noise ratio. Of course, what is happening in this hypothetical experiment is that our quasi-static assumption does not apply, even for a "small" $6°$ incidence angle, so one must be quite careful in the use of such a reduced theory.

Also, as we will see in Section 7.3, there will always be some slippage between the probe beam and the plasma wave due to "photon acceleration" (Wilks et al. 1989).

Let us look at another example where the quasi-static assumption clearly breaks, in this case when the probe enters the plasma such that it is not even approximately collinear with the direction of the relativistic plasma wave, perhaps coming in at $\approx 90°$ as in Figure 13. If, at a particular z-position, the probe enters the plasma wave at a phase where $\tilde{n}$ is, say, at a minimum, it will exit the plasma wave after a transit time $\tau \approx d/v_{ph,i}$ and corresponding phase advance of $\phi(\eta_p) = \phi_o(L) \approx \omega_i \tau$. If by some chance the probe exits the plasma wave where $\tilde{n}$

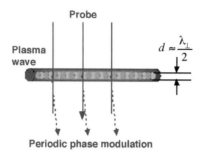

Figure 13. *Schematic of a side-on (probe beam at $\approx 90°$), small angle (or near-forward) collective Thomson scattering experiment. If the relativistic plasma wave is traveling from left to right, then the scattered light, indicated by the dashed lines, will be blue-shifted; e.i, $\omega_s = \omega_i + \omega_p$. For a transverse size of d, the corresponding transverse Fourier wavenumber $k_\perp \approx \pi/d$ or $d \approx \lambda_\perp/2$ where $\lambda_\perp \equiv 2\pi/k_\perp$. (From Clayton (2004). With permission.)*

has just had its next minimum, the net phase change due to the plasma wave itself is $\approx 2\pi$, as in our hypothetical 6° angle experiment above. For the duration of the plasma wave, this phase advance would be essentially constant; i.e., independent of t. Moreover, the phase change would be 2π for all z-positions as well! Although this probe would still be truly useful for average-density interferometry, as in Figure 10, would not seem to carry any spatial information about the plasma wave itself. However, this reasoning, based solely on the refractive-index description, is once again false, as discussed below.

The great experimental advantage of a setup such as suggested by the dashed lines in Figure 13 is that one could potentially dump the unscattered probe beam and produce an *image* of the plasma wave. We will argue below that, under most (common) conditions, spatial information *is* retained. Suppose the net phase change was, say, $< \pi$, then even though we are far from quasi-static, our intuition tells us that information about the plasma wave amplitude will be imprinted on the probe beam.

Consider two "rays" of the probe beam entering the plasma at z_{low} and z_{high} where $z_{low} - z_{high} = \lambda_p/2$. The ray entering at z_{low} enters at a density minimum and the other at a density maximum of $\tilde{n}(z, t)$. If the phase change is $< \pi$, then the z_{low} ray will exit before the next maximum of $\tilde{n}(z_{low}, t)$ and the z_{high} ray will exit before the next minimum of $\tilde{n}(z_{high}, t)$. The phase-fronts of that portion of the probe beam that traversed the plasma wave will thus have a periodic phase modulation as indicated in Figure 13. In other words, $E_S(t, z) = E_i(t', z)$ but this time with $t' = t - <\eta_p(z)> d/c$ where $<\eta_p(z)> = \int_0^{d/c} \eta(t, z)\, dt$. In turn, this periodic phase modulation in z leads to scattered "orders" (m) at deviation angles of $\hat{i} \cdot \hat{s}_m = \partial\theta_{y,m} = m\lambda_i/\lambda_p$ where $m = \pm 1, \pm 2, ...$ If one builds an optical system to collect, say, the $m = +1$ blue-shifted order, then this could be imaged onto a slit of a spectrometer resulting in an image of the relativistic plasma wave amplitude (squared) vs. position and frequency.

Figure 14 shows just such an image where the horizontal axis is position in the gas jet target (0 mm is the center of the 5.0 mm diameter jet) and the vertical axis is wavelength relative to the 532 nm probe beam wavelength (Gordon et al. 1998). The feature at $\triangle\lambda \approx -35$ nm located between -1.8 mm and -1.2 mm from the center of the gas jet is the image of the plasma wave. Note that the field-of-view in this Thomson scattering image is about 6 mm. To ensure that the Thomson scattering would capture the image regardless of where the plasma wave was, the pulse duration of the probe was set to twice the field-of-view; i.e., to a pulse length of

2×0.6 cm $\div (3 \times 10^{10}$ cm/sec$) = 40$ ps. On this particular shot, electrons from SM-LWFA were observed with energies ~ 100 MeV, which, given the length of the plasma wave image, would imply $E_z \sim 170$ GeV/m. However, given the plasma density, the energy gain was more than the "dephasing-limited" energy gain, which was attributed to a finite beam-loading effect (Gordon et al. 1998).

The relationship between our heuristic argument that the phase advance in crossing the plasma wave be $< \pi$ is related to the microscopic treatment and k-matching in the following way: as indicated in Figure 13, for a transverse size of d, the corresponding transverse Fourier wavenumber is $k_\perp \approx \pi/d$. Thus $d \approx \lambda_\perp/2$ where $\lambda_\perp \equiv 2\pi/k_\perp$. For a phase change $< \pi$, we must have the transit time $\tau = d/c < \tau_p/2$; i.e., less than half of a plasma period. Eliminating d and using $\tau_p = 2\pi/\omega_p = 2\pi/(ck_p)$ we have the requirement on the transverse portion of the vector k-matching as $k_\perp > k_p$. This sets rough limit on the maximum width of the plasma wave along the direction of the probe: that $d \sim \lambda_p$. Thus, in Equation (9), $\vec{k} = (\omega_p/c)\hat{z} + k_\perp\hat{y}$ where $\hat{z}$ and $\hat{y}$ are unit vectors in the direction of the plasma wave and the incident probe beam, respectively. Even if $\tau = \tau_p$ (which gave zero scattered power in the refractive description), the scattered power would be finite if $\tilde{n}(k_\perp = 2k_p) > 0$,

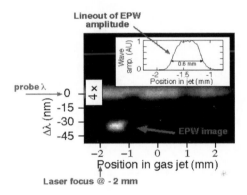

Figure 14. *Image of a relativistic plasma wave from small-angle Thomson scattering with the probe beam incident at $\approx 90°$ from the direction of the plasma wave. The inset shows a lineout of the electron plasma wave image. The white box labeled $4\times$ indicates the rows of the image where the CCD counts were divided by a factor of 4. This region is "stray light" (or noise) at the probe frequency. For this particular shot, the maximum electron energy observed was ~ 100 MeV. (From Gordon et al. (1998).)*

Perhaps it is intuitively clear that a phase modulation leads to scattered orders, as Lord Rayleigh (Rayleigh 1907) realized. But we have no intuition at this stage as to why the orders are shifted by $\pm m\omega_p$. We would have this intuition had we carried out the microscopic-theory based computation: we would have arrived again at the simultaneous ω- and k-matching results of Section 6.1.1. There is a nice chapter in Born and Wolf (1980) (Chapter 12) which highlights the fairly complex theory of near forward scattering for probe beams impinging at $\approx 90°$ onto ultrasound waves in transparent materials. If we apply the "Raman-Nath" diffraction theory described in that chapter to plasma waves, we find that $\partial\theta_m = \pm m\lambda_i/\lambda_p$ with corresponding frequency shifts of $\pm\omega_p$ and, incredibly, the power scattered into either of the $m = \pm 1$ orders as the coefficient of Equation (15) (within a factor of two)! Moreover, the ω- and $\vec{k}$-matching of Equation (9) is also recovered. All of these findings would be obtained by treating the geometry of Figure 13 using the microscopic description for Thomson scattering that we used in Section 6.1.

The reason that Raman-Nath diffraction from, say, a germanium crystal, agrees with the microscopic description for Thomson scattering from a plasma is likely due to the fact that the underlying physics of forward scattering is essentially the same. The mathematical treatment presented by Born and Wolf (1980) is quite similar to the microscopic treatment (where we summed over all electrons) except that the field giving rise to the dipole radiation for a particular atom or molecule (see Equation (5)) is the sum of the mean field $\vec{E}_i$ and the "effective field" $E' = (4\pi/3)\vec{P}$ seen by the radiating atom or molecule within the germanium where $\vec{P}$ is the electric dipole moment per unit volume. In the Thomson scattering theory, we have neglected the effective field since the densities we deal with are 10^3–10^7 times lower than solid or liquid densities.

7.3 Photon acceleration

One way to measure the amplitude of the plasma wave is to inject an *electron* bunch with γ_b equal to γ_{ph} of the wave, placed at exactly the maximum-field location in the wave, and measure their acceleration integrated over some distance $L \ll L_{dp}$ where L_{dp} is the dephasing length. This gives the maximum E_z, and from Gauss' law, one can recover the wave amplitude $\tilde{n}$. As was pointed out by Wilks et al. (1989), if you don't happen to have such a source of electrons in your laboratory, you can do essentially the same measurement using a photon bunch and measuring the integrated *photon* acceleration!

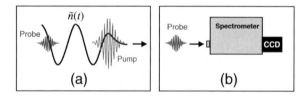

Figure 15. *(a) A single, short probe pulse rides along with the density wave driven by the pump pulse.(b) After dumping the pump pulse, the probe pulse is analyzed by a CCD camera coupled to a spectrometer. (From Clayton (2004). With permission.)*

This is the Idea in the Experiment represented in Figure 15 which was apparently first proposed by a group at the University of Texas at Austin, as reported by Joshi (1995). The photon bunch is located at the zero-crossing of $\tilde{n}$. Since the electron density rises behind this point and the unshielded ion density rises forward of this point, this is where E_z is maximum (actually it is minimum, but since we don't have positrons to accelerate, we will stick to this syntax which this author prefers). The photon bunch propagates at the group velocity $v_{gr} = c(1 - \omega_p^2/\omega_i^2)^{1/2}$. Thus, for the photon bunch to accelerate (to propagate at a higher v_{gr}), the frequency of the photons must increase beyond ω_i. This is just what happens in Figure 15(a). We can see physically how this happens: recalling that the phase velocity of this probe beam is $v_{ph,i} = \omega_i/k_i$ is given by $c/\eta_p = c(1 - \omega_p^2/\omega_i^2)^{-1/2} \approx c(1 + 0.5\,\omega_p^2/\omega_i^2)$ we see that where the electron density is lower, say, at the front of the photon bunch, $v_{ph,i}$ is slower. However, at the back of the photon bunch, the electron density is higher and thus those phase fronts towards the rear of the bunch will travel faster. After a finite propagation distance, the planes of constant phase representing, for example, the peaks in the waveform, will move towards each other as in the press of a concertina. Since the separation of the peaks of the waveform become smaller with increasing propagation distance, by definition, the frequency is increasing. Note that from causality arguments, the number of oscillations of the laser field entering and leaving this short length of plasma must be conserved (assuming no self-phase modulation). Thus the photon number is conserved. The photons gain their energy (one plasmon quanta $\hbar\omega_p$ at a time) at the expense of the plasma wave. Alternatively, one can envision the wakefield launched by the *probe* pulse as "canceling" that of the pump pulse (through a superposition argument) and transferring the plasma-wave energy to the probe pulse in the process (Mori 1997).

A careful examination of Figure 11 shows that the phase fronts of the probe beam have bunched up towards the "downhill" side of $\tilde{n}(z)$ (forward of the peak electron density) and, where the slope of $\tilde{n}(z)$ is opposite, the phase fronts have stretched apart. Therefore we anticipate that we can use the same formalism used to derive Equation (15) (which also describes Figure 11) to solve for this photon acceleration. The main difference here is that we will describe the incident field as a short wave packet which samples only a small region of the sinusoidal electron density variation. For a $\tilde{n}(t) \propto \sin(\omega_p t)$, we will take the centroid of the wave packet to be at $t = \pi\omega_p^{-1} \equiv t_1$ as in Figure 15(a) and Taylor expand the phase of the scattered wave at the exit of the plasma around this time. Note that this is the time where the density perturbation crosses zero and hence the point of maximum E_z. From the Taylor expansion of the bottom line of Equation (14), the light-wave phase at the plasma exit $\phi_S(L)$ is

$$\phi_S(L,t) = \phi_i(L,t') \approx \omega_i t - \phi_o(L,t_1) + \epsilon_1 \frac{d}{dt}\left[\sin(\omega_p t)\right]_{t=\pi\omega_p^{-1}} t \approx \omega_i t - \phi_o(L,t_1) - \omega_p\epsilon_1 t, \quad (16)$$

where $\phi_o(L,t_1)$ and ϵ_1 have the same meaning as in Equation (14) except that clearly $\phi_o(L,t_1)$ depends on where in time t_1 the probe pulse rides.

The net effect on the probe beam is to modify $E_i \exp(-i\omega_i t)$ to $E_i \exp(-\phi_o(L,t_1)) \times \exp(-i(\omega_i t + \omega_p\epsilon_1 t)$. Recalling (Equations (2) and (8)) that the measured quantity on the CCD camera in Figure 15(b) is the magnitude squared of the Fourier transform of the detected field, we have, for the phase modulation described in Equation (16)

$$I_S(\omega) = \frac{1}{2\pi}\left|e^{-i\phi_o(L,t_1)}\int_{-\infty}^{+\infty} e^{-i(\omega+\omega_p\epsilon_1)t} E_S^T(t)dt\right|^2 = I_i(\omega + \omega_p\epsilon_1), \quad (17)$$

where the last equality follows from the "shift theorem" of Fourier transforms. The frequency of the detected light has been up-shifted by an amount $\Delta\omega = \omega_p\epsilon_1 = \pi\frac{\tilde{n}}{n_{cr}}\frac{L}{\lambda_p}\omega_i$. This could have been immediately obtained from Equation (16) by finding the "instantaneous frequency" $\omega_S \equiv d\phi_S(L)/dt$. Note that as $\omega_S(z,t)$ increases with z, the resulting photon acceleration will move the pulse forward in the frame of the wave and hence $t' > \pi\omega_p^{-1}$ and this continuous phase slippage will reduce the rate of increase of $\omega_S(z,t)$ with z. Plugging in some numbers, for $n_e = 4 \times 10^{18}\text{cm}^{-3}$ we have $\lambda_p = 17$ μm and thus a maximum probe pulse duration of $\tau_i = \lambda_p/c/2 = 28$ fs. For $\tilde{n}/n_e = 0.3$ and $L = 200$ μm, the shift in the centroid wavelength is about $\Delta\lambda_{shift} = 20$ nm. Although this is certainly measurable since it is on the order of the bandwidth of the laser itself, the linear expansion around $t = \pi\omega_p^{-1}$ is not valid for our chosen τ_i, and even less valid if wave-steepening is significant due to strong harmonic content or radial motion of the plasma electrons.

For very large frequency shifts, as in ionization-induced blue shifts – see (Siders et al. 1996a) and references therein – centroid frequency shifts or large "wings" on the blue side carry information on the ionization dynamics while making clear centroid frequency shifts due solely to wakefields difficult to measure. Ultimately, even in a background-free experiment, Equation (17) is depressing if one wants to map out the variation of the plasma wave vs delay time of the probe with respect to the pump. The frequency shift requires photon acceleration and thus becomes useless near the peaks and valleys of the plasma wave where $d\sin(\omega_p t)/dt \sim 0$. However, since the "average" density in this experiment depends on where the short probe pulse rides on the wake, $\phi_o(L,t_1)$ will oscillate with delay time t_1. This term, which is "just" an overall phase shift in Equation (17), has well-separated extrema at densities of $n_e \pm \tilde{n}$. The difference in phase-advance from the top to the bottom of the wave is $\Delta\phi_o(L) = 2\pi(L/\lambda_i)(\tilde{n}/n_{cr}) \approx 0.01$ radians for the example given above. It is now clear that, for a *short* probe pulse (where the pulse postion t_1 has some meaning), the $\phi_o(L, t_1)$ term in Equation (14) is the most useful if it can be measured somehow. This is the subject of Section 7.4 below, with the details in Section 7.4.3.

7.4   Spectral interferometry

Using the centroid shift via photon acceleration as in Section 7.3 is analogous to doing interferometry without background fringes! Without the background fringes in the interferograms of Figure 10, the static (constant in space) or "DC" phase change (the first term in the integrand of Equation (13)) would have to be measured by the far-field (k-space) perturbation imposed onto the probe beam as represented by the spatial Fourier transform $\mathcal{F}_{x,y}$ of the probe beam at the plasma exit $E(x, y) = \widetilde{E}(x, y) \exp[i\phi(x, y)]$

$$E(k_x, k_y) \equiv \mathcal{F}_{x,y}\{E(x, y)\} = \mathcal{F}_{x,y}\{\widetilde{E}(x, y) \exp[\int_0^L \eta_p(\vec{r}) 2\pi/\lambda_i dz]\}.$$

However, this approach is sensitive only to the spatial variation of the imparted phase rather than the imparted phase itself. The far-field approach is thus the basis for the techniques of Schlieren, shadowgraphy, dark-field, etc. which can all be classified as angular changes to the probe beam, destined to be observed in (k_x, k_y) space. Similarly, for the photon acceleration analysis (with its lack of a reference beam) of Section 7.3, we also had to resort to the equivalent "far field" transform of the temporal phase $\phi_S(t) = \omega_i t - \phi(\eta_p)$

$$I_S(\omega) = |\mathcal{F}\{E_i \exp[\phi_S(t)]\}|^2 = |\mathcal{F}\{E_i(t) \exp[-i(\omega_p \epsilon_1 t)] \exp[-i\phi_o(L, t_1)]\}|^2$$

and, as in the spatial case described above, it was the derivatives in $\phi_S(t)$ (Equation (16)) that contributed to $I_S(\omega)$ in Equation (17) where the most useful term $\phi_o(L, t_1)$ ends up as just an undetected phase shift.

As was pointed out in Section 6.2, by adding the reference beam to the interferometer, the "DC" *spatial phase* information from the plasma $\phi(x, y)$ is placed onto a "carrier wave"; i.e., the spatial fringes due to the angular (k-space) interference of the two beams. In effect, *all* the information, including the otherwise DC spatial phase, is now available in derivative form (by the placing it on a carrier wave) and therefore accessible through the Fourier transform process.

So what is the analogous process for extracting the DC *temporal phase* $\phi_o(L, t_1)$? Recall that this (very informative) phase term was thrown out of the Equation (17) result. If you, the reader, are prone to think that Nature is symmetric in the conjugate-pair variables time/frequency and space/anglular-frequency, then you may already have an intuition as to the answer.

Figure 16 shows an experimental setup identical to that of Figure 15 except that there are a two beam splitters (not shown); one to split off a "reference" pulse $E_{ref}(t)$, and the other to recombine it back onto the same optical path as the probe beam $E_{pr}(t - \tau)$ with an adjustable time delay τ. The pair of initially identical pulses may be timed relative to the pump pulse such that the reference pulse traverses the interaction region *before* the pump pulse (as shown in Figure 16) and thus before any plasma wave is generated. In a sense, it "bypasses" the plasma (in time) just as the reference beam in spatial interferometry bypassed the plasma (in space), as in Figure 10(a). The reference beam's temporal phase is just that in the original pulse plus any accumulated in traversing the apparatus with the plasma off. Also shown on the right side of Figure 16 are "spectral fringes", the very same fringes that our intuition predicted as necessary to recover the DC temporal phase factor $\phi_o(L, t_1)$. These fringes follow from the mathematics of Fourier transforms which appear to be the result of the spectrograph being able to do (all by itself) a definite integral! In Section 7.4.1 below, we will attempt to justify this optical calculus of the spectrograph. In Section 7.4.2 we will use the optical calculus to derive spectral fringes in the absence of plasma while the analysis leading to the recovery of $\phi_o(L, t_1)$ will be presented in Section 7.4.3.

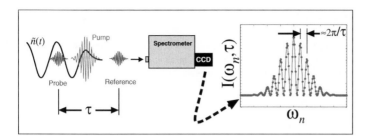

Figure 16. *The basic setup for frequency-domain interferometry (or longitudinal interferometry). A reference pulse precedes the pump pulse (no phase modulation) and interferes with the probe pulse (with plasma-induced phase modulation) at the exit of a spectrometer. A typical lineout from the camera (inset at right) shows the "spectral fringes". (From Clayton (2004). With permission.)*

7.4.1 The calculus of spectrographs

By now, we are quite familiar with the Fourier-transform properties of the spectrograph. We have argued that if all the light enters the spectrograph in a time shorter than the integration time of the spectrograph, we can take the time-integration limits to $\pm\infty$ in the resulting Fourier transform. But why does a spectrograph perform a time integration? We saw evidence for this in the experimental forward-scattered spectrum of Figure 12(b) where a single reflection off of a highly-tilted grating in the notch filter caused the the data to be stretched out in time. Each "color" or spectral interval $\omega_a \pm 0.5\delta\omega_a$ is stretched by an amount $\Delta\tau_a = (L_{a,max} - L_{a,min})/c$ where $L_{max,a}$ ($L_{min,a}$) is the longest (shortest) optical path between the slit of the spectrometer and the pixel which detects the spectral band $\omega_a \pm 0.5\delta\omega_a$ on the CCD camera. Here, L_a is the path length of the light hitting the *center* of the grating to the ω_a pixel on the CCD camera. If the grating was set to "zero order", it is equivalent to a mirror ($\theta_{in} = \theta_{exit}$) and, by Fermat's principle (light takes the shortest path), $\Delta\tau_a$ would be zero; *all* paths have the length L_a. But for a spectrograph to produce the intensities of the spectral components of the incoming pulse, we operate at a diffraction order $m > 0$. In this case, the grating equation states that $\sin(\theta_{in}) + \sin(\theta_{exit}) = m\lambda_a/d$ and as a result, $\Delta\tau_a > 0$. Here d is the groove spacing for the grating in use and λ_a is the wavelength associated with ω_a. If the grating is fully illuminated with light, then $L_{max,a}$ and $L_{min,a}$ are just $L_a \pm X_g \sin(\theta_{grating})/2$ where X_g is the size of the grating in the dispersion direction, $\theta_{grating}$ is the angle that the grating needs to be rotated to move from $m = 0$ to $m = 1$. If the grating size is, say, 40mm and $\theta_{grating} = 20°$, then $\Delta\tau_a = \sin(20°) \times 4\text{cm}/c = 46\text{ps}$! This is an overestimate since one usually does not "overfill" the grating but rather chooses an "f-number" ($\approx$ the full angle of the beam exiting the slit) such that the intensity at the edges of the grating is near zero. With this weighting factor, $\Delta\tau_a$ would be closer to 10ps for the above example.

From the above discussion we conclude that in practice $\Delta\tau_a$ is many ps. How is this related to an integration in time? Let us consider a particular case where $\theta_{in} = 0$ and $\theta_{exit} = 20°$. For this case, the entire used surface of the grating is illuminated at exactly the same time. Suppose the pulse is 60 fs in duration. The ω_a portions of the first 20 fs of the pulse (we'll call this the pulse "head") will already be traveling towards the ω_a pixel before the final 20 fs of the pulse (the pulse "tail") even strikes the grating. However, the $\approx$ 40 fs time advantage given to the pulse head is negligible compared to the now $\approx$ 10 ps duration of the pulse head. How can the diffracted pulse head suddenly become 500× longer? It is because the wavefront is tilted: the pulse came in shaped like a pancake (or possibly a crêpe) with the group velocity vector $\hat{g}$ normal to the surfaces of constant intensity but exits the grating with, in this case, $\hat{g}$ rotated about $20°$ from the constant-intensity surfaces. Note that the angle $\hat{g}$ within the spectrograph will, by our grating choice and appropriate setting of $\theta_{grating}$, necessarily take all the ω_a light to the ω_a pixel. In effect, the ω_a portions of the entire pulse will be detected by the ω_a pixel; not by "first-come, first-served", but all times within the pulse *integrated* together into one long photon stream.

7.4.2   The origin of spectral fringes

In the calculus of spectrographs of Section 7.4.1, we discussed how the grating stretches a diffracted *single* pulse. For simplicity, let us suppose that the temporal phase of this pulse, perhaps the reference pulse in Figure 16 with pulse length τ_{ref}, can be described by the first two terms in the temporal-phase expansion $\phi_{ref}(t) = \phi_{0,ref} + \phi_{1,ref}t + \cdots$. In this case, the "instantaneous frequency" $d\phi_{ref}(t)/dt = \phi_{1,ref}$ is a constant and the pulse is said to be "transform limited". If the pulse is derived from a mode-locked laser (as is usually the case), all n modes, with temporal phases $\phi_n(t) = \psi_n + \omega_n t$, contributing to the pulse will have a the same $\phi_n(t)$ at a particular time $t = t_{peak}$ corresponding to the peak-intensity of the pulse, i.e., such that $I_{ref}(t = t_{peak})$ is maximum. At this particular time, all the modes add *in phase* resulting in an intense, short pulse. As $|t - t_{peak}|$ approaches τ_{ref}, the phases of the n modes begin to deviate from each other and the net field drops. The phases $\phi_n(t)$ need not vary rapidly with n, but since $\log(n) \gg 1$ (a huge number of modes, in other words), any deviation from $t = t_{peak}$ leads to rapid 'phase mixing' within the addition of the multitude of terms. This is one mathematical explanation of a short, mode-locked pulse (without the math!).

Let $t = t_{peak} = 0$ be the time when the peak of the pulse $I_{ref}(t)$ is just entering the spectrograph. Furthermore, let the constant-phase term $\phi_{0,ref} = 0$. Following the example give earlier, a pulse with $\tau_{ref} = 60$ fs diffracts off the grating and pixel a detects the light within a bandwidth $\omega_a \pm 0.5\delta\omega_a$. Similarly, any arbitrary pixel $b = a \pm k$ (with $a \pm k$ in the range 1 to the total number of pixels) detects light in the spectral band $\omega_b \pm 0.5\delta\omega_b$. The grating equation is expressed in terms of wavelength (not frequency) so that, while the wavelength span $\delta\lambda_b$ covered by a single pixel is $\approx$ independent of the pixel number b, the *frequency* span of pixel depends strongly on pixel number since $\delta\omega_b \approx c\delta\lambda_b/\lambda_b^2$ and thus $\delta\omega_b$ is a function of b. Each spectral band of the pulse is stretched by a large amount: i.e., by $\Delta\tau_b \sim 10$ps $\gg \tau_{ref}$. Since each "color" ω_b in our single reference pulse has a fixed or slowly-varying phase within $\pm 0.5\delta\omega_b$, the energy detected by each pixel is simply the integrated spectral power at that color.

We now add an identical pulse $E_{pt}(t)$, the *probe* pulse or second pulse in Figure 16, separated in time by τ from the reference pulse. In this case, the phase relationships of the frequency components of the *total* integrated spectral spectral power changes dramatically as we move from, say, pixel a to pixel $b = a \pm 1, 2, 3, ..., k$. The time when the peak of the pulse $I_{pr}(t)$ just enters the spectrograph is $t = \tau$. Since it is unlikely that $c\tau$ is exactly an integer multiple of the laser wavelength, there must be in general a constant-phase term $\phi_{0,pr} = \psi$. If $\tau \approx 1.5$ ps (a typical value), $\tau \ll \Delta\tau_b$ and so both the reference and probe pulses will be time-integrated by the spectrograph. Now the meaning of time-integration is more clear; although each pulse entered the spectrograph at different times, they will *overlap in time* while being detected by the CCD camera pixels. Both pulses travel at c to the grating. We consider the phase advance of various colors in propagating from the grating to the various pixels. Suppose $z = 0$ is defined to be the center of the grating so that the distance to the ω_a pixel is L_a, as discussed before. We can choose ω_a as a color where, at the ω_a pixel, the probe and reference electric fields (which we are considering as identical and equal to $E_i(t)$) add in phase; i.e. one of the peaks in Figure 16

$$
\begin{aligned}
E_{ref}(t) + E_{pr}(t) &= [\widetilde{E}_{ref}(t)\cos(\omega_a(t - L_a/c) \\
&\quad + \widetilde{E}_{pr}(t)\cos(\omega_a t(1 - L_a/c) - \omega_a \tau + \psi)] \\
&= 2E_i(t),
\end{aligned}
\tag{18}
$$

where, as discussed in Section 7.4.3 below, the $E(t)$ are, in general, complex and the $\widetilde{E}(t)$ are real. Likewise, we can choose ω_b as a color where, at the ω_b pixel, the probe and reference electric fields are $180°$ out of phase. The total field is

$$
\begin{aligned}
E_{ref}(t) + E_{pr}(t) &= [\widetilde{E}_{ref}(t)\cos(\omega_b(t - L_b/c) \\
&\quad + \widetilde{E}_{pr}(t)\cos(\omega_b t(1 - L_b/c) - \omega_b \tau + \psi)] \\
&= 0.
\end{aligned}
\tag{19}
$$

For Equation (18), we need $\omega_a \tau = \psi$ to obtain constructive interference of the two pulses and for destructive interference in Equation (19), we need $\omega_b \tau = \psi \pm \pi$. Defining $\Delta\omega_{half} = \omega_b - \omega_a$, we have $\Delta\omega_{half} = \pm\pi/\tau$ as the distance between a bright fringe and a the nearest dark fringe. The distance between consecutive bright fringes $\Delta\omega$ is thus $\pm 2\pi/\tau$, as indicated on the right side of Figure 16.

Experimentally, the distance between consecutive bright fringes $= 2(\omega_b - \omega_a)$ can be express in terms of pixels. The change was from pixel a to pixel $b = a \pm 1, 2, 3....k$ so the difference is [pixel $b = a \pm 1, 2, 3, ..., k$]−[pixel a] $= \pm k$. The minimum $|k|$ to "resolve" this fringe period comes from the Nyquist sampling limit which states that we need at least 2 samples per period or $|k| > 2$. The pixel-spacing of fringes in practice depends on the number of pixels/mm on the CCD camera, the "focal length" of the spectrograph, the number of lines/mm on the grating, and the pulse separation τ. We assume that the slit width, when imaged onto the CCD camera, is less than the size of a pixel.

Given this physical picture of how two short pulses separated in time can produce spectral fringes, we can go ahead and derive these fringes mathematically. Back in Equation (1), we supposed that the pulse to be measured was fully described by its electric field $E(t)$ which is generally true in vacuum. In Sections 6 and 7, we described the probe beams as essentially plane waves with temporal variations given by $\cos(\omega_i t)$. For short pulses we must allow for the envelope of the pulse to vary and so the general laser field is

$$
\begin{aligned}
E(t) &= \widetilde{E}(t)\cos(\phi(t)) \\
&= \widetilde{E}(t)\tfrac{1}{2}\{\exp[i\phi(t) + \exp[-i\phi(t)]]\} \\
&= \mathrm{Re}\,\{\widetilde{E}(t)\exp[i\phi(t)]\} \\
&\rightarrow \widetilde{E}(t)\exp[i\phi(t)],
\end{aligned}
\tag{20}
$$

where $\widetilde{E}(t)$ is the (real) amplitude or envelope which varies slowly on an optical-cycle timescale and $\phi(t)$ is (again) the temporal phase which is real and a slowly-varying function of time such that it can be represented by the first few terms of the expansion

$$
\phi(t) = \phi_0 + \phi_1 t + \phi_2 t^2 + ...
\tag{21}
$$

In the final form of Equation (20), we implicitly define a complex representation for $E(t)$. For nonlinear operations on complex functions, such as $|E(t)|^2 = E(t)E^*(t)$ (where $E^*(t)$ is the complex conjugate of $E(t)$), we must remember to use the second form of Equation (20) to evaluate the operation. With the Equation (21) representation of the temporal phase, the FROG data of Figure 6 can be seen to have a temporal phase which is essentially quadratic; that $|\phi_2| > 0$

such that the instantaneous frequency $\omega(t) = d\phi(t)/dt$ changes linearly in time: the pulse indeed has a linear chirped. The Fourier transform $\mathcal{F}\{E(t)\}$ has the complex representation

$$E(\omega) = \widetilde{E}(\omega)\exp[i\varphi(\omega)], \tag{22}$$

where $\widetilde{E}(\omega)$ is the (slowly varying) real spectral amplitude and the *spectral phase* $\varphi(\omega)$ can be expressed as

$$\varphi(\omega) = \varphi_0 + \varphi_1\omega + \varphi_2\omega^2 + ... \tag{23}$$

Again, we must use the real part of $E(\omega)$ for any nonlinear mathematical operation. Ultimately, all results are real, but it is convenient to use the complex representations for linear operations during the derivation of these results, especially in the Fourier transforms.

We can now reconsider the situation in Figure 16 where there are two pulses entering the spectrograph. The magnitude squared of the Fourier transform of the detected field (the signal on the CCD camera, as we have stated many times in this chapter) is

$$\begin{aligned} I_S(\omega,\tau) &= \frac{1}{2\pi}\left|\int_{-\infty}^{+\infty} e^{-i\omega t}\left[E_{ref}(t) + E_{pr}(t-\tau)\right]dt\right|^2 \\ &= \left|E_{ref}(\omega) + E_{pr}(\omega)e^{-i\omega\tau}\right|^2. \end{aligned} \tag{24}$$

The electric fields in both time- and frequency-domains in Equations (24) are all still in their complex forms. Using $|a+b| = |a|^2 + |b|^2 + 2\mathrm{Re}\{ab^*\}$, we have

$$I_S(\omega,\tau) = |E_{ref}(\omega)|^2 + |E_{pr}(\omega)|^2 + 2\mathrm{Re}\{E_{ref}(\omega)E_{pr}^*(\omega)e^{-i\omega\tau}\}. \tag{25}$$

Equation (25) is the temporal equivalent of a interferogram. To get a feeling for this, let us consider what $I_S(\omega,\tau)$ would look like if the pump in Figure 16 were to be blocked so that there would be no plasma or plasma wave. The only light entering the spectrometer in this case is from two identical pulses $E_i(t)$ and $E_i(t-\tau)$ separated in time by τ. In this case, Equation (25) reduces to

$$I_i(\omega,\tau) = 2|E_i(\omega)|^2[1 + (\cos\omega\tau)] \tag{26}$$

where $E_i(\omega) = \mathcal{F}\{E_i(t)\}$ and we have used $\mathrm{Re}\{e^{-i\omega\tau}\} = \cos(\omega\tau)$. Equation (26) shows clearly the promised "spectral fringes" (which we also obtained the hard way in the interpretation of Equations (18) and (19)). These fringes will act as the carrier wave that we concluded is necessary to obtain the (otherwise invisible) phase shift $\phi_o(L,t_1)$ that is imparted to the probe beam upon transiting the plasma wave. A lineout of the image (the "spectrogram") on the CCD camera would look similar to that on the right of Figure 16. The $[1 + (\cos\omega\tau)]$ says that the spacing should be $2\pi/\tau$ (for identical pulses) with 100% modulation. Of course, if either beam is blocked, the lineout would be simply the spectrum $I_i(\omega) = |E_i(\omega)|^2$.

7.4.3 Recovery of the temporal phase

To recover the temporal phase, we must first take our data, described in Equation (24), and extract the *change* in the probe beam spectral phase (relative to that of the reference beam) which is given by $\varphi_{ref}(\omega) - \varphi_{pr}(\omega)$. This change is due to the interaction of the probe with the plasma and is therefore related to $\phi(\eta_p(t))$, the entire temporal phase (DC and AC components). The entire process will involve: (1) finding a decent representation of the spectral interferogram; (2) analyze the "spectrum" of ωs in the data via a forward Fourier transform of the data; (3) extracting the phase variation within the fringes (the variation in spacing vs ω) by filtering the "spectrum" of ωs and returning to ω-space via an inverse Fourier transform; (4) realize that the extracted phase variation within the fringes is actually the change in the probe beam spectral phase; and (5) return to t-space with the inverse Fourier transform after adding the extracted spectral changes to the phase of the *reference* beam. This assumes that the spectral phase of the reference beam is identical to that of the plasma-off probe beam. All this and some other minor details follow.

Starting from the bottom line of Equation (24), we have

$$I_S(\omega, \tau) = \left| \widetilde{E}_{ref}(\omega) \exp[i\varphi_{ref}(\omega)] + \widetilde{E}_{pr}(\omega) \exp[i\varphi_{pr}(\omega) - \omega\tau] \right|^2,$$

where we have used Equation (22) to expand both $E(\omega)$s. Again using $|a + b| = |a|^2 + |b|^2 + 2\mathrm{Re}\{ab^*\}$, we have

$$I_{(\omega, \tau)} = \widetilde{E}^2_{ref}(\omega) + \widetilde{E}^2_{pr}(\omega) + \widetilde{E}_{ref}(\omega)\widetilde{E}_{pr}(\omega)P_{cross}(\omega), \tag{27}$$

where

$$\begin{aligned} P_{cross}(\omega) &= \{\exp[i(\varphi_{ref}(\omega) - \varphi_{pr}(\omega) + \omega\tau)] + \exp[-i(\varphi_{ref}(\omega) - \varphi_{pr}(\omega) + \omega\tau)]\} \\ &= 2\cos(\varphi_{ref}(\omega) - \varphi_{pr}(\omega) + \omega\tau). \end{aligned} \tag{28}$$

Equations (27) and (28) together fully describe the shape of the spectrogram lineout shown on the right of Figure 16. In particular, Equation (28) shows that the addition of the reference pulse which, as described physically in Section 7.4.2, has created a carrier wave (the spectral fringes) and thereby make the entire function $\varphi_{pr}(\omega)$ available. Thus the desired DC *temporal* phase term $\phi_o(L, t_1)$ is, in principle, available. Note that $\varphi_{ref}(\omega) - \varphi_{pr}(\omega)$ is typically small (we estimated $\sim 10\mathrm{mrad}$ earlier) and varies slowly with ω relative to the carrier wave $\omega\tau$. Thus, the spacing ($\approx 2\pi\mathrm{rad}$) and/or location of the spectral fringes will be slightly perturbed after the probe beam has accumulated the temporal phase imprint described by $\phi(\eta_p(t))$ given by Equation (11). The perturbed fringe spacing is equivalent to a carrier wave (the fringes) with a slight phase variation. This is what we are after.

A very practical Fast Fourier Transform (FFT) numerical technique was developed by Takeda et al. (1982) to isolate the phase variation $[\varphi_{pr}(\omega) - \varphi_{ref}(\omega)]$ from 1-D or 2-D interferograms as long as the phase variation is very slow with respect to the carrier wave. As discussed above, this is generally the case. Starting now from the last term in Equation (25), we can write

$$2\mathrm{Re}\{E_{ref}(\omega)E^*_{pr}(\omega)e^{-i\omega\tau}\} = E_{ref}(\omega)E^*_{pr}(\omega)e^{-i\omega\tau} + E^*_{ref}(\omega)E_{pr}(\omega)e^{i\omega\tau} \tag{29}$$

where we have used $\mathrm{Re}\{a\} = (a + a^*)/2$. Let $a(\omega) \equiv (|E_{ref}(\omega)|^2 + |E_{pr}(\omega)|^2)$ and $c(\omega) \equiv E_{ref}(\omega)E^*_{pr}$. Note that $c(\omega)$ is complex and contains the desired spectral phase $\exp[i(\varphi_{pr}(\omega) - \varphi_{ref}(\omega))]$. Using these definitions of $a(\omega)$ and $c(\omega)$, we can now rewrite Equation (25) as

$$I_S(\omega, \tau) = a(\omega) + c(\omega)e^{i\omega\tau} + c^*(\omega)e^{-i\omega\tau}. \tag{30}$$

Equation (30) is equivalent to Equation (27) and Equation (28) and, in addition to being compact, it is very useful for understanding the FFT numerical technique that is used to extract $[\varphi_{pr}(\omega) - \varphi_{ref}(\omega)]$.

Since the phase variation is encoded within the carrier wave, we must perform a "spectral analysis" of the carrier wave. This requires the Fourier transform of the signal $I_S(\omega, \tau)$ of Equation (30) giving

$$\begin{aligned} \mathcal{F}\{I_S(\omega, \tau)\} &\equiv \frac{1}{2\pi} \int_{-\infty}^{+\infty} e^{-i\omega\check{t}} I_S(\omega)\mathrm{d}\omega \\ &= A(\check{t}) + C(\check{t} - \tau) + C^*(\check{t} + \tau), \end{aligned} \tag{31}$$

where $\check{t}$-space represents the "spectral" space of functions of ω and $A(\check{t})$ and $C(\check{t})$ are the Fourier transforms of $a(\omega)$ and $c(\omega)$, respectively. Note: an *inverse* Fourier transform would bring us back to the physical t-space—not what we want just yet. Equation (31) has peaks at $\check{t} = \pm\tau$ and at $\check{t} = 0$. To spectrally analyze just the *carrier wave*, we can choose either of the off-zero peaks. Assuming that these sidebands are well separated from the $\check{t} = 0$ feature (which will be the case if the $|E_{ref,pr}(\omega)|^2$ has very little structure on the scale of $2\pi/\tau$ or, equivalently, if τ is large enough) we can "window" the $\check{t}$ data with a smooth (apodized) band-pass filter function $G(\check{t}-\tau)$ centered at $\check{t} = +\tau$. The width of our apodization function $G(\check{t}-\tau)$, perhaps a "Hanning" function or "super-Gaussian" function, must be wide enough to collect all the information around $\check{t} = +\tau$ but not so wide as to pick up information from the DC term. Applying this linear filter, we have

$$C(\check{t} - \tau) \approx G(\check{t} - \tau)\mathcal{F}\{I_S(\omega, \tau)\}.$$

$C(\check{t})$ is obtained by translating $C(\check{t}-\tau)$ along the $\check{t}$-axis by an amount τ: i.e., $C(\check{t}) = e^{i\omega\tau}C(\check{t}-\tau)$. We can measure τ by "turning off the physics" (blocking the pump pulse) and taking the centroid of the resulting $C_{pump-off}(\check{t} - \tau)$ as τ. Alternatively, if the spectrogram is 2-D and the x-width of the reference and probe beams is wider than the pump-perturbed region $(2x_{max})$, τ may be obtained from the centroid of $C(\check{t} - \tau, x > x_{pump})$.

The function $c(\omega)$ is obtained by taking the inverse Fourier transform (performed numerically using the IFFT, the inverse fast fourier transform) of $C(\check{t})$

$$
\begin{aligned}
c(\omega) &= \int_{-\infty}^{+\infty} e^{i\omega\check{t}}C(\check{t})\mathrm{d}\check{t} \\
&= \int_{-\infty}^{+\infty} e^{i\omega\check{t}}e^{i\omega\tau}G(\check{t} - \tau)\mathcal{F}\{I_S(\omega, \tau)\}\mathrm{d}\check{t} \\
&= \widetilde{E}_{ref}(\omega)\widetilde{E}_{pr}(\omega)exp[i(\varphi_{ref}(\omega) - \varphi_{pr}(\omega))],
\end{aligned}
\tag{32}
$$

where the last equality in Equation (32), which follows from the definition of $c(\omega)$, is put here just as a reminder. Here both amplitude functions $\widetilde{E}_{ref,pr}(\omega)$ and both phase functions $\varphi_{ref,pr}(\omega)$ are real. The desired quantity $\Delta\varphi(\omega) \equiv \varphi_{ref}(\omega) - \varphi_{pr}(\omega) = \mathrm{Arg}\{c(\omega)\}$ can be mathematically obtained from Equation (32) by $\mathrm{Im}\{\ln(c(\omega))\}$ (since $\ln(a)\exp(ib) = \ln(a)+ib$). Since the Fourier transform in Equation (31) and the inverse Fourier transform in Equation (32) are done by a numerical FFT and IFFT, the function $\Delta\varphi(\omega)$ is returned in the range $\pm\pi$ so, in general, there will be phase jumps of $\approx \pm 2\pi$ within the pixel-sampled data. Fortunately, many programming languages have a command to produce a continuous phase function by "concatenation" of the "wrapped" phase returned by the IFFT. Failing that, there is an excellent description of this proceedure in Takeda et al. (1982).

One can rightfully question as to why we would need to concatenate the phase function when we said that at the end of Section 7.3 that the peak-to-valley phase change in a plasma wave of $\tilde{n}/n_e = 0.3$ was only ~ 10 mrad (for $\lambda_i = 800$ nm, $L = 200$ μm, and $n_e = 4 \times 10^{18}$cm^{-3}). However, the probe beam sees the background plasma density and thus has an additional phase shift of $\varphi_{pr,plasma} \approx 0.5k_i L n_e/n_{cr} \approx 3.7$ rad for the same parameters. Thus $\Delta\varphi(\omega) \equiv \varphi_{ref}(\omega)-\varphi_{pr}(\omega) = \mathrm{Im}\{\ln(c(\omega))\}$ is non-trivial and we must "unwrap" the phase prior to the inverse Fourier transform needed to return to t-space (step (5) of the 5-step process discussed earlier).

By the definition of $\Delta\varphi(\omega)$, we make the substitution $\varphi_{pr}(\omega) = \varphi_{ref}(\omega) - \mathrm{Im}\{\ln[c(\omega)]\}$ to obtain $\mathcal{F}^{-1}\{E_{pr}(\omega)\}$

$$E_{pr}(t) = \widetilde{E}(t)e^{i\phi_{pr}(t)} = \int_{-\infty}^{+\infty} e^{i\omega t}\widetilde{E}_{pr}(\omega)\exp[i\varphi_{ref}(\omega) - \mathrm{Im}\{\ln[c(\omega)]\}]\mathrm{d}t \tag{33}$$

and $\mathcal{F}^{-1}\{E_{ref}(\omega)\}$ is

$$E_{ref}(t) = \widetilde{E}(t)e^{i\phi_{ref}(t)} = \int_{-\infty}^{+\infty} e^{i\omega t}\widetilde{E}_{ref}(\omega)\exp[i\varphi_{ref}(\omega)]\mathrm{d}t. \qquad (34)$$

Finally, noting that $\phi_{pr}(t) = \mathrm{Im}\{\ln[E_{pr}(t)]\}$ and $\phi_{ref}(t) = \mathrm{Im}\{\ln[E_{ref}(t)]\}$, the temporal phase due solely to the plasma $\phi(\eta_p(t)) = \phi_{pr}(t) - \phi_{ref}(t)$ is obtained by combining Equations (32), (33), and (34) and can be written in the form (Kim et al. 2002a)

$$\phi(\eta_p(t)) = \mathrm{Im}\left\{\ln\left[\frac{\int_{-\infty}^{+\infty} e^{i\omega t}\widetilde{E}_{pr}(\omega)\exp\left[i\varphi_{ref}(\omega) - \mathrm{Im}\{\ln[c(\omega)]\}\right]\mathrm{d}t}{\int_{-\infty}^{+\infty} e^{i\omega t}\widetilde{E}_{ref}(\omega)\exp[i\varphi_{ref}(\omega)]\mathrm{d}t}\right]\right\} \qquad (35)$$

The amplitude $\widetilde{E}_{pr(ref)}$ is obtained from the measured probe (reference) spectrum as $(|\widetilde{E}_{pr(ref)}|^2)^{1/2}$. The only quantity not measured in Equation (35) is $\varphi_{ref}(\omega)$. This can be calculated from the FFT of a measurement of E_{ref} from a FROG trace or a SPIDER (Iaconis and Walmsley 1998) trace. Alternatively, $\varphi_{ref}(\omega)$ can be obtained by sending a transform-limited (TL) pulse ($\varphi_{TL}(\omega) =$ constant) and the reference pulse into the spectrograph. Since the TL pulse has a "flat" spectral phase, the fringe analysis gives $\varphi_{ref}(\omega)$ directly. Actually, there are even more ways to obtain $\varphi_{ref}(\omega)$ which we will discuss in the next Section.

Let us review. Recalling Equation (28) or the last equality in Equation (32), it has been our intension (so far) to extract only the DC temporal phase term $\phi_o(L, t_1)$ from the full phase term given, for example, in Equation (14) which also contained an AC temporal phase term $\epsilon_1 \sin(\omega_p t)$. Photon acceleration was somewhat sensitive to this AC term, but we ultimately decided that the sensitivity was poor and we would rather measure the DC term directly. This brought us into the idea of spectral interferometry as an analogous technique to spatial interferometry.

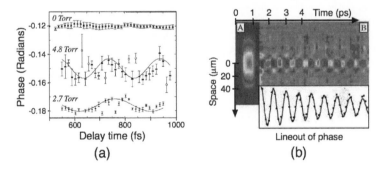

Figure 17. *(a) Results from longitudinal interferometry showing the measured phase shift $\phi_o(L)$ vs. delay for 0T, 2.7T and 4.8T of He. The data for 4.8T is not offset vertically while the 0T and 2.7T data are offset from their nominal value of 0rad for clarity. (b) Data similar to (a) but with imaging in radial direction and $\approx 0.7T$ He. Each vertical column of the image is a single shot. The time scale is at the top of the image and the radial space-scale is at the left. See text for details. ((a) From Siders et al (1996b). With permission. (b) From Marquès et al. (1997). With permission.)*

Figure 17(a) shows the first experimental results for $\phi_o(L, t_1)$ from a laser wakefield (Siders et al. 1996b). The setup was similar to Figure 16. The twin pulses were 800nm but polarized $90°$ with respect to the pump. $\Delta\varphi(\omega)$ for each point was obtained from a 100 shot average with and without the pump (≈ 40 sec for each data point). The reference phase $\varphi_{ref}(\omega)$ was obtained occasionally during the run by observing the spectral fringes by essentially "replacing" the probe pulse with a near-transform-limited pulse as discussed above. For the middle curve, the vacuum chamber was back-filled to 4.8 Torr of He and the pump and twin pulses were focused to 3.6 μm (giving a Rayleigh length of 52 μm). The reference pulse was *ahead* of the pump and the timing of the probe pulse was varied. In this "absolute" mode, since $\phi_o(L, t_1)$ includes the plasma itself, the fringes oscillate around $\phi_o(L, t_1) \approx -15$ mrad due to the background density of 3×10^{18} cm^{-3}. The oscillation period is ≈ 1–1.1 times that for a linear plasma wave. For such a tight focus, theses are radial plasma waves (the transverse ponderomotive force was estimated to be about 10$\times$ larger than the longitudinal ponderomotive force). Comparing the measured phase shifts with a model of the radial wakefield, the peak wave amplitude was estimated to be $\widetilde{n}_e/n_e \approx 0.8$. The lower curve for 2.7 Torr of He used the "differential" mode (Geindre et al. 1994) where both the probe and reference pulses were behind the pump pulse. The separation of the twin pulses, $t_1 - t_2 = \tau$ was fixed at 3/2 plasma periods and both twin pulses were simultaneously scanned in time away from the pump. In this case, $\phi_o(L, t_1 - t_2)$ oscillates about zero. The top curve shows the null scan (no pump) corresponding to the absolute-mode scan. The $\pm\sigma$ error bars in the, say, 4.8 Torr data indicate that the stability of the spectral fringes was not high. Experimental issues in this work included leakage of the intense pump pulse into the spectrograph (due to depolarization), pointing jitter of the pump and twin pulse, and fluctuations in the mode-locked lasers center wavelength and spectral shape.

Another group, working independently on a similar experiment, used frequency-doubled twin pulses which eliminated the pump-leakage issue (Marquès et al. 1996). This group also used the direction along the slit to image a slice of the radial plasma wake. This required focusing the twin pulses more weakly. For the very nice data is shown in Figure 17(b) (Marquès et al. 1997), the pump was focused to a spot size of 6μm and the twin pulses to a spot size of 140μm. The chamber was back-filled with He producing a fully-ionized density of 2.47×10^{16} cm^{-3}. This data was taken in the "differential" mode with the probe and reference pulses separated by 3/2 plasma periods. Each vertical column in the image is a single shot. For delays near 0.7ps, the reference is still in front of the pump and $\phi_o(L, t_1 - t_2)$ obtains a large ($\approx$ -80 mrad) contribution from the plasma. For times greater than about 1.3 ps, the entire reference pulse is in the radial wakefield and the phase oscillates about zero (as in the 2.7 Torr case of Figure 17(b)) with a $\phi_o(L, t_1 - t_2)$ excursion of $\approx \pm 7$ mrad. Also shown is a lineout down the axis. It is interesting to note (although difficult to see in this image) that when the on-axis density is lower than the background density, the off-axis density is a bit higher than background; consistent with radial wakes. For the data shown, the oscillation period is ≈ 1.05 times that for a linear plasma wave and $\widetilde{n}_e/n_e \approx 0.8$. The frequency, amplitude, and damping rate of the radial wake were measured as a function of density.

Although the multi-shot 2-D data of Figure 17(b) provides a powerful "image" of the plasma wave, variations in the temporal and/or radial structure cannot be correlated with, say, accelerated electrons on any given shot. One can imaging adding a second probe pulse (triplet pulses!) or a fourth probe pulse or perhaps a total of N probe pulses. The Idea is to view N longitudinal locations in the wakefield, not just one. One *did* image this and took $N \to \infty$ (Siders et al. 1996a). This is the concept behind frequency-domain holography.

7.5   Frequency-domain holography

As mentioned above, the original Idea for frequency-domain holography (FDH) was to extend the probe pulse to a longer temporal window. If this can be done, then $\varphi_{ref}(\omega) - \varphi_{pr}(\omega)$ can be measured to higher order. This implies that $\phi(\eta_p(t))$ can be extracted with the temporal dependence intact, not just the *local* $\phi_o(L, t_1)$ (where the probe beam resided) obtained in the multi-shot approach of the previous Section. In fact, there is no need to differentiate between the DC $\phi_o(L, t_1)$ term and the AC $\widetilde{n}(t)$ term. With the longer temporal window, if we measure

$\phi(\eta_p(t))$, we measure all terms; DC and AC. By adding *imaging* we can obtain $\phi(\eta_p(t,y))$ where $\hat{y}$ is the direction parallel to the slit.

Frequency-domain holography with a long temporal window was first achieved (Le Blanc et al. 2000) by sending the originally 70 fs, 800 nm probe pulse through its own frequency-doubling crystal; a 2 mm-thick Lithium Iodate (LiIO$_3$) crystal. The small phase-matching bandwidth (≈ 0.45nm) converts only a small fraction of the incident bandwidth, effectively reducing the number of phase-locked modes contributing to the pulse and thereby producing a long, $\approx$ 1ps 400 nm probe pulse. To maintain good temporal resolution, the reference pulse must still be short. Thus it is doubled in a 150 μm potassium dihydrogen phosphate (KDP) crystal where the phase-matching bandwidth is wide enough to produce an $\approx$ 70 fs 400 nm reference pulse. To produce these separate paths (to separate crystals) requires a modified Michelson interferometer shown in the upper left of Figure 18(a). The rest of the experimental setup is similar to that described in reference to Figure 17(b) in that the interaction region is imaged onto the slit of an imaging spectrometer producing a 2-D spectrogram. However, now there is a continuum of delay times (reference-probe separations) spanning about 1 ps! Take a look at Figure 19 to get a feeling for a 2-D spectrogram, but realize that due the the narrow bandwidth of the probe beam in Figure 18(a), the interference fringes will be most visible in the spectral overlap region of the reference and probe pulses.

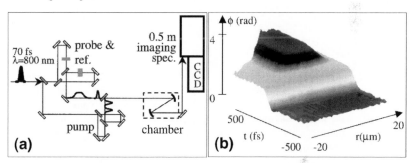

Figure 18. *(a) Schematic of the setup for the first frequency domain holography. (b) Spatially-and temporally-resolved phase variation due to an ionization front in 120-Torr air, measured in a single shot. (From Le Blanc et al. (2000). With permission.)*

The analysis of the 2-D spectrograms is done in much the same way as discussed in Section 7.4.3 and summarized in Equation (35) except that now $\phi(\eta_p(t,y))$ is a function of y and so each individual row with constant y is individually analyzed. One could replace the ω-dependent terms in the integrals of Equation (35) with their 2-dimensional counterparts; i.e., $\phi(\eta_p(t,y))$ depends on $\widetilde{E}_{pr}(\omega,y)$, $\widetilde{E}_{pr}(\omega,y)$, $\varphi_{ref}(\omega,y)$, and $c(\omega,y)$ with the desired $\varphi_{ref}(\omega,y)$ given by $\varphi_{ref}(\omega,y)-\text{Arg}\{c(\omega,y)\}$. However, lurking in our analysis and evident in Equation (35) is the assumption that the twin pulses are truly identical. After all, the left-hand side is $\phi(\eta_p(t)) = \phi_{pr}(t) - \phi_{ref}(t)$. This assumption is not necessary, since what we really want to know is $\Delta\phi_{pr} = \phi_{pr}^{off}(t) - \phi_{pr}^{on}(t)$ where the "off" and "on" refer to the presence or absence of the pump beam, respectively. Also, when the twin pulses are heavily chirped, we do not need to use FROG or SPIDER to find $E_{ref}(t)$ and Fourier transform the result to obtain $\varphi_{ref}(\omega)$. A fairly straight-forward cross-phase modulation measurement for the twin chirped pulses (Kim et al. 2002a) gives ϕ_2 in Equation (21) from which one can estimate φ_2 in given by Equation (23). The φ_1 term in Equation (23) results in a group delay and is thus not a very important term in $\varphi_{ref}(\omega)$.

The system in Figure 18(a) was checked by producing twin pump pulses, both of which

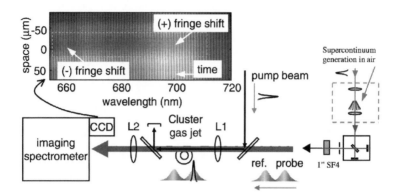

Figure 19. *Schematic of the experimental setup used to measure the time-dependent refractive index of laser-heated Ar clusters. The method for producing the supercontinuum light for the twin beams (insert at right). A typical spectrogram revealing, via the inherent chirp on the probe beam, the time-dependent phase shift of the probe beam due to the exploding Ar clusters. (From Kim et al. (2003). With permission.)*

resided *within* the probe pulse as shown by the modified Michelson interferometer in the lower center of Figure 18(a). The twin pump pulses induced two temporally separated phase changes onto the long probe pulse through cross phase modulation (XPM) via the twin-pump-induced nonlinear susceptibility χ^3 in a thin piece of glass. The result of this system check was that the temporal features from the recovered $\phi(\eta_{glass(I)}(t,y))$ had the correct temporal tracking of the pump pulses and the magnitude of the phase changes was in agreement with those expected from the twin-pump intensities. Note: the phase changes can now be measured with respect to the unperturbed portion of the spectrogram at large transverse distances compared to the pump spot size. Next, using a single pump beam, spatio-temporal ionization fronts in various pressures of air were measured. The pump was focused with $f/7$ optics to a maximum intensity of $\approx 10^{15}$ W/cm^2. Figure 18(b) shows the recovered phase $\phi(\eta_p(t,y))$ for 120 Torr of air. There are two temporal steps in $\phi(\eta_p(t,y=0))$ and two transverse step in $\phi(\eta_p(t=500 \text{ fs},y))$.

Each step is consistent with the phase change given in Equation (14); $\Delta(\phi_o(L)) \equiv 0.5k_i L n_e/n_{cr}$ where L is roughly a Raleigh range for $f/7$ focusing; $L \approx 2Z_r \equiv 2\pi w_o^2/\lambda_i$. For a focused spot size $w_o \approx f\lambda_i$, $L \approx \pi f^2 \lambda_i = 246$ μm giving $\Delta(\phi_o(2Z_r)) \approx 2.6$ rad for 120 Torr ($f = 7$ is the f-number). For doubly-ionized air, the observed 4 rad phase change is close to the 5.6 rad just estimated. The clear step-wise ionization observed in Figure 18(b) is "intuitively" obvious given the highly nonlinear variation of ionization rate with laser intensity as in the ADK model (Ammosov et al. 1987). However, as pointed out by Kim et al. (2002b), the accumulated phase acquired by the probe beam in propagating through the plasma depends on the full integral of $n_e(r,z)$ along the mutual path of probe and pump beams. In a thick target of gas, this means that the probe will experience much more phase accumulation due to singly-ionized gas than through doubly-ionized gas. This will tend to "wash out" the step-like structure anticipated to occur near the focus. Indeed, for the data of Figure 18(b), it was noted (Le Blanc et al. 2000) that for pressures lower than 120T, the data showed much more subtle step-wise ionization. It was concluded that ionization-induced refractive-divergence of the pump and probe beams (Monot et al. 1992) for pressures > 100 Torr may have limited the integrated $n_e(r,z)$ to some point where the steps were somehow clearly visible.

Single-shot frequency-domain holography took on a twist (actually, a stretch) when the con-

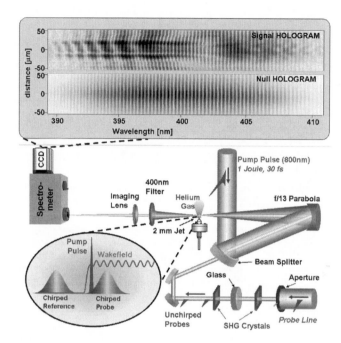

Figure 20. *Schematic of the setup for obtaining single-shot frequency-domain holograms. The relative timing of the reference, pump, and probe pulses (lower insert). Typical interferograms with and without the pump beam (upper insert). (From Matlis et al. (2006). With permission.)*

cept of twin pulses returned as twin *supercontinuum* pulses (Kim et al. 2002a). Supercontinuum (SC) generation in gases (see, for example, Bellini 2004) is readily observed for sub-100fs pulses for focused intensities $\sim 10^{13}$ W/cm^2; the threshold for self-focusing. The small inset at the right of Figure 19 shows the setup for producing the SP pulses. A portion of the main 800 nm, 80 fs pulse is focused into 1 bar of air producing the SC light. Due to the conical-dependent blue shift of the SC light (Corkum et al. 1986), the resulting SC light is apertured prior to the production of twin pulses in the Michelson interferometer. The twin pulses emerging from the interferometer are chirped in 2.5 cm of high-disperion leaded glass (SF4) to ≈ 1.5 ps in length and recombined with the 800 nm pump with the pump located in the second (reference) pulse. Also shown in Figure 19 is a schematic of the setup to measure the time-dependent complex refractive index of the pump-heated Ar cluster jet (Kim et al. 2003); i.e., of the "exploding clusters". Due to the aperturing of the SC beams, the spot size of the reference and probe are much larger than that of the pump beam. The Idea here was to see if a clustered gas could be efficiently ionized at a very low average density.

Some of the experimental benefits of using SC pulses for FDH are: (1) the wavelength is close to the 800 nm pump wavelength, limiting the ω-dependent group-velocity "walk-off" between the pump and the probe; (2) much of the large bandwidth of the SC pulse can be transported through the two 800nm mirrors (see Figure 19) allowing a bandwidth ($\Delta\omega_{SC}$) of ≈ 65 nm to enter the spectrograph. In principle, the temporal resolution can be a low as the corresponding coherence time $\tau_c \approx 2.8/\Delta\omega_{SC} \approx 10$ fs; (3) the central frequency of the SC light is about 700nm which allows for efficient transmission through the high-reflecting pump mirror, similar to the use of frequency-doubled probe pulses; and (4) the nearly linear chirp of $\omega_i(t)$ provides a visual clue of the time-dependent refractive index from the raw data. This latter benefit can be immediately seen in fringes shown in Figure 19. Since the (ordinary) refractive index in glass is larger for blue light than for red light, the red-to-blue chirp is correlated with increasing time.

In the spectrogram, we see that the exploding clusters (EC) have a positive $\phi(\eta_{EC})$ and changing sign later in the pulse. The retrieved phase $\phi(\eta_{EC}(t, y))$ (not shown) shows the early-time positive phase shift varies transversely like a focusing optic leading to self-guiding of the leading edge of the pump pulse through the exploding clusters. Later in time, the cluster po-larizability takes on the distinctive plasma appearance as shown by the negative $\phi(\eta_{EC})$. The plasma density at long delays is indeed quite low ($\approx 5 \times 10^{17}$ cm^{-3}) as desired. The ionization state of the Ar atoms is $Z \approx 10$. One of the future challenges is to reproduce these results in a low-Z gas such as H_2.

The most recent manifestation of single-shot frequency-domain holography is shown in Fig-ure 20 (Matlis et al. 2006). As oppose to the setup of Figure 18 (short and long transform-limited pulses), this uses twin chirped pulses, but formed in a very different way than those of Figure 19 (a supercontinuum pulse replicated by a Michelson interferometer). Looking at the bottom-right corner of Figure 20, the incoming 800 nm, ≈ 30 fs pulse is first apertured and sent through three optics. The first is a 200 μm-thick KDP crystal where a portion of the incoming pulse is converted to 400nm. This and the remaining 800 nm light then enter an ≈ 13 mm-thick piece of glass where, due to the large group-velocity difference between these two colors, the residual 800 nm pulse exits the glass ≈ 3 ps before the 400 nm pulse. An identical KDP crystal follows in which a portion of the residual 800 nm light is frequency-doubled. This setup produces exactly coaxial twin 400 nm pulses without the use of a Michelson interferometer. Such a simplification in the optical arrangement needed to produce twin pulses not only can make initial alignment easier, but can be more stable and thus the experimentalist need not wonder if "something is wrong" with the Michelson interferometer when problems arise!

After the second KDP crystal, the twin pulses are essentially unchirped. However, in travers-ing the 2.5 cm-thick mirror for the pump beam at 45°, a chirp is imparted and each of the twin pulses is stretched to about 1ps while maintaining the 3 ps separation. Due to the aperture at the beginning, the off-axis parabolic mirror focuses the twin pulses to a spot much larger than that of the f/13 focused pump beam. Upon exiting the plasma, a filter blocks out the 800 nm light and a lens images the interaction region onto the slit of an imaging spectrometer, as usual. The timing of the twin pulses with respect to the pump pulse is illustrated in the inset at the lower-left corner of Figure 20. The probe beam sees the ionization front in front of the pump beam as well as the plasma wave behind the pump beam as it co-propagates with the pump. The target is a He gas jet operated at quite low densities of $0.5 - 6 \times 10^{18}$cm^{-3}. The gas jet does not, however, produce a uniform-density plume of He. The longitudinal variation of the plasma density was measured with interferometry (not shown) similar to that shown in Figure 10 and it was found that the density rises to its maximum in about 1mm and falls over about 1.5 mm.

Two typical spectrograms are shown at the top of Figure 20 where the upper of these is with the pump on and the lower is with the pump off. Because the twin pulses are chirped, to zero-order one imagine time moving from long wavelengths to short, as in Figure 19. However, by performing the full phase-retrival algorithm discussed in Section 7.4.3, data on the 2-D structure of the plasma wave is reveal, as shown in Figure 21(a) (Matlis et al. 2006). The vertical axis is the transverse dimension (say, y) that was imaged onto the spectrometer slit. The horizontal axis is time. The gray-scale image is $\Delta\phi_{pr}(t, y) = \phi_{pr}^{off}(t, y) - \phi_{pr}^{on}(t, y)$ in radians with the laser traveling to the right. For this laser shot, the laser power was 30 TW and the density was 2.2×10^{18}cm^{-3}.

There are several features that are clear in the data of Figure 21(a): (1) An ionization front near $t = 0$; (2) A transverse fall off of the plasma density. In fact, careful examination shows the transition from He^{+2} to He^{+1} at $y \approx \pm 35$ μm; and (3) a plasma wave centered at $y = 0$ with curved wavefronts, especially towards later times. Due to the inhomogeneous plasma, the data is an integral over the longitudinal variation of the plasma density (as in Equation (13)) and so the measured $\Delta\phi_{pr}(t, y)$ cannot be read out directly as the magnitude of the density fluctuation $\tilde{n}$. However, as evidenced by the curved wavefronts, this is a strongly driven plasma wave. Shots at the same density but at 10 TW of laser power showed planar wave fronts. The reason the wave fronts curve is related to the quiver energy of the electrons that participate in the plasma wave. The quiver energy is high enough that the relativistic correction to the plasma frequency has a radial dependence. This is more clearly seen in the simulation in Figure 21(b). The rate at which the radial-dependent oscillation frequencies dephase (leading to the curvature) can be related to the wave amplitude (Bulanov et al. 1997). For these data, the wave amplitude is estimated to be $\tilde{n}/n_e^m ax \approx 0.5$.

The ability to obtain single shot images of such a plasma accelerator structure is the first step in being able to control the quality of the structure! Moreover, this information can be combined with the angular and energy spread of accelerated electrons, potentially allowing for the correlation of the electron beam parameters with the microscopic details of the plasma wave. Finally, if the accelerated charge is high enough, direct observation of beam loading and/or self-trapping could be "photographed", bringing almost real-time experimental observations of highly nonlinear phenomena to a point well beyond the capability of simulations!

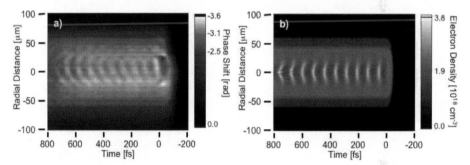

Figure 21. *(a) Contour or color plot of the plasma-induced phase shift $\Delta\phi_{pr}(t, y)$ of the probe beam from an analysis of a single, frequency-domain hologram taken with the apparatus described in Figure 20 for a relatively high-power (30 TW) laser shot. (b) Two-dimensional (r, z) simulation of $\Delta\phi_{pr}(t, r)$ for parameters close to those used to obtain the experimental data in (a). A series of such simulations confirm the analytically-derived variation of the wavefront curvature as a function of the amplitude of the plasma wave. (From Matlis et al. (2006). With permission.)*

8 Electron diagnostics

In this section, we will touch on *just a few* of the existing short bunch electron diagnostics. In addition to being of interest to the plasma accelerator community, these techniques are also being developed for diagnosing linacs and storage rings and for studying, for example, the bunching that occurs in free-electron lasers.

Before we discuss how to measure properties of accelerated (or injected) electrons, we must first define the properties that we wish to measure. All the information about a distribution of electrons (from which "properties" are assigned) could be derived from a simple matrix with six columns and N_b rows, where N_b is the number of electrons we are studying. The six columns may have the headings

$$x \quad p_x \quad y \quad p_y \quad z \quad p_z$$

where at particular time the coordinate of the, say, jth electrons is $\vec{r}_j = (x_j, y_j, z_j)$ and $\vec{p}_j = (p_{x,j}, p_{y,j}, p_{z,j})$ in physical space and momentum space, respectively. Thus each electron is described fully by its unique $\vec{r}$ and $\vec{p}$. If we let $\hat{z}$ be the direction that the ensemble $j = 1, 2, 3, ..., N_b$ of electrons (the beam or bunch or bunches) is moving, we can define $f_z(z, p_z)$ as the *longitudinal phase space* distribution; i.e., the distribution resulting from the projection of all N_b electrons, with a distribution function $F(x, y, z, p_x, p_y, p_z)$, onto the (z, p_z) plane. Similarly, the projection of all electrons onto the orthogonal planes (x, p_x) and (y, p_y) gives the *transverse phase space* distributions $f_x(x, p_x)$ and $f_y(y, p_y)$. The distribution functions F, f_x, f_y and f_z are not analytic functions, but "scatter plots" with one point per electron. Nevertheless, for $\log(N_b) \gg 1$, there are statistical representations of these distribution functions, and properties of the statistical representations are normally used to describe the corresponding physcial properties within the three, 2-D projections. For example, the *mean* values of x_j and y_j tells us where the centroid of the beam/bunch is while the standard deviation (about the mean)of x_j is related to the size of the beam/bunch in the $\hat{x}$-direction.

8.1 Electron beam or bunch parameters

Below is a list of various electron beam/bunch related quantities and their definitions.

Electron kinetic energy $E_{kin} = (\gamma - 1)mc^2$

The (invariant) mass of an electron in its rest frame is $m = E_o/c^2$ and is a property of the electron and therefore frame-invariant. Here, E_o is referred to as the electron rest-mass energy. Somehow, under Einstein's postulate that the speed of light c would be measured the same in any inertial (non-accelerating) reference frame, the more general expression for the invariant mass of an electron moving at v_z is $m \equiv (E^2 - (p_z c)^2)^{1/2}/c^2$ where E the *total* electron energy (rest-mass energy + kinetic energy E_{kin}) and $p_z = |\vec{p}|$ is the momentum (we neglect the assumed small contribution of p_x and p_y to the total momentum). Defining the "relativistic mass" as $M \equiv E/c^2$ conveniently allows for expressing momentum in a familiar way; that is $p_z = Mv_z$. Substituting these latter two equations involving M into the general expression of the invariant mass, we get $M = \gamma m$ where $\gamma \equiv (1 - v_z^2/c^2)^{-1} = (1 - \beta_z^2)^{-1/2}$ where $\beta_z \equiv v_z/c$. These definitions lead to the relationships

$$E = \gamma E_o \qquad \text{and} \qquad E_{kin} = E - E_o = (\gamma - 1)mc^2 \approx \gamma mc^2$$

as well as

$$c^2 p^2 = E^2 - E_o^2 \qquad \text{and} \qquad p = mc(\gamma^2 - 1)^{1/2} \approx \gamma mc$$

where the last approximations are for $\gamma^2 \gg 1$ and we have dropped the subscript z. We implicitly assume that the electrons are primarily moving in a particular direction (defined as the "longitudinal" $\parallel$ or z-direction) and that the velocities in the transverse ($\perp$) directions are such that $p_\perp/\gamma mc \ll 1$. Note that for $\gamma \gg 1$, $E_{kin} \approx p/c$. Thus, one will often see *momentum* as expressed as "MeV/c".

Electron bunch charge $Q_b = -N_b e$

This may seem trivial, but the electrons may be spread out over more than one accelerating bucket. A typical charge monitor connected to an oscilloscope or to an analog-to-digital (ACD) converter will integrate the total charge over closely spaced bunches. We will attempt to focus on the charge per accelerating bucket since this is what is ultimately is limited by "beam-loading". If N_b is the number of electrons in a single bunch, the the charge is $Q_b = -N_b e$ where e is the charge of a single electron. This is a scalar quantity of the electrons.

Bunch length $\sigma_\parallel$ or σ_z

For $\gamma_\parallel \gg 1$, this is a strictly a longitudinal property of the electron beam. The bunch length is usually expressed as σ_z where σ_z is the standard deviation of $f_z(z)$ about the mean $\langle z \rangle$. The bunch length σ_z is commonly referred to at the "RMS" bunch length. To calculate the standard deviation, one takes the square **root** (the R in RMS) of the **mean** value (the M in RMS) of the square (the S in RMS) of $z_j - \langle z \rangle$. In beam physics, everything is assumed to be gaussian (or so it seems!) and, if one expresses the statistical probability-function $f_z(z) = \exp[-(z - \langle z \rangle)^2/(2\sigma_z^2)]$ (note the factor of two in the denominator), then the "RMS" value of z is exactly σ_z. This is related to the above discussion of the bunch charge Q_b in two ways. First, simple current monitors (peak bunch current $\approx \beta_z c Q_b/\sigma_z$) cannot not resolve short (ps or fs) bunch lengths σ_z and second, since it will be the electric or magnetic field of the electron bunch that will ultimately enable a measurement of this quantity, more charge per bunch will result in a cleaner measurement.

Energy spread $\sigma_{E_{kin}}$

The spread in energies of electrons from a plasma accelerator can be due to the energy spread in a single bunch and/or the total energy spread if there are multiple bunches. The energy spread in a single bunch is typically due to the longitudinal variation of the accelerating field E_z but this variation can be substantially modified if N_b is large enough to beam-load the plasma wake. There are very few measurements that can definitively correlate the energy spread $\sigma_{E_{kin}}$ with the longitudinal spread of electrons. As in the bunch length σ_z, this number is often quoted as an RMS value.

Transverse spot size $\sigma_\perp$ or σ_x or σ_y

This has already been discussed in the introduction to this section as a property of the statistical variation of the electrons about the transverse centroid location. If a CCD camera views a phosphorecent, fluorescent, scintillating, or otherwise optically-emitting material placed in the beam path, the recorded intensity variation with x or y shows up as, in principle (the material must emit linearly with N_b and used below its saturation level), the transverse variation of the number of electrons in the beam/bunch. As discussed above, if the distribution seen on the CCD camera seems to be $\approx$ gaussian, then the RMS of lineout of the projected image in the x- or y-direction gives the desired quantity: σ_x or σ_y. Here, as in all beam/bunch parameter measurements, deviation from a gaussian distribution must be noted and sometimes the full-width at half-maximum (FWHM) is a more appropriate description of the measured projection of the phase space.

Transverse emittance

The transverse emittance is closely related to the scatter plots of $f_x(x, p_x)$ and $f_y(y, p_y)$ discussed in the introduction to this Section. Suppose that the distribution functions, say, $f_x(x)$ and $f_x(p_x)$ were peaked at roughly $\langle x \rangle$ and $\langle p_x \rangle$, respectively. Suppose further that, in a single electron bunch, we have $N_b = 1 \times 10^{10}$ and we plotted all these data onto the coordinate system

$(x - \langle x \rangle, p_x - \langle p_x \rangle)$. In this case, the scatter plot would be centered around $(x = 0, p_x = 0)$ and fall off from the origin with a scale length on the order of σ_x in x and σ_{px} in p_x. Here, σ_{px} is, as is hopefully obvious by this time, the standard deviation of p_x about $\langle p_x \rangle$.

In Item 1 of this list, we assumed that $p_\perp/\gamma mc \ll 1$ so that $p_{x,j}/p_{z,j}$ is just the (small) angle that electron j makes with the $\hat{z}$-axis. Defining this angle as $\theta_{x,j}$, we can scale the momentum-component of the transverse distribution functions to $\langle p_z \rangle$ giving a equivalent transverse distribution functions $f_x(x, \theta_x)$ and $f_y(y, \theta_y)$ and, for the example of the x-projection, the new coordinate system $(x - \langle x \rangle, \theta_x - \langle \theta_x \rangle)$. Again, all N_b electrons will be centered at $(x = 0, \theta_x = 0)$ and fall off from the origin with a scale length on the order of σ_x in x and σ_{θ_x} in θ_x. Since $\theta_x \approx dx/dz \equiv x'$, from here on we will use the alternative symbol x' for θ_x. This latter notation is similar to that for optical ray-tracing.

Returning to our imagined scatter plot in the plane (x, x') (actually, the plane $(x - \langle x \rangle, x' - \langle x' \rangle)$, but we assume now that all beam quantities are measured with respect to their centroid values), we supposed that with a large enough number of points (like 1×10^{10}), we could construct a curve which follows (approximately) a constant-density contour of the scatter-plot points in the distribution $f_x(x, x')$. Three such contours are shown in the upper row of Figures 22(a), (b), and (c) representing the *same* contour of the distribution function $f_x(x, x')$ at three consecutive values of z; namely, z_a, z_b and z_c, respectively ($z_a < z_b < z_c$). The lower row of Figure 22 shows the projection of $f_x(x, x')$ onto the x-axis or $f_x(x)$ (the integration of $f_x(x, x')$ over x') at z_a, z_b and z_c, respectively. These are, of course, the transverse profiles of the beam and represent the "spot size" vs z. Note that the integral $f_x(x)$ over x gives the total bunch charge Q_b.

For an ensemble of electrons, the *transverse emittance* is defined by

$$\epsilon_{x,rms} = \sqrt{\langle x^2 \rangle \langle x'^2 \rangle - \langle xx' \rangle}. \tag{36}$$

This can be simply interpreted if we look at Figure 22(b) where the contour is upright and the spot size of the beam is a minimum (the beam is at a "waist"). In this case, there is no correlation between x and x' and so Equation (36) reduces to $\epsilon_{x,rms} = \sqrt{\langle x^2 \rangle}\sqrt{\langle x'^2 \rangle} = \sigma_x \sigma'_x$ where the latter equality is true if $f_x(x, x')$ is gaussian in both x and x'; i.e. if $f_x(x, x') \propto \exp[-(x)^2/(2\sigma_x^2)] \times \exp[-(x')^2/(2\sigma'^2_x)]$ (see Item 3 above). In addition, since both projections have the same functional form (gaussian, in this case), the contour is a general ellipse.

Finally, if the contour was chosen such that the x-intercepts are $\pm\sigma_x$, then the ellipse necessarily lies on the $\exp[-1/2]$ contour of $f_x(x, x')$ and the x'-intercepts are necessarily $\pm\sigma'_x$. Since $\epsilon_{x,rms} = \sigma_x \sigma'_x$ for this choice of contour, the RMS emittance $\epsilon_{x,rms}$ is just the area of the ellipse divided by π. For example, suppose that the $+x$-intercept in Figure 22(b) was 10 μm and the $+x'$-intercept was 150 mrad. Then the emittance would be 0.0 1mm × 150 mrad or 1.5 π mm-mrad. Here, the π is not a "unit" but a decoration or, more precisely, a *declaration* that the quoted emittance is only the product of x and x' (at a waist) and not the total area of the ellipse.

The three panes of Figure 22 represent an ensemble of electrons simply drifting in vacuum with no forces applied. Since no forces were applied, neither $\langle p_z \rangle$ nor p_x have been changed. Thus x'_j is constant for each of the N_b electrons. In Figure 22(a) at $z = z_a$, most of the electrons with $x(z_a) < 0$ have a positive $x' = dx/dz$ and so, for $z > z_a$, these electrons will move to the right; towards larger x. Similarly, those electrons with $x(z_a) > 0$ have negative values of x' and so the ellipse will rotate into an upright position after a predictable drift $z_b - z_a$. So, the transverse phase-space ellipse at $z = z_a$ (Figure 22(a)) represents a beam that is focusing while the phase-space ellipse at $z = z_c$ shows a diverging beam and the beam at $z = z_b$ is at a "waist" or a focus. This is clearly seen in the spot size and intensity of the x-projection shown in the bottom row of Figure 22; the beam becomes smaller and more intense at $z = z_b$ and subsequently increases in size.

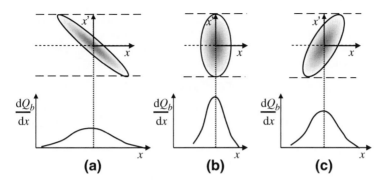

Figure 22. *Upper row: transverse phase-space diagrams of a beam drifting in vacuum (no forces applied). Lower row: projections of the phase-space diagrams (from the upper row) onto the x-axis. The three panels show: (a) a beam approaching a waist in x; (b) the same beam as in (a) but at a waist; and (c) the same beam diverging from the waist in x. The transverse emittance, equal to the area of the ellipses in the upper row (divided by π) is a conserved quantity over the short drift space implied in the figure.*

In an accelerator or in a long drift tube, the beam will need to be refocused many times to keep it within the beam pipe. In other words, we need some device which will transform the ellipse at $z = z_c$ back into something that resembles the ellipse at $z = z_a$. Such a device would need to change the sign of the angle x' for each electron and do it in proportion to the x location. This is just what a lens of focal length f does for optical beams; $x'_{out} = -1/f \times x_{in}$. This is "linear focusing"; the angle-kick is linear with x. The equivalent electron optic is a "doublet" or "triplet" of quadrapole magnets. A single quadrapole magnet can focus in, say, the x-plane but it will defocus in the y-plane. A typical quadrapole triplet will have, for example, two x-focusing optics at the entrance and exit of the triplet with focal length f and a y-focusing optic in the center of focal length $f/2$ (stronger focusing). If the thickness of the x-focusing quadrapole is L_x, then a gap of roughly L_x between the outer "quads" and the inner quad will produce a optical assembly that can focus a wide variety of incoming beams.

Strangely, in Figure 22(b), electrons at each x have a range of x'. In this sense, $\epsilon_{x,rms}$ is a measure of the "quality" of the electron beam. Suppose we want to focus this beam in the x-plane. Using a "perfect" lens with the perfect angular kick $\delta\theta_x \propto x$ and focal length f_b, all the electrons in a zero-emittance beam would cross $x = 0$ at a distance $z = z_b + f_b$. However, for a finite $\epsilon_{x,rms}$, the distance at which each electron crosses $x = 0$ will be at $z(x) = z_b + f_b(1 \pm \theta_x(x))$ where $\theta_x(x)$ is the distribution of angles at the location of the lens. Thus the spot size of the focused beam is degraded by the transverse emittance.

8.2   Basic measurements

The most basic measurement of interest (it seems) in plasma accelerator research is the energy of the accelerated electrons. A dipole magnet is the equivalent to an optical prism or grating and thus is the basic tool for determining the energy (or more properly, the momentum) of an electron. The radius of curvature ρ of an electron in a constant magnetic field B is $\rho = p/(eB)$ (for $\gamma \gg 1$) which has the engineering formula

$$\rho(\text{m}) = \frac{3.33\ p(\text{GeV}/c)}{B(\text{T})},$$

where p is the momentum in units of GeV/c (see Item (1) in the list of Section 8.1 for these strange units of momentum) and B is in units of Tesla. This transverse force $F_x = -e\vec{v_z} \times \vec{B_y}$ causes the electrons to bend and for small bend angles, the impulse Δp_x imparted during a transit time $t_{tr} = l/v_z$ is $\Delta p_x \approx F_x t_{tr} = eB_y l = pl/\rho$ which gives a deflection angle $\Delta\theta_x$

$$\Delta\theta_x \equiv \Delta p_x/p \approx l/\rho = 0.3 B_y(\mathrm{T})\, l/p(\mathrm{GeV/c}),$$

where we have assumed that the deflection is in the $\hat{x}$-direction, the dominant momentum is in the $\hat{z}$-direction and l is the length of the B_y-field in the $\hat{z}$-direction. For larger bend angles (where the above approximation breaks down) it can be shown that a "sector magnet" which may have a "sweep angle" (input angle - output angle) of (in practice) $> 50°$ or so, can accept a monoenergetic beam with a range of θ_x and bend them such that they will all intercept at approximately the same point in the "energy-imaging plane". Moreover, if these monoenergetic electrons enter the sector magnet with a variation of angles θ_y in the non-dispersed $\hat{y}$-plane (suppose this is the vertical plane), the "fringing fields" of the sector magnet can be used to provide vertical kicks to the electrons and cause them all to roughly intercept at the same point in the "vertical-imaging plane" of the sector magnet. This can occur if the direction to the normal of the "pole face" at the entrance to the sector magnet is rotated with respect to the $\hat{z}$-direction. In this case, there will be a B_x component to the fringing field which changes sign top-to-bottom and the $\vec{v_z} \times \vec{B_x}$ will focus the beam vertically: i.e., provide a downward kick if the beam enters near the top and an upward kick if the beam enters near the bottom. If this "pole-face rotation" angle is θ_{pfr} and the radius of curvature in the dipole is again ρ, then the vertical focal length f_y is roughly $\rho/\tan(\theta_{\mathrm{prf}})$ (Enge 1977). If the electrons exit the sector magnet again with an angle with respect to the pole face, another vertical kick occurs, bringing the vertical imaging plane closer to the magnet. In practice, it is possible to overlap the energy-imaging and vertical-imaging planes providing a *fully* imaging spectrometer. There are many designs of such "double-focusing" sector magnets that extend well into the GeV energy range (Livingood 1969) and can be constructed relatively cheaply for energies below 100MeV (Clayton et al. 1995). By choosing the shape of the poles, the imaging properties can be roughly maintained over a factor of 2–3 in energy.

For the experiments with the 28.5 GeV beam at SLAC, any ordinary dipole will have a very large radius of curvature ρ which results in small bend angles and essentially no fringing-field focusing. A cartoon of the final incarnation of PWFA experiments done in the (now dismantled) FFTB is shown in Figure 23. This is similar to Figure 4 except the imaging electron spectrometer (composed of quadrapole magnets in addition to a dipole magnet) was not used. In order to observe the energy gain and loss over a *very* wide range of energies (over a factor of ≈ 8) in a single shot, the fact that quadrapole-imaging optics have a strong "chromatic" aberration or energy-dependent focusing strength forbade the use of a *imaging* electron spectrometer. Recall the discussion in Section 3.2 where it was stated that the plasma exit is the "slit" of the spectrometer and finite energy resolution can be obtained if *and only if* this exit plane (the slit) is imaged onto the detector plane. Now, without the possibility of imaging, how can the changes in position at the detector plane possibly be interpreted as energy changes? The answer is that they cannot be so interpreted! The energy-dependent kick in angle provided by just dipole magnet cannot be distinguished from a kick at the plasma exit unless the absolute angle that the electrons have in the dispersion plane is also measured. This is the purpose of the two CCD cameras in Figure 23. If a bunch of electrons is seen on both cameras, then these two measurements uniquely determine the kick at the plasma exit and the kick due to the dipole, the latter revealing the momentum of the electrons.

Several other beam diagnostics are also shown in Figure 23. Moving from left to right in the figure, there is a beam-position monitor (BPM) (Hinkson 1998). One cannot have enough of these! A BPM generally has four electrodes ("button" or "stripline") which pickup the beams electromagnetic field. The signal processing usually provides the difference-over-sum of opposing electrodes where, for example, the difference is that from the top and bottom electrodes or from the left and right electrodes. This signal is proportional to the position of the beam and, for intense bunches, can easily have sub-μm accuracy. The sum of all four electrodes, proportional to the bunch charge, is also acquired. The actual charge requires pre-calibration of the BPM electronics. This author has had success with home-made button-type BPM's with ≈ 100 μm resolution. A much more robust charge measurement come from a current-pulse monitor or "toroid" in the jargon. These devices sense the magnetic field of the bunch and thus need to be placed on an non-conducting section of beam pipe. Ideally, these transformers will integrate the current over the bunch length. For sub-ps pulses, the capacitance of the windings is enough to perform the integration. Home-made toroids can be equipped with an second winding to calibrate the charge. Store-bought integrating current transformers with fixed calibrations are also available.

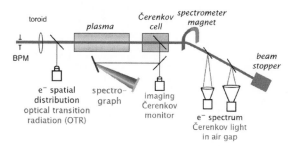

Figure 23. *Cartoon of some of the diagnostics used in the recent PWFA experiments in the FFTB. (From Ischebek, R., Stanford Linear Accelerator Center. With permission.)*

The next diagnostic shown in Figure 23 is a camera imaging the optical transition radiation (OTR) emitted when the bunch strikes a 1 μm-thick Ti foil. One description of this source of radiation is that the field of an electron-induce polarization in the foil that is highly time-dependent as the electron traverses the dielectric transition (in this case, the first dielectric is vacuum and the second is metal). Since the electron is a point charge, the frequency components of the time-changing polarization contain all frequencies for which the foil has an AC polarizability. This is generally finite in the optical and thus OTR is emitted. The OTR light comes off the foil at the specular angle, but in a cone peaking at an angle of $\approx 1/\gamma$ from pure specular. The camera records the OTR with a spatial resolution of better than 10 μm. Thus the beam size in the two transverse directions can be recorded for every shot. This diagnostic is repeated downstream of the plasma (not shown) to observe how the plasma affects the transverse properties of the incoming beam. The radiated energy per unit frequency interval for transition radiation $dW/d\omega$ is roughly independent of ω (Carr and Wiedemann 1998) which tends to favor shorter wavelengths and is linear in the number of electrons. However, as we will see in Section 8.3, when the electrons arrive in a short bunch, the frequency-dependence of the radiation can change dramatically due to coherent effects when the emitted wavelength is greater than the bunch length. These coherent effects are possible for all forms of radiation that single-electrons are capable of radiating (transition, synchrotron, Cherenkov, etc.).

Next in line in Figure 23 is radiation produced in a Cherenkov (Čerenkov) cell. Cherenkov light is emitted when the electron propagates with a velocity $\beta_z c > c/\eta_{cell}$ where η_{cell} is the refractive index of the gas within the cell. Since η_{cell} can be varied from zero to some finite number depending on the pressure, the Cherenkov cell can act as a threshold detector; that is, only β_z values greater than $1/\eta_{cell}$ will emit. In practice, since variations in β_z are negligible for energies beyond 1 GeV, this threshold technique is more suited to electron energies below, say,

500 MeV. Thus, this diagnostic was mainly used to study low-energy electrons that were trapped from the background gas. Although it was not discussed above in regards to the electron energy measurements, the two cameras which record the vector of the accelerated (and deaccelerated) electrons view Cherenkov light from the surrounding air. A "air gap" is created by placing two pairs of twin ≈ 350 μm Si wafers at 45° to the beam axis with a roughly 2 cm gap between each pair. The first wafer blocks all previously created Cherenkov light and the second wafer reflects just that Cherenkov light produced in the gap. The small gap is necessary for good spatial resolution.

8.3   General properties of radiation from electrons

In the previous Section, we discussed transition radiation for a single electron but suggested that for emission wavelengths > the bunch length, coherent effects would occur. In this section we briefly justify this statement in a general way.

Let $e(t)$ be the radiated field from transition radiation or synchrotron radiation or Cherenkov radiation, etc. for one electron. The spectrum is $\mathcal{F}\{e(t)\} = \mathbf{e}(\omega)$. The total field for N electrons is in some well-defined bunch is

$$E(t) = \sum_{k=1}^{N} e(t - t_k)$$

with the spectrum

$$E(\omega) = \int_{-\infty}^{+\infty} E(t)e^{i\omega t}dt = e(\omega) \sum_{k=1}^{N} e^{i\omega t_k},$$

where t_k is the (random) timing of the k-th electron within the bunch. The spectral intensity of the radiation is $I(\omega) = |E(\omega)|^2$ given by

$$I(\omega) = E(\omega)E^*(\omega) = e(\omega)e^*(\omega) \sum_{k=1}^{N} e^{i\omega t_k} \sum_{m=1}^{N} e^{i\omega t_m} = |e(\omega)|^2 \sum_{k,m=1}^{N} e^{i\omega(t_k-t_m)}. \qquad (37)$$

Let $g(t)dt$ be the probability of finding an electron between t and $t + dt$ with $\int_{-\infty}^{\infty} g(t)dt = 1$. Thus $g(t)$ is the normalized distribution function of electrons within the bunch. Averaging over all possible values of t_ks and t_ms given by the distribution function $g(t_j)$ gives

$$\langle I(\omega) \rangle = |e(\omega)|^2 \sum_{k,m=1}^{N} \int_{-\infty}^{+\infty} \int_{-\infty}^{+\infty} dt_k dt_m g(t_k)g(t_m)e^{i\omega(t_k-t_m)}.$$

In the double summation $(\sum_{k,m}^{N})$, there is the possibility that $k = m$ and that $k \neq m$. The former possibility will occur N times (for N electrons) and the latter case will occur $N - 1$ times in *each* summation. Separating these two possibilities, we have

$$\langle I(\omega) \rangle = |e(\omega)|^2 \sum_{k=m}^{N} \int_{-\infty}^{+\infty} \int_{-\infty}^{+\infty} dt_k dt_m g(t_k)g(t_m)e^{i\omega(t_k-t_m)}+$$

$$|e(\omega)|^2 \sum_{k\neq m}^{N} \int_{-\infty}^{+\infty} \int_{-\infty}^{+\infty} [g(t_k)e^{i\omega t_k}dt_k][g(t_m)e^{-i\omega t_m}dt_m].$$

For $g(t)$ real and $\mathcal{F}\{g(t)\} \equiv g(\omega)$, the first term $(k = m)$ results in $N|e(\omega)|^2$ since $(t_k - t_m) = 0$ for all k, m and the second term (for $k \neq m$) contains the the product of two Fourier transforms of $g(t)$; i.e., equal to $|e(\omega)|^2(N-1)^2g(\omega)g^*(\omega)$. For $N-1 \approx N$, we have $\langle I(\omega) \rangle \approx$ $|e(\omega)|^2 \left(N + N^2 |g(\omega)|^2 \right)$.

Let $I_{incoh}(\lambda) \equiv N|e(\omega)|^2$ represent the contribution to $\langle I(\omega) \rangle$ from the N pseudo-randomly timed electrons; the incoherent part of the measured spectrum. We have changed variables from measured ω to measured $\lambda \equiv 2\pi c/\omega$. Furthermore, let $f_{bunch}(\lambda) \equiv |g(\omega)|^2$. We then have

$$\langle I(\lambda) \rangle \approx I_{incoh}(\lambda) \left(1 + N f_{bunch}(\lambda) \right). \tag{38}$$

The function $f_{bunch}(\lambda)$ is the so-called "form factor" which is strictly a property of the envelope $g(t)$ of the bunch. The form-factor might be more intuitively defined as $f_{bunch}(\lambda) \equiv g(\omega)$; i.e., not with $|g(\omega)|^2$. Unfortunately, both representations appear in the literature so one must be careful. This function will become important if, for a particular measured λ_m, the (square) of the Fourier transform of the *bunch shape* $g(t)$ has significant spectral component at λ_m. For a gaussian bunch with a RMS bunch length of σ_z, the total energy $\int dt \int d\Omega \langle I(\lambda) \rangle$ is dominated by the energy in wavelengths $> \sigma_z$.

For example, if the bunch length σ_z is 1 mm and we are measuring transition radiation (TR) in the optical wavelengths ($\lambda_m \sim 0.5\ \mu$m), then it is intuitively obvious that $f_{bunch}(\lambda_m)$ would be vanishingly small such that $N f_{bunch}(\lambda_m) \ll 1$. Thus, for such long bunches, the only emission contributing to this OTR signal $\langle I(\lambda_m) \rangle$ is from $I_{incoh}(\lambda_m)$. On the other hand, if σ_z was, say 1 μm, then $f_{bunch}(\lambda_m)$ is quite finite and, since N tends to be a large number $\sim 10^9$–10^{10}, then $N f_{bunch}(\lambda_m) \gg 1$ and subsequently $\langle I(\lambda_m) \rangle$ will be enhanced by many orders of magnitude compared to $I_{incoh}(\lambda_m)$! In this case, the TR signal is dominated by the *coherent addition* of the individual $e_k(t)$ fields. Effectively, for observation wavelengths $\sim \sigma_z$ or longer, the electron bunch looks like one super-electron and the fields $e_k(t)$ add in phase resulting in, for this example, coherent TR or CTR.

Note that the above discussion of radiation from electrons was done in only one dimension; the longitudinal dimension using $g(t,z) = g(z - \beta_z ct)$. In general, the finite spot size of the beam can modify some of the above conclusions (Shibata et al. 1994). For example, using the more general $g(t, \vec{r})$, we would find that this idea of a super-electron emitting all the $e_k(t)$ fields is phase is incorrect. Instead of t, we must used the transverse-postion-dependent retarded time $t'(x,y)$. For example, if the spot size at the TR foil is $\gg \sigma_z$, then even if $\sigma_z < \lambda_m$ (the measured wavelength), then the enhancement of $\langle I(\lambda_m) \rangle$ will be smaller than for a narrow beam because of the phase-mixing due to $t'(x,y)$.

8.4   Radiation-based electron measurements

With electrons being charged particle, there is seemingly no end to the ways in which a electron can radiate when placed in various environments. We have already discussed transition radiation and Cherenkov radiation. A more full list is given by Carr and Wiedemann (1998). Wiggler and undulator radiation are common in storage rings and free-electron lasers. These are connected to "synchrotron radiation" (SR) which seems to be a "catch-all" term for any radiation due to acceleration occurring far away from materials and boundaries. In this sense, Equation (5) describes SR since we neglected the material aspect of the dilute plasma, as discussed at the end of Section 7.2.

We will start out this Section describing two SR-based measurements; using bending magnet SR to map the transverse spatial distribution of an electron beam onto a photon-sensitive screen and using the spectral properties of spontaneous emission from a magnetic wiggler to estimate the longitudinal bunch length. This section will end with a description of the use of CTR to estimate the bunch length.

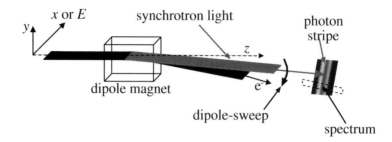

Figure 24. *Setup for measuring the transverse distribution (or energy spectrum) of electrons using synchrotron radiation.*

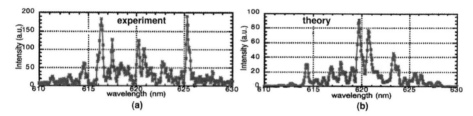

Figure 25. *Single-shot spectra from a 44MeV beam traversing a 60-period wiggler. (a) A typical experimental spectrum for a 1.5ps, 100nC electron bunch. (b) A calculated spectrum for a 1.5ps, 100nC electron bunch. (From Catravas et al. (1999). With permission.)*

Figure 24 shows a setup used in the PWFA experiments in the FFTB (see the diagnostic labeled "Energy Spectrum, X-ray" in Figure 4). The dipole magnet was located in a section of the FFTB beamline where the x-dispersion D_x was high and the horizontal- or x-size of the beam $\sigma_{x,\delta E}$ due to an energy spread δE, given by $\sigma_{x,\delta E} \approx D_x \delta E / E_0$ where E_0 is the centroid energy. The inherent spot size of the beam at this high-disperion location was such that the emittance-limited spot size σ_x as in Figure 22(b) was such that $\sigma_x \ll \sigma_{x,\delta E}$.

In this limit, the transverse distribution of electrons entering the dipole in Figure 24 represents an energy spectrum of the beam! As the beam bends in the dipole, the acceleration due to $F_t = -e\vec{v_z} \times \vec{B_x}$ causes the beam to radiate SR continuously over the entire duration of the "dipole sweep" (the continuous change in angle given by $\vec{\beta}(t) \cdot \hat{y}$). The result continuous emission of SR photons over the dipole sweep angle. A scintillating crystal, sensitive to ionizing radiation is the ~ 10 keV range, is "painted" with a photon stripe. A 12-bit CCD camera (not shown) imaged the scintillation photons. By summing over a region of of this stripe, as shown at the right of Figure 24 gave a array of numbers which is a scaled copy of the electron spectrum. In this way, the spectrum of electrons prior the PWFA plasma was recorded for every shot. The beam was put back on the $\hat{z}$-axis after this measurement by the use of two more dipoles; a center dipole of opposite polarity to bring the beam back to $y = 0$ and a final dipole (a mirror of the first dipole) to remove the sweep angle. This combination of dipole magnets is called a "chicane"; however, for this application, it is better called a half-cycle undulator.

When an electron beam enters a magnetic wiggler, the periodic transverse acceleration of the electron will cause it to emit SR photons along the entire trajectory. Due to the wiggling, the average velocity in the $\hat{z}$-direction must be substantially less than c, even if $\gamma \gg 1$ ($\langle \beta_z c \rangle < (1 - \gamma^{-2})^{1/2} c < c$). For a SR photon wavelength λ_c such that the longitudinal slippage of the relativistic electrons is one wavelength per wiggler period, there is a possibility for "in-phase"

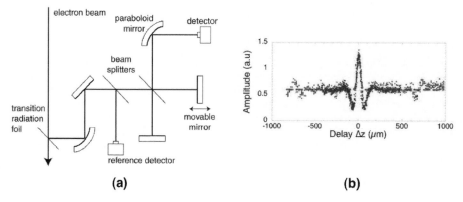

Figure 26. *(a) A schematic of a CTR interferometer. The movable mirror provides the mechanism for recording an autocorrelation trace. (b) A typical raw autocorrelation trace showing the spectral filtering of the optics. (From Muggli et al. (2005). With permission.)*

addition of this wavelength at the end of the wiggler. This requires that $\lambda_c = \lambda_w(1 + K_w^2)/(2\gamma^2)$ where λ_w is the wiggler period and K_w is the wiggler strength. Typically the bandwidth $\Delta\omega_c$ of the wiggler radiation around this central wavelength of λ_c is $\Delta\omega_c/\omega_c \approx 1/N_w$ where N_w is the number of wiggler periods. Note that this spontaneous emission is incoherent in that the finite longitudinal extent of the beam is $\gg \lambda_c$; we are dealing with the first (incoherent) term in Equation (38).

In some cases, the full bandwidth is not observed on a single shot, as shown in Figure 25(a). This is the spectrum observed when a 44MeV electron beam passes through a 60-period wiggler with $K_w = 0.34$ and $\lambda_w = 8.8$mm. There is a distribution around $\lambda_c = 620$ nm, but it is not "filled in". For the 1.5 ps bunch with ≈ 100 pC of charge used in this experiment, a single shot does not appear to produce enough photons to fill in the spectrum. But rather than being simply a weak spectrum, there are distinct spikes with apparently a characteristic width. On a shot-to-shot basis, the location and number of these spikes are random and if one adds up *many* shots, the spectrum indeed looks like that expected for wiggler radiation. For a single shot, it turns out that the number of spikes and the characteristic width is intimately related to the bunch length σ_z (Catravas et al. 1999).

In Section 8.3, we calculated some properties of electron radiation. In particular, Equation (38) shows the bunch-averaged spectrum with its incoherent and coherent terms. However, if the bunch-avering is not done and we begin again at Equation (37), we see that the summation over single electrons goes like $\sum_{k,m=1}^{N} \exp[i\omega(t_k - t_m)]$. Just as the sum in Equation (24) produced spectral fringes with period $2\pi/\tau$, each term in the summation of Equation (37) is also oscillatory with with a spectral period of $2\pi/(t_k - t_m)$ (Sajaev 2004). The highest frequency (or narrowest spikes) will be for $(t_k - t_m) \sim \sigma_z/c$. If the bunch is longer (wider span of possible $(t_k - t_m)$), the spikes will be closer together and there will thus be more of them in the bandwidth of the spectrum. Using a statistical model for the electron beam, the the characteristics of the observed spectra can be reproduced, as shown in Figure 25(b), and fitted to a bunch length with a fairly small uncertainty. In fact, to correctly describe the observed spectra, the beam emittance is also used as a fitting parameter (Catravas et al. 1999), thus giving a measure of σ_z and ϵ_x on a single shot! Even more impressive to this author, by examining many such single-shot spectra, the bunch *shape* can be extracted!! It is said that the variance of the Fourier transform of the spectra is proportional to the convolution of bunch current profile and there are experimental results using incoherent SR (not wiggler radiation) which suggest that the bunch profile can indeed be extracted (Sajaev 2004).

The final radiation-based electron measurement we will discuss is bunch-length estimates based on coherent transition radiation. Recalling that $f_{bunch}(\lambda)$ is $|g(\omega)|^2$, and accepting that the magnitude or "modulus" squared discards any *phase* information, it would be surprising if the *shape* of the bunch could be determined by a measurement of $\langle I(\lambda) \rangle$. However, this seems to be the case if one considers causality! Hidden in the mathematics of Section 8.3 is the notion that the distribution function is centered at $t = 0$ through the integration limits $\int_{-\infty}^{\infty}$. If we had chosen $t = 0$ as the timing t_k of the first electron in the bunch, then the integral of Equation (37) would be over $\int_{0}^{\infty}$ and, as discussed by Lai and Sievers (1997), one can use the Kramers-Kroning relation to connect the modulus and phase of $g(\omega)$. If $\langle I(\lambda) \rangle$ is measured over a bandwidth $\gg 2\pi c/\sigma_z$, then, by 'reasonable' asymptotic extensions of of the form factor for $\omega \to 0$ and $\omega \to \infty$, the spectral phase associated with $g(\omega)$ can be recovered giving the bunch shape up to a time-reversal in $g(t)$ (Lai and Sievers 1997). In practice, experimentalists are limited not by the available analysis techniques, but by the material science involved in measuring frequency spectra ranging from the near infrared to the THz frequencies.

For most of the current electron-radiation-based bunch-length measurements, the experimental goal is simply to measure the form factor $f_{bunch}(\lambda)$ as expressed in Equation (38). Since this term is in the coherent contribution to the measured spectrum, CTR is a good way to estimate the bunch length. In principle, the phase-retrieval method described above is always available in the post-processing of the data, but in practice, the material science limitations make this very challenging. A great deal of CTR work has been done at the University of Tokyo (Uesaka et al. 2000) where the bunch shape was measured directly with a 200 fs-resolution streak camera and the resulting $f_{bunch}(\lambda)$ compared to two independent measurements: (1) a direct spectral measurement of $\langle I(\lambda) \rangle$ using a "polychromator" which is similar to a spectrograph but with only a few (~ 10) discrete infrared detectors; and (2) CTR interferometry, which we will discuss below. Detailed comparisons of the streak camera measurements and the spectral measurements, done properly, are quite good and this research has lead to some best-practice techniques (Uesaka et al. 2000).

As in Section 6.2 on spatial interferometry and in Section 7.4 on spectral interferometry, the interference patterns rely on the coherence of the probe and reference beams/pulses. Moreover, the resulting fringes were visualized by a 2-D detector (a CCD camera). In the CTR interferometry setup shown in Figure 26(a), there is no reference beam/pulse. Also, there is only one "pixel" for the detector; either a Golay cell or a pyroelectric detector which ideally would have a flat response from the near infrared to the THz frequencies. The CTR pulse must interfere with itself on this single detector. For $E(t)$ the field at the detector, the measured interference energy $I_{CTR}(\tau)$ is

$$I_{CTR}(\tau) = \int_{-\infty}^{+\infty} |E(t) + E(t+\tau)|^2 \, dt \qquad (39)$$
$$= 2 \int_{-\infty}^{+\infty} |E(t)|^2 \, dt + 2 \, \text{Re} \int_{-\infty}^{+\infty} E(t) E^*(t+\tau). dt$$

The first term in the bottom line of Equation (39) represents the (non-interfering) contribution from the two arms of the interferometer for delays $|\tau| \gg \sigma_z$. The variation of $I_{CTR}(\tau)$ is given by the second term in the bottom line of Equation (39) which is an "autocorrelation". Using the Wiener-Khinchin theorem (see http:// mathworld.wolfram.com/ Autocorrelation.html) which states that $\mathcal{F}\{|E(\omega))|^2\} = E(t) \star E(t) = E^*(-t) * E(t)$ (where the operator $\star$ denotes cross-correlation and the operator $*$ denotes convolution), we can rewrite $I_{CTR}(\tau)$ as

$$I_{CTR}(\tau) = A + 2 \text{ Re} \int_{-\infty}^{+\infty} |R(\omega)T(\omega)E(\omega)|^2 e^{i\omega\tau} d\omega$$

where, in the frequency domain, we can apply the total linear refection and transmission coefficients $R(\omega)$ and $T(\omega)$, respectively, of the vacuum window and beam splitters so that now $E(\omega)$ is the Fourier transform of $E(t)$ emitted by the CTR foil rather than the field at the detector. The quantity A is the delay-independent term (first term) in Equation (39) appropriately scaled for the transmission and reflection of the optics. Note the difference from Equation (24) which shows $I_S(\omega,\tau)$ for longitudinal interferometry. Since there is no grating, the time-integration occurs for *all frequency components* on the same detector! For the same reason, there is no possibility of spectral fringes and thus a single value of τ provides only one data point; $I_{CTR}(\tau)$!

Figure 26(b) shows the the interference trace obtained when τ is manually changed on a shot-by-shot basis (Muggli et al. 2005). If we normalize the curve so that the maximum *possible* $I_{CTR}(\tau) = 1$, then for $|\tau| \gg \sigma_z$, the maximum $I_{CTR}(\tau) = 0.5$. This is clear is we look again at bottom line of Equation (39) for $\tau = 0$. There is an equal contribution from each of the two terms for $\tau = 0$. For long delays, the second term drops out and only the first term contributes. Thus the data should all lie between 1.0 and 0.5. What happens as we move from away from $\tau = 0$? The coherent addition of the CTR signal in each arm 'phase mixes' and drops on the scale of $\tau \sim \sigma_z$. The fact that the data goes below 0.5 in Figure 26(b) comes back to the material science issues. By incorporating the expected frequency filtering functions $R(\omega)$ and $T(\omega)$ for the beam splitters and vacuum window used for these data, the measured autocorrelation trace is quite reasonable and can, indeed, result in a good measure of σ_z (Muggli et al. 2005) for $5 < \sigma_z < 30$ μm. However, the material limitations become worse at longer bunch lengths and more broadband materials need to be used, but at the expense of the ability to align the interferometer with visible light (Muggli et al. 2005).

9 Conclusions

The Experimental capabilities of plasma accelerator laboratories around the world are currently undergoing rapid expansion through creative and innovative Ideas. Through journal articles, proceedings, and the personal interactions at conferences, workshops, and Summer Schools, we not only learn what is happening at a particular Laboratory, but also get a feeling of how things are done. However, when we get back into our laboratories, we find it hard enough just to know if the laser or linac is running properly, let alone do anything with it! Often it is a very long road from having a pump beam to having a working Experiment. Hopefully, this path is not via trial and error but rather via quantify and correlate! We hope that this article has given the reader an understanding as to how much of the laboratory Apparatus operates and can thus develop an intuition that will ultimately shorten the Experimentalist's path to the Unknown.

Acknowledgments

I thank the organizers of this SUSSP for their warm welcome and assistance in all thing big and small. Appreciation goes to the students for their enthusiasm and interest. I myself found wisdom and inspiration from my fellow lecturers at this School. My gratitude also goes to Dino Jaroszynski; his comments and patient(!) encouragement helped enormously in the writing of this manuscript. Nicholas Matlis, Mike Downer, and Howard Milchberg contributed to my own understanding of certain ω-space details which helped in this writing. Patric Muggli and Art

Pak did some needed markups on the drafts. Thanks must go as well to Alan Cairns and Bob Bingham for disclosing several fairway hazards from the tees. Finally, a very special thanks to Gisele Mackenzie Smith and David Jones for their extraordinary hospitality and friendship throughout my stay in Scotland.

Bibliography

Ammosov, M. V., Delone, N. B. and Krainov, V. P., 1987, Tunnel ionization of complex atoms and of atomic ions in an alternating electromagnetic field *Soviet Physics JETP* **64** 1191.

Andreev, N. E., Gorbunov, L. M., Kirsanov, V. I., Pogosova, A. A. and Ramazashvili, R. R. 1992, Resonant excitation of wakefields by a laser-pulse in a plasma, *JETP Letters* **55** 571.

Bellini, M. 2004 Supercontinuum and high-order harmonics: "extreme" coherent sources for atomic spectroscopy and attophysics, in *Femtosecond laser spectroscopy* 29–60 editor Hannaford, P (Springer Verlag, Heidelberg-Berlin-New York).

Blumenfeld, I., Clayton, C. E., Decker F.-J., Hogan, M. J., Huang, C., Ischebeck R., Iverson, R., Joshi, C., Katsouleas, T., Kirby, N., Lu, W., Marsh, K. A., Mori, W. B., Muggli, P., Oz. E, Siemann, R. H., Walz, D. and Zhou, M. 2006 Energy Doubling of 42 GeV Electrons in a Meter Scale Plasma Wakefield Accelerator, *submitted for publication.*

Born, M. and Wolf, E. 1980 *Principles of Optics* (Pergamon Press, Oxford).

Bulanov, S. V., Pegoraro, F., Pukhov, A. M. and Sakharov, A. S. (1997), Transverse-wake wave breaking, *Physical Review Letters* **78** 4205.

Carr, R. and Wiedemann, H. Other radiation sources in *Handbook of Accelerator Physics and Engineering* Chapter 3.1.6, pg 190 editors Chao A. and Tigner, M. (World Scientific Publishing Co. Pte. Ltd.).

Catravas, P., Leemans, W. P., Wurtele, J. S., Zolotorev, M. S., Babzien, M., Ben-Zvi, I., Segalov, Z., Wang, X.-J. and Yakimenko, V. 1999, Measurement of electron-beam bunch length and emittance using shot-noise-driven fluctuations in incoherent radiation, *Physical Review Letters* **82** 5261.

Chen, P., Dawson, J. M., Huff, R. W. and Katsouleas, T. 1985, Acceleration of Electrons by the Interaction of a Bunched Electron-Beam With a Plasma, *Physical Review Letters* **54** 693.

Clayton, C. E, Joshi, C., Darrow, C. and Umstadter, D. 1985, Relativistic plasma-wave excitation by collinear optical mixing, *Physical Review Letters* **54** 2343.

Clayton, C. E., Marsh, K. A., Dyson, A., Everett, M., Lal, A., Leemans, W.P., Williams, R. and Joshi, C. 1993, Ultrahigh-gradient acceleration of injected electrons by laser-excited relativistic electron-plasma waves, *Physical Review Letters* **70** 37.

Clayton, C. E., Marsh, K. A., Joshi, C., Darrow, C. B., Dangor, Modena, A., Najmudin, Z. and Malka, V. 1995, A broadband electron spectrometer and electron detectors for laser accelerator experiments, in *Proceedings of the 1995 Particle Accelerator Conference, Dallas, TX, USA* 637–639 (IEEE).

Clayton, C. E. 2004, Diagnostics for Laser Accelerators in *Advanced Accelerator Concepts: Eleventh Advanced Accelerator Concepts Workshop* 137–159 editor Yakimenko, V (AIP Vol 737).

Cohen, B. I., Kaufman, A. N. and Watson, K, M. 1972, Beat heating of a plasma, *Physical Review Letters* **29** 581.

Corkum, P. B., Rolland, C. and Srinivasan-Rao T. 1986, Supercontinuum generation in gases, *Physical Review Letters* **57** 2268.

Coverdale, C. A., Darrow, C. B., Decker, C. D., Mori, W. B., Tzeng, K-C., Marsh, K. A., Clayton, C. E and Joshi, C. 1995, Propagation of intense subpicosecond laser pulseshrough underdense plasmas, *Physical Review Letters* **74** 4659.

Drake, J. F., Kaw, P. K., Lee, Y. C., Schmidt, G., Liu, C. S. and Rosenbluth, M. N. 1974, Parametric instabilities of electromagnetic waves in plasmas, *Physics of Fluids* **17** 778.

Durfee, C. G. and Milchberg, H. M. 1993, Light pipe for high-intensity laser-pulses, *Physical Review Letters* **71** 2409.

Enge , H. A. 1977, Deflecting magnets, in *Focusing of Charged Particles* Chapter 4.2 editor Septier, A. (Academic, Ney York).

Esarey, E., Sprangle, P., Krall, J. and Ting, A. 1996, Overview of plasma-based accelerator concepts, *IEEE Transactions On Plasma Science* **24** 252.

Esarey, E., Hubbard, R. F., Leemans, W. P., Ting, A. and, Sprangle, P. 1997, Electron injection into plasma wake fields by colliding laser pulses, *Physical Review Letters* **79** 2682.

Evans, D.E., von Hellermann, M. and Holzhauer, E. 1982, Fourier optics approach to far forward scattering and related refractive index phenomena in laboratory plasmas, *Plasma Physics* **24** 819.

Everett, M. J., Lal, A., Clayton, C. E., Mori, W. B., Johnston, T. W. and Joshi, C 1995, Coupling Between High-Frequency Plasma-Waves In Laser-Plasma Interactions *Physical Review Letters* **74** 2236.

Faure, J., Rechatin, C., Norlin, A., Lifschitz, A., Glinec, Y. and Malka, V. 2006, Controlled Injection and Acceleration of Electrons in Plasma Wakefields by Colliding Laser Pulses *Nature* Accepted for publication.

Filip, C. V., Narang, R., Tochitsky, S. Y., Clayton, C. E., Musumeci, P., Yoder, R. B., Marsh, K. A., Rosenzweig, J. B., Pellegrini and C. Joshi, C. 2004, Nonresonant beat-wave excitation of relativistic plasma waves with constant phase velocity for charged-particle acceleration, *Physical Review E* **69** 026404.

Forslund, D. W. Kindel, J. M. and Lindman, E. L. 1975, Theory of Stimulated Scattering Processes in Laser-Irradiated Plasmas, *Physics of Fluids* **18** 1002.

Geindre, J. P., Audebert, P., Rousse, A., Fallies, F., Gauthier, J. C., Mysyrowicz, A., Santos, A., Dos Hamoniaux, G. and Antonetti, A. 1994, Frequency-domain interferometer for measuringhe phase and amplitude of a femtosecond pulse probing a laser-produced plasma, *Optics Letters* **19** 1997.

Gordon, D., Tzeng, K. C., Clayton, C. E., Dangor, A. E., Malka, V., Marsh, K. A., Modena, A., Mori, W. B., Muggli, P., Najmudin, Z., Neely, D., Danson and C. Joshi, C. 1998, Observation of electron energies beyond the linear dephasing limit from a laser-excited relativistic plasma wave *Physical Review Letters* **80** 2133.

Hinkson. J 1998, Beam position monitors, in *Handbook of Accelerator Physics and Engineering* Chapter 7.4.6, pg 555 editors Chao A. and M. Tigner, M. (World Scientific Publishing Co. Pte. Ltd.).

Hsieh, C. T., Huang, C. M., Chang, C. L., Ho, Y. C., Chen, Y. S., Lin, J. Y., Wang, J. and Chen, S. Y. 2006 Tomography of injection and acceleration of monoenergetic electrons in a laser-wakefield accelerator, *Physical Review Letters* **96** 095001.

Iaconis, C., and Walmsley, I. 1998, Spectral phase interferometry for direct electric-field reconstruction of ultrashort optical pulses, *Optics Letters* **23** 792.

Joshi, C., Mori, W. B., Katsouleas, T., Dawson, J. M., Kindel, J. M. and Forslund, D. W. 1984, Ultrahigh gradient particle-acceleration by intense laser-driven plasma-density waves, *Nature* **311** 525.

Joshi, C 1995, Working group on plasma accelerators: status report in *Proceedings Of The Advanced Accelerator Concepts, Fontana, Wisconsin, June 1994* 271–227, editor Schoessow, P (AIP Vol. 335).

Kane, D.J. and Trebino, R. 1993, *Optics Letters* **18** 823.

Kim, K. Y., Alexeev, I. and Milchberg, H. M. 2002a, Single-shot supercontinuum spectral interferometry, *Applied Physics Letters* **81** 4124.

Kim, K. Y., Alexeev, I. and Milchberg, H. M. 2002b, Single-shot measurement of laser-induced double step ionization of helium, *Optics Express* **10** 1563.

Kim, K.Y., Alexeev, I., Parra, E. and Milchberg, H.M. 2003, Time-resolved explosion of intense-laser-heated clusters, *Physical Review Letters* **90** 023401.

Krall, J., Ting, A., Esarey, E. and Sprangle, P. 1993, Enhanced acceleration in a self-modulated-laser wake-field accelerator, *Physical Review E* **48** 2157.

Kroll, N. M., Ron, A. and Rostoker, N. 1964, Optical mixing as a plasma density probe, *Physical Review Letters* **13** 83.

Lai, R. and Sievers, A. J. 1997, On usinghe coherent far ir radiation produced by a charged-particle buncho determine its shape: I analysis, *Nuclear Instruments and Methods in Physics Research Section A: Accelerators, Spectrometers, Detectors and Associated Equipment* **397** 221.

Lal, A., K. Gordon, D., Wharton, K., Clayton, C. E., Marsh, K. A., Mori, W. B., Joshi, C., Everett, M. J. and Johnston, T. W. 1997, Spatio-temporal dynamics of the resonantly excited relativistic plasma wave driven by a CO_2 laser, *Physics of Plasmas* **4** 1434.

Le Blanc, S. P., Gaul, E. W., Matlis, N. H., Rundquist, A. and Downer, M. C. 2000, Single-shot measurement of temporal phase shifts by frequency-domain holography, *Optics Letters* **25** 764.

Leemans, W. P., Nagler, B., Gonsalves, A. J., Toth, Cs., Nakamura, K., Geddes, C. G. R., Esarey, E., Schroeder, C. B. and Hooker, S. M. 2006, GeV electron beams from a centimetre-scale accelerator, *Nature Physics* **2** 696.

Livingood, J. J. 1969, *The Optics of Dipole Magnets* (Academic, New York).

Marquès, J. R., Geindre, J. P., Amiranoff, F., Audebert, P., Gauthier, J. C., Antonetti, A. and Grillon, G. 1996, Temporal and spatial measurements ofhe electron density perturbation produced in the wake of an ultrashort laser pulse, *Physical Review Letters* **76** 3566.

Marquès, J. R., Dorchies, F., Audebert, P., Geindre, J. P., Amiranoff, F., Gauthier, J. C. and Hammoniaux, G. 1997, Frequency increase and damping of nonlinear electron plasma oscillations in cylindrical symmetry, *Physical Review Letters* **78** 3463.

Matlis, N. H., Reed, S., Bulanov, S. S., Chvykov, V., Kalintchenko, G., Matsuoka, T., Rousseau, P., Yanovsky, V., Maksimchuk, A., Kalmykov, S., Shvets and G. Downer, M. C. 2006, Snapshots of laser wakefields, *Nature Physics* **2** 749.

Monot, P., Auguste, T., Lompre, L. A., Mainfray and G. Manus, C. 1992, Focusing limits of a terawatt laser in an underdense plasma, *Journal of the Optical Society of America B* **9** 1579.

Muggli, P., Hogan, M. J., and Barnes, C. D., Walz, D., Krejcik, P., Siemann, R. H., Schlarb, H. and Ischebek, R. 2005, Coherent transition radiation to measure the SLAC electron bunch length, in *Proceedings of the 2005 Particle Accelerator Conference* **C15** 4102.

Mori, W.B., The physics of the nonlinear optics of plasmas at relativistic intensities for short-pulse lasers, *IEEE Journal of Quantum Electronics* **33** 1942.

Patterson, F. G., Gonzales, R. and Perry, M. D. 1991, Compact 10-TW, 800-fs Nd:glass laser, *Optics Letters* **16** 1107.

Rayleigh, Lord 1907, On the dynamical theory of gratings, *Proceedings of the Royal Society of London, Series A* **79** 399.

Rosenbluth, M.N. and Liu, S.C. 1972, Excitation of plasma waves by two laser beams, *Physical Review Letters* **29** 701.

Rosenzweig, J. B., Cline, D. B., Cole, B., Figueroa, H., Gai, W., Konecny, R., Norem, J., Schoessow, P. and Simpson, J. 1988, Experimental-Observation of Plasma Wake-Field Acceleration, *Physical Review Letters* **61** 98.

Sajaev, V. 2004, Measurements of Bunch Length Using Spectal Analysis of Incoherent Radiation Fluctuations, in *Beam Instrumentation Workshop 2004: Eleventh Beam Instrumentation Workshop, Knoxville, Tennessee, 3-6 May 2004* 73–87 editors Shea, T. and Sibley III, C. (AIP Vol. 732).

Sheffield, J. 1975, *Plasma Scattering of Electromagnetic Radiation* Chapter 2 (Academic Press, New York).

Shibata, Y., Takahashi, T., Kanai, T., Ishi, K., Ikezawa, M., Ohkuma, J., Okuda, S. and Okada, S. 1994, Diagnostics of an electron beam of a linear accelerator using coherent transition radiation, *Physical Review E* **50** 1479.

Siders, C. W., Le Blanc, S. P., Babine, A., Stepanov, A., Sergeev, A., Tajima, T. and Downer, M. C. 1996a, Plasma-based accelerator diagnostics based upon longitudinal interferometry with ultrashort optical pulses, *IEEE Transactions on Plasma Science* **24** 301.

Siders, C. W., Le Blanc, S. P., Fisher, D., Tajima, T., Downer, M. C., Babine, A., Stepanov, A. and Sergeev, A. 1996b, Laser wakefield excitation and measurement by femtosecond longitudinal interferometry, *Physical Review Letters* **76** 3570.

Siegman, A. E. 1986 *Lasers* Chapter 9 (University Science Books).

Slusher, R. E. and Surko, C. M. 1980, Study of density fluctuations in plasmas by small-angle CO_2 laser scattering, *Physics of Fluids* **23** 472.

Spence, D. E., Kean, P. N. and Sibbett, W. 1991, 60-fsec pulse generation from a self-mode-locked Ti:sapphire laser, *Optics Letters* **16** 42.

Strickland, D. and Mourou, G. 1985, Compression of amplified chirped optical pulses, *Optics Communications* **56** 219.

Tajima, T. and Dawson, J. M. 1979, Laser electron accelerator, *Physical Review Letters* **43** 267.

Takeda, M., Ina, H. and Kobayashi, S. 1982, Fourier-transform method of fringe-pattern analysis for computer-based topography and interferometry, *Journal of the Optical Society Of America* **72** 156.

Trebino, R. 2002 *Frequency-Resolved Optical Gating: The Measurement of Ultrashort Laser Pulses* (Kluwer Academic).

Uesaka, M., Kinoshita, K., Watanabe, T., Sugahara, J., Ueda, T., Yoshii, K., Kobayashi, T., Hafz, N., Nakajima, K., Sakai, F., Kando, M., Dewa, H., Kotaki, H. and Kondo, S. 2000, Experimental verification of laser photocathode rf gun as an injector for a laser plasma accelerator, *IEEE Transactions on Plasma Science* **28** 1133.

Umstadter, D., Kim, J. K. and Dodd, E. 1996, Laser injection of ultrashort electron pulses into wakefield plasma waves, *Physical Review Letters* **76** 2073.

Wilks, S. C., Dawson, J. M., Mori, W. B., Katsouleas, T. and Jones, M. E. 1989, Photon accelerator, *Physical Review Letters* **62** 2600.

Zgadzaj, R., Gaul, E. W., Matlis, N. H., Shvets, G. and Downer, M. C. 2004 Femtosecond pump-probe study of preformed plasma channels, *Journal of the Optical Society of America B* **21** 1559-1567.

3
Waveguides for High-Intensity Laser Pulses

Simon Hooker

Department of Physics, University of Oxford

1 Introduction

In these lectures we will discuss the physics underlying several techniques for guiding high-intensity laser pulses. After a brief discussion of the processes which limit the distance over which a beam of radiation maintains an intensity close to that at its focus, we describe some general principles underlying the operation of waveguides, before discussing in detail how these principles may be applied to guiding intense laser pulses.

2 Why are waveguides necessary?

The interaction of intense laser pulses with plasma leads to a rich variety of important applications, including high-harmonic generation, soft x-ray lasers, and laser-driven particle accelerators. In most cases the useful output — whether it be photons or accelerated particles — increases with the distance over which the laser-plasma interaction occurs, at least up to some upper limit which might, for example, be set by the onset of phase mismatching. The strength of any such interaction will depend on the intensity of the driving radiation, and will usually require the intensity to be greater than some threshold, or to lie within a certain range.

2.1 Diffraction

The distance over which the intensity of a beam of radiation lies within a given range is limited fundamentally by diffraction. For example, consider the propagation of a cylindrical beam of light of wavelength λ and diameter w. As is well known from the theory of diffraction (see, for example, the textbook by Brooker (2003)), in the Fraunhofer limit the beam will diverge with a half-angle of order λ/w. Consequently the distance z_R over which the diameter of the beam will double may be estimated from $(\lambda/w)z_R \approx w/2$, and hence $z_R \approx w^2/2\lambda$.

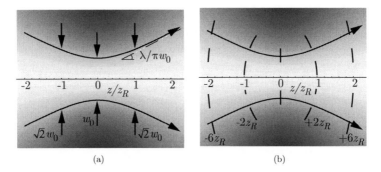

(a)   (b)

Figure 1. *Variation of the spot size (a) and the radius of curvature of the wavefronts (b) of a lowest-order Gaussian beam in the region near the beam waist. The greyscale background gives the variation of the intensity of the beam, lighter regions being of higher intensity.*

To put these ideas onto a firmer footing we consider the case of a lowest-order Gaussian beam (Brooker, 2003) propagating along the z-axis, the amplitude $U(z, r, t)$ of which may be written in the form,

$$U(z, r, t) = \frac{iU_0}{z_R} \frac{w_0}{w(z)} e^{-[r/w(z)]^2} \exp[i(kz - \omega t)] e^{ikr^2/2R(z)} e^{-i\psi(z)}, \tag{1}$$

where r is the radial distance from the z-axis, $R(z)$ is the radius of curvature of the wavefronts, and $\psi(z)$ is a propagation-dependent phase shift. The *intensity* of the beam is proportional to $|U|^2$, and hence the intensity varies with distance from the propagation axis as $\exp\{-2[r/w(z)]^2\}$. The quantity $w(z)$ is known as the **spot-size**, and we see that it is equal to the radius at which the intensity of the beam falls to $1/e^2$ of its value on axis. It may be shown that the spot size varies with propagation distance according to,

$$w(z) = w_0 \sqrt{1 + \left(\frac{z}{z_R}\right)^2} \tag{2}$$

$$w_0 = \sqrt{\frac{\lambda z_R}{\pi}}. \tag{3}$$

In this case the focus of the beam — that is, where the spot size is smallest — occurs at $z = 0$. The focus is often referred to the **beam waist**, and the spot size at this point, w_0, is known as the **waist size**. As shown in Fig. 1 the beam propagates away from the waist its spot size increases, and consequently the intensity of the beam on axis decreases. A useful measure of this behaviour is the **Rayleigh range** $z_R = \pi w_0^2/\lambda$, which corresponds to the distance from the beam waist at which the spot size increases by a factor of $\sqrt{2}$, and hence the intensity is decreased by a factor of 2, compared to the values at focus.

As an example, in order to reach the high intensities of present interest ($10^{15} - 10^{19}$ W cm^{-2}), it is typically necessary to focus the laser beam to a waist size of order $w_0 = 10$ μm. In this case, radiation of wavelength $\lambda = 1$ μm has a Rayleigh range of only 0.3 mm. Consequently we expect diffraction to restrict the length of any interaction with the laser to a few millimetres at most.

2.2 Refractive defocusing

In the discussion above we assumed that the refractive index of the medium through which the beam propagated was uniform. This will rarely be the case in practice, and consequently *refraction* of the beam may also be important in determining the interaction length. In fact, very frequently it is more important than diffraction.

Consider the propagation of an initially-plane wave through a medium for which the refractive index increases with r. In this case the phase velocity of the wavefronts will be slower at points away from the axis, and consequently the wavefronts will develop a curvature which is concave with respect to the source: the wave will diverge. The position $\boldsymbol{R}$ of a ray propagating through a medium of refractive index η is described by (Born and Wolf, 1980),

$$\frac{\mathrm{d}}{\mathrm{d}s}\left(\eta\frac{\mathrm{d}\boldsymbol{R}}{\mathrm{d}s}\right) = \nabla\eta, \tag{4}$$

where s is the distance along the ray from some reference point.

Let us consider the idealized case of propagation through a region of cylindrical symmetry in which the refractive index increases linearly with distance from the propagation axis: $\eta(r) = \eta_0 + \eta' r$, such that $\nabla\eta = \eta'\hat{\boldsymbol{r}}$. We may then write the position of the ray as $\boldsymbol{R} = z\hat{\boldsymbol{k}} + r\hat{\boldsymbol{r}}$, and since the gradient of the refractive index has only one non-zero component the equation describing the position of a ray becomes

$$\eta_0\frac{\mathrm{d}^2 r}{\mathrm{d}z^2} = \eta', \tag{5}$$

where we have used the fact that for a paraxial ray $s \approx z$ and $\eta \approx \eta_0$. The solution to this equation is simply $r(z) = r(0) + (\eta'/2\eta_0)z^2$, and hence if η' and η_0 are positive the ray will diverge — a process known as **refractive defocusing**. For example, a ray initially a distance r_0 from the axis will double its distance from the axis after propagating a distance $z_{\mathrm{ref}} = \sqrt{2r_0\eta_0/\eta'}$. Hence we may estimate of the distance over which a Gaussian beam will be defocused significantly by letting $r_0 = w_0$.

Refractive defocusing frequently occurs when an intense laser pulse propagates through a gas or a partially ionized plasma, and in such cases it is commonly referred to as **ionization-induced defocusing**. In order to estimate the strength of this defocusing we first note that in the non-relativistic regime the refractive index of a plasma may be written,

$$\eta = \sqrt{1 - \frac{N_e e^2}{m_e \epsilon_0 \omega^2}} \approx 1 - \frac{1}{2}\frac{N_e e^2}{m_e \epsilon_0 \omega^2}, \tag{6}$$

where N_e is the electron density, ω the angular frequency of the radiation, and m_e the rest mass of the electron. The approximation on the right-hand-side of eqn (6) holds in a dilute plasma such that the plasma frequency $\omega_p = \sqrt{N_e e^2/m_e \epsilon_0}$ is small compared to ω. Hence for a dilute plasma the radial gradient in the refractive index depends on the radial electron density gradient as:

$$\frac{\partial\eta}{\partial r} \equiv \eta' \approx -\frac{1}{2}\frac{e^2}{m_e \epsilon_0 \omega^2}\frac{\mathrm{d}N_e}{\mathrm{d}r} = -r_e\frac{\lambda^2}{2\pi}\frac{\mathrm{d}N_e}{\mathrm{d}r}, \tag{7}$$

where $r_e = e^2/4\pi\epsilon_0 m_e c^2$ is the classical electron radius. Imagine an intense laser beam of spot size w_0 propagating through a partially-ionized plasma, and let us suppose that the beam is sufficiently intense to double the electron density on axis (corresponding, for example, to ionization of He^+ to He^{2+}). Since the intensity profile of the beam decreases rapidly with radius,

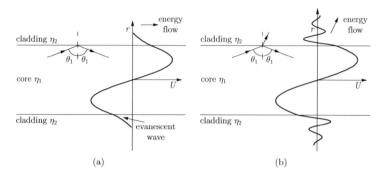

Figure 2. *Schematic diagram showing both ray and wave propagation in a generalized waveguide formed by layers of material of refractive index η_1 and η_2. Representative behaviour of the rays and amplitudes, U, of the electromagnetic fields are shown for: (a) a bound mode in a waveguide for which $\eta_1 > \eta_2$; (b) a lossy mode in a waveguide for which $\eta_1 < \eta_2$.*

the beam will be too weak to further ionize the plasma at radii greater than of order w_0. Hence we may estimate $\mathrm{d}N_\mathrm{e}/\mathrm{d}r \approx -N_\mathrm{e}/w_0$. For a plasma with $N_\mathrm{e} = 10^{19}$ cm^{-3} we find that for a beam with $w_0 = 10$ μm and $\lambda = 1$ μm, $z_\mathrm{ref} \approx 0.2$ mm, i.e. about equal to the Rayleigh range for diffraction.

2.3   General properties of waveguides

In order to overcome the limitations imposed by diffraction and refraction it is necessary to channel or guide the laser beam in some way. For low laser intensities this is routinely achieved using optical fibres, or related waveguide structures, in which radiation is guided through a core surrounded by a cladding layer of lower refractive index. This approach cannot be directly applied to lasers with the high intensities of present interest since the laser radiation would destroy the material of the core! However, the techniques which have been developed for high intensity laser pulses are very closely related and therefore it is worth briefly reviewing the general properties of waveguides.

Figure 2 shows a generalized waveguide formed by a layer of refractive index η_1 surrounded by layers of refractive index $\eta_2 < \eta_1$. Let us consider first the propagation in terms of rays. A ray striking the boundary between the core and cladding layers with an angle of incidence θ_1 will suffer total internal reflection provided $\sin \theta_1 > \eta_2/\eta_1$. Since the reflection is total the ray will, in principle, then propagate through the structure without loss.

Of course the ray picture is incomplete and we should consider the problem from the point of view of the electromagnetic field within the core and cladding layers. We will not do this here (see, for example, the book by Lorrain et al. for solution of a one-dimensional dielectric waveguide), but instead outline the approach and comment on the general form of the solutions found.

In essence the problem is one of solving Maxwell's equations for the core and cladding regions and ensuring that the solutions satisfy the boundary conditions at the boundary between the core and cladding layers. This is usually achieved by assuming solutions of the form,

$$\begin{aligned}
\boldsymbol{E}(x,y,z,t) &= \boldsymbol{E}_0(x,y)\exp[\mathrm{i}(\beta z - \omega t)] & (8)\\
\boldsymbol{H}(x,y,z,t) &= \boldsymbol{H}_0(x,y)\exp[\mathrm{i}(\beta z - \omega t)] & (9)
\end{aligned}$$

where $\boldsymbol{E}(x,y,z)$ and $\boldsymbol{H}(x,y,z)$ are the electric and magnetic field respectively. Although the solutions (8) and (9) look rather like the usual plane-wave solutions for an electromagnetic wave, they are rather different. In particular:

- The amplitude of the waves depends on the transverse co-ordinates (x,y).

- The propagation constant β is *not* equal to the usual value $\omega\eta/c$ for either the core or cladding layer. Instead β satisfies a **dispersion relation** that will depend on the refractive index in both layers, the frequency of the radiation, *and* the geometry of the waveguide.

- For a given frequency ω only certain values of β are allowed by the boundary conditions. Each allowed value corresponds to a **transverse mode** of the waveguide.

- Each transverse mode of the waveguide is associated with particular transverse variations of the fields, described by $\boldsymbol{E}_0(x,y)$ and $\boldsymbol{H}_0(x,y)$.

- In general the waves are not purely transverse, although they may be nearly so.

- For $\eta_1 > \eta_2$ a set of **bound modes** exist for which the amplitudes of the fields in the cladding do not oscillate with distance from the core, but decrease exponentially. As a consequence the electromagnetic wave in the cladding — known as an evanescent wave — does not carry energy away from the core, and consequently these modes may propagate without loss.

- For $\eta_1 > \eta_2$ a set of higher-order **lossy modes** also exist for which the amplitudes of the fields in the cladding *do* oscillate with distance from the core, and hence energy is carried away from the core.

- For $\eta_1 < \eta_2$ no bound modes exist, and consequently all modes are lossy to some extent.

Now using the ray picture we found that a ray may propagate without loss for *any* angle such that $\sin\theta_1 > \eta_2/\eta_1$, and consequently the z-component of the wavevector, β, could take any value within an allowed range. This is quite different from the conclusions of an analysis using the wave, or Maxwell, equations. The two pictures may be reconciled once it is realized that in propagating between two equivalent points in the waveguide the optical phase accumulated by the ray must equal an integer times 2π. This restriction leads to only certain angles θ_m being allowed, each allowed angle corresponding to a mode labelled by m. Further, solutions for which θ_m is greater than the critical value for total internal reflection are expected to experience significantly higher reflection losses at the core-cladding boundary, and hence may be associated with the lossy modes of the waveguide. The connections between the ray and wave pictures is explored in more detail in the book by Senior (1985).

The discussion above applies to a so-called **step-index** waveguide in which the refractive index varies sharply between two values, η_1 and η_2. For high-intensity laser pulses it is not possible to use a solid core since it would be destroyed early during the laser pulse. Instead the core must be vacuum, a low density gas, or a plasma, and consequently the cladding material will almost always have a *higher* refractive index than the core. Such waveguides must have finite propagation losses. Step-index waveguides for high-intensity laser pulses are usually referred to as **hollow** or **grazing-incidence** waveguides.

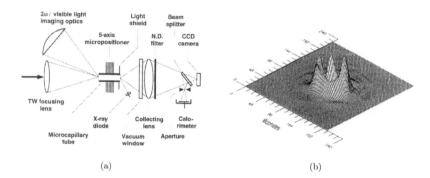

(a)   (b)

Figure 3. *The experiments of Jackel et al. on multimode guiding of high-intensity laser pulses through hollow capillaries. (a) Schematic diagram of the experimental arrangement used. (b) The measured transverse intensity profile of laser pulses transmitted through a 30 mm long, 100 μm diameter capillary.*

A second class of waveguide, **gradient refractive index waveguides**, may be formed in which the refractive index varies smoothly with radial distance. For high-intensity laser pulses the only suitable material in which a refractive index gradient may be formed is a plasma, and waveguides of this type are known as **plasma waveguides**.

We will discuss these two classes of waveguide in turn.

3   Grazing-incidence waveguides

Perhaps the most obvious way of attempting to guide high-intensity laser pulses is to focus the beam into an evacuated, or "hollow" capillary and to use the reflections from the inner wall to transport the beam. From the discussion above it is clear that since $\eta_1 < \eta_2$ the waveguide modes must be lossy, but without further analysis it is not clear *how* lossy the modes will be.

We may obtain a simple estimate of the losses from the ray picture. Away from its waist a Gaussian beam will diverge with a half-angle of $\Delta\theta = w_0/z_R = \lambda/\pi w_0$. For a beam with $w_0 = 10$ μm and $\lambda = 1$ μm we find $\Delta\theta \approx 30$ mrad. The rays striking the inner wall of the capillary therefore do so at a grazing-incidence, for which the reflectivity is high. Using the Fresnel equations (Brooker, 2003) the reflectivity R may be calculated to be $80 - 90\%$, depending on the polarization of the radiation. It is straightforward to calculate the number of wall reflections that a ray will suffer in propagating through a length ℓ of capillary and hence to show that the fraction of energy transmitted by the capillary is $T = R^{\ell/2a\tan\theta_1}$, where a is the inner radius of the capillary. For example, using the numbers above we estimate $T \approx 20\%$ for a 30 mm long capillary of inner diameter 100 μm.

This approach was first investigated experimentally by Jackel et al. in 1995. In that work 1 J, 0.9 ps pulses from a hybrid Ti:sapphire / glass amplifier chain were focused to a peak input intensity of 2×10^{17} W cm^{-2} at the entrance to hollow capillaries up to 126 mm long and with a diameter of 100 μm or 266 μm. The experimental arrangement is shown in Fig. 3(a). Relatively high pulse energy transmission was observed. For example, for 30 mm long, 266 μm diameter capillaries the energy transmission was measured to be 26%. This work was later extended to a peak input intensity of up to 10 TW by Borghesi et al.

Whilst relatively good energy transmission was achieved in this early work, the transverse

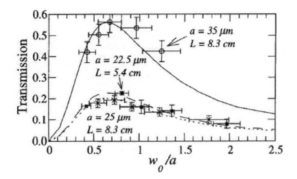

Figure 4. *The pulse energy transmission measured by Dorchies et al. (1999) as a function of w_0/a for capillaries of various lengths and diameters.*

profiles of the transmitted radiation typically showed large variations in intensity. For example, Fig. 3(b) shows the transverse intensity profile measured by Jackel et al. at the exit of a 30 mm long, 100 μm diameter capillary. The output shows several sharp features, and the intensity on axis is relatively low. These undesirable properties of the transmitted beam arise from the fact that for such experimental configurations the input laser pulse excites several transverse modes of the waveguide which then propagate with different phase velocities. For the same reason, **multimode guiding** of this type will also in general cause the propagating pulse to be distorted in temporally. Further, the losses of a mode typically increase as the order of the mode is raised. It is therefore desirable to guide in the lowest-order mode only; this is frequently done in optical fibres in telecommunications for exactly the same reasons.

In order to achieve **monomode** guiding it is necessary to match more closely the transverse intensity profile of the incident laser beam with the lowest-order mode of the waveguide. The mode structure of hollow capillaries, and particularly the transmission losses per unit length and the relative field strength at the wall of the capillary, have been studied by Cros et al. Those authors investigated the set of modes known as "hybrid modes", denoted EH_{1s}, which are linearly polarized and can therefore couple strongly with a linearly-polarized incident laser beam. It had previously been shown (Abrams, 1972) that for an incident Gaussian beam focused to a waist of size w_0 at the entrance to a capillary of radius a, optimum coupling into the lowest-order mode, EH_{11}, occurs when $w_0 = 0.645a$. Under these conditions the coupling into the lowest-order mode is 98%, leading to essentially monomode guiding.

For practical applications it is important that the capillary is not damaged by the propagating laser pulse, which requires that the power per unit area flowing into the capillary wall does not exceed the intensity damage threshold of the wall material. Cros et al. have calculated that for the EH_{11} mode the ratio of the power flux directed into the capillary wall to the that propagating along the capillary on axis is approximately 10^{-4} for a glass capillary. Given the damage threshold of glass of order 10^{14} W cm^{-2}, theory suggests that it should be possible for pulses with a peak axial intensity of order 10^{18} W cm^{-2} to be guided without damage to the capillary wall.

The same group have also performed experiments to investigate the conditions for optimum coupling (Dorchies et al., 1999). Figure 4 shows the measured pulse energy transmission through capillaries of various lengths and diameters as a function of w_0/a. It is clear that in all cases the highest transmission occurs when $w_0/a \approx 0.645$, in agreement with the theoretical expectation that the transmission will be highest when the greatest possible fraction of the input energy is coupled into the lowest-order mode. The solid lines are plots of $C_{\exp}\exp(-\alpha\ell)$ where $C_{\exp}$ is an experimentally determined coupling constant, and $\alpha = 2/L_{\mathrm{d},1}$ in which $L_{\mathrm{d},1}$ is the calculated damping coefficient for the *amplitude* of the lowest-order mode. We note that the damping coefficient depends on the capillary radius, a, and that the agreement with the experimental data reflects the fact that the transmitted radiation is predominantly in the lowest-order mode provided w_0 is not too far from $0.645a$.

This approach has been used to demonstrate guiding of intense laser pulses. Monomode guiding of pulses with a peak intensity of up to 5×10^{16} W cm^{-2} has been demonstrated through capillaries of 50 μm diameter and lengths of up to 105 mm (Dorchies et al., 1999). The observed damping coefficient $\alpha = 0.17$ cm^{-1} was in good agreement with theory, and corresponded, for example, to a measured transmission of approximately 40% for 50 mm long capillaries.

These results are for evacuated capillaries. Clearly in any application a target gas or plasma will be introduced into the capillary. If the contents of the capillary is further ionized by the guided laser radiation the losses of the waveguide will increase. This may be explained within the ray picture by the fact that this ionization-induced defocusing will decrease the angle of incidence θ_1 of rays incident on the capillary wall, and hence increase the proportion of radiation transmitted into the capillary wall and reduce the reflection coefficient. Alternatively, the increase in the losses may be considered from the point of view of the waveguide modes. Within this picture, ionization-induced defocusing couples the lowest-order mode to higher-order modes (Courtois et al., 2001). These modes have higher losses and hence the rate of attenuation of the guided radiation will be increased. In addition to increased *propagation* losses, introduction of a gas or partially-ionized plasma into the capillary may increase the *coupling* losses owing to defocusing of the incident radiation in the plume of gas or plasma leaving the entrance of the capillary.

Increased losses of this type have been observed experimentally. For example, it has been observed that the addition of 20 mbar of He gas decreases the energy transmission of a 40 mm long capillary by a factor of 6 from that achieved under vacuum (Dorchies et al., 1999). As a consequence, this technique is likely to be most useful for guiding intense pulses in low density, low atomic number gases, corresponding to plasma densities less than approximately 5×10^{17} cm^{-3} (Courtois et al., 2001). Indeed, guiding of laser pulses with a peak intensity of greater than 10^{17} W cm^{-2} has been demonstrated for 60 mm long capillaries filled with 2.2 mbar of helium; at this intensity the helium is expected to have been ionized to form a plasma with an electron density of approximately 5×10^{16} cm^{-3} (Cros et al., 2004).

It should also be mentioned that monomode guiding in hollow capillaries requires the incident laser pulse to have a very well-controlled and consistent transverse intensity profile. This requirement arises from the fact that when the condition for optimum coupling into the lowest-order mode is met the transverse dimensions of the incident beam are comparable to the diameter of the capillary. As a consequence, unwanted wings in the laser transverse profile can cause damage to the front face of the capillary and can destroy the capillary in only a few laser shots (Courtois et al., 2000). These problems could be overcome if techniques were developed for controlling the transverse intensity profile of high-power laser systems, or for manufacturing capillaries with tapered entrances.

4   Plasma waveguides

We consider now in more detail the operation of gradient refractive index waveguides. In this approach a channel is formed in which the refractive index decreases with radial distance r from the axis. A refractive index profile of this form will cause the phase velocity to increase with

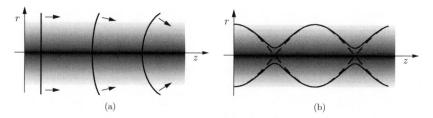

Figure 5. *Gradient refractive index guiding. (a) Schematic diagram showing how an initially plane wavefront becomes convex relative to the source after propagating through a medium which has a refractive index that decreases with radius. (b) Calculated propagation in a parabolic channel with a refractive index profile $\eta = \eta_0 - (1/2)\eta''r^2$. The dashed lines show solutions of the parabolic ray equation (11); the solid lines are solutions of eqn (20).*

r and hence the wavefronts of an initially-plane wave will become convex when viewed from the source, as shown schematically in Fig. 5(a). In other words, the channel will counteract diffraction and refractive defocusing.

4.1 Analysis of ideal parabolic channels

4.1.1 Ray picture

We may use the ray picture to investigate such a channel in a straightforward way. Let us suppose that the refractive index of the channel has the form,

$$\eta(r) = \eta_0 - \frac{1}{2}\eta''r^2. \tag{10}$$

In the paraxial approximation eqn (4) becomes,

$$\frac{\mathrm{d}^2 r}{\mathrm{d}z^2} = -\frac{\eta''}{\eta_0}r, \tag{11}$$

which has solutions of the form $r(z) = r(0)\cos(2\pi z/L_{\mathrm{osc}}^{\mathrm{ray}})$ for rays which travel exactly parallel to the z-axis at $z = 0$, and where,

$$L_{\mathrm{osc}}^{\mathrm{ray}} = 2\pi\sqrt{\frac{\eta_0}{\eta''}}. \tag{12}$$

We see that the distance of the ray away from the axis oscillates as it propagates through the channel with a period $L_{\mathrm{osc}}^{\mathrm{ray}}$, as shown in Fig 5(b). The period of this oscillation decreases as η'' increases. Finally in this section we also note that if η'' is negative, the solutions to eqn (11) would be rays which diverge away from the axis exponentially. In other words the rays would be refractively defocused.

4.1.2 Wave picture

If the refractive index does not vary with radial distance too quickly the electromagnetic modes of the channel will be almost transverse. In this case we may reduce Maxwell's equations to the scalar wave equation, which we may write as,

$$\nabla^2 E = \frac{\eta^2}{c^2}\frac{\partial^2 E}{\partial t^2}, \tag{13}$$

where E is one of the transverse components of the electromagnetic field.

A mode of the channel must be such that the modulus of its amplitude is independent of z. Hence we seek solutions of the form,

$$E(x, y, z, t) = E_0(x, y)\exp[\mathrm{i}(\beta z - \omega t)], \tag{14}$$

i.e. a wave propagating towards positive z with a phase velocity ω/β, and an amplitude that depends on the transverse co-ordinates only. Substitution of this trial solution and eqn (10) into the wave equation yields,

$$\left\{\nabla_\perp^2 - k^2\frac{\eta''}{\eta_0}r^2\right\}E_0 = \left(\beta^2 - k^2\right)E_0, \tag{15}$$

where $k = (\omega/c)\eta_0$ is the wave vector of light of angular frequency ω propagating through a *uniform* medium of refractive index η_0 and $\nabla_\perp^2$ is a Laplacian containing derivatives of the transverse components only (e.g., for a Cartesian co-ordinate system $\nabla_\perp^2 \equiv \partial^2/\partial x^2 + \partial^2/\partial y^2$). Note that in evaluating η^2 we have retained terms up to quadratic only.

Notice that this equation has the same form as the Schrodinger equation for a 2-D harmonic oscillator, where the term $(\beta^2 - k^2)$ takes the role of an eigenvalue. The solution of eqn (15) may be written in cylindrical co-ordinates as.

$$E(r, z, \phi, t) \propto \mathrm{e}^{-r^2/w^2}\,r^m L_p^m(2r^2/w^2)\mathrm{e}^{\pm \mathrm{i}m\phi}\exp[\mathrm{i}(\beta z - \omega t)], \tag{16}$$

where L_p^m is a Laguerre polynomial and $p = 0, 1, 2, 3, \ldots$ and $m = 0, 1, 2, 3, \ldots$ are the radial and azimuthal mode numbers. The constant w corresponds to the spot size of the lowest-order mode and is related to the properties of the channel by,

$$w = \left(\frac{2}{k}\right)\left(\frac{\eta_0}{\eta''}\right)^{1/4}. \tag{17}$$

The solution also requires that the following dispersion relation holds:

$$\beta^2 = k^2 - 2k\left(\frac{\eta''}{\eta_0}\right)^{1/2}(2p + m + 1). \tag{18}$$

Notice that the modes of the channel show **intramodal dispersion**. That is, even for a single mode the phase velocity ω/β is a function of frequency owing to the frequency-dependence of $k = (\omega/c)\eta_0$, and dispersion will also arise from any frequency-dependence of the refractive index (either η_0 or η''). In addition, of course, **intermodal dispersion** arises from differences in the phase velocity for different modes.

Notice that the transverse profile of the lowest-order ($p = 0, m = 0$) mode of the channel is identical to that at the waist of a lowest-order Gaussian beam of spot size $w_0 = w$. Hence, if such a beam is focused at the entrance of the channel the beam will excite the lowest-order mode of the channel only, and hence will propagate with a constant transverse intensity profile. The input beam is then said to be matched to the channel, the matching condition being that the waist size of the Gaussian beam is equal to w_M where,

$$w_M = w = \left(\frac{2}{k}\right)^{1/2}\left(\frac{\eta_0}{\eta''}\right)^{1/4}. \tag{19}$$

4.1.3 Source-dependent expansion method

Esarey et al. have solved eqn (15) using an alternative approach known as the source-dependent expansion method. In this approach the amplitude of the propagating beam is assumed to have a Gaussian profile of the form $\exp\{-[r/w(z)]^2\}$, and eqn (15) is then used to derive a differential equation for the beam spot size. For our parabolic channel, this differential equation may be written in the form,

$$\frac{\mathrm{d}^2 W'}{\mathrm{d}z^2} = \frac{1}{z_{\mathrm{RM}}^2} \frac{1}{W'^3} \left(1 - W'^4\right), \tag{20}$$

where $W' = w(z)/w_{\mathrm{M}}$, and $z_{\mathrm{RM}} = \pi w_{\mathrm{M}}^2/\lambda$ is the Rayleigh range of a Gaussian beam with a spot size equal to the matched spot size of the channel. The first term on the right-hand side of eqn (20) gives rise to diffraction of the beam; the second term describes the focusing action of the channel. It should be clear that when $W' = 1$ the beam will propagate with a constant spot size w_{M}.

Figure 5(b) shows the calculated variation of the spot size of a beam which is *not* matched to the channel. It is seen that the spot size oscillates as the beam propagates, a process known as **beam scalloping**. Notice that away from the axial region the behaviour of the beam spot size follows quite closely that of a ray initially travelling parallel to the axis at a distance w_0; as we would expect, however, this agreement does not hold at small values of r where diffraction becomes important. The reason for this scalloping is straightforward: if the spot size is greater than the matched spot size of the channel, the action of the channel will be to focus the beam to a smaller spot size; however once the spot size is smaller than the matched spot size, diffraction will cause the spot size to increase.

The period of this scalloping may be shown to be,

$$L_{\mathrm{osc}}^{\mathrm{Gauss}} = \pi \sqrt{\frac{\eta_0}{\eta''}}. \tag{21}$$

Notice that this is equal to exactly half of $L_{\mathrm{osc}}^{\mathrm{ray}}$, as we might expect given that the beam spot size cannot take negative values.

Of course, beam scalloping may equally well be described in terms of the modes of the channel. If a beam is introduced into the channel with a spot size which is not equal to the matched spot size, it will excite more than one of the channel modes. Since the modes propagate with different phase velocities the excited modes will periodically come in and out of phase with each other, leading to periodic variations in the transverse profile of the beam.

4.2 Techniques for generating plasma waveguides

Until now we have considered the operation of gradient refractive index waveguides in a general way. Gradient refractive index guiding of high-intensity laser pulses must employ a core formed from plasma — the transverse gradient of the refractive index being achieved by a gradient in the plasma density — since the ionization-induced defocusing that would occur in a gaseous core would be too strong to be counteracted by the waveguide. Gradient refractive index waveguides based on a preformed, shaped plasma are generally known as **plasma waveguides** or **plasma channels**.

In an ideal parabolic plasma channel the electron density can be described by,

$$N_e(r) = N_e(0) + \Delta N_e \left(\frac{r}{r_{\text{ch}}} \right)^2, \tag{22}$$

where ΔN_e is the increase in electron density at $r = r_{\text{ch}}$. It should be noted that r_{ch} is *not* necessarily equal to the size of any physical component of the channel (such as a capillary radius); but together with ΔN_e it is a convenient way of specifying the curvature of the channel. In other words it is the *ratio* $\Delta N_e / r_{\text{ch}}^2$ that determines the properties of the channel as a waveguide.

From eqn (6) the refractive index of the plasma channel may be written as,

$$\eta(r) = 1 - \frac{1}{2} \frac{N_e(0)e^2}{m_e \epsilon_0 \omega^2} - \frac{1}{2} \frac{e^2}{m_e \epsilon_0 \omega^2} \Delta N_e \left(\frac{r}{r_{\text{ch}}} \right)^2, \tag{23}$$

from which we may identify the relations:

$$\eta_0 = 1 - \frac{1}{2} \frac{N_e(0)e^2}{m_e \epsilon_0 \omega^2} \approx 1 \tag{24}$$

$$\eta'' = \frac{e^2}{m_e \epsilon_0 \omega^2} \frac{\Delta N_e}{r_{\text{ch}}^2}. \tag{25}$$

These results may be used in the formulae derived above. For example, of particular importance is the matched spot size. Substitution of eqns (24) and (25) into eqn (19) yields,

$$w_M = \left(\frac{r_{\text{ch}}^2}{\pi r_e \Delta N_e} \right)^{1/4}. \tag{26}$$

Notice that the matched spot size of a given channel is independent of the wavelength of the radiation. In fact the transverse extent of *all* the modes of a parabolic plasma channel are independent of the wavelength, as can be seen from eqns (16) and (19). This interesting result is important in some applications of plasma waveguides, particularly those involving the generation of radiation within the channel of some wavelength λ by pump radiation of wavelength λ_{pump}.

A very wide variety of techniques has been investigated for creating plasma waveguides, some of these are described below.

4.2.1 Plasma waveguides formed by hydrodynamic expansion

The first technique for creating a plasma waveguide suitable for high-intensity laser pulses was developed by Durfee and Milchberg (1993). In this approach a laser pulse is used to create a hot, approximately cylindrical plasma in a gas. Radial expansion of this plasma column drives a shock wave into the surrounding cold gas, to form a plasma channel behind the shock front.

Figure 6(a) shows the experimental arrangement employed. The initial plasma is formed by focusing a laser pulse with an axicon lens to form a longitudinal line focus in the gas. The duration of the laser pulse must be sufficiently long for heating, and further collisional ionization of the plasma formed early in the laser pulse. However, the plasma column expands in a timescale of order 1 ns which therefore forms an upper limit to the duration of the laser pulse that may be used. Typical values for the plasma-forming laser pulse are a pulse energy of 100 - 500 mJ and a duration of order 100 ps. This forms a cylindrical plasma of with a radius of order 10 μm and a temperature of approximately 100 eV. Figure 6(b) shows the temporal evolution of the plasma column formed in an argon gas jet. The formation of a plasma channel some 1.4 ns after the creation of the plasma column is evident.

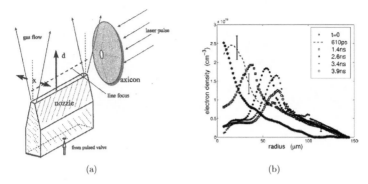

Figure 6. *Formation of a plasma channel by hydrodynamic expansion. (a) the experimental arrangement for generating the initial cylindrical plasma column in a gas jet. (b) Transverse electron density profiles of a plasma column in argon gas measured at various delays after its formation.*

The same group has demonstrated guiding of 50 mJ, 110 fs laser pulses with a peak input intensity of approximately 1×10^{17} W cm^{-2} through 15 mm plasma channels, with a pulse energy transmission of greater than 50% (Nikitin et al., 1999).

Plasma channels formed by this technique are typically not fully ionized, and can therefore only be used for laser pulses below the threshold for further ionization of the channel. For example, the plasmas formed in Ar are typically ionized to Ar^{8+} by the initial laser pulse. The threshold intensity for optical field ionization (OFI) of this ion is approximately 1.6×10^{18} W cm^{-2}, and consequently ionization-induced defocusing will occur for laser intensities above this value. Further, gases with a high atomic number are likely to increase the coupling losses of the channel owing to defocusing in regions of unionized gas near the channel entrance, as well as distorting the laser pulse through phase modulation (Gaul et al., 2000).

The obvious solution to these problems is to use a gas with a low atomic number, such as hydrogen or helium, and to ensure that it is fully ionized. The difficulty in achieving this arises from the conflicting requirements that the laser pulse forming the channel should be: (i) of sufficiently high intensity ($> 10^{14}$ W cm^{-2}) to ionize neutral H or He by OFI; but yet of low intensity ($< 10^{13}$ W cm^{-2}) and long duration so that the plasma is heated sufficiently by inverse bremsstrahlung. Several extensions of the hydrodynamic expansion technique have been developed to overcome these problems.

In the ignitor-heater method (Volfbeyn et al., 1999) a short (< 100 fs), intense ($> 5 \times 10^{14}$ W cm^{-2}) "ignitor" pulse produces initial ionization by OFI. This initial plasma is then heated and further ionized by a long (of order 100 ps) "heater" pulse of relatively low intensity. In practice this approach is implemented by overlapping the line foci produced by focusing the ignitor and heater beams with cylindrical mirrors. Approximately cylindrical channels are formed if the ignitor and heater beams propagate perpendicular to each other and the common line focus.

Recently this approach has been used to guide very intense laser pulses (Geddes et al., 2004a). A significant feature of this work is that the intensity of the guided laser pulses was sufficiently high for relativistic self-focusing to occur (see section 4.2.3). For example, pulses with a peak power of 4 TW, focused to a waist of 7 μm at the entrance of the channel — corresponding to a peak input intensity of 7×10^{18} W cm^{-2} — were guided through 2.5 mm long plasma channels, equivalent to more than 10 Rayleigh ranges of the input beam. The pulse energy transmission

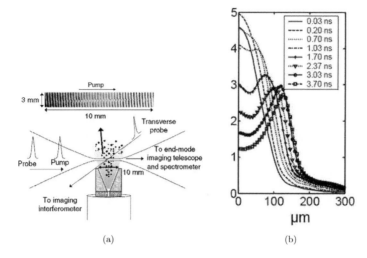

(a)   (b)

Figure 7. *Formation of a plasma channel by hydrodynamic expansion in a clustered gas jet. (a) the experimental arrangement for generating the initial cylindrical plasma column in a clustered gas jet using an ordinary lens. Inset: transverse interferogram showing formation of a plasma channel over approximately 8 mm. (b) Measured transverse electron density profiles (in units of 10^{18} cm^{-3}) generated in argon clusters for a jet temperature and backing pressure of 113 K and 20 bar respectively.*

was measured to be 35%, but a significant fraction of the absorbed energy was deposited in the (useful) form of longitudinal plasma waves. The onset of relativistic self-focusing was confirmed by the fact that if the peak power of the laser pulses was increased to 9 TW, optimum guiding was achieved for plasma channels with a smaller curvature of the transverse electron density profile; in other words in the presence of relativistic self-focusing the refractive power of the plasma channel needed to be reduced. At these higher powers trapping and acceleration of plasma electrons to form a quasi-monoenergetic electron beam of 86 MeV energy was observed (Geddes et al., 2004b)

An alternative extension of the hydrodynamic expansion scheme has been investigated by Gaul et al. In this approach a simple discharge is employed to ionize He gas partially before heating and further ionization by a 0.3 J, 100 ps laser pulse. The same group has also found that stretching the laser pulse to 400 ps and increasing its energy to 0.6 J gave sufficient time for avalanche ionization and inverse bremsstrahlung heating to build up from OFI of trace impurities in the He gas. The same group have used this approach to guide laser pulses with a peak input intensity of 2×10^{17} W cm^{-2} through 10 mm long plasma channels (Zgadzaj et al., 2004)

Despite the significant advances that have been achieved with this guiding technique, several areas remain to be addressed. First, the channels formed have a relatively high electron density on axis, typically above 5×10^{18} cm^{-3}, and this is too high for many applications. A high plasma density is necessary for the ionization and heating of the initial plasma column to be fast relative to the time for hydrodynamic expansion. Second, the efficiency of heating the initial plasma column is relatively low, of order 10%, meaning that relatively large and complex laser systems are required to generate the plasma column. Third, narrowing of the plasma channels can occur at their entrance and exit owing to gradients in the gas density in the gas jets typically used.

Some of these problems have recently been addressed by Milchberg's group (Kumarappan et al., 2005) by forming the plasma channel in clustered gases. A gas cluster is an assembly of atoms bonded by van der Waals forces, and can be formed following the rapid cooling induced by expansion of gas through a gas jet. The use of gas clusters provides two key advantages: extremely efficient absorption of the pump laser energy, and self-guiding of the laser pulse used to provide the initial plasma column.

The reason for the efficient absorption of the pump laser radiation may be understood with the aid of a 1-D model of the laser interaction with a gas cluster (Ditmire et al.1997). In this model initial low-level ionization of the gas cluster occurs through OFI. The electrons so produced are heated by the laser pulse and drive rapid and efficient collisional ionization *within each cluster*, which has a density near that of the solid state. The hot, highly-ionized clusters then explode to form an essentially uniform plasma in 10 - 100 ps; subsequent radial expansion of the plasma column so formed then creates a plasma channel by the mechanisms discussed above. A key advantage of using a clustered medium is that the efficiency of the absorption of the pump laser radiation is determined primarily by the cluster size, whereas the final density of the plasma column produced depends on the initial density of clusters. It is therefore possible to decouple, to a significant extent, the requirement for efficient heating of the plasma from its final density.

The second advantageous feature of clustered gas targets, self-guiding of the pump laser pulses, arises from the temporal dependence of the polarizability of the ionizing cluster. Early during the pump laser pulse each ionized cluster behaves as a very small ball of plasma above the critical density ($\omega_p > \omega$), and so excludes the laser field. At such times the real part of the susceptibility of the cluster is positive, and hence the refractive index of the ensemble of ionized clusters is greater than unity. For a pump laser beam with a peak intensity on axis, the degree of cluster ionization will also be greatest on axis, leading to a transverse refractive index profile with an axial maximum: a gradient refractive index waveguide. This feature allows long plasma columns to be produced by longitudinal pumping with an ordinary lens, rather than using an axicon lens. This geometry increases the efficiency of the production of the plasma column since with an axicon lens the incident rays make a relatively large angle with respect to the axis of the system, and therefore pass through a relatively thin region of plasma.

Milchberg's group (Kumarappan et al., 2005) have used this approach to demonstrate guiding of high-intensity laser pulses using the experimental arrangement shown in Fig. 7(a). The measured transverse electron density profile generated by a 30 mJ, femtosecond pump laser pulse interacting with argon clusters is shown in Fig. 7(b). The pump energy required is some ten times lower than typically required to form plasma waveguides in gases with axicon lenses. Further, the axial electron density can be as low as 1×10^{18} cm^{-3}, significantly lower than is possible with conventional gas jets. The length of the plasma channel produced (8 mm) is some 40 times longer than the 200 μm Rayleigh range of the focused pump laser pulse, illustrating the importance of self-guiding of the pump laser radiation. By injecting a second laser pulse into the plasma channel, guiding of laser pulses with a peak intensity of approximately 3×10^{17} W cm^{-2} was achieved with a coupling efficiency of 50%. This method has also been used to generate fully-ionized channels in hydrogen, but to date the length of the channels formed has been limited to 3 mm owing to difficulties in generating hydrogen clusters in a gas jet with a slit nozzle.

4.2.2 Plasma waveguides formed by discharges

It is also possible to create plasma channels by the reverse process, (magneto)hydrodynamic compression, using fast capillary discharges. In this approach a capillary of a few millimeters diameter is filled with gas at low pressure and ionized by a discharge current with a rise time of

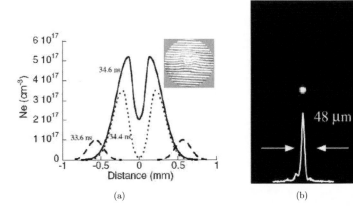

(a)   (b)

Figure 8. *Formation of a plasma channel by hydrodynamic compression in a fast capillary discharge in Ar. (a) Measured temporal evolution of the transverse electron density profile. The times indicated are with respect to the beginning of the current pulse, and the inset shows an interferogram corresponding to the 34.4 ns profile. (b) The transverse intensity profile of a laser pulse with a peak input intensity of 2×10^{17} W cm^{-2} measured at the exit of a 55 mm long capillary.*

10 - 50 ns, and a peak of 5 - 20 kA. Since the current rises rapidly, the skin effect ensures that the initial ionization occurs close to the capillary wall. The large magnetic field generated by the current then compresses the plasma through the $\boldsymbol{J} \times \boldsymbol{B}$ force to drive a strong shock towards the capillary axis. A plasma channel is formed just before the rapidly collapsing annulus of plasma reaches the axis, and since the compression is close to 100%, the axial plasma density is much greater than the initial density of the gas. This process is illustrated in Fig 8(a) which shows interferometric measurements of the transverse electron density profile formed in a fast capillary discharge in Ar.

This approach has been used (Luther et al., 2004), (Hosokai et al., 2000) to guide intense laser pulses channels formed by fast capillary discharges in He and Ar. For example, Luther et al. have demonstrated guiding of pulses with a peak input intensity and spot size of 2×10^{17} W cm^{-2} and 30 μm respectively through 55 mm long Ar^{8+} plasma channels with an axial electron density of approximately 2×10^{17} cm^{-3}. Fig 8(b) shows the measured transverse intensity profile of the transmitted beam. The pulse energy transmission was measured to be 75%.

Plasma channels have also been formed in slow capillary discharges, an approach pioneered by Zigler and his group (1996). In that technique a discharge pulse with a rise-time of order 100 ns and a peak current of a few hundred amperes is passed through an initially evacuated capillary formed in a soft material such as polypropylene. The discharge current ablates and ionizes the wall material to fill the capillary with plasma and forms a plasma channel by radiative and collisional heat transfer to the capillary wall. This causes the temperature of the plasma to be greater on axis and, since the pressure across the capillary is uniform, an axial minimum in the plasma density.

The disadvantage of this technique is that the plasma channels formed are typically only partially ionized. For example, discharge-ablated polypropylene capillaries form a plasma comprised of fully-ionized hydrogen and carbon ionized to a variety of stages up to C^{4+}. As a consequence, laser-induced ionization can cause severe modulation of the spot size of an intense laser pulse propagating along the plasma channel (Spence and Hooker, 2000).

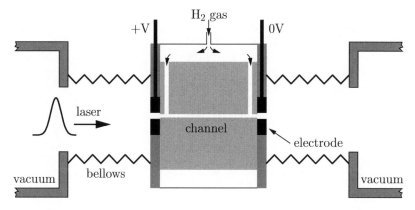

Figure 9. *Schematic diagram of a gas-filled capillary discharge waveguide.*

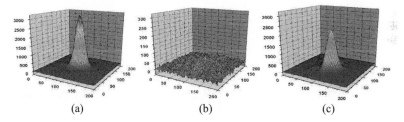

Figure 10. *Demonstration of guiding of high-intensity laser pulses in a gas-filled capillary discharge waveguide. Shown are the transverse intensity profiles: (a) at the entrance; and at the exit of a 33 mm long, square cross-section waveguide filled with 120 mbar of H_2 for laser pulses injected (b) before, and (c) 115 ns after the onset of the discharge. All plots have the same spatial scale (in microns); plots (a) and (c) have the same intensity scale, whilst that in (b) has been magnified by a factor of 10.*

This problem can be overcome using the hydrogen-filled capillary discharge waveguide developed by the author and his group (Spence and Hooker, 2001). In this approach the plasma channel is formed by discharge ionization of a low atomic number gas injected into a capillary channel. In early designs the capillary was formed from a refractory material, such as alumina, to avoid ablation of the wall. In more recent experiments, however, the capillary channels were formed by laser-machining two half-channels (Jaroszynski et al., 2006) in sapphire plates and then contacting the plates together, as shown schematically in Fig. 9. Channels made in this way are extremely robust; and the capillary wall material is transparent which allows transverse observation of the discharge. A slight modification of this design allows channels of square cross-section to be formed which allows a transverse beam to interact with the plasma without the distortion of the probe beam that would be caused by the refraction at a cylindrical wall. Square cross-section capillaries of this type are presently being used to measure the transverse electron density profile in the plasma channel.

The mechanisms leading to the formation of the plasma channel have been investigated in magnetohydrodynamic simulations of the capillary discharge (Bobrova et al., 2002). These show that for a 300 μm diameter capillary filled with 67 mbar of H_2 gas, and with a half-sinusoidal discharge pulse of full width 200 ns and 300 A peak, the gas is fully ionized approximately 60 ns after the onset of the discharge. By some 80 ns after the onset of the discharge current the plasma is in a quasi-equilibrium in which the pressure is uniform across the diameter of the capillary, and Ohmic heating of the plasma is balanced by conduction of heat to the wall. The electron density profile so formed is found to be approximately parabolic with a matched spot size in good agreement with longitudinal interferometry of the plasma channel (Spence and Hooker, 2001).

This approach has been used to guide laser pulses with a peak input intensity of 1.2×10^{17} W cm^{-2} through 30 mm and 50 mm channels with a pulse energy transmission of approximately 90% and 80% respectively (Butler et al., 2002). Fig. 10 shows recent results obtained with 33 mm long, 400 μm square cross-section capillaries for laser pulses with a peak input intensity of approximately 5×10^{16} W cm^{-2}. It is seen that laser pulses injected prior to the onset of the discharge current are strongly defocused by ionization-induced refraction in the neutral hydrogen, and have a low pulse energy transmission, as shown in (b). The pulse energy transmission is found to increase rapidly during the discharge, reaching approximately 90% once the hydrogen gas becomes essentially fully ionized. This is accompanied by a reduction in the transverse dimensions, and a corresponding increase in intensity, of the transmitted laser pulse as a plasma channel is formed.

4.2.3 Relativistic guiding

An electron in a laser field will be driven in an oscillatory motion known as the **quiver motion**. For laser intensities of order 10^{18} W cm^{-2} the quiver motion becomes relativistic, modifying the refractive index of the plasma from eqn (6) to $\eta = \sqrt{1 - N_e(r)e^2/\gamma m_e \epsilon_0 \omega^2}$, where γ is the usual relativistic factor.

In order to understand how the relativistic motion leads to self-focusing, we consider the weakly relativistic limit; a more general consideration may be found in a paper by Esarey et al. In a laser field $E(t) = E_0 \cos \omega t$ the electron will be driven with a velocity $v(t) = v_0 \sin \omega t$, where $v_0 = -eE_0/m_e \omega$. The relativistic factor may be written $\gamma^{-1} = \sqrt{1 - (v/c)^2} \approx 1 - (1/2)(v/c)^2$. Since the square of the quiver velocity is proportional to the intensity of the laser radiation, the refractive index of the plasma becomes,

$$\eta \approx 1 - \frac{1}{2} \frac{N_e e^2}{\gamma m_e \epsilon_0 \omega^2} \equiv \eta_0 + \eta_I I, \tag{27}$$

where η_I is a positive constant, and I is the laser intensity. We see, therefore, that a laser with a transverse intensity profile which is peaked on axis will be self-focused by the gradient in the refractive index profile. This **relativistic self-guiding** exactly balances diffraction of the laser pulse at a critical power $P_c^{\text{rel}} = 17.4(\omega/\omega_p)^2$ (Esarey et al., 1997).

In addition to channeling arising from modification of the refractive index by the quiver motion of the electrons, guiding can also occur through changes induced in the plasma density. The energy of the quivering electrons and ions is proportional to the intensity of the laser radiation, and the gradient of this quiver energy corresponds to a force – the **ponderomotive force** – which expels electrons and ions radially. Owing to their much greater mass, the motion of the ions under the ponderomotive force is much slower than that of the electrons, and hence for *short* laser pulses the modification to the plasma density arises solely from expulsion of electrons from the axial region, reducing the critical power for relativistic self-focusing to $P_c^{\text{pon}} = 16.2(\omega/\omega_p)^2$ (Sun et al., 1987).

Ions are expelled by the ponderomotive force over a much longer time scale, and hence ion motion is only important for self-channeling of long laser pulses. Ion motion *is* important, however, in forming plasma channels *after* the passage of a short, intense laser pulse. In this

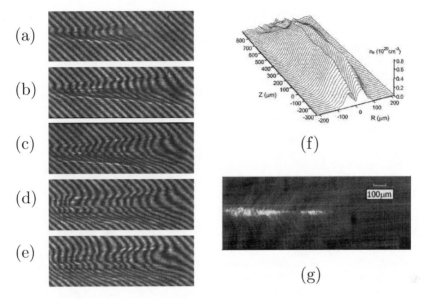

Figure 11. *Interferograms of a plasma in the regime of relativistic self-focusing: (a) 0 ps, (b) 5 ps, (c) 15 ps, (d) 35 ps, and (e) 45 ps after the passage of a 4.3 TW laser pulse through a He gas jet with an atomic density of 4×10^{19} cm^{-3}. In (f) is shown the reconstructed electron density profile of the channel formed 35 ps after the passage of the pulse. An image of the Thomson scattered radiation from the passage of the driving laser pulse through the He gas jet is shown in (g).*

case the transverse momentum acquired by the electrons from the ponderomotive force continues to carry electrons away from the axial region after the laser pulse has passed; ions are dragged with the electrons via the Coulomb force to form a channel in which the total plasma density is reduced on axis.

Relativistic and ponderomotive self-channeling has been reported by several groups (Borisov et al., 1992), (Monot et al., 1995), (Clayton et al., 1998). For example, Monot et al. have demonstrated channeling of 6.8 TW laser pulses over a length of approximately 2 mm in a hydrogen gas jet. Comparison of the observed Thomson scattered light with theory suggested that the peak laser intensity in the channel was greater than 10^{19} W cm^{-2}. The temporal evolution of plasma channels formed ponderomotively during and after the passage of an intense laser pulse has been investigated by Sarkisov et al. as shown in Fig. 11. In that work 400 fs, 4.3 TW laser pulses with a wavelength of 1.053 μm were focused into a He gas jet with an atomic density of approximately 4×10^{19} cm^{-3}. Fig.11(a) shows the formation of a plasma channel during the laser pulse, and extended propagation over a length of approximately 1 mm is evident in Fig. 11(g) which shows Thomson scattered radiation from the pulse. The development of a plasma channel on a picosecond timescale after the passage of the driving pulse is shown clearly in Fig. 11(b)-(e).

Ponderomotively-formed channels of this type have been used to guide a second laser pulse injected after an intense channel-forming pulse. For example, Krushelnick et al. have observed guiding of pulses with a peak intensity of 5×10^{16} W cm^{-2} through a 2.5 mm long channel formed by a pulse with a peak intensity of approximately 6×10^{18} W cm^{-2}.

Technique	Electron density (cm^{-3})	Length (mm)	Peak intensity (W cm^{-2})
Grazing-incidence	$0 - 10^{17}$	100	10^{17}
Hydrodynamic expansion	$10^{18} - 10^{19}$	15	10^{19}
Magnetohydrodynamic compression	10^{17}	100	10^{17}
Gas-filled capillaries	$10^{18} - 10^{19}$	50	10^{17}
Relativistic guiding	10^{19}	5	10^{18}

Table 1. *Summary of the performance of some techniques used to guide high-intensity laser pulses through plasmas. The figures given are indicative of what has been achieved to date, and do not necessarily represent fundamental limitations.*

5 Summary

It will be apparent from the above that a wide variety of techniques for guiding laser pulses with intensities as high at 10^{19} W cm^{-2} have been developed in the last decade or so. However, as yet there is no "ideal" technique able to guide laser pulses of arbitrary intensity over any desired length through a plasma density which may be freely chosen. As such, the particular conditions of an experiment will be the main factor determining the most appropriate guiding technique. Table 1 summarizes the guiding performance achieved to date with the methods discussed in these lectures. It is perhaps useful, however, to comment further on each method.

Grazing-incidence guiding has been demonstrated to be capable of guiding laser pulses with a peak intensity of order 10^{17} W cm^{-2} over more than 100 mm, and extension to intensities of order 10^{18} W cm^{-2} would seem to be feasible. This technique is likely to be most useful at low gas densities owing to the rapid increase in propagation and coupling losses caused by ionization-induced defocusing when gas is introduced to the capillary. It should also be noted that the transverse spatial profile of the beam to be guided must be well controlled if damage to the capillary is to be avoided.

Many techniques for creating plasma channels have been, and continue to be, investigated. In the ideal case the plasma channel will be fully ionized, in which case it may be used — at least in principle — to guide pulses with intensities at least up to the onset of relativistic self-focusing. If the channel is not fully ionized, ionized-induced defocusing will cause scalloping of the guided beam for intensities above that required to ionize the channel.

Plasma channels formed by hydrodynamic expansion have many advantages. These include: a small matched spot size (of order 10 μm), an essentially infinite shot lifetime, and good optical access for diagnostics. Until recently the channel formed with this technique were not fully ionized, and the axial electron density was limited to the range 10^{18}–10^{19} cm^{-3}. However, with the development of channels formed in clustered gases it has become possible to create *fully-ionized* channels at relatively low densities. To date, the length of plasma channels formed by hydrodynamic expansion has not exceeded 15 mm. This limitation partly reflects the fact that it is difficult to make very long initial plasma columns by laser-induced ionization, and that the energy of the laser pulse (and cost of the laser system) required to generate the channel increases linearly with the length of the plasma.

Several types of capillary discharge have been used to create plasma channels. These all have the advantages that no auxiliary laser system is required to form the channel, long plasma channels can be created, and it may be possible to generate tapered or structured channels.

However, at least to date, the matched spot size of the plasma channels produced by capillary discharges are relatively large, approximately 30 μm, and in many cases transverse optical access to the channel is restricted.

The plasma channels formed by fast capillary discharges can be fully ionized, and have a reasonably small matched spot size at low axial electron densities. On the other hand, the discharge circuitry is relatively complex and the channels evolve on a nanosecond time-scale, which means that the discharge must be triggered with very low timing jitter. It should also be noted that this approach is likely to be limited to reasonably low values of the axial electron density, owing to difficulties in driving a controlled shock wave at high initial gas pressures.

The major advantage of using slow capillary discharges to generate a plasma channel is its simplicity. Fully ionized channels with axial electron densities in the range 10^{18}–10^{19} cm^{-3} have been demonstrated, and long shot lifetimes achieved. However, formation of plasma channels at densities outside this range may prove to be difficult. In particular, it is likely to be difficult to use this approach to generate plasma channels with low values of the electron density whilst maintaining a reasonably small matched spot size.

Relativistic and ponderomotive guiding are, by their nature, able to guide very intense laser pulses. However, these guiding mechanisms are ineffective at guiding short laser pulses with a duration less than $2\pi/\gamma\omega_p$ owing to the fact that the refractive index of the plasma is modified on the time-scale of the plasma, rather than that of the laser pulse (Esarey et al., 1997). On the other hand, long laser pulses are prone to developing modulation instabilities which are undesirable in some, but not all, applications. It should be appreciated, of course, that relativistic and ponderomotive guiding are intrinsic effects and will therefore occur whenever the incident laser power is above a certain threshold. As such an understanding of these mechanisms is important even if they are not used as the primary guiding mechanism.

References

Abrams, R. L.: 1972, *IEEE Journal of Quantum Electronics* **QE-8**, 838

Bobrova, N. A., Esaulov, A. A., Sakai, J.-I., Sasorov, P. V., Spence, D. J., Butler, A., Hooker, S. M., and Bulanov, S. V.: 2002, *Physical Review E* **65**, 016407

Borghesi, M., Mackinnon, A. J., Gaillard, R., and Willi, O.: 1998, *Physical Review E* **57**, R4899

Borisov, A. B., Borovskiy, A. V., Korobkin, V., Prokhorov, A. M., Shiryaev, O. B., Shi, X. M., Luk, T. S., McPherson, A., Solem, J. C., Boyer, K., and Rhodes, C. K.: 1992, *Physical Review Letters* **68**, 2309

Born, M. and Wolf, E.: 1980, *Principles of Optics: Electromagnetic Theory of Propagation Interference and Diffraction of Light*, Pergamon Press, Oxford, sixth edition

Brooker, G. A. *Modern Classical Optics*, Oxford University Press, Oxford, first edition, 2003.

Butler, A., Spence, D. J., and Hooker, S. M.: 2002, *Physical Review Letters* **89**, 185003

Clayton, C. E., Tzeng, K. C., Gordon, D., Muggli, P., Mori, W. B., Joshi, C., Malka, V., Najmudin, Z., Modena, A., Neely, D., and Dangor, A. E.: 1998, *Physical Review Letters* **81**, 100

Courtois, C., Couairon, A., Cros, B., Marquès, J. R., and Matthieussent, G.: 2001, *Physics of Plasmas* **8**, 3445

Courtois, C., Cros, B., Malka, G., Matthieussent, G., Marquès, J. R., Blanchot, N., and Miquel, J. L.: 2000, *Journal of the Optical Society of America* **17**, 864

Cros, B., Courtois, C., Godiot, J., Matthieussent, G., Maynard, G., Marquès, J. R., David, P. G., Amiranoff, F., Andreev, N. E., Gorbunov, L. M., Mora, P., and Ramazashvili, R. R.: 2004, *Physica Scripta* **T107**, 125

Cros, B., Courtois, C., Matthieussent, G., Bernardo, A. D., Batani, D., Andreev, N., and Kuznetsov, S.: 2002, *Physical Review E* **65**, 026405

Ditmire, T., Smith, R. A., Tisch, J. W. G., and Hutchinson, M. H. R.: 1997, *Physical Review Letters* **78**, 3121

Dorchies, F., Marquès, J. R., Cros, B., Matthieussent, G., Courtois, C., Vélikoroussov, T., Audebert, P., Geindre, J. P., Rebibo, S., Hamoniaux, G., and Amiranoff, F.: 1999, *Physical Review Letters* **82**, 4655

Durfee, III, C. G. and Milchnerg, H. M.: 1993, *Physical Review Letters* **71**, 2409

Esarey, E., Sprangle, P., Krall, J., and Ting, A.: 1997, *IEEE Journal of Quantum Electronics* **33**, 1879

Gaul, E. W., Blanc, A. P. L., Rundquist, A. R., Zgadzaj, R., and Langhoff, H.: 2000, *Applied Physics Letters* **77**, 4112

Geddes, C., Toth, C., Tilborg, J. V., Esarey, E., Schroeder, C., Bruhwiler, D., Nieter, C., Cary, J., and Leemans, W.: 2004a, in V. Yakimenko (ed.), *AIP Conference Proceedings*, Vol. 37, American Institute of Physics

Geddes, C. G. R., Toth, C., vanTilborg, J., Esarey, E., Schroeder, C. B., Bruhwiler, D., Nieter, C., Carey, J., and Leemans, W. P.: 2004b, *Nature* **431**, 538

Hosokai, T., Kando, M., Dewa, H., Kotaki, H., Kondo, S., Hasegawa, N., Horioka, K., and Nakajima, K.: 2000, *Nuclear Instruments and Methods in Physics Research A* **455**, 155

Jackel, S., Burris, R., Ting, A., Manka, C., Evans, K., and Kosakowski, J.: 1995, *Optics Letters* **20**, 1086

Jaroszynski D., et al.: 2006, *Phil. Trans. R. Soc.* **A 364**, 689

Krushelnick, K., Ting, A., Moore, C. I., Burris, H. R., Esarey, E., Sprangle, P., and Baine, M.: 1997, *Physical Review Letters* **78**, 4047

Kumarappan, V., Kim, K. Y., and Milchberg, H. M.: 2005, *Physical Review Letters* **94**, 205004

Lorrain, P., Corson, D., and Lorrain, F.: 1996, *Electromagnetic Fields and Waves*, W. H. Freeman and Company, New York, third edition

Luther, B. M., Wang, Y., Marconi, M. C., Chilla, J. L. A., Larotonda, M. A., and Rocca, J. J.: 2004, *Physical Review Letters* **92**, 235002

Monot, P., Auguste, T., Gibbon, P., Jakober, F., Mainfray, G., Dulieu, A., Louisjaquet, M., Malka, G., and Miquel, J. L.: 1995, *Physical Review Letters* **74**, 2953

Nikitin, S. P., Alexeev, I., Fan, J., and Milchnerg, H. M.: 1999, *Physical Review E* **59**, R3839

Sarkisov, G. S., Bychenkov, V. Y., Novikov, V. N., Tikhonchuk, V. T., Maksimchuk, A., Chen, S. Y., Wagner, R., Mourou, G., and Umstadter, D.: 1999, *Physical Review E* **59**, 7042

Senior, J. M.: 1985, *Optical Fibre Communications Principles and Practice*, Prentice-Hall International, London, first edition

Spence, D. J. and Hooker, S. M.: 2000, *Journal of the Optical Society of America* **17**, 1565

Spence, D. J. and Hooker, S. M.: 2001, *Physical Review E* **63**, 015401

Sun, G. Z., Ott, E., Lee, Y. C., and Guzdar, P.: 1987, *Physics of Fluids* **30**, 526

Volfbeyn, P., Esarey, E., and Leemans, W. P.: 1999, *Physics of Plasmas* **6**, 2269

Zgadzaj, R., Gaul, E. W., Matlis, N. H., Shvets, G., and Downer, M. C.: 2004, *Journal of the Optical Society of America* **21**, 1559

Zigler, A., Ehrlich, Y., Cohen, C., Krall, J., and Sprangle, P.: 1996, *Journal of the Optical Society of America* **13**, 68

4

Harnessing Plasma Waves as Radiation Sources and Amplifiers

D.A. Jaroszynski, Albert Reitsma and Bernhard Ersfeld

Department of Physics, University of Strathclyde, Glasgow

1   Introduction

The interaction of intense ultra-short laser pulse with plasma provides the basis for developing new technologies such as ultra-compact wakefield accelerators (Tajima and Dawson 1979), compact light sources (Jaroszynski et al. 2002, 2006) and high power laser amplifiers (Shvets et al 1998). In this lecture we will explore how laser driven plasma waves can be harnessed to accelerate particles and produce useful electromagnetic radiation for various applications. As we have seen in this series of lectures, lasers can be used to drive fission and fusion reactions, the latter of which has potential for energy production. However, lasers are also creating new opportunities for studying new phenomena by realising astrophysical-like conditions in the laboratory (Bingham 2005), creating particles from vacuum and studying nonlinear optics of plasma (Joshi 1990). Plasma has long been recognised as providing acceleration gradients that are three orders of magnitude larger than conventional radio frequency (RF) accelerators that are limited by electrical breakdown of the accelerating cavity structures (Tajima and Dawson 1979). Laser and electron driven wakefield acceleration is creating a revolution in accelerator technology by miniaturising the accelerator module. Researchers have demonstrated gradients exceeding 100 GV/m in plasma using intense laser pulses (Modena et al. 1995, Coverdale et al. 1995, Nakajima et al. 1995). Low density plasma in the range of $n_p = 10^{17} - 10^{19}$ cm^{-3} can potentially accelerate to GeV energies in a few millimetres. Initially, electron beams were produced with a 100 % Maxwellian energy spread due to plasma wavebreaking (Malka et al. 2002), which limited their usefulness and was seen as a major impediment to controllable acceleration. In 2004 several groups (Mangles et al. 2004, Geddes et al. 2004, Faure et al. 2004) accelerated relatively high charge (20 - 500 pC) *monoenergetic* electron bunches to more than 100 MeV in several millimetres. These experiments demonstrated gradients in excess of 100 GV/m and effectively put laser-driven wakefield accelerators on the map. A very surprising aspect of these observations was that they did not require a separate injector because electrons were injected into the plasma wake from the background plasma and that the electron beams were produced with a relatively small energy spread, $\delta\gamma_0/\gamma_0 \approx 1$ - 3%, where $\gamma_0 = (1 - \beta^2)^{-1/2}$ is the Lorentz factor and $\beta = v/c$

(c and v are the respective velocities of light and the electron bunch in vacuum). Simulations by Martins et al. (2005) show that injection is rapidly shut off by beam loading (Trines et al. 2001), which results in extremely short electron bunches ($<$ 10 fs) with a small energy spread (Pukhov and Meyer-ter Vehn 2002).

The surprisingly high quality of the electron beams produced in wakefield accelerators should make them suitable as drivers of compact light sources (Jaroszynski et al. 2006). Coherent and incoherent sources have become key research tools for scientists and industrialists. X-ray radiation from synchrotron light sources are indispensable probes used by very large user community. A new "fourth" generation of facilities providing coherent radiation based on free-electron lasers (FELs) are being developed and now reach wavelengths as short short as a few nanometres (Ayvazyan et al. 2002, 2006). Excellent progress is being made to extend the spectral range of these self amplification of spontaneous emission (SASE) FELs to the ngstrom range where they can be used to study biological relevant structures and follow chemical reactions in real time (Zewail 1999). Light sources are some of teh largest instruments in the world. Their high cost is determined by the size of conventional RF accelerator structures on which they are based and their large radiation shielding infrastructure. The compactness of laser-driven plasma wakefield accelerators provides and opportunity to reduce both the size and cost of both synchrotron sources and SASE FELs, which would make them widely available to universities and medium sized research establishments. Furthermore, as will be discussed below, the ultra-short electron bunches from wakefield accelerators would make light sources unique and enable the production of brilliant femtosecond duration electromagnetic pulses over a wide spectral range extending to hard x-rays. The ultra-short "pre-bunched" electron bunches should also enable the sources to emit coherently at longer wavelengths thus opening up the possibility of producing high power electromagnetic pulses down to a single cycle in the mid-infrared and terahertz frequencies. Because the high current electron bunches from wakefield accelerators are very short ($<$ 10 fs) (Pukhov 2002) they could potentially be used to produce attosecond pulses at shorter wavelengths through a superradiant FEL process (Bonifacio et al. 1989, 1991, 2005, Jaroszynski 1997, 1997a). Wakefield based sources combined with synchronised femtosecond lasers would provide a compact suite of powerful tools for time resolved two-colour and two-type (electron-photon) pump-probe studies. The wide availability of radiation sources based on wakefield technology could revolutionise the way science is done.

In the mid-1980s, the application of intense laser pulses changed dramatically with the advent of chirped pulse amplification (CPA) (Strickland and Mourou 1985). CPA is a technique for generating short, high-intensity laser pulses by the stretching, amplification and recompression of short, low-intensity laser pulses. Stretching and compression is achieved using a pair of gratings in antiparallel configuration. Since the optical path off a grating depends on the wavelength, propagation through the stretcher results in a long pulse with a frequency chirp. In the time domain, the frequencies are separated in such a way that either the low frequency part (negative chirp) or the high frequency part (positive chirp) of the pulse comes first. After recompression in a second grating, the frequencies overlap again and the pulse is compressed. The stretched pulse is amplified by a series of amplifiers giving a total gain of 10^7 or give an intensity of focused beams as high as $10^{19} - 10^{21}$ W cm^{-2} (Danson et al. 1993). Because of the relatively small size and high peak power, these CPA systems are known as T^3 (table top terawatt) lasers.

2   Laser-plasma interactions

In laser-plasma interaction, ultrahigh intensities produce a wealth of nonlinear phenomena, including strong wakefield excitation (Akhiezer and Polovin 1956, Lu et al. 2006), self-focusing (Sun et al. 1987, Hafizi et al. 2000), and wave-wave interactions such as stimulated Raman scattering (McKinstrie and Bingham 1992, Darrow et al. 1992). A short survey of these nonlinear phenomena in laser-plasma interaction is given here. An important parameter for ultrahigh intensity lasers is the amplitude a_0 of the dimensionless vector potential $e\vec{A}/mc^2$. In terms of the peak intensity I_0, it is given by

$$a_0 \simeq 0.85 \cdot 10^{-9} \, \lambda_0 \, [\mu m] \, I_0^{1/2} \, [W \, cm^{-2}] \tag{1}$$

where λ_0 is the laser wavelength in vacuum. Since the amplitude of the quiver momentum is $a_0 mc$, the condition $a_0 \geq 1$ implies that the plasma electron motion is relativistic and nonlinear.

A laser pulse can propagate in a plasma if the plasma density n_p is below the *critical density* $n_{cr} = \epsilon_0 m \omega_0^2 / e^2$, where ω_0 denotes the central (carrier) frequency of the laser pulse. Plasmas with $n_p > n_{cr}$ are called overdense, plasmas with $n_p < n_{cr}$ are called underdense.

The refractive index of the plasma is

$$ck_0/\omega_0 = \sqrt{1 - n/(\gamma n_{cr})}, \tag{2}$$

where k_0, ω_0 denote the wavenumber, resp. the frequency of the laser pulse, n is the density and γ the Lorentz factor of the plasma electrons. The pulse is self-focused if the refractive index is peaked on axis: this can be achieved by relativistic mass increase associated with the quiver motion (as this gives γ a radial dependence) or it is due to wakefield excitation (which results in density perturbation, giving n a radial dependence). There is a power threshold for relativistic self-focusing, given by $\mathcal{P} > \mathcal{P}_{cr}$, where

$$\mathcal{P}[GW] \approx 43 \, (a_0 r_0/\lambda_0)^2 \tag{3}$$

is the *peak power* for a laser pulse with a Gaussian transverse intensity profile $a^2 \propto \exp(-r^2/r_0^2)$, and

$$\mathcal{P}_{cr}[GW] \approx 17 \, (\lambda_0/\lambda_p)^2 \tag{4}$$

is the *critical power* for self-focusing.

High power pulsed lasers which drive the plasma waves rely CPA, a technique originally developed to avoid damage to optical components. CPA involves amplifying a stretched, frequency chirped, "seed" pulse in a solid state broadband amplifying medium excited by a monochromatic "pump" beam. After amplification, the seed pulse is compressed by dispersive optical elements, to ultra-short durations and high powers, currently up to several petawatts (Edwards 2003). Because CPA is limited to several petawatts by damage to optical components, plasma has been suggested as a possible medium for amplification. Stimulated Raman back-scattering in plasma (Shvets et al. 1997, 1998) occurs when two slightly detuned laser beams collide in plasma to produce a beat wave with a frequency equal to the plasma frequency, $\omega_p = \sqrt{n_p^2 e^2/\epsilon_0 m}$ (SI units) or $= \sqrt{4\pi n_p^2 e^2/m}$ (CGS units), where n_p is the plasma density, e the electron charge, ϵ_0 the permittivity of free space, and m the electron mass. The intrinsically narrow bandwidth of the Raman instability (Kruer 1998) and its convective instability (Bobroff 1967), due to the excitation of long lived plasma waves, limit the gain bandwidth achievable in the *linear regime* thus reducing its usefulness as a short pulse amplifier. Linear Raman amplification (Kruer 1998, Forslund et al. 1973), is susceptible to undesirable noise amplification because of its intrinsically high gain (Ping 2003, Shvets et al. 1997, 1998). To avoid these limitations, Shvets et al (Shvets et al. 1997, 1998, Malkin et al. 1999) proposed taking advantage of the *non-linearity* of the medium at high intensities to produce soliton-like pulses through pump-depletion or, at higher intensities, by operating in the Compton or superradiant regime where the plasma density echelon is dominated by the ponderomotive force associated with the beat wave (Shvets et al. 1998, Malkin et al. 2000). By using a chirped pump pulse, distributed amplification occurs for different spectral components of the seed at different positions in the plasma, through the creation of a "chirped" plasma density echelon. This behaves as a long "chirped mirror" which simultaneously backscatters and compresses the chirped pump pulse and effectively broadens the gain bandwidth to that of the pump (Ersfeld and Jaroszynski 2005). We will show that the gain and the bandwidth of the amplifier depend on ω_p and the chirp rate and spectral bandwidth of the pump. This contrasts with conventional CPA where the seed is chirped while the pump is monochromatic. The chirped pulse Raman amplifier has potential as a high fidelity ultra-short pulse high power *linear* amplifier or as a compressor of high energy chirped pulses from a conventional CPA amplifier, avoiding the requirement for extremely large compressors in large vacuum chambers. Furthermore, because chirped pulse Raman amplification is a three wave parametric interaction it provides a means of eliminating pre-pulses and pedestals which usually limit the usefulness of conventional solid state CPA.

In the following section we will explore laser-driven plasma waves as accelerators, radiation sources and amplifiers.

3   The laser-driven plasma wakefield accelerator

Electrostatic forces of a plasma with a density of $n_p \approx 10^{18}$ cm^{-3} are of the order of 100 GV/m (Tajima and Dawson 1979), which is three orders of magnitude larger than that attainable with conventional RF cavities. The ponderomotive force (Kruer 1988) of an intense laser pulse drives a plasma wave in the form of a wake travelling at the group velocity, v_g ($\equiv c(1 - \omega_p^2/\omega_0^2)^{1/2} \approx c(1 - \omega_p^2/2\omega_0^2)$), of the laser pulse in the plasma, where ω_p is the plasma frequency and ω_0 is the radiation frequency. The Lorentz factor associated with the wake velocity is $\gamma_g = \omega_0/\omega_p$. As the wavelength of these plasma waves is very short, (≈ 33 μm for $n_p \approx 10^{18}$ cm^{-3}), they provide an opportunity to harness their electrostatic forces to produce a very compact travelling wave accelerating "structure". These structures have dimensions that are approximately λ_p^3, where $\lambda_p \approx 2\pi c/\omega_p$, which is $\approx 4 \times 10^{-5}$ cubic millimetres for the above density. Furthermore, as the accelerating medium is fully ionised i.e. already broken down, it is extremely robust. As with conventional RF accelerator cavities, the accelerating potential of a wakefield accelerator has both longitudinal forces, which give rise to acceleration, and transverse forces, which cause focussing or defocussing of the electron bunches, thus many of the techniques of conventional RF accelerator physics can be applied to plasma based accelerators. However, the small dimension of the plasma "structures" implies several significant challenges for harnessing the waves to accelerate charged particles. The most stringent of these is the requirement that electron bunches must be injected into the accelerating part of the electrostatic wave. This in turn demands bunches to be much shorter than 100 fs for the above density. Furthermore, electrons should also be injected precisely into the structure at a velocity close to the phase velocity of the wake i.e. at the group velocity of the laser pulse. This implies that initially $\gamma_g \approx \gamma_0$ (typically $\gamma_g = 10$ - 50). For a plasma wave of relative amplitude, $\delta n_p/n_p$, electrons of the correct phase velocity are accelerated until they gain an energy $\gamma_0 = 2\gamma_g^2 \delta n_p/n_p$ over a distance $l_d = \gamma_0 \lambda_p = 2\gamma_g^2 \lambda_p/\pi$, after which deceleration occurs. This sets the dephasing length for acceleration and limits the useful accelerator length for a particular density. Rapid longitudinal variation of the accelerating potential also increases the resulting energy spread and can cause fluctuations of the mean energy of accelerated electrons. The production of high quality electron beams with a reproducible mean energy and small energy spread implies bunch durations and synchronism between plasma wake of around 10 fs, for $\delta\gamma_0/\gamma_0 < 1\%$, where γ_0 and $\delta\gamma_0$ are the respective mean and variance of the electron bunch energy. The maximum energy acquired by an injected electron bunch is inversely proportional to the plasma density and thus it is an advantage to accelerate particles over longer lengths using low density plasma. The maximum charge that can be accelerated is proportional to the background plasma density (Chiou et al. 1998, Reitsma 2000, 2005, Wilks 1987), giving a maximum of between ≈ 100 pC and 1 nC for 10^{18} cm^{-3}. At this density the dephasing length is of the order of one centimetre. Because diffraction limits the interaction length, it is usually necessary to rely on self-guiding of the laser or use some form of plasma waveguide to extend the length beyond the Rayleigh range to fully utilise the accelerating potential (Butler et al. 2002, Spence et al. 2001). To achieve sufficiently high intensities to drive a substantial plasma wave the normalised vector potential of the laser should be $a \geq 1$, which implies a laser spot size of the order of 50 μm for a 20 terawatt 800 nm laser pulse. The optimum pulse duration for driving the wake is $\tau_l \approx \lambda_p/2c$ (Fritzler 2004), the resonant regime, though it has become clear that laser pulses with a duration longer than $\lambda_p/2c$ will evolve to an optimum dimension for driving the wake because of the space and time varying permittivity of the wake. The laser power rapidly evolves to the threshold for self focussing, $P_{crit}[\text{GW}] \approx 17(\omega_0/\omega_p)^2$, which leads to a further increase in the laser intensity and self-guiding.

4 Plasma-based acceleration

This section contains a brief overview of plasma-based acceleration and an explanation of the main concepts.

A plasma-based accelerator harnesses high-amplitude plasma waves produced by an intense laser pulse propagating in plasma which drives collective oscillations of plasma electrons in a background of plasma ions (Esarey et al. 1996). The driving pulse is assumed to propagate at a speed close to the speed of light so that ions can be considered immobile. The plasma electron motion induces charge separation, which results in high electric fields and a strong restoring Coulomb force that drives the oscillation of the plasma electrons. The oscillation frequency of the plasma electrons is the *plasma frequency* ω_p. The *phase velocity* v_φ of the plasma wave is equal to the propagation velocity of the source. Together, these quantities determine the *plasma wavelength* $\lambda_p = 2\pi/k_p$, where $k_p = \omega_p/v_\varphi$ is the plasma wavenumber.

In *laser wakefield acceleration*, a single short, intense laser pulse is the source for creating a high-amplitude plasma wave (Tajima and Dawson, 1979). For a given laser wavelength $\lambda_0 = 2\pi c/\omega_0$, the critical density is found to be $\pi/r_e\lambda_0^2$, where $r_e = e^2/4\pi\epsilon_0 mc^2$ is the classical electron radius. For a typical solid-state laser with a wavelength of 800 nm often used in laser wakefield acceleration, the critical density is $1.75 \cdot 10^{21}$ cm^{-3}. The typical plasma density for laser wakefield acceleration is 2 to 4 orders of magnitude lower ($10^{17} - 10^{19}$ cm^{-3}). The propagation velocity of the laser pulse, which is the group velocity $v_g = c(1 - n_p/n_{cr})^{1/2}$, is close to c, the speed of light in vacuum.

If a laser pulse propagates in a plasma, the plasma electrons perform oscillations at the laser frequency ω_0 in response to the electric fields of the laser. This motion of plasma electrons is called the *quiver motion*. The rapidly oscillating momentum (*quiver momentum*) of the plasma electron is given by

$$\vec{p} \approx e\vec{A}/c, \tag{5}$$

where $\vec{A}$ denotes the vector potential that describes the laser electromagnetic fields. In a strongly underdense plasma ($n_p \ll n_{cr}$) the quiver motion can be clearly separated from the response on the much slower plasma timescale, which is driven by the *ponderomotive force* of the laser pulse. The ponderomotive force is due to the finite amplitude of the quiver oscillation, which is proportional to the laser field amplitude, and the gradients of the laser pulse envelope. On the slow timescale these effects combine into a net force that expels the plasma electrons from the laser pulse region. The ponderomotive force is given by

$$\vec{F}_p = -mc^2\nabla\gamma \approx -\nabla\sqrt{m^2c^4 + e^2 < \vec{A}^2 >}, \tag{6}$$

where $\gamma = (1 + \vec{p}^2/m^2c^2)^{1/2}$ is the Lorentz factor of the electron and the brackets $< \dots >$ denote the average over the fast laser oscillation timescale. The ponderomotive excitation of a plasma wave is sketched in Fig. 1. The ponderomotive force excites a plasma wave by pushing the plasma electrons aside, which induces a charge separation. Due to this charge separation, strong restoring Coulomb forces are excited in the wake of the laser pulse. The laser pulse and the induced wakefields are stationary in the co-moving frame that propagates with velocity v_g, as sketched in Fig. 1, where the pulse propagates in the z-direction. For effective wakefield excitation, the pulse must be short: the maximum amplitude of the accelerating wakefield is found when the pulse length is about half a plasma wavelength, depending on the particular shape of the laser pulse envelope. For this reason, acceleration in the wakefield of a short laser pulse is also known as *resonant* laser wakefield acceleration.

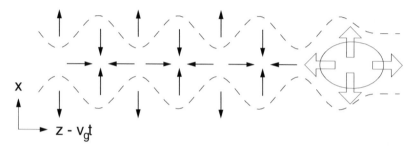

Figure 1. *Sketch of ponderomotive wakefield excitation: ponderomotive force (big arrows) pushes plasma electrons aside (dashed curve is trajectory of plasma electrons, white oval is region of laser pulse); as a result, strong electrostatic forces (small arrows) are excited.*

In laser wakefield acceleration, the energy gain of the electrons is limited by *phase slippage* (Hubbard et al. 2000), which results from the difference between v_g and the actual velocity v_z of an electron. This is illustrated in Fig. 2, in which a sketch of laser wakefield acceleration of a short electron bunch is given. Due to phase slippage, the electron eventually reaches a decelerating region in the wave. This process is called *dephasing* and it limits the acceleration distance to the *dephasing length*. During most of the acceleration the electron velocity v_z is larger than v_g and can be approximated with c. With this approximation, the dephasing length is estimated to be $(\lambda_p/2)/(c/v_g - 1)$. Therefore, the strongly underdense regime ($n_p \ll n_{cr}$) is preferred for laser wakefield acceleration, because in this regime a long dephasing length is found from the group velocity $v_g = c(1 - n_p/n_{cr})^{1/2}$ of the laser pulse, which is close to c.

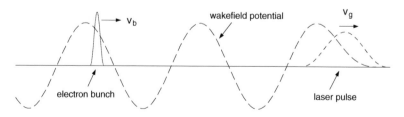

Figure 2. *Sketch of laser wakefield acceleration: the laser pulse that propagates with velocity v_g excites a plasma wave on which an electron bunch with velocity v_b is accelerated.*

As an alternative method for driving the plasma wave with a single laser pulse, a scheme involving two long, overlapping laser pulses with slightly different frequencies has been tried.

This scheme is called *beat-wave acceleration* (BWA - Rosenbluth and Liu 1972), because here the maximum amplitude is determined by the condition that the beat distance is equal to the plasma wavelength, which can be realised by taking the frequency difference exactly equal to the plasma frequency. Plasma waves can also be driven by a relativistic electron bunch (Chen et al. 1985). In this case it is the Coulomb force of the electron bunch instead of the ponderomotive force that induces charge separation. Acceleration in the wakefield of an electron bunch is known under the somewhat confusing name *plasma wakefield acceleration* (PWFA). For acceleration, a trailing electron bunch can be placed in the wakefield of the driving bunch. This scheme works best if both bunches are short compared to the plasma wavelength. Alternatively, one bunch that is somewhat longer than a plasma wavelength can be used, such that the wakefield of the leading edge accelerates the electrons in the tail of the bunch. In the early 1980s, when powerful short-pulse lasers were not yet available, plasma-based acceleration was investigated using both the BWA and PWFA schemes. These lines of research are still continued today, in particular at SLAC (PWFA - Hogan et al. 2005) and at UCLA (BWA - Tochitsky et al. 2004).

In principle, for a given pulse energy the wakefield amplitude can be maximized by compressing the laser pulse as much as possible (both in longitudinal and transverse direction). Thus the ideal laser pulse for wakefield excitation is a very short "light bullet" focused tightly into an underdense plasma. For sufficiently high a_0, the wakefield excitation can be so strong that the pulse expels all plasma electrons close to the axis, leaving behind a region of unshielded ions. This regime of wakefield excitation is called the "blowout" or "bubble" regime (Kostyukov and Pukhov 2004) and it also exists for dense electron bunch drivers (Rosenzweig et al. 1991).

Nonlinear laser-plasma interaction enables coupling of waves of different frequencies. This coupling is most effective if the frequencies and wavenumbers satisfy a *resonance* condition. An electromagnetic pump wave (laser pulse) with frequency ω_0, wavenumber k_0 can couple resonantly to a copropagating plasma wave with frequency ω_p, wavenumber k_p and copropagating electromagnetic waves (sidebands) with frequency $\omega_0 \pm \omega_p$, wavenumber $\approx k_0 \pm k_p$, a process known as Raman forward scattering. The excitation of a counterpropagating electromagnetic wave with frequency $\omega_0 - \omega_p$, wavenumber $-k_0 + k_p$ and a plasma wave with frequency ω_p, wavenumber $2k_0 - k_p$ is called Raman backward scattering. Raman scattering can severely affect the propagation of the laser pulse and lead to *parametric instability*. Such an instability is driven by the feedback loop sketched in Fig. 3. The presence of a plasma wave, e.g., excited through Raman scattering, leads to density perturbations. Since the refractive index of the plasma depends on the electron density, as given by (2), the laser pulse amplitude tends to increase in regions with low electron density and to decrease in regions with high electron density. In this way, the laser pulse can eventually break up in "beamlets" of length $\simeq \lambda_p/2$. Such a modulated pulse can drive the plasma wave very efficiently, especially if it extends over many plasma wavelengths. Finally the loop is closed due to enhancement of density perturbations associated with a stronger plasma wave.

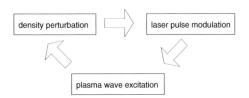

Figure 3. *Feedback loop of parametric instability.*

In the presence of scattered waves and a strong wakefield, the motion of plasma electrons becomes highly nonlinear and relativistic. In particular, Raman forward scattering can lead to fast electron generation by *self-trapping* and acceleration of background electrons in the plasma wave (Tzeng et al. 1999). Alternatively, electron trapping may be induced by *wave breaking* of a high-amplitude plasma wave (Bulanov et al. 1997). Fast electrons have been found in several *self-modulated laser wakefield* (SMLW) experiments (Modena et al. 1995, Umstadter et al. 1996, Wagner et al. 1997, Malka et al. 1997, Ting et al. 1997). In these experiments, an ultraintense laser pulse is used to ionize a gas jet and create a plasma. Inside the plasma, the laser pulse undergoes relativistic self-focusing and, through Raman scattering and wave breaking, accelerates electrons up to multi-MeV energies. The typical plasma density for SMLW experiments is $10^{19} - 10^{20}$ cm^{-3}. For these densities, the regime of strongly nonlinear interaction is easily accessible with T^3 lasers, because the critical power (4) for self-focusing is around 1 TW and the typical laser pulse length (100 fs - 1 ps) contains multiple plasma wavelengths. In SMLW acceleration large amounts of fast electrons can be produced: up to 8 nC in a single bunch has been reported. The energy distribution shows a large spread, with most electrons at low energy (a few MeV) and only a small fraction at high energy (up to a few 100 MeV). From the short acceleration length in these experiments (typically a few mm), it is deduced that the accelerating gradients inside the plasma are very large (of order 100 GV m^{-1}).

At slightly lower densities (between $10^{18} - 10^{19}$ cm^{-3}) another regime of LWFA with a quasi-monoenergetic spectrum of accelerated electrons has been found recently (Mangles et al. 2004, Geddes et al. 2004, Faure et al. 2004). Because of the lower density, the pulse covers at most about 1 plasma wavelength, and will be compressed due to self-steepening in the plasma. Subsequently, the laser-plasma interaction leads to efficient wakefield excitation, formation of a "bubble" and electron injection through plasma wave breaking. The injection stops when *beam loading* prevents further trapping of the electrons in the plasma. A rough definition of beam loading (Katsouleas et al. 1987) is the modification of the plasma wave due to the presence of trapped electrons, as illustrated in Fig. 4. In the figure, the amplitude of the plasma wave behind the electron bunch is seen to be much lower than in front of it, indicating that the bunch extracts energy from the plasma, which is an expression of energy conservation. After the injection has terminated, the electrons are accelerated to high energy with a small energy spread due to the bunch being very short. The typical energy is around 100 MeV and the charge is several 10s to several 100s of pC/bunch.

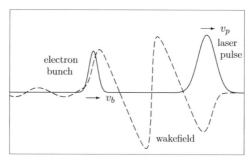

Figure 4. *Sketch of beam loading effect: electron bunch takes energy from the plasma wave.*

In comparison with SMLW experiments, self-focusing is less effective because the critical power is higher at low density. Consequently, long focal length optics must be used to ensure that the interaction length is sufficiently long for the electrons to reach maximum energy (i.e. Rayleigh length comparable to dephasing length). However, this comes at the expense of intensity, as a long focal length implies a relatively large spot size. Therefore, it is beneficial for the further development of laser wakefield acceleration to find other methods of laser pulse guiding, to be able to reach a higher energy by scaling down to lower plasma densities. The two main options are plasma channel guiding (Durfee and Milchberg 1993, Ehrlich et al. 1996, Spence and Hooker 2001), and guiding with capillary tubes (Jackel et al. 1995, Dorchies et al. 1999).

A plasma channel provides local changes in the refractive index (2) through a transverse profile of the plasma density. In order to guide a laser pulse, the refractive index on axis must be higher than off axis, so the density on axis must be lower than off axis. One particular realization is a hollow plasma channel (Chiou, Katsouleas and Mori 1996), which is completely evacuated around the axis and has a constant density beyond a certain radius. In order to support plasma waves of sufficient amplitude, the hollow channel radius must be small compared to a plasma wavelength. Another possibility is the use of parabolic channels, which have a density profile $n_i(r) = n_p(1 + r^2/w_c^2)$, with n_i the ion (plasma) density, n_0 the on-axis density, r the distance to the axis (cylindrical symmetry is assumed), and w_c the channel radius. These profiles support certain eigenmodes for the laser, of which the fundamental mode is a Gaussian with $a^2 \propto \exp(-r^2/w_l^2)$ with pulse width (spot size) w_l. The *matching condition* is $w_c = k_p w_l^2$, where $k_p^2 = 4\pi n_p e^2/mc^2$. This condition must be applied to find the matched spot size, i.e. the width of the Gaussian laser profile that is supported by a given channel radius or, alternatively, to find the matched radius of the channel that can be used to guide a pulse with given spot size. A slight mismatch between w_c and w_l leads to spot size oscillations. These oscillations can be thought of as the beat wave of the different supported eigenmodes that the mismatched pulse consists of. The effect of small-amplitude spot size oscillations on the excited wakefields is generally assumed to be of minor importance.

One way of creating a parabolic plasma profile is the manipulation of a gas jet with additional laser pulses. This technique was pioneered by Milchberg et al. (Durfee and Milchberg 1993) and later refined by Leemans et al. (Volfbeyn, Esarey and Leemans 1999) The scheme relies on creating the plasma via inverse bremsstrahlung heating, which leads to plasma expansion through a hydrodynamic shock wave. At a particular point in the expansion, the density profile is nearly parabolic around the axis and it has walls with a density about 5 times the on-axis value. Still further away from the axis, the density decreases to zero. Strictly speaking, this type of density profile can only support decaying laser modes because the laser light can penetrate through the wall (Clark and Milchberg 2000) - the channel is sometimes called "leaky". Also because the shock expansion is dynamic, it is necessary to tune the delay between the main pulse that creates the wakefield and the laser pulses used for creating the channel. Experimental results show that these issues can be dealt with and laser pulse guiding over cm distances has been demonstrated by various groups. A limitation of this scheme is that the on-axis density is usually not lower than about 10^{18} cm^{-3} (a lower density is desirable as it can provide a higher final energy).

A parabolic density profile can also be created in capillary discharge waveguides of plastic (Ehrlich et al. 1996) or aluminum oxide (Spence and Hooker 2000). In such waveguides, a plasma channel is formed on such a long timescale that the pressure is constant inside the tube. A hollow density profile is formed, because the plasma temperature, which is in the eV-range, is higher on-axis than off-axis. The density limitations of this scheme is comparable to the limitation of Milchberg's scheme ($n_p \geq 10^{18}$ cm^{-3}).

Another possibility of extending the interaction length is to guide the laser pulse in a capillary tube. Such a tube supports laser eigenmodes of which the fundamental has a large overlap with a Gaussian profile. These eigenmodes are decaying with a decay distance that can be several 10s of cm. A plasma can be introduced by filling the tube with gas and using a laser pulse that is strong enough to create a fully ionized plasma inside the tube. Therefore this scheme does not require tuning of the delay between different laser pulses.

Besides diffraction there are other limitations on the propagation length of a laser pulse in a plasma. If the laser pulse is sufficiently long and/or intense, instabilities such as Raman scattering and modulational instability may develop, as discussed above. The instabilities are usually suppressed if the pulse length l_l is short compared to the plasma wavelength λ_p and if the pulse amplitude is in the linear or weakly nonlinear regime ($a_0 \leq 1$). For channel-guided laser pulses of sufficiently low intensity and sufficiently short pulse length to suppress instabilities, one finds that the dynamics is governed by group velocity dispersion with dispersion length $k_0 (\lambda_p/\lambda_0)^2 l_l^2$ and energy loss of the driving pulse to the plasma wave, for which the typical length is the pump depletion length $(6/a_0^2) (\lambda_p/\lambda_0)^2 l_l$. For a short, relativistic pulse ($l_l \approx \lambda_p/2$, $a_0 \geq 1$) the length scales for pump depletion and dephasing become comparable, leading to efficient coupling between the laser pulse and the plasma wave (Gordon et al. 2003).

In order to make plasma-based acceleration competitive to other techniques, it is paramount to find a completely controlled way of acceleration. This is because all applications require particular electron source characteristics. As an example, high energy gain is required in particle physics experiments (e.g. the TeV collider). For high energy gain one needs a large accelerating gradient, which plasma-based acceleration obviously provides, and/or a sufficiently long acceleration distance. An obvious requirement is that the source produces enough electrons, where "enough" strongly depends on the particular application. Other criteria can be summarized as *bunch quality* requirements, which typically involve things like small energy spread and small angular spread of the electron bunch. Some applications require very high quality bunches: to mention one, the ideal electron bunch for an X-ray free-electron laser (*see e.g.* LCLS Conceptual Design report) has a charge of about 1 nC at 1 GeV energy with 0.1% energy spread and a transverse emittance of 1 μm. Although the latest LWFA experiments have demonstrated good progress towards these specs, there are still many technical issues to be resolved, related e.g. to shot-to-shot reproducibility and wall-plug efficiency of laser systems.

5   Acceleration dynamics

In this section, we describe the laser wakefield accelerator at its most basic level by discussing the dynamics of a test electron in a prescribed wakefield. Already this simple model reveals some interesting aspects of wakefield acceleration, especially when we examine the transverse motion. The scalar and vector potentials (ϕ and $\vec{A}$) are assumed to depend only on the comoving coordinate $\zeta = z - v_g t$ and the transverse coordinates (x, y), where v_g denotes the phase velocity of the wakefield, which is equal to the group velocity of the laser pulse.

Since the electrons move predominantly in the forward direction (z-direction), the *paraxial approximation* can be used for analyzing the dynamics of accelerated electrons. For evaluating the Lorentz force on an electron with momentum $\vec{P}$, the approximation $\vec{v} \approx v_g \hat{e}_z$ can be used to give

$$\frac{d\vec{P}_\perp}{dt} = -e\vec{E}_\perp - ev_g\hat{e}_z \times \vec{B} = e\nabla_\perp(\phi - v_g A_z) \tag{7}$$

$$\frac{dP_z}{dt} = -eE_z = e\frac{\partial}{\partial\zeta}(\phi - v_g A_z). \tag{8}$$

The approximation $\vec{v} \approx v_g \hat{e}_z$ is correct for evaluating the Lorentz force, but it does *not* imply that the electron energy γmc^2 is constant. This is because the difference between the forward velocity v_z and the group velocity v_g is always small. The evolution of the energy $\gamma = (1 + \vec{P}^2/m^2c^2)^{1/2}$ follows from (7)-(8), which can be written in Hamiltonian form with Hamiltonian

$$\mathcal{H} = \gamma mc^2 - v_g P_z - e\Psi \tag{9}$$

where the comoving coordinate $(\vec{r}_\perp, \zeta)$ and momentum $(\vec{P}_\perp, P_z)$ are canonical coordinates. The quantity Ψ, defined by

$$\Psi = \phi - v_g A_z \tag{10}$$

is the *wakefield potential*. Since $|P_\perp| \ll P_z$ in the paraxial approximation, it is convenient to expand the Hamiltonian in a Taylor series around $\vec{r}_\perp = 0$, $\vec{P}_\perp = 0$. To leading order, this gives

$$\mathcal{H}_0 = \gamma_0 mc^2 - v_g P_z - e\Psi_0 \tag{11}$$

with $\gamma_0 = (1 + P_z^2/m^2 c^2)^{1/2}$ and $\Psi_0(\zeta) = \Psi(\vec{r}_\perp = 0, \zeta)$. With canonical coordinates ζ, P_z the lowest order equations are

$$\frac{dP_z}{dt} = F_z(\vec{r}_\perp = 0, \zeta) = -\frac{\partial \mathcal{H}_0}{\partial \zeta} \tag{12}$$

$$\frac{d\zeta}{dt} = \frac{P_z}{\gamma_0 m} - v_g = \frac{\partial \mathcal{H}_0}{\partial P_z}. \tag{13}$$

These equations describe the momentum transfer and phase slippage between the electron and the wave. Now consider the case of electron acceleration in a sinusoidal plasma wave

$$e\Psi_0(\zeta) = e_0 mc^2 \cos\zeta, \tag{14}$$

where e_0 is the dimensionless amplitude of the wave. The approximation of a sinusoidal waveform (14) is correct in the low-amplitude limit $e_0 \ll 1$ (linear regime). The nonlinear regime, with $e_0 = \mathcal{O}(1)$, is of course more interesting because increasing the wakefield amplitude yields higher energy gain (Esarey and Pilloff 1995). However, no analytical expression for the wakefield is available for the general nonlinear case (14). However, the linear regime is used here to illustrate the acceleration mechanism. For the value of e_0 chosen in the examples below, namely $e_0 = 0.1$, the linear approximation is valid.

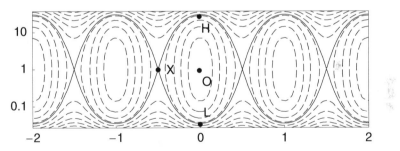

Figure 5. *Phase diagram* $(\zeta/\lambda_p, \gamma_0/\gamma_g)$ *for* $\mathcal{H}_0$ *with O-point (O), X-point (X), highest (H) and lowest (L) point of separatrix.*

For the Hamiltonian $\mathcal{H}_0$ (11) there are 3 types of orbits, as can be seen in Fig. 5, which shows the $(\zeta/\lambda_p, \gamma_0/\gamma_g)$-phase space diagram for $e_0 = 0.1$, $\gamma_g = 50$. The quantity $\gamma_g = (1 - v_g^2/c^2)^{-1/2}$ is called the *resonant energy*. As seen in Fig. 5, there are closed orbits inside the separatrix and open orbits both *above* and *below* the separatrix. The orbits below the separatrix describe the motion of electrons that are too slow to be captured in the wave. The orbits above the separatrix correspond to the motion of electrons that outrun the wave. The orbits inside the separatrix describe motion of electrons that are trapped inside the wave. Trapped electrons oscillate back and forth in the moving potential well: these oscillations are called *synchrotron* oscillations. For acceleration, it is unwanted to let the electrons perform multiple synchrotron oscillations: it is sufficient to inject electrons at low energy and extract them when they have reached a maximum average γ_0 after about half a synchrotron oscillation.

Stable equilibrium points (O-points) in Fig. 5 are seen to exist at $\zeta = n\lambda_p$, $\gamma_0 = \gamma_g$ and unstable equilibrium points (X-points) at $\zeta = (n+1/2)\lambda_p$, $\gamma_0 = \gamma_g$ for all $n \in Z$. For orbits inside the separatrix, one can define the *turning points* by the condition $\partial\mathcal{H}_0/\partial P_z = d\zeta/dt = 0$, i.e. points at which the backward phase slip of the electron changes to forward slip or vice versa. In Fig. 5 these points are seen to be at $\gamma_0 = \gamma_g$. Points of minimum and maximum P_z, defined by $\partial\mathcal{H}_0/\partial\zeta = 0$, are found at $\zeta = n\lambda_p/2$ for $n \in Z$. The minimum and maximum energy for the separatrix (points indicated as H and L in phase diagram 5) are denoted by $\gamma_0 = \gamma_{max(min)}$ and given by

$$\gamma_{max} = 2\gamma_g + 4e_0\gamma_g^2 \tag{15}$$

$$\gamma_{min} = e_0 + \frac{1}{4e_0} \tag{16}$$

in the approximation $\gamma_g \gg 1$.

The energy gain is proportional to the amplitude of the wakefield and the acceleration distance L_a, which is limited by phase slippage. The acceleration distance is equal to $v_g T$, where T denotes the time during which the electron can remain in the accelerating region (i.e. half a synchrotron period). The maximum acceleration distance is the *dephasing length* L_d. Since for a large part of the acceleration the approximation $\gamma_0 \gg \gamma_g$ is valid, the phase slippage can be taken constant:

$$\frac{d\zeta}{dt} \approx c - v_g \approx \frac{c}{2\gamma_g^2}. \tag{17}$$

The dephasing length corresponds to a phase slippage distance of half a plasma wavelength. With (17) it is found that

$$L_d \approx cT = \int c\,dt \approx 2\gamma_g^2 \int d\zeta = \lambda_p\gamma_g^2 \tag{18}$$

and this formula explains the scaling of $\gamma_{max} \propto e_0 L_d$ with γ_g^2.

The energy at the lowest point of the separatrix is the minimum energy for trapped electrons, which is the *trapping threshold*. The formula for γ_{min} has a minimum at $e_0 = 1/2$, suggesting that beyond $e_0 = 1/2$ the trapping threshold increases with amplitude. This is not the case, however, because these γ_0-values correspond to electrons with $v_z < 0$. This means that, above $e_0 = 1/2$, the wave is strong enough to trap counterpropagating electrons.

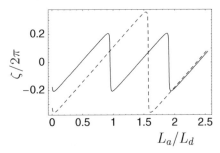 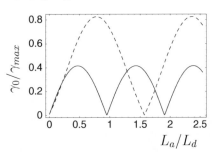

Figure 6. *Phase ζ/λ_p (left) and energy γ/γ_{max} (right) as functions of acceleration distance for initial conditions $(\zeta, \gamma_0) = (-0.15\lambda_p, 0.2\gamma_g)$ (solid lines) and $(\zeta, \gamma_0) = (-0.3\lambda_p, 0.2\gamma_g)$ (dashed lines).*

To illustrate the dynamics further, results of numerical integration of the lowest-order equations of motion are shown in Fig. 6. Two different initial conditions inside the separatrix have been chosen: $(\zeta, \gamma_0) = (-0.15\lambda_p, 0.2\gamma_g)$, $(-0.3\lambda_p, 0.2\gamma_g)$ for $e_0 = 0.1$, $\gamma_g = 50$. The time variable is multiplied by v_g to get the acceleration distance L_a, which is expressed as a fraction of the dephasing length L_d. From Fig. 6 the approximation of constant phase slippage is seen to hold for a large part of the motion. It fails only over a short period, when the electron rapidly slips backward. This leads to typical sawtooth oscillations for ζ. Orbits near the O-point have a shorter synchrotron oscillation period than orbits close to the separatrix. The maximum energy scales approximately linearly with the synchrotron period.

The transverse motion follows from the second-order expansion of the Hamiltonian (9)

$$\mathcal{H} \approx \mathcal{H}_0 + \mathcal{H}_2, \tag{19}$$

with

$$\mathcal{H}_2 = \frac{1}{2\gamma_0 m} \vec{P}_\perp^2 - \frac{1}{2} e\Psi_2 \, \vec{r}_\perp^2 \tag{20}$$

where the function Ψ_2 denotes the curvature of the potential Ψ in the vicinity of the propagation axis, assuming cylindrical symmetry. The function Ψ_2 is given by

$$\Psi_2(\zeta) = \frac{\partial^2 \Psi(r, \zeta)}{\partial r^2}(\zeta, r = 0). \tag{21}$$

Because of the cylindrical symmetry, for now the analysis can be restricted to one of the transverse directions, say x. For the y-direction, the same formulas apply. The transverse equations of motion are

$$\frac{dP_x}{dt} = e\Psi_2 x \tag{22}$$

$$\frac{dx}{dt} = \frac{P_x}{\gamma_0 m} \tag{23}$$

The transverse forces are focusing in regions with $\Psi_2 < 0$ and defocusing in regions with $\Psi_2 > 0$. In focusing regions, $\mathcal{H}_2$ is the Hamiltonian of a harmonic oscillator with a time-dependent mass $\gamma_0 m$ and focusing strength $-e\Psi_2$. The time dependence enters through the dependence of γ_0 and Ψ_2 on P_z and ζ, resp. The transverse oscillations are called *betatron* oscillations. If the typical timescale for the transverse motion is much shorter than for the longitudinal motion, the betatron frequency ω_β of the transverse oscillation is given by

$$\omega_\beta^2 = -\frac{e\Psi_2}{\gamma_0 m}. \tag{24}$$

In this case, the (ζ, P_z)-dependence is adiabatically slow and the area $\mathcal{A}_x$ in (x, P_x) phase space

$$\mathcal{A}_x = \oint P_x \, dx \tag{25}$$

is an adiabatic invariant of the motion. Note that the requirement that the longitudinal timescale is much longer than the transverse timescale may fail during the rapid backward slip of the electron (see Fig. 6). In this case there is no adiabatic invariant.

For the harmonic oscillator, the adiabatic invariant can be calculated explicitly and it is

$$\mathcal{A}_x = \pi x_0 P_{x0} \tag{26}$$

where x_0, P_{x0} denote the slowly varying amplitudes (envelopes) of the oscillations for x and P_x. From the condition that $\mathcal{H}_2$ evolves on the longitudinal timescale, it is found that

$$x_0 = (\mathcal{A}_x/\pi)^{1/2} \, (-e\Psi_2\gamma_0 m)^{-1/4} \tag{27}$$

$$P_{x0} = (\mathcal{A}_x/\pi)^{1/2} \, (-e\Psi_2\gamma_0 m)^{1/4} \tag{28}$$

The amplitude x_0 of the betatron oscillation is seen to decrease with increasing focusing strength ($-\Psi_2$) and with increasing energy (γ_0).

By inserting the expressions for x_0, P_{x0}, y_0, P_{y0} in the second-order Hamiltonian $\mathcal{H}_2$ one finds the second order corrections to the longitudinal motion. Including these corrections, the equations for ζ, P_z can be written as

$$\frac{dP_z}{dt} = e\frac{\partial\Psi_0}{\partial\zeta} + \frac{\alpha}{2}\left[\frac{-e}{\Psi_2\gamma_0 m}\right]^{1/2}\frac{\partial\Psi_2}{\partial\zeta} = -\frac{\partial\mathcal{H}_\alpha}{\partial\zeta}, \tag{29}$$

$$\frac{d\zeta}{dt} = \frac{P_z}{\gamma_0 m}\left(1 - \frac{\alpha}{2c^2}\left[\frac{-e\Psi_2}{\gamma_0^3 m^3}\right]^{1/2}\right) - v_g = \frac{\partial\mathcal{H}_\alpha}{\partial P_z} \tag{30}$$

where

$$\mathcal{H}_\alpha = \mathcal{H}_0 + \mathcal{H}_2 = \gamma_0 mc^2 - v_g P_z - e\Psi_0 + \frac{\alpha}{2}(-e\Psi_2/\gamma_0 m)^{1/2} \tag{31}$$

is the Hamiltonian that depends on the adiabatic constants through $\alpha = (\mathcal{A}_x + \mathcal{A}_y)/\pi$. The Hamiltonian $\mathcal{H}_\alpha$ is defined only in the focusing region, where $\Psi_2 < 0$. Equations (29) - (30) describe the phase slippage and momentum transfer averaged over the betatron timescale, including the effect of a finite amplitude of the transverse oscillation, characterized by a nonzero value of α. For example, it includes the effect of the radial variation (r-dependence) of the accelerating field F_z on the energy gain.

As an example, consider again a sinusoidal plasma wave and take a Gaussian transverse profile with spot size w_l:

$$e\Psi(r,\zeta) = e_0\,mc^2\,\cos\zeta\,e^{-r^2/w_l^2}. \tag{32}$$

The focusing strength is

$$-e\Psi_2(\zeta) = 2\,\frac{e_0}{w_l^2}\cos\zeta \tag{33}$$

which means that only 1/4 of the wave is both focusing and accelerating ($-\lambda_p/4 < \zeta < 0$). Because the defocusing region is not accessible, the maximum and minimum energy are not on the separatrix in fig. 5, but on the orbit through ($\zeta = -\lambda_p/4$, $\gamma_0 = \gamma_g$):

$$\gamma_{max} = 2\gamma_g + 2e_0\gamma_g^2 \tag{34}$$

$$\gamma_{min} = \frac{1}{2}\left(e_0 + \frac{1}{e_0}\right) \tag{35}$$

where again the approximation $\gamma_g \gg 1$ has been used.

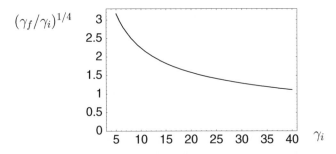

Figure 7. *Adiabatic focusing ratio as a function of initial energy for $\gamma_g = 50$.*

If an electron slips forward in the region $-\lambda_p/4 < \zeta < 0$, it is strongly focused, since both $-\Psi_2$ and γ_0 are increasing. To estimate the magnitude of the focusing effect, consider injection with energy $\gamma_i < \gamma_g$ and extraction at energy $\gamma_f > \gamma_g$. The injection phase and the extraction phase are taken identical, so that the value of Ψ_2 is the same. In this case, the *adiabatic focusing factor*, defined as the ratio of initial to final x_0, equal to the ratio of final to initial P_{x0}, is found to be $(\gamma_f/\gamma_i)^{1/4}$. With (34)-(35) it is found that

$$(\gamma_f/\gamma_i)^{1/4} \leq (\gamma_{max}/\gamma_{min})^{1/4} \approx (2e_0\gamma_g)^{-1/2}, \qquad (36)$$

where $e_0 \ll 1$, $e_0\gamma_g \gg 1$ have been used. For $e_0 = 0.1$, $\gamma_g = 50$, this results in an upper limit of about 3.16. Note that acceleration leads to a decrease of opening angle, roughly given by $P_{x0}/\gamma \propto (\gamma_f/\gamma_i)^{-3/4}$. A plot of the adiabatic focusing factor for $\gamma_g = 50$ is given in Fig. 7, where the value of γ_f has been found by taking the zero-order Hamiltonian (11) as a constant of the motion.

The condition for separation of betatron and synchrotron timescales $\omega_\beta/\omega_p \gg 1/\gamma_g^2$ gives

$$w_l \ll \gamma_g^2 (2 e_0 \cos\zeta/\gamma_0)^{-1/2} < \gamma_g (\cos\zeta)^{-1/2} \qquad (37)$$

where the expression for the maximum energy γ_{max} has been used. This condition is violated if the electron slips too close to a defocusing region (i.e. when $\cos\zeta \to 0$) or if the potential well is too wide (i.e. if w_l is too large).

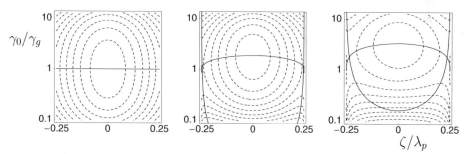

Figure 8. *Phase diagrams for $\mathcal{H}_\alpha$ with $\alpha/w_l = 0$ (left), $mc/2$ (middle), and $3mc/2$ (right).*

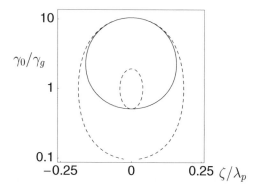

Figure 9. *Selected orbits for $\alpha = 0$ (dashed lines) and $\alpha = w_l mc$ (solid line).*

The influence of the transverse motion on the longitudinal dynamics is illustrated in Fig. 8, which shows phase diagrams of $\mathcal{H}_\alpha$ for $\alpha/w_l = 0$, $mc/2$, $3mc/2$. Also indicated are contours of $\partial\mathcal{H}_\alpha/\partial\zeta = 0$ (points of maximum or minimum energy) and $\partial\mathcal{H}_\alpha/\partial P_z = 0$ (turning points).

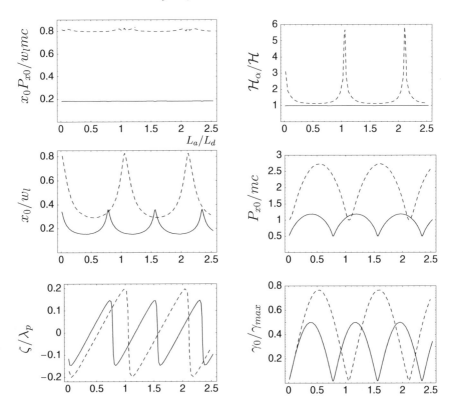

Figure 10. *Simulation results for initial conditions (a) $(x, P_x) = (0, mc/2)$ (solid lines) and (b) $(x, P_x) = (0, mc)$ (dashed lines) showing influence of betatron oscillation on longitudinal dynamics.*

For $\alpha > 0$, energy maxima and minima are found around $\zeta = \pm\lambda_p/4$, which are absent in the case $\alpha = 0$. The turning points, which are always at $\gamma_0 = \gamma_g$ for $\alpha = 0$, are seen to occur at $\gamma_0 > \gamma_g$, where the value of γ_0 depends on the orbit. In Fig. 8, X-points are seen to exist near $\zeta = \pm\lambda_p/4$, $\gamma_0 = \gamma_g$. The area inside the separatrix decreases with increasing α in such a way that γ_{min} increases with α and γ_{max} decreases with α.

The influence of transverse motion on longitudinal dynamics is further illustrated in Fig. 9. This figure shows one orbit for $\alpha = w_l mc$ and two orbits for $\alpha = 0$, chosen such that one of them has the same maximum energy and the other one has the same minimum energy. The main difference between the $\alpha = 0$-orbits and the $\alpha > 0$-orbit is seen to be in the low energy ($\gamma_0 < \gamma_g$) part, where the electron rapidly slips backward. In the high energy ($\gamma_0 > \gamma_g$) part, the $\alpha > 0$-orbit is barely different from the large $\alpha = 0$-orbit, indicating that the radial variation of the accelerating field has only little effect on energy gain. This is because the electron moves close to the axis as a result of strong adiabatic focusing during the rapid backward slip - see also (27).

To check the validity of the paraxial approximation, it is instructive to look at results of numerical integration of the full equations of motion (7)-(8). These simulations are presented in Fig. 10, which shows $x_0 P_{x0}$, $\mathcal{H}_\alpha/\mathcal{H}$, x_0, P_{x0}, ζ and γ_0 as functions of the acceleration distance. The following initial conditions have been chosen: $(\zeta, P_z) = (0, 0.2\gamma_g)$, $(y, P_y) = (0, 0)$. For

(x, P_x), the cases $(0, mc/2)$ (*a*) and $(0, mc)$ (*b*) are compared. The wakefield parameters are $e_0 = 0.1$, $w_l = \lambda_p$, $\gamma_g = 50$.

The quantity $x_0 P_{x0}$ is seen to be nearly constant for electron a, while there are some fluctuations for electron b. The change of $x_0 P_{x0}$ is most pronounced during the backward slip of electron b, when the assumption that the transverse dynamics is much faster than the longitudinal dynamics is violated. This is confirmed by the behaviour of $\mathcal{H}_\alpha/\mathcal{H}$, which indicates that the second-order approximation breaks down during the backward slip. There is a considerable difference in longitudinal dynamics for the two electrons: electron b is seen to slip closer to the defocusing regions $\zeta < -\lambda_p/4$, $\zeta > \lambda_p/4$ and reaches a higher energy than electron a. The influence of phase slippage and acceleration on the betatron motion is seen in the graphs of x_0 and P_{x0} as functions of acceleration distance. The maximum of x_0 is at the point of minimum energy, the maximum of P_{x0} is at the point of maximum energy, indicating that the influence of γ_0 on x_0, P_{x0} dominates over the influence of Ψ_2.

6   Radiation source development

Laser driven plasma wakefield accelerators are about one thousand times more compact that conventional accelerators based on microwave cavities, which are usually limited to about 20 MV/m for superconducting cavities and about 100 MV/m for room temperature devices. To achieve very high energies long and expensive multi-staged accelerators are required. Their high cost is due to the numerous microwave power sources required, the highly engineered accelerator systems and the huge shielding infrastructures needed to house them. Furthermore, the duration of electron bunches from conventional accelerators are restricted to between 100 fs and several tens of picoseconds. Achieving shorter durations and high peak currents requires bunch compressors, which adds to the complexity and cost of state-of-the art accelerators.

Apart from the well known use of accelerators to study the sub-atomic structure of matter, such as searching for the Higgs Boson, high energy accelerators are central components of light sources. Light sources are some of the most useful tools available to scientists and technologists for mapping out the molecular and solid state structure of matter. They have become essential tools for developing drugs and new materials, and are powerful sources for imaging matter on all length scales, offering unique windows into biological function. Light sources, such as synchrotrons, which are often called third generation light sources, have become ubiquitous brilliant sources of incoherent radiation covering an extremely broad spectral range, from terahertz frequencies (millimeter wavelengths) to hard X-rays. Free-electron lasers, which are coherent sources based on high peak current, high brightness accelerators, and undulators (Marshall 1985), are being developed as fourth generation light sources (Ayvazyan et al. 2002, 2006). FELs produce brilliant ultra-short pulses of coherent radiation that is tuneable over a very wide spectral range, similar in extent to synchrotrons but restricted to photon energies of the order of several keV because of the extreme demands on the electron beam quality and peak current required to develop microbunching on an ngstrom wavelength. Synchrotrons require high repetition rate electron bunches with high average currents and energies in the range of 0.5 GeV to more than 10 GeV, whereas FELs usually require lower energy, high peak current electron beams with durations less than 100 fs to achieve high peak currents with modest totsl charge. Infrared FELs, on the other hand, are based on 10's MeV electron beams and are usually low gain oscillators consisting of undulators contained in an optical cavity to allow trapped radiation to build up over many repetitive micropulses. However, VUV to X-ray FELs are usually high gain single-pass amplifiers because of the lack of suitable cavity mirrors for oscillators in this wavelength region.

6.1   Synchrotron sources and free-electron lasers

The FEL (Marshall, 1985, Bonifacio et al., 1989, 1991, Colson, 2001) is a unique source of coherent electromagnetic radiation because of its simplicity: the amplifying medium consists of an electron beam in vacuum subject to a spatially periodic magneto-static field (undulator or wiggler) which enables the transfer of energy between electrons and electromagnetic wave. A ponderomotive force arising from the Lorentz force of the combined magnetic fields of electromagnetic wave and undulator gives rise to bunching of the electron beam, and results in coherent radiation at a Doppler up-shifted frequency. The absence of a solid or gaseous amplifying medium allows the FEL to attain extremely high powers and broad tuneability. Tuning of the FEL wavelength, which is given by the expression $\lambda = \lambda_u/2\gamma^2(1 + a_u^2/2)$ (where a_u is the deflection parameter and λ_u the periodicity, for a planar undulator), is achieved by varying either γ or the undulator parameters. Several x-ray FELs are being developed as 4th generation light sources at centres throughout the world. The ultimate goal of these projects is to reach the water window and beyond using a self-amplification of spontaneous emission (SASE) FEL amplification driven by a multi-GeV electron beams. When complete, these x-ray sources will produce bright and coherent x-ray pulses with durations of the order of the electron bunch duration. However, a drawback of SASE amplifiers is that they are essentially noise amplifiers and have spiky and fluctuating outputs (Bonifacio et al., 2005, 2005a). New strategies are being adopted to improve the FEL properties, such as mode locking or amplifying a signal from high harmonic emission from laser-gas interaction.

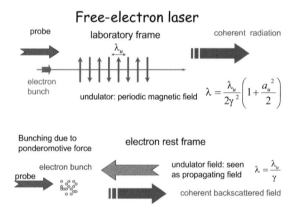

Figure 11. *FEL schematic*

The radiator structure of both FELs and synchrotron sources usually consists of an array of magnets arranged in a regular array to periodically deflect the electron beam as it traverses the magnetic fields. When the deflection angle is smaller than the synchrotron cone, $\vartheta = 1/\gamma$, the array is called an undulator. Deflections larger than $1/\gamma$ produces highly non-sinusoidal motion and results in emission of numerous harmonics. To understand how synchrotron or FEL radiation is produced, we need to consider the motion of the electrons in the magnetic field of the undulator or wiggler, of length L_u (Luchini and Motz, 1990). As an example, consider a periodic magnetic field in a planar undulator polarised in the y-z plane with the form $B = B_u \sin(k_u z)\hat{e}_y$, where B_u is the undulator peak magnetic field and $k_u = 2\pi/\lambda_u$ is the wave number of the undulator. An electron travelling in the z direction with velocity $v_z = v_{z0}\hat{e}_z$, will be subject to a Lorentz

$(\vec{v} \times \vec{B})$ force in the x-z plane, giving rise to periodic acceleration with the velocity varying as $v_x = v_u \cos(k_u z)\hat{e}_x$ to give an electron trajectory of the form

$$x = x_0 + X \sin(k_u z), \quad y = y_0, \quad z = z_0 + v_z t, \tag{38}$$

where (x_0, y_0, z_0) is the initial position of the electron, and X is the amplitude of oscillation which depends on the undulator field strength. For an observer in the electron frame of reference, the undulator period appears contracted by the Lorentz factor γ. The forced oscillations produce radiation with a wavelength contracted by approximately $1/2\gamma^2$, in the laboratory frame. The undulator magnetic field is transformed to $E'_x = \gamma_z(E_x + \beta_z B_y)$ in the electron reference frame and the electron is accelerated as $\dot{v}' = -e\beta_z B_y \gamma/m_e$ giving rise to dipole radiation with an energy radiated in the laboratory frame

$$W = \frac{1}{6\pi\epsilon_0} \left(\frac{e^2}{m_e c} \right)^2 \beta_z \lambda_u \gamma_z^2 B_u^2 \tag{39}$$

on every undulator period that the electron passes. Radiation is emitted into a narrow cone of $1/\gamma$ around the z-axis as a series of harmonics with the wavelength varying as

$$\lambda = \frac{\lambda_u}{2\gamma_z^2 h} = \frac{\lambda_u}{2\gamma^2 h}(1 + a_u^2/2), \tag{40}$$

where h is the harmonic number, $a_u = |e| B_u/m_e c k_u$ ($\approx 0.93 B_u[\text{Tesla}]\lambda_u[\text{cm}]$) is the undulator or wiggler constant and $k_u = 2\pi/\lambda_u$ is the undulator wavenumber. The undulator deflection parameter is a measure of the transverse velocity $\beta_\perp \approx a_u/\gamma$. The slight increase of the wavelength by a factor $1 + a_u^2/2$, arises from the extended (sinusoidal) path of the electrons giving $1/\gamma_z^2 \approx 1/\gamma^2(1 + a_u^2/2)$ from $1/\gamma^2 = 1 - \beta^2 = 1 - \beta_z - \beta_\perp$. The radiator is called a wiggler when $a_u > 1$ and the searchlight-like radiation periodically deflects out of the $1/\gamma$ Lorentz cone resulting in a broad, rich spectrum consisting of many harmonics. For $a_u >> 1$, the spectrum becomes quasi-continuous with a peak intensity occurring at

$$\lambda_{\text{peak}} = \frac{2}{3} \frac{\lambda_u}{\gamma^2 a_u}, \tag{41}$$

which is the wavelength above which half the power is radiated. The intensity of a wiggler is a factor of $2N_u$ times larger than a single bending magnet with the same strength magnetic field, where N_u is the number of periods of the magnetic poles. The wavelength of radiation emitted depends not only on the electron beam energy and the deflection parameter, a_u, but also on the angle of the electron beam with respect to the undulator axis and/or the angle of observation relative to the undulator axis, as

$$\lambda = \frac{\lambda_u}{2\gamma_z^2 h} = \frac{\lambda_u}{2\gamma^2 h}(1 + a_u^2/2 + \gamma^2 \vartheta^2), \tag{42}$$

which is slightly larger off-axis because the undulator periodicity appears longer when viewed off-axis at angle ϑ. The Lorentz contraction restricts the emitted radiation to a narrow cone $\vartheta_r = 1/\gamma_z$; and λ doubles when the undulator is viewed at an angle of $\vartheta = 1/\gamma_z$. Given the known trajectory of an electron in the undulator, the radiation fields can be calculated by directly evaluating the retarded field of the Liénard-Wiechert potential (Landau, Liftshitz and Pitaevski, 1984). However, for an ideal undulator or wiggler, an approximate solution in the ultra-relativistic and far field paraxial approximations can be evaluated . To help characterise the beam we consider the total brightness of the beam of radiation impinging on a surface is the power per unit area from a unit solid angle. The brilliance, which is defined by $S_\omega = 2\epsilon_0 c |E|^2 \times \text{area}$, is the power radiated (per unit frequency, beam area and solid angle) by a collimated electron

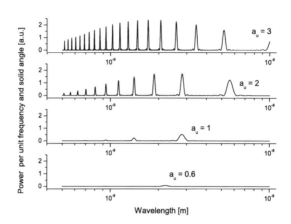

Figure 12. *Spectral brightness of undulator harmonics for $a_u = 0.6, 1, 2$ & 3, $N_u = 100$, $\lambda_u = 1.5$ cm periodicity undulator.*

beam of current density j_b (where ϵ_0 is the permittivity of free space). This is given by (Luccini and Motz, 1990)

$$S_\omega(\omega, \vartheta_{we}) \approx \frac{k^2 L_u^2 e j_b}{32 c \epsilon_0 \pi^2 \gamma^2} \sum_{n=0}^{\infty} |A_{u,n}(\vartheta_{we})|^2 \left[\frac{\sin(\nu_n)}{\nu_n}\right]^2 \tag{43}$$

where ϑ_{we} is the angle between observation and motion directions and the normalised detuning parameter, ν_n, is given by

$$\nu_n = \pi N_u \left(\frac{\lambda_u}{2\lambda\gamma^2}\left(1 + \frac{a_u^2}{2} - \gamma^2\vartheta_{we}^2\right) - 1\right), \tag{44}$$

where the coefficients $A_{u,n}$ are given by (Luccini and Motz, 1990)

$$A_{u,n}(\vartheta_{we}) = \frac{1}{\lambda_u}\int_0^{\lambda_u}(\vec{a}_u e^{-k_u z} + \vec{a}_u^* e^{-k_u z} + 2\gamma\vartheta_{we}) \times \tag{45}$$

$$\exp[-i\left(nk_u z + \mathrm{Re}\left[\frac{k\vec{a}_u \cdot \vec{\vartheta}_{we}}{i\gamma k_u}(e^{ik_u z} - 1) + \frac{k\vec{a}_u^2}{8i\gamma^2 k_u}(e^{2ik_u z} - 1)\right]\right)]dz,$$

and $\vec{a}_u = -\hat{e}_z \times \vec{B}_u/ik_u = a_{u,x}\hat{e}_x + a_{u,y}\hat{e}_y$ and $a_{u,x}$ and $a_{u,y}$ are the vector potentials along the x and y axes, respectively. A circularly polarised undulator has $a_{u,y} = \pm i a_{u,x}$. From this expression it can be seen that the spectrum consists of equally spaced odd harmonics on axis and both even and odd harmonics off axis, for a planar undulator. The spectrum of each harmonic, h, is a sync function, $(\sin \nu^2/\nu^2)$, giving a full width at half maximum (FWHM) spectral width of

$$\frac{\delta\lambda_{\mathrm{FWHM}}}{\lambda} \approx \frac{0.9}{Nh}. \tag{46}$$

Radiation over a spectral width $\delta\lambda_{\mathrm{FWHM}}$ is emitted within an angle of approximately $\delta\vartheta_{we,\mathrm{FWHM}}$. For a particular harmonic, the solid angle of radiation emitted into the first lobe is

$$\delta\vartheta_{we,\mathrm{FWHM}} \approx \frac{0.9}{(N_u h)^{1/2}} \cdot \frac{1 + a_u^2/2}{\gamma}, \tag{47}$$

with all harmonics emitted into the solid angle $\vartheta_{we,\mathrm{FWHM}} \approx 1/\gamma_z = (1 + a_u^2/2)/\gamma$. For angles less than $N^{1/2}\delta\vartheta_{we,\mathrm{FWHM}}/2$, $A_{u,n}(\vartheta_{we}) \approx A_{u,n}(0)$ and for odd harmonics

$$A_{u,2p+1}(0) \approx a_u G_p, \tag{48}$$

where

$$G_p = \left[J_p\left(\frac{(2p+1)a_u^2}{4 + 2a_u^2}\right) - J_{p+1}\left(\frac{(2p+1)a_u^2}{4 + 2a_u^2}\right) \right], \tag{49}$$

whereas for a circularly polarised undulator, $A_{u,1}(0) \approx a_u$ and $A_{u,2p+1}(0) = 0$ when $p > 1$, i.e. only first harmonic emission occurs on-axis.

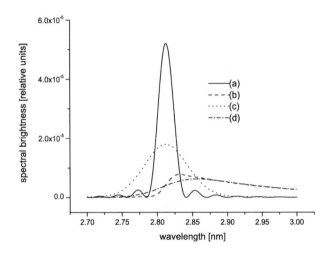

Figure 13. *Comparison of synchrotron radiation with (a) $\delta\gamma/\gamma = 0$ and $\sigma_\vartheta = 0$, (b) $\sigma_\gamma/\gamma = 0$ and $\sigma_\vartheta = 10^{-4}$, (c) $\sigma_\gamma/\gamma = 5 \times 10^{-3}$ and $\sigma_\vartheta = 0$, (d) $\sigma_\gamma/\gamma = 5 \times 10^{-3}$ and $\sigma_\vartheta = 10^{-4}$, for $N_u = 100$ with $\lambda_u = 1.5$ cm, $\gamma = 2000$ and $a_u = 1$*

From (49) we find that for $a_u > 1$, i.e. wiggler motion, many harmonics are radiated with the intensity of each harmonic growing as $p^{-1/2}a_u$ until a peak at harmonic number

$$h_{peak} \approx \frac{3a_u}{4}\left(1 + \frac{a_u^2}{2}\right). \tag{50}$$

Figure 12 shows the spectral brightness for various values of a_u for undulators and wigglers with, $a_u = 0.6, 1, 2$ & 3.

A finite divergence and/or energy spread of the electron beam reduces the brilliance of the beam and smears out the spectrum as

$$S_\omega(\omega, \varphi_w) = \int S_{w0}(\omega, \gamma, \varphi_w - \varphi_e) F(\gamma, \vartheta_e) d\gamma d\vartheta_e, \tag{51}$$

where ϑ_e is the direction of electrons and S_{w0}, is the distribution for a perfect beam with $\delta\gamma = 0$ and $\delta\vartheta = 0$. $F(\gamma, \vartheta_e)$ is the energy and angular probability distribution of the beam. Figure (13) shows the effect of energy spread and finite divergence of the electron beam on the spectrum.

The brilliance of radiation emitted from the undulator is reduced by the energy and transverse momentum spread. The total spectral width of the radiation is the sum of contributions from broadening due to the energy, angular and natural spread components:

$$\left(\frac{\delta\lambda}{\lambda}\right)^2 = \left(\frac{2\sigma_\gamma}{\gamma}\right)^2 + \left(\vartheta^2\gamma^2\right)^2 + \frac{1}{N_u^2}. \tag{52}$$

To minimise the spectral width and thus maximise the brilliance, the second term, $\left(\vartheta^2\gamma^2\right)^2$, can be reduced by matching the electron beam to the undulator, which minimises the beam divergence ϑ. To minimise the divergence the betatron wavelength, λ_β, which is analogous to the Rayleigh range of a laser beam, should be made approximately equal to the undulator length. The un-normalised r.m.s. emittance, $\epsilon_{\text{r.m.s.}}$, of an electron beam of radius r_e is analogous to the wavelength of a laser beam (i.e. has units of length), is given by

$$\epsilon_{\text{r.m.s.}} = \pi\sigma_{\beta_\perp} r_e = k_\beta r_e^2, \tag{53}$$

where $\sigma_{\beta_\perp}$ is the variance of the (normalised) transverse velocities and $k_\beta = 2\pi/\lambda_\beta = a_u k_u/2\gamma$ is the betatron wavenumber. The r.m.s. emittance is defined as the transverse momentum/position phase-space area of the beam (Lawson, 1977):

$$\epsilon_{\text{r.m.s.}} = \left(\langle x^2\rangle \left\langle\left(\frac{p_x}{m_e c}\right)^2\right\rangle - \left\langle\frac{xp_x}{m_e c}\right\rangle^2\right)^{\frac{1}{2}}, \tag{54}$$

where p_x and x are respective the transverse momenta and coordinates of an electron. The edge or envelope emittance is four times as large. The normalised emittance, $\epsilon_n = \beta_z\gamma\epsilon_{\text{r.m.s.}}$ governs the broadening i.e. through $\gamma\vartheta$. The minimum beam divergence that is consistent with the smallest average beam radius i.e. $\lambda_\beta \approx L_u$. When the Fresnel number $F = r_e/\lambda_\beta L_u \approx 1$ and the electron beam divergence (given by $\vartheta \approx \epsilon_{\text{r.m.s.}}/r_e$) matches the diffraction angle of radiation emitted by the undulator ($\vartheta \approx \lambda/r_e$) i.e. when $\vartheta^2 = 2k_\beta\epsilon_{\text{r.m.s.}}$. The brilliance is maximised when the divergence angle of the beam is

$$\vartheta_{\text{matched}} \approx 2\sqrt{\frac{\lambda_u\epsilon_n}{a_u}}. \tag{55}$$

Fig. 13(b) shows the spectrum of a matched beam with $\epsilon_n \approx 1\pi$ mm mrad giving a divergence angle of $\vartheta \approx 10^{-4}$ rad, which slightly increases the spectral width and therefore reduces the brilliance of the undulator radiation.

The brilliance of radiation from an undulator can be enhanced significantly if the electron beam is bunched on a wavelength scale and coherent emission occurs. This can be achieved either by pre-bunching the beam on a wavelength scale or producing a density modulation or bunching with a periodicity close to the resonance wavelength (40) of the undulator. Bunching is the main mechanism of the free electron laser. It arises from the ponderomotive force due to interaction of either an externally injected field, or spontaneously emitted radiation, with the undulator. Bunching leads to field growth which in turn increases the bunching i.e. an instability leading to cooperative emission. The resulting amplification, starting from noise, as in SASE, or an externally injected probe field from another source, can produce tuneable radiation in regions of the spectrum where conventional high power sources are not available.

The sum of the fields $E(k,r)$ radiated by an ensemble of electrons in an undulator or wiggler gives an intensity

$$\left\langle|E(k)|^2\right\rangle = N_e \int f(k,r)E(k,r)^2 dr + N_e(N_e - 1)\left(\int f(k,r)E(k,r)dr\right)^2, \tag{56}$$

where $f(k, r)$ is a form or bunching factor giving the probability of an electron being found at location r and N_e is the total number of electrons in the bunch. The first term on the l.h.s is the usual spontaneous emission term while the second term is a measure of the electron correlation or bunching and gives rise to coherent emission. Intense radiation is produced only when the Fourier components with significant amplitude are close to the resonance wavelength. Constructive interference leads to an enhancement of the intensity by a factor of N_e when $\int f(k, r) dr > 1/N_e$ which makes a contribution only when $k \approx k_r$ (k_r is the resonance wavenumber) (40). This implies that if the bunch length is smaller than the resonance wavelength it will radiate coherently and intensely (Jaroszynski et al., 1993, 1997, 1997a, 2000a, Wiggins et al., 2000, McNeil et at., 1999, 2000). Coherent mid-infrared radiation implies that the bunch duration should be shorter than about 10 microns or 30 fs! Thus ultra-short electron bunches in long period undulators can be very effective radiators of infrared radiation.

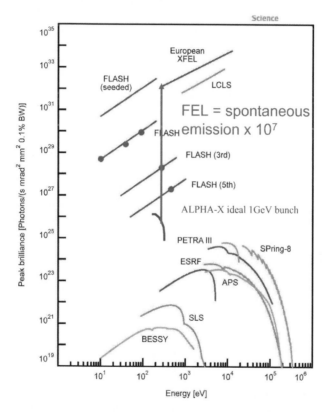

Figure 14. *Comparison of synchrotron radiation with FEL radiation showing the 7 orders of magnitude increase due to coherent bunching for the ALPHA-X project parameters (matched beam with $\gamma = 2000$, $\epsilon_n = 1\pi$ mm mrad, $I_{pk} = 10$ kA, $\lambda_u = 1.5$ cm, $\tau_e = 10$ fs)*

The build up of bunching and coherence leads to an exponential growth in intensity. Energy is exchanged between the radiation field at frequencies close to the resonance frequency. Following Bonifacio et al. (1989, 1991, 2005, 2005a), we consider a circularly polarised (helical) undulator, where a_u is a factor of $\sqrt{2}$ larger for the same peak magnetic field of a planar undulator. The motion of an electron in the combined fields of probe and undulator is governed mainly by the magnetic field of the undulator. The Lorentz equations giving the motion is

$$\frac{d(\gamma m \vec{v})}{dt} = e \left[\vec{E} + \vec{v} \times (\vec{B}_u + \vec{B}) \right] \tag{57}$$

and

$$\frac{d(\gamma m c^2)}{dt} = e\vec{E} \cdot \vec{v}_\perp. \tag{58}$$

The transverse motion is given by $\beta_\perp = -\vec{a}_{tot}/\gamma$. Thus, in terms of the normalised vector potentials, the motion equations reduce to

$$\frac{d(\gamma \vec{\beta}_\perp)}{dt} = -\frac{d}{dt}(\vec{a}_u + \vec{a}) = -\frac{d}{dt}\vec{a}_{tot} \tag{59}$$

and

$$\frac{d\gamma}{dt} = \frac{1}{2\gamma}\frac{\partial |\vec{a}_{tot}|^2}{\partial t}, \tag{60}$$

where $\vec{a}_{tot} = \vec{a}_u + \vec{a}$, with

$$\vec{a}_u(z) = \frac{1}{\sqrt{2}}(\hat{e}a_u e^{-ik_u z} - c.c.) \tag{61}$$

and

$$\vec{a}(z,t) = -\frac{i}{\sqrt{2}}(\hat{e}a(z,t)e^{i(kz-\omega t)} - c.c.), \tag{62}$$

where $\hat{e} = (\hat{e}_x + i\hat{e}_y)/\sqrt{2}$ for a helical undulator. The total field intensity, is the sum of the field amplitudes:

$$|\vec{a}_{tot}|^2 = a_u^2 - ia_u[a(z,t)e^{i\theta(z,t)} - c.c.] + |a(z,t)|^2, \tag{63}$$

where $\theta = (k_u + k)z - \omega t$ is the phase of an electron in the ponderomotive potential propagating at phase velocity $v_\phi = \omega/(k_u + k)$. (For a linear undulator we replace a_u by $a_u G_p$ from (49).) The changes in electron energy and phase are given by

$$\frac{d\gamma}{dz} = -\frac{a_u k}{2\gamma}(a(z,t)e^{i\theta(z,t)} + c.c.) \tag{64}$$

and

$$\frac{d\theta}{dz} = k_u + k \left(1 - \frac{1}{\beta_z} \right) \tag{65}$$

which gives, for $\gamma \gg 1$,

$$\frac{d\theta}{dz} = k_u \left[1 - \left(\frac{\gamma_r}{\gamma} \right)^2 \right], \tag{66}$$

where $\gamma_r = \sqrt{(1 + a_u^2)(k/2k_u)}$ is the resonant energy. On-resonant electrons, with $\gamma = \gamma_r$, travel at the phase velocity, v_ϕ, of the ponderomotive potential. The ponderomotive force, $\propto aa_u/\gamma \propto \vec{E}.\vec{v}_\perp$, gives the electron-radiation coupling via the undulator field. $d\gamma/dz > 0$ leads to electron acceleration while $d\gamma/dz < 0$ leads to deceleration and amplification. In the Compton limit, when $\gamma - \gamma_r \ll \gamma_r$, the equations of motion reduce to

$$\frac{d\gamma}{dz} = -\frac{a_u k}{2\gamma_r} \left(a(z,t)e^{i\theta(z,t)} + c.c. \right) \tag{67}$$

and

$$\frac{d\theta}{dz} = 2k_u \left[\frac{\gamma}{\gamma_r} - 1 \right]. \tag{68}$$

Space-charge forces due to electron-electron interactions due to Coloumb fields result in variations in the electron beam energy:

$$\left(\frac{d\gamma_j}{dz}\right)_s = -ik\left(\frac{\omega_p}{\omega}\right)^2\left(\left\langle e^{-i\theta}\right\rangle e^{i\theta_j} - c.c\right). \tag{69}$$

Electrons are bunched by the ponderomotive force and give rise to a transverse current that results in a coherent field. The growth of the field amplitude is governed by the wave equation (in the slowing varying envelope approximation $|\partial a/\partial z| << |ka|$ and $|\partial a/\partial t| << |\omega a|$):

$$\left(\frac{\partial}{\partial z} + \frac{1}{c}\frac{\partial}{\partial t}\right)a = \frac{k}{2}\left(\frac{\omega_p}{\omega}\right)^2\left(a_u\left\langle\frac{e^{-i\theta}}{\gamma}\right\rangle - ia\left\langle\frac{1}{\gamma}\right\rangle\right). \tag{70}$$

The right-hand term is the bunching term describing the transverse current in the undulator field. Dropping the insignificant radiation term, $ia\left\langle\frac{1}{\gamma}\right\rangle$, the main bunching contribution governing the field is given by

$$b = \left\langle e^{-i\theta}\right\rangle = \frac{1}{N_e}\sum_{j=1}^{N_e}e^{\theta_i}, \tag{71}$$

where θ_i is the phase of the ith electron in a bunch of N_e electrons. The equations can be cast in a "universal" dimensionless form, following Bonifacio et al. (1989, 1991, 2005, 2005a), by defining

$$\rho = \frac{1}{\gamma_r}\left(\frac{a_u\omega_p}{4ck_u}\right)^{2/3}$$

$$z_1 = \frac{z - v_z t}{L'_g}$$

$$z_2 = \frac{z}{L'_g} \tag{72}$$

$$L'_g = \frac{\lambda_u}{4\pi\rho}p_j = \frac{\gamma_j - \gamma_r}{\rho\gamma_r}$$

$$A = \frac{\omega a}{\omega_p\sqrt{\rho\gamma_r}}$$

where ρ is sometimes called the universal FEL or Pierce parameter and, in real units, for a helical undulator, is given as

$$\rho = \frac{1.136B_u\lambda_u^{4/3}I_{pk}^{1/3}}{\gamma_r} \tag{73}$$

and $L'_g = L_g p_j$, where $L_g = \lambda_u/4\pi\rho$ is a gain length as we shall see below. The system of equations describing radiation growth and bunching reduces to a simple form:

$$\frac{\partial}{\partial z_2}p_j = -\left(A(z_2, z_1)e^{i\theta_j} + c.c.\right), j = 1, ..., N_e, \tag{74}$$

$$\frac{\partial}{\partial z_2}\theta_j = p_j, j = 1, ..., N_e, \tag{75}$$

and

$$\left(\frac{\partial}{\partial z_2} + \frac{\partial}{\partial z_1}\right)A = \left\langle e^{-i\theta}\right\rangle = b. \tag{76}$$

A linear stability analysis following linearisation of the equations, assuming a solution of the form $A \propto e^{i\lambda z}$ results in the following dispersion relationship (Bonifacio et al., 1987):

$$\lambda^3 - \delta\lambda^2 + \rho\lambda + 1 = 0, \tag{77}$$

where the detuning δ is given by

$$\delta = \frac{\bar{\gamma}_0^2 - \gamma_r^2}{\gamma_r^2}\frac{1}{\rho}. \tag{78}$$

This admits one real and two complex roots i.e. $\lambda = \lambda_2 \pm i\lambda_3$. The field growth on resonance, $\delta = 0$, gives $\lambda_3 = \sqrt{3}/2$, and

$$A(z_2) = C_1 e^{i\lambda_1 z_2} + C_2 e^{i\lambda_2 z_2} + C_3 e^{i\lambda_3 z_2}, \tag{79}$$

which gives a growth in intensity:

$$|A(z_2)|^2 \approx \frac{|A(0,0)|^2}{9}\left(4\cosh^2\left(\frac{\sqrt{3}}{2}z_2\right) + 4\cos\left(\frac{3}{2}z_2\right)\cosh^2\left(\frac{\sqrt{3}}{2}z_2\right) + 1\right). \tag{80}$$

When $z < L_g$ (i.e. $z < L_g$) then $|A(z)|^2 \approx (1 + (3z/2L_g)^6)|A_0|^2$; this is the so-called "lethargy" period. During this phase the electrons are bunched and the radiation phase shifts, but without any real growth in field amplitude. For $z > L_g$, $|A(z)|^2 \approx (|A_0|/3)^2 e^{\sqrt{3}z/L_g}$ i.e. exponential growth with a rate given by $\sqrt{3}L_g$, the gain length. This is accompanied by a growth in bunching

$$|b(z)| \approx \frac{|b(0)|}{3}\exp\left[\frac{\sqrt{3}z}{L_g}\right]. \tag{81}$$

Saturation occurs when $|b(z_{\mathrm{sat}})| \approx 1$. Because the initial bunch distribution is random, the initial bunching contribution is $|b_0| \approx 1/\sqrt{\bar{N}_\lambda}$, where $\bar{N}_\lambda$ is the average number of electrons which contribute to the initial field. The current density is $J_e = n_e e c$, therefore $N_\lambda = J_e \lambda_r s_e/ec = I_e \lambda_r/ec$, where I_e and s_e are respectively the current and beam area. Given this value of $|b_0|$ and requiring $|b(z_{\mathrm{sat}})| \approx 1$, we find the saturation length $z_{\mathrm{sat}} \approx \ln(3/|b_0|)L_g/2\sqrt{3} \approx L_g(1 + 2\ln(ec/I_e\lambda_r))/\sqrt{3}$. This can be expressed in terms of a noise equivalent power, P_{nep}, since the saturated power is given by the efficiency ρ: $P_{\mathrm{sat}} \approx \rho P_{\mathrm{e-beam}}$, where $P_{\mathrm{e-beam}}$ is the electron beam power. Thus $P_{\mathrm{nep}} = \rho P_{\mathrm{e-beam}} ec/I_e\lambda_r \approx \rho\gamma_r m_e c^3/\lambda_r$ (i.e. about 1 kW for $\rho \approx 10^{-3}$, $\gamma \approx 2000$, $\lambda_r \approx 1$ nm, which is about one photon per electron emitted in the first gain length, resulting in about 20 GW peak power at saturation at the end of the undulator).

Energy conservation is expressed as $\langle p \rangle + |A|^2 = \mathrm{constant}$, which gives a maximum efficiency $\eta = 1/\rho$, occurring when $|\langle p \rangle| = \sigma_\gamma/\gamma \cdot 1/\rho \approx 1$, i.e. when $b \approx 1$, thus the power is $P(z) \approx P_{\mathrm{nep}}\exp[\sqrt{3}z/L_g]/3^2$.

In the limit of small signal amplitudes ($b << 1$) the electron phases exhibit harmonic motion (for weakly coupled pendula via the build up of correlations, $b > 0$), and vary as

$$\frac{d^2\theta_j}{dz_2} = -2|A_0|\cos\theta_j. \tag{82}$$

The field will only grow when δ is within the gain bandwidth i.e. $\delta \leq 2$ therefore it sets a limit to the acceptable energy spread

$$\frac{\sigma_\gamma}{\gamma_r} \leq 2\rho \tag{83}$$

and similarly, the beam angular spread, which is a function of the emittance, produces a longitudinal energy spread and thus also limits the gain analogous to increasing the spectral width of spontaneous undulator radiation (52)

$$\frac{\sigma_{\gamma(\epsilon_r)}}{\gamma_r} = \frac{k_\beta k_r \epsilon_r}{2k_u} = \frac{\lambda_u \epsilon_r}{2\lambda_\beta \lambda_r}, \tag{84}$$

which for a matched undulator $\lambda_\beta \approx L_g \approx \lambda_u/4\pi\rho$, reduces to

$$\frac{\sigma_{\gamma(\epsilon_r)}}{\gamma_r} = \frac{2\pi\rho\epsilon_r}{\lambda_r}. \tag{85}$$

This sets a limit to the emittance:

$$\frac{\sigma_{\gamma(\epsilon_r)}}{\gamma_r} < 2\rho \;\rightarrow\; \pi\epsilon_r < \lambda_r \;\rightarrow\; \pi\epsilon_n < \gamma_r\lambda_r. \tag{86}$$

When the inhomogenous broadening due to energy and emittance spread is small the field will grow exponentially until $|b| \approx 1$.

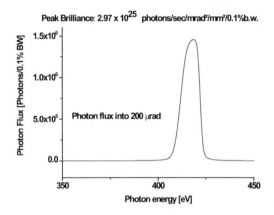

Figure 15. . *Synchrotron spectrum (Matched beam with $\gamma = 2000$, $\epsilon_n = 1\pi$ mm mrad, $I_{pk} = 10$ kA, $\lambda_u = 1.5$ cm, $\tau_e = 10$ fs)*

However, superradiance (Bonifacio et al., 1989, 1991, Jaroszynski et al., 1997, 2000, 1997a), self-amplification of coherent spontaneous emission (SACSE) (McNeil et al., 1999, 2000) and amplification of an injected signal are ways of improving the temporal characteristics of the x-ray pulses. As x-ray SASE FELs are extremely expensive devices it is very important that they produce useful output. If ways are found to produce an electron bunch microstructure with Fourier components at the resonance frequency, a large stable "spontaneous" coherent signal will act as an (intrinsic) injection source in the FEL amplifier (i.e. SACSE FEL) (Wiggins, 2000).

A potentially revolutionary way of driving an FEL is to use an electron beam from a plasma wakefield accelerator because their predicted electron bunch durations can approach one femtosecond or less. The growth in intensity of an injected or spontaneous field in a FEL amplifier is given by $I = I_0 \exp(gz)$, where z is the propagation distance, $g = 4\pi\rho 3^{1/2}/\lambda_u$ (Bonifacio et al., 1989, 1991), which is a function of the beam energy γ, peak current I_{pk} and normalised emittance ϵ_n. For a matched electron beam, the FEL gain parameter is given by $\rho = 1.1\gamma^{-1}B_u\lambda_u^{4/3}I_{pk}^{1/3}\epsilon_n^{-1/3}$. The matched electron beam radius for electron beams from laser-plasma accelerators ($\epsilon_n < 1$ mm mrad) is of the order of the plasma wake wavelength, giving $\rho \approx 0.01$ to 0.001, for the electron beam parameters expected from a laser-plasma accelerator, and a gain length $L_g = \lambda_u/(2\pi\sqrt{3}\rho)$ of between 10 and 100 undulator periods, which is sufficient to obtain saturation in a 200-period for the largest ρ value and about 1000 periods for $\rho = 10^{-3}$. With a $\lambda_u = 1.5$ cm undulator saturation should be just reached with a 100 m long undulator. However, the main advantage of using a laser-driven plasma wakefield accelerator to drive a FEL is that the peak current could be between 10 kA and 30 kA. With injection of an external field from high harmonic generation in a gas the saturation length could be decreased further.

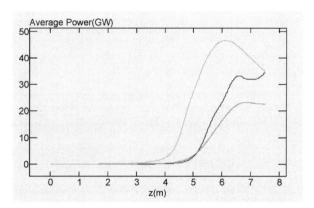

Figure 16. *FEL growth. (Matched beam with* $\gamma = 2000$, $\epsilon_n = 1\pi$ *mm mrad,* $I_{pk} = 10$ *kA,* $\lambda_u = 1.5$ *cm,* $\tau_e = 10$ *fs)*

To obtain growth we need $\delta\gamma/\gamma < \rho$ and $\epsilon_n < 4\lambda\beta\gamma\rho/\lambda_u$ or $\epsilon_n < \gamma\lambda$ (matched). A 4 nm source assuming a 1 GeV beam with 100 pC charge and a duration of 10 fs we obtain a peak current of $I_{pk} = 10$ kA and a $\rho \approx 0.005$, which gives a gain length of $10\lambda_u$ and a constraint on the energy spread of $\delta\gamma/\gamma \approx 0.1 - 0.5\%$, which may be achievable using a wakefield accelerator. FEL sources at x-ray wavelengths are less compact because the dependence of the gain on electron energy, $\rho \propto \gamma^{-1}$ leads to a lower gain and therefore the requirement of a longer undulator to achieve saturation. The enhancement over the expected synchrotron radiation brilliance is shown in Fig. 14.

To significantly shorten the undulator length SACSE could be used to enhance the start-up power. This has the additional benefit that the non-linear regime is entered promptly and the superradiant pulses should evolve self-similarly leading to very high efficiencies and extremely short, smooth and stable pulses. Pulses as short as several attoseconds should be feasible in future x-ray FEL sources because the gain bandwidth is automatically increased in this regime. An FEL operating in the water window would require a beam of short duration electron bunches with an energy of about 1 GeV. For these conservative estimates the challenge is to obtain an emittance $\epsilon_n < 1$ mm mrad (Fritzler, 2004) and an energy spread $\delta\gamma/\gamma \leq 0.01$ (Bonifacio, 1989, 1991). The synchrotron radiation peak brilliance has been calculated for a 6 fs electron bunch for a 17 kA beam with an emittance of 1 π mm mrad and a relative energy spread $\delta\gamma/\gamma \approx 1\%$ and shown in Fig. 15.

6.2 Betatron radiation from plasma wakefield accelerator

The magnetic field of an undulator or wiggler is not the only way to exert a transverse force on a relativistic charged beam. Electrostatic forces or propagating electromagnetic fields are alternatives. Compton or Thomson backscattering from a laser beam, where the laser field acts as an undulator, has long been considered as a means of converting laser photons to very high photon energies using relatively low energy electron beams. Compton Scattering occurs when the scattered photon energy is greater than the rest energy of the electron, $\hbar\omega > mc_e^2$. A high frequency laser beam is Doppler shifted by a factor of $4\gamma^2$ and appears to the electron as an effective undulator, thus all the theory discussed above on undulator and synchrotron radiation applies, with the undulator period L_u replaced by $\lambda/2$. When $a_0 > 1$ the characteristic wiggler type of radiation is often called non-linear Thomson or Compton scattering. An undulator

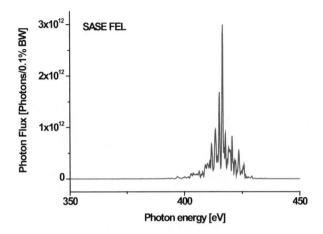

Figure 17. *FEL output. (Matched beam with $\gamma = 2000$, $\epsilon_n = 1\pi$ mm mrad, $I_{pk} = 10$ kA, $\lambda_u = 1.5$ cm, $\tau_e = 10$ fs)*

parameter of $a_u \approx a_0 \approx 1$ requires an intensity of around 10^{18} Wcm^{-2} for 1μm radiation. It is thus challenging to obtain a large diameter laser beam so that the field seen by the electron beam appears as a plane wave, and sufficiently long and flat topped, so that Compton backscattered spectrum is narrow. Furthermore, the interaction length is limited by the Rayleigh and/or the betatron length, which again limits the interaction length and results in a radial variation of a_0. However, Compton backscattering is a very effective means of producing gamma rays. As example scattering 800 nm radiation from a 1 GeV beam results in 23 MeV photons. A 100 MeV beam produces photons of around 230 KeV. There have been several studies exploring the potential of using a laser-plasma accelerator to produce Compton back-scattered photons at x-ray energies (Ta Phuoc, 2003, Schwoerer et al., 2006). These studies indicate the potential of producing femtosecond duration x-ray pulses.

In addition to static and electromagnetic wave undulators it is also possible to use plasma waves exited in structures as an effective undulator. One approach with low energy electron beams is to use a periodic dielectric or metal structure (Smith and Purcell, 1953, Luo et al., 2003), or even a medium (or photonic structure) where the electron velocity exceeds the phase velocity of the wave in the medium causing Čerenkov radiation (Jelley, 1958). A related method that is used to produce intense pulses of terahertz radiation is based on coherent transition radiation (CTR). Transition radiation occurs when an electron bunch passes through a transition between two media with different refractive indices (Ginsburg and Frank, 1946). It is very closely related to Čerenkov radiation, Smith Purcell radiation and diffraction radiation, which occurs when an electron beam exits a small hole in a conducting or dielectric surface (Ter-Mikaelian, 1972). Coherent emission occurs when the electron bunch is very short and intense radiation pulses can be produced over a wide spectral range from terahertz to optical frequencies (Leemans et al., 2003). Because the emission very closely reflects the electron beam properties such as divergence, bunch duration and cross-section, it provides a powerful means of characterising the electron beam.

The huge forces in perturbed plasma can also be harnessed to produce undulating or wiggler motion. In the first section of this lecture we showed that the plasma wave or wake behind an intense laser beam can have a very large radial restoring force, equal in magnitude to the accelerating forces which drive the injected electron beam to high energies, as depicted schematically in Fig. 1. Because the phase velocity of the plasma bubble structure is equal to the group velocity of the laser beam in the plasma, the accelerated electrons experience a relatively constant transverse force.

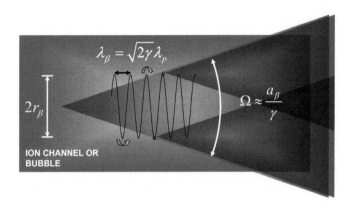

Figure 18. *Diagram showing searchlight-like betatron radiation into $\Omega \approx a_u/\gamma$ cone.*

A bubble structure is approximated, which is reminiscent of the cavity of a conventional accelerator, exerts a force which results in a transverse oscillation at the betatron frequency ω_β, from (24), is given by

$$\omega_\beta^2 = -\frac{e\Psi_2}{\gamma_0 m}. \tag{87}$$

thus from Gauss' Law gives

$$\omega_\beta = \frac{\omega_p}{\sqrt{2\gamma}} \tag{88}$$

and an effective undulator/wiggler parameter

$$a_\beta = \frac{\sqrt{2\gamma}\pi r_\beta}{\lambda_p}, \tag{89}$$

where γ is the Lorentz factor of the electron and r_β is the betatron oscillation amplitude. r_β varies as the electrons are accelerated but for the parameters used in the examples, it is found to be ≈ 1 μm at maximum energy. For an emittance of $\epsilon_n = 1\pi$ mm mrad the beam is matched when $2\pi r_\beta^2 = \epsilon\lambda_\beta$, which for a 1 GeV beam at the dephasing length gives $r_\beta \approx 1$ μm. It is then easy to calculate all the properties of the radiation using standard synchrotron wiggler or undulator theory, as described above. It should be noted that the undulator parameter depends on r_β and thus for an electron beam of r.m.s. radius σ_β, the emission will be the sum of radiation arising from the individual wiggling motions for each electron, each with a distinct radial amplitude $r_{\beta,i}$ and an undulator parameter, $a_{\beta,i}$, and wavelength given by (42)

$$\lambda_i = \frac{\lambda_\beta}{2\gamma_{z,i}^2 h} = \frac{\lambda_\beta}{2\gamma^2 h}(1 + \frac{a_{\beta,i}^2}{2} + \gamma^2\vartheta^2) = \frac{\sqrt{2}\lambda_p}{\gamma^{3/2}h}(1 + \frac{a_{\beta,i}^2}{2} + \gamma^2\vartheta^2), \tag{90}$$

where h is the harmonic number. For $a_\beta > 1$, i.e. wiggler motion, many harmonics are produced. As in conventional wiggler radiation the emission peaks at harmonic number

$$h_c = \frac{3a_{\beta,i}}{4}\left(1 + \frac{a_{\beta,i}^2}{2}\right), \tag{91}$$

which (from equation 90) gives a peak in photon energy of $E_c = \hbar 2\pi c/e\lambda_i$, in eV, the same as that shown in Fig. 12. The power radiated every betatron period is proportional to both γ^2 and r_β^2 i.e.

$$P \propto \gamma^2 r_\beta^2 \tag{92}$$

therefore most of the radiation is emitted from electrons with the largest amplitude betatron orbit when electrons reach their highest energy, $\gamma \approx 2\omega_o^2/\omega_p^2$, i.e. over a distance equal to the dephasing length L_d. Therefore, significant emission occurs only in the last 12% of acceleration i.e. $\delta_z \approx 0.12L_d$, at the highest energy, which is of the order of one betatron oscillation. E.g. for a plasma density of 1×10^{18} cm^{-3}, $L_d \approx 37$ mm, $\gamma_{\max} \approx 3500$ and $r_\beta = 1$ μm, $\lambda_\beta \approx 2.7$ mm, $\delta_z \approx 4.4$ mm then only the two last betatron oscillations result in significant emission. The large undulator/wiggler parameter, $a_\beta \approx 7.9$, results in harmonics peaking at the 187th harmonic. The first harmonic occurs at a relatively low photon energy (340 eV) because of the high value of a_β and the on-axis spectrum peaks at 64 keV, with significant power beyond 150 keV. As in conventional wiggler radiation, radiation is emitted into a cone of $\Omega \approx a_\beta/\gamma$. The betatron radiation wavelength at the peak of the spectrum is given by

$$\lambda_{\text{peak}} \approx \frac{2\lambda_p^2}{3\pi\gamma^2 r_\beta} \approx \frac{\lambda_0^4}{6\pi\lambda_p^2 r_\beta}(\text{at dephasing}), \tag{93}$$

thus to obtain higher photon energies, γ and/or r_β should be increased and/or the plasma density should be decreased. One consequence of this scaling is that the peak of the spectrum becomes

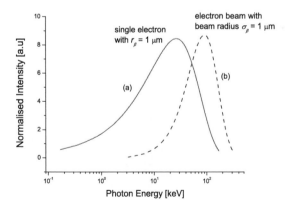

Figure 19. *Betatron spectrum for (a) single electron and (b) electron beam, for a plasma density of $n_e = 1 \times 10^{18}$ cm^{-3}, $L_d \approx 37$ mm, $\gamma_{\max} \approx 3500$ and $r_\beta = \sigma_\beta = 1$ μm.*

very sensitive to the radial distribution of the electron beam. Fig. 19 compares the normalised betatron spectra for a single electron with $r_\beta = 1$ μm with that for an electron beam with a beam with of $\sigma_\beta = 1$ μm, where the electron radial distribution is Gaussian, centered on the bubble axis, with a variance σ_β. Betatron radiation is a promising means of producing intense ultra-short pulses of hard x-rays and even gamma rays with more than 10^8 photons per pulse in a relatively small spectral bandwidth.

7   Raman amplification

When a photon interacts with a medium, it can be scattered either elastically, thus retaining its incident energy (Rayleigh scattering), or inelastically, exchanging energy with elementary excitations of the medium; this inelastic scattering is named after Sir Chandrasekhera Venkata Raman, who was awarded the 1930 Nobel Prize in physics for the discovery and exploration of this phenomenon. Energy can be exchanged either way: if the photon loses energy to the medium this is called Stokes scattering, the case in which the photon gains energy is called anti-Stokes scattering.

In plasma, the excitations are Langmuir waves: space charge oscillations of electrons against the ion background. Their natural frequency is the plasma frequency ω_p. If the frequency of an incident photon is ω_0 (corresponding to energy $\hbar\omega_0$), scattered photons can emerge at the Stokes frequency $\omega_S = \omega_0 - \omega_p$, if energy is transferred from the electromagnetic wave to the plasma oscillation, or the anti-Stokes frequency $\omega_a = \omega_0 + \omega_p$ in the opposite case.

Raman scattering can occur spontaneously, when the incident wave is scattered by, *e.g.*, thermal fluctuations, or it can be stimulated by a seed pulse at the Stokes frequency, which is amplified in the process. The latter case offers potential to amplify laser pulses to powers exceeding the breakdown threshold of current state-of-the-art solid state amplifiers.

8   Three-wave interaction

The description of Raman scattering in plasma involves three waves or modes: incident and scattered transverse electromagnetic waves, and longitudinal oscillations of the plasma electrons. (The latter only becomes a propagating wave if, at sufficiently high temperatures, the electron pressure plays a role.) In the case of stimulated scattering, the incident wave is called the pump wave, and the scattered wave the probe. We adopt a hydrodynamic description of the plasma, characterising the electrons by their number density $n(\vec{r}, t)$ and average velocity $v(\vec{r}, t)$ or momentum $p(\vec{r}, t)$.

We describe the transverse waves by their vector potential $\vec{A}(\vec{r}, t)$, so that the electromagnetic fields are given by $\vec{E} = -\partial\vec{A}/\partial t$ and $\vec{B} = \nabla \times \vec{A}$. In these fields, the electron momentum follows the vector potential: $\vec{p}_\perp(\vec{r}, t) = e\vec{A}_\perp(\vec{r}, t) + \vec{p}_0(\vec{r})$, where $\vec{p}_0$ is determined by the initial conditions, and can in many cases be set to zero. As previously, it is convenient to use the normalised (dimensionless) vector potential $\vec{a} = e\vec{A}_\perp/(mc)$. Taking into account the current corresponding to the electron motion, Maxwell's equations yield a Klein-Gordon type equation for the vector potential:

$$[\partial^2/\partial t^2 - c^2\nabla^2 + \omega_p^2(1 + \delta n/n_0)]\vec{a} = 0, \tag{94}$$

where we allow for deviations $\delta n = n - n_0$ of the electron density n from its equilibrium value n_0 (which cancels out the ion charge density, thus corresponds to neutral plasma).

We use these deviations to characterise the longitudinal plasma wave. The electric field due to the charge separation exerts a restoring force on the electrons, so that

$$(\partial^2/\partial t^2 + \omega_p^2)\delta n/n_0 = -\nabla \cdot \vec{F}/m, \tag{95}$$

where $\vec{F}$ is an external force (*i.e.* not due to the charge separation). In the case of interacting transverse waves in plasma, the force acting on the electrons is the $\vec{v} \times \vec{B}$ or ponderomotive force (the latter term is often used for the slowly varying part only), which may be written as $\vec{F}_p = -\nabla \Phi_p$. The ponderomotive potential is the kinetic energy of the electrons in the forced transverse oscillations: $\Phi_p = (\gamma_\perp - 1)mc^2$, $\gamma_\perp = \sqrt{1 + \vec{a}^2}$. For small amplitudes (non-relativistic electron motion), $\Phi_p \approx mc^2 \vec{a}^2/2$.

9   Dispersion of scattered wave

Let us first address the case of a weak scattered field, and derive its dispersion relationship. Consider a transverse incident or pump wave, $\vec{a}^{(0)}$ (*e.g.* from a laser), being scattered at a small perturbation δn in the electron density, and determine the scattered wave or probe, $\vec{a}^{(1)}$:

$$\vec{a}(\vec{r}, t) = \vec{a}^{(0)} + \vec{a}^{(1)}.$$

The velocity and current density are

$$\vec{v} = \vec{v}^{(0)} + \vec{v}^{(1)}, \quad \vec{j}^{(1)} = -e(n_0 \vec{v}^{(1)} + \delta n \vec{v}^{(0)}),$$

respectively, and thus the wave eqn. becomes

$$\partial^2 \vec{a}^{(1)}/\partial t^2 - c^2 \nabla^2 \vec{a}^{(1)} + \omega_p^2 \vec{a}^{(1)} = -e^2 \delta n \, \vec{a}^{(0)}/(\epsilon_0 m).$$

Choosing $\vec{a}^{(0)} = \vec{a}_0 \cos(\omega_0 t - \vec{k}_0 \cdot \vec{r})$ (where ω_0 and $\vec{k}_0$ satisfy the plasma dispersion relation $D_\perp(k_0, \omega_0) = \omega_0^2 - c^2 k_0^2 - \omega_p^2 = 0$) we may Fourier-analyse $\vec{a}^{(1)}$ and δn,

$$\{\vec{a}^{(1)}(\vec{r}, t), \, \delta n(\vec{r}, t)\} = \int d^3k \, d\omega \, \{\vec{a}_{\vec{k},\omega}, \, \delta n_{\vec{k},\omega}\} e^{i\omega t - i\vec{k}\cdot\vec{r}},$$

to find

$$D_\perp(k, \omega)\vec{a}_{\vec{k},\omega} = \omega_p^2 \vec{a}_0 \, [n_{\vec{k}-\vec{k}_0, \omega-\omega_0} + n_{\vec{k}+\vec{k}_0, \omega+\omega_0}]/(2n_0),$$

$$D_\parallel(\omega) n_{\vec{k},\omega} = c^2 k^2 n_0 \, \vec{a}_0 \cdot [\vec{a}_{\vec{k}-\vec{k}_0, \omega-\omega_0} + \vec{a}_{\vec{k}+\vec{k}_0, \omega+\omega_0}]/2,$$

with $D_\parallel(\omega) = \omega^2 - \omega_p^2$.

The dominating Fourier components of the density perturbation will be those near the plasma resonance, $\omega \approx \omega_p$. Combining both eqns., and neglecting nonresonant components at frequencies $\omega \pm 2\omega_0$ we find a dispersion relation for the frequency of the density perturbation:

$$D_\parallel(\omega) = c^2 k^2 \omega_p^2 a_0^2/4 \, (1/D_+ + 1/D_-),$$

where $D_\pm = D_\perp(k_\pm, \omega_\pm)$ correspond to waves with up- and downshifted frequencies $\omega_\pm = \omega_0 \pm \omega$, respectively, and similarly shifted wave vectors $\vec{k}_\pm = \vec{k}_0 \pm \vec{k}$. Close to the plasma resonance, these frequencies will be close to the Stokes and anti-Stokes frequencies, ω_S and ω_a, respectively.

The most interesting case is when the Stokes wave can propagate, *i. e.* D_- is small, and $1/D_+$ can be neglected:

$$D_\parallel(\omega) D_\perp(k_-, \omega_-) = c^2 k^2 \omega_p^2 a_0^2/4. \tag{96}$$

To derive the dispersion relation we choose the wave number k_- of the scattered wave. The corresponding frequency in the absence of the pump wave would be $\omega' = \sqrt{c^2 k_-^2 - \omega_p^2}$, and the frequency which the scattered wave will adopt with the pump wave present is ω_-. We denote the differences of these and the Stokes frequency by $\Delta = \omega' - \omega_S$, and $\delta = \omega_- - \omega_S$, respectively; thus $\omega_- = \omega' + \delta - \Delta$, and the frequency of the plasma wave is $\omega = \omega_p - \delta$. For $|\Delta|, |\delta| \ll \omega_p$, $D_\parallel(\omega) \approx -2\omega_p\delta$, and similarly, $D_\perp(k_-, \omega_-) \approx 2\omega'(\delta - \Delta)$, and the dispersion relation becomes:

$$\delta(\delta - \Delta) + \Gamma_0^2 = 0, \quad \delta = \Delta/2 \pm \sqrt{\Delta^2/4 - \Gamma_0^2}. \tag{97}$$

We find that for detuning $\Delta/2$ within the bandwidth $\pm\Gamma_0$, the system exhibits an instability, *i. e.* exponential growth, at a rate

$$\Gamma(\Delta) = \sqrt{\Gamma_0^2 - \Delta^2/4}, \tag{98}$$

which reaches its maximum $\Gamma_0 = \sqrt{\omega_p/\omega'}\, c|\vec{k}_0 - \vec{k}_-|a_0/4$ at the plasma resonance, $\Delta = 0$. This growth rate is highest for backscattering. In this case, assuming dilute plasma, $\vec{k}_- \approx -\vec{k}_0$, $k_0 \approx \omega_0/c$, we find $\Gamma_0 \approx \sqrt{\omega_0\omega_p}\, a_0/2$. Looking at the real part of δ, within the instability bandwidth, the group velocity of the scattered wave is $v_g = d\omega_-/dk_- = (d\mathrm{Re}\,(\delta)/d\Delta)(d\omega'/dk_-) = v_g'/2$, half the group velocity without interaction.

If energy dissipation by electron-ion collisions is taken into account, the dispersion relation is modified to

$$\delta = \Delta/2 - is + \sigma(\Delta)\sqrt{(\Delta/2 + id)^2 - \Gamma_0^2}, \tag{99}$$

with $s = (\nu_\parallel + \nu_\perp)/4$, $d = (\nu_\parallel - \nu_\perp)/4$, where $\nu_\perp$ is the average momentum dissipation rate at the probe frequency ω_1 (for transverse oscillations), multiplied with $(\omega_p/\omega_1)^2$, and $\nu_\parallel$ is the rate at the plasma frequency (for longitudinal oscillations). The dissipation introduces a threshold for the pump intensity, below which there is no Raman instability: $\mathrm{Im}(\delta) > 0$ requires $\Gamma_0^2 > \nu_\perp\nu_\parallel/4$.

10   Slowly varying envelope approximation

In order to gain a more intuitive picture of the Raman interaction, we shall look at it in the time domain. We choose the special case of Raman backscattering, and reduce the second order partial differential equations to first order ones by using the slowly-varying envelope approximation. We write the vector potential and density modulations as

$$\vec{a}(z, t) = \vec{u}\,(a_0 e^{i\varphi_0} + a_1 e^{i\varphi_1})/2 + c.c, \tag{100}$$
$$\delta n/n_0 = n e^{-i\Delta\varphi}/2 + c.c., \tag{101}$$

where $\vec{u}$ is a unit vector (e. g. $\vec{u} = \vec{u}_x$ for linear, and $\vec{u} = (\vec{u}_x + i\vec{u}_y)/\sqrt{2}$ for circular polarization), and *c.c.* denotes the complex conjugate of the preceding expression. We make the assumption that the variation of the envelopes a_0, a_1, n with position and time is much slower than that of the corresponding phases

$$\varphi_0 = \omega_0 t + k_0 z, \quad \varphi_1 = \omega_1 t - k_1 z, \quad \Delta\varphi = \varphi_0 - \varphi_1 = \Delta\omega\, t + \Delta k\, z, \tag{102}$$

with $\Delta\omega = \omega_0 - \omega_1$, $\Delta k = k_0 + k_1$. This allows us to neglect second derivatives of the envelopes. With this approximation, the second time derivatives $\ddot{} \equiv \partial^2/\partial t^2$ are:

$$\ddot{\vec{a}} \approx \vec{u}\,[(-\omega_0^2 a_0 + 2i\omega_0\dot{a}_0)e^{i\varphi_0} + (-\omega_1^2 a_1 + 2i\omega_1\dot{a}_1)e^{i\varphi_1}]/2 + c.c.,$$
$$\ddot{\delta n}/n_0 \approx (-\Delta\omega^2 n - 2i\Delta\omega\,\dot{n})e^{-i\Delta\varphi}/2 + c.c.,$$

and analogous for the z-derivatives $\vec{a}'' \equiv \partial^2\vec{a}/\partial z^2$; thus

$$(\partial^2/\partial t^2 - c^2\nabla^2 + \omega_p^2)\vec{a}$$
$$\approx \vec{u}\{[(c^2 k_0^2 - \omega_0^2 + \omega_p^2)a_0 + 2i\omega_0\dot{a}_0 - 2ic^2 k_0 a_0']e^{i\varphi_0}$$
$$+[(c^2 k_1^2 - \omega_1^2 + \omega_p^2)a_1 + 2i\omega_1\dot{a}_1 + 2ic^2 k_1 a_1']e^{i\varphi_1}\}/2 + c.c.$$
$$= i\vec{u}\,[(\omega_0\dot{a}_0 - c^2 k_0 a_0')e^{i\varphi_0} + (\omega_1\dot{a}_1 + c^2 k_1 a_1')e^{i\varphi_1}] + c.c., \tag{103}$$

where we have chosen the wave numbers to satisfy the plasma dispersion relation for the respective frequencies. Similarly, choosing $\Delta\omega = \omega_p$:

$$(\partial^2/\partial t^2 + \omega_p^2)\delta n/n_0 \approx -i\omega_p \dot{n}e^{-i\Delta\varphi} + c.c.. \tag{104}$$

We now have to find the envelopes of the product $\delta n/n_0\, \vec{a}$, and of the ponderomotive potential Φ_p:

$$
\begin{aligned}
\delta n/n_0\,\vec{a} &= (n\,e^{-i\Delta\varphi}/2 + c.c.)\,[\vec{u}\,(a_0 e^{i\varphi_0} + a_1 e^{i\varphi_1})/2 + c.c.] \\
&= \vec{u}\,(na_0 e^{i\varphi_1} + na_1 e^{2i\varphi_1 - i\varphi_0} + n^* a_0 e^{2i\varphi_0 - i\varphi_1} + n^* a_1 e^{i\varphi_0})/4 \\
&\qquad\qquad\qquad\qquad\qquad\qquad\qquad\qquad\qquad +c.c., \\
\Phi_p &= mc^2\vec{a}^2/2 = mc^2\,[\vec{u}(a_0 e^{i\varphi_0} + a_1 e^{i\varphi_1}) + c.c.]^2/8 \\
&= mc^2\,[\vec{u}^{\,2}(a_0^2 e^{2i\varphi_0} + 2a_0 a_1 e^{i\varphi_0 + i\varphi_1} + a_1^2 e^{2i\varphi_1}) \\
&\qquad +|\vec{u}|^2(|a_0|^2 + 2a_0^* a_1 e^{-i\Delta\varphi} + |a_1|^2) + c.c.]/8.
\end{aligned}
$$

We collect terms according to their phase factors, neglect the non-resonant ones (as before; for circular polarization, $\vec{u}^{\,2} = 0$ in the ponderomotive potential anyway), and the slowly varying ones $|a_{0,1}|^2$, and arrive at the evolution equations for pump, probe, and density envelopes:

$$
\begin{aligned}
\omega_0 \dot{a}_0 - c^2 k_0 a_0' &= i\omega_p^2/4\,n^* a_1, & (105) \\
\omega_1 \dot{a}_1 + c^2 k_1 a_1' &= i\omega_p^2/4\,na_0, & (106) \\
\omega_p \dot{n} &= -ic^2\Delta k^2 a_0^* a_1/4. & (107)
\end{aligned}
$$

The left-hand sides of the first two equations are time derivatives of the fields in their respective co-moving frames of reference. For simplicity, we assume dilute plasma, $ck_0 \approx ck_1 \approx \omega_1 \approx \omega_0$:

$$
\begin{aligned}
\dot{a}_0 - ca_0' &= i\omega_p^2/(4\omega_0)\,n^* a_1, & (108) \\
\dot{a}_1 + ca_1' &= i\omega_p^2/(4\omega_0)\,na_0, & (109) \\
\dot{n} &= -i\omega_0^2/\omega_p\,a_0^* a_1. & (110)
\end{aligned}
$$

10.1   Low scattered amplitude

First, we shall look again at the case of low scattered amplitude, in which the pump amplitude remains constant and we can disregard the first equation. We introduce new variables $\zeta = z/c$ and $\tau = t - z/c$. The derivatives are $\partial/\partial t = \partial/\partial\tau$, $c\partial/\partial z = \partial/\partial\zeta - \partial/\partial\tau$, thus

$$\partial a_1/\partial\zeta = i\omega_p^2/(4\omega_0)\,a_0 n, \quad \partial n/\partial\tau = -i\omega_0^2/\omega_p\,a_0^* a_1. \tag{111}$$

Next, we eliminate the scattered field and find

$$\partial^2 n/\partial\zeta\partial\tau = \Gamma_0^2\,n, \tag{112}$$

with $\Gamma_0 = \sqrt{\omega_0\omega_p}\,a_0/2$ as before (choosing a_0 real). This type of equation has special, self-similar, solutions $n(\zeta,\tau) = N(\xi)$ which depend only on the product $\xi = \zeta\tau$. The derivatives are $\partial/\partial\zeta = \tau\,d/d\xi$, $\partial/\partial\tau = \zeta\,d/d\xi$, $\partial^2/\partial\zeta\partial\tau = \xi d^2/d\xi^2 + d/d\xi$:

$$(\xi d^2/d\xi^2 + d/d\xi)N = \Gamma_0^2\,N, \tag{113}$$

with solution

$$N(\xi) = N_0 I_0(2\Gamma_0\sqrt{\xi}), \tag{114}$$

where I_0 is the zero-order Bessel function. From the second equation (111) we find

$$a_1 = \zeta A(\xi), \quad A(\xi) = iN_0 \sqrt{\omega_p^3/(4\omega_0^3)}\, I_1(2\Gamma_0\sqrt{\xi})/\sqrt{\xi}, \tag{115}$$

with I_1 is the first-order Bessel function.

As written above, the self-similar solution describes the evolution of the instability starting from density modulations with homogeneous amplitude N_0. Alternatively, one can introduce a short seed pulse (idealized as a δ-function) propagating into the plasma from the surface (taken to be $z = 0$):

$$a_1 = S\delta(\tau) + \zeta A(\xi)\Theta(\xi), \quad n = N(\xi)\Theta(\xi), \tag{116}$$

where $\Theta(\xi)$ is the Heaviside step function. The discontinuity N_0 at $\xi = 0$ is then determined by the amplitude of the seed: $N_0 = -2i\Gamma_0\sqrt{\omega_0/\omega_p}^{-3} S$. This solution describes a scattered field that extends from the plasma edge to the seed position, and grows approximately eponentially at every point that the seed has passed. The maximum lies near the centre of this region, which explains why the group velocity is half that of the seed, as we have found earlier.

Although the idealization of the seed as a δ-function violates, strictly speaking, the assumption of a slowly-varying envelope, the result can be interpreted as the system's Green's function: since the equations are linear (with respect to A, N), the response to arbitrary seeds may be found as its convolution with the Green's function. The Green's function also represents an approximate solution for short (but finite) seeds.

10.2   Pump depletion

However, this rapid growth will be limited by two factors: Firstly, as the amplitude of the density modulations approaches the background density, their evolution will become anharmonic, leading to scattering into waves which are down-shifted by multiples of the plasma frequency, and eventually the plasma wave will break. And secondly, the energy taken up by the plasma wave and the probe will deplete the pump energy, thus its amplitude is no longer constant. We shall take this latter effect into account, and try to find a self-similar solution for this case. We start again with (108) and the following two equations, and try a solution of the form: $a_0(\zeta, \tau) = a_{00}B(\xi)$, $a_1 = \zeta a_{00} A(\xi)$, $n = -2ia_{00}\sqrt{\omega_0/\omega_p}^{-3} N(\xi)$, where a_{00}, which can be chosen real, is the pump amplitude before any interaction, and the scaling has been introduced to simplify the coupling constants. While the second and third equation are still consistent, the derivatives in the first transform into $\partial/\partial t - c\partial/\partial z = 2\partial/\partial\tau - \partial/\partial\zeta = (2\zeta - \tau)d/d\xi$, thus the validity of the solution seems to be restricted to $z \approx ct$, close to the position of the seed. On the other hand, this is where the pump depletion is expected to happen, thus:

$$2dB/d\xi = -\Gamma_{00}AN^*, \tag{117}$$
$$\xi dA/d\xi + A = \Gamma_{00}NB, \tag{118}$$
$$dN/d\xi = \Gamma_{00}B^*A, \tag{119}$$

where the growth rate Γ_{00} corresponds to the undepleted pump, a_{00}. Since all the coefficients in these coupled equations are real, the functions A, B, and N can be chosen real (if they are real initially, they will remain so).

By multiplying the first equation with B, the third with N, and adding both we find that $2B^2 + N^2$ is conserved, and we may write

$$B(\xi) = \cos(s), \quad N(\xi) = \sqrt{2}\sin(s). \tag{120}$$

To derive the evolution of the new variable s, we substitute B and N into the first, and the result into the second equation, yielding

$$\Gamma_{00}A = \sqrt{2}ds/d\xi, \quad \xi d^2 s/d\xi^2 + ds/d\xi = \Gamma_{00}^2 \sin(2s)/2. \tag{121}$$

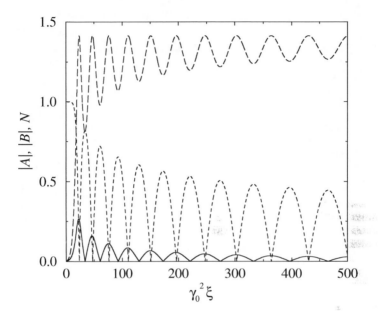

Figure 20. *Self-similar solution of sine-Gordon equation for amplitudes $|A|$ of scattered wave (solid), $|B|$ of pump wave (dotted), and N of density (dashed).*

The variable s is thus a self-similar solution $f(xy)$ of the sine-Gordon equation $\partial_{xy}^2 f(x,y) = \sin f$. Before the interaction, $B = 1$ and $N = 0$, corresponding to $s = 0$. Starting close to this value, which is an unstable fixpoint, the solution initially grows exponentially, but then oscillates about the stable fixpoint $s = \pi/2$, with decreasing amplitude. Correspondingly, $A(\xi)$ has a maximum for small ξ, and oscillates with decreasing amplitude as ξ increases. For large ξ, both A and B tend to zero, while N tends to $\sqrt{2}$ (Fig. 20). The scattered wave, a_1, has a maximum close to $z = ct$ which grows proportional to $\zeta \approx t$; its width Δz decreases proportional to $1/t$ (Fig. 21). This is called superradiant scaling.

10.3   Chirped pump, low scattered amplitude

In experimental setups for Raman amplification of a short seed pulse by a long pump pulse, often the latter is not monochromatic, but chirped, *i. e.* its frequency changes with time. To find the effect of a linear frequency chirp α, we modify the phase φ_0, equation (102), to

$$\varphi_0 = \omega_0(t + z/c) + \varphi_{ch}, \quad \varphi_{ch} = \alpha(t + z/c)^2/2, \tag{122}$$

where we have approximated $k_0 = \omega_0/c$ for dilute plasma. We restrict our attention to the case of low scattered amplitude. The imprint of this chirped phase on the probe pulse and density modulations is taken into account in their respective envelopes, for which the evolution equations now read:

$$\partial a_1/\partial\zeta = i\omega_p^2/(4\omega_0)\, a_0 e^{i\varphi_{ch}} n, \quad \partial n/\partial\tau = -i\omega_0^2/\omega_p\, a_0^* e^{-i\varphi_{ch}} a_1, \tag{123}$$

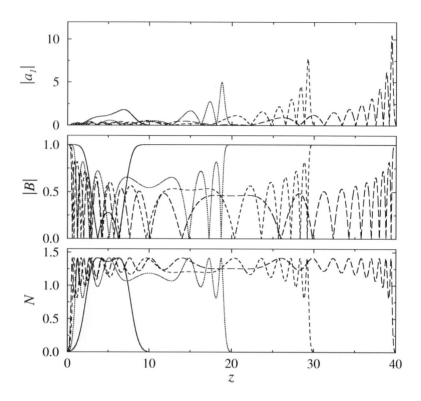

Figure 21. *Snapshots of amplitudes $|a_1|$ of scattered wave (top), $|B|$ of pump wave (middle), and N of density (bottom), at $\Gamma_0 t = 10, 20, 30, 40$; z-axis scaled with Γ_0/c.*

with $\varphi_{ch} = \alpha(\tau + 2\zeta)^2/2$ in these variables. The coupling of these equations is simplified by absorbing parts of the chirped phase in modified envelopes $\tilde{a}_1 = a_1 e^{-i\alpha\tau^2/2}$, $\tilde{n} = n e^{2i\alpha\zeta(\zeta+\tau)}$:

$$\partial\tilde{a}_1/\partial\zeta = i\omega_p^2/(4\omega_0)\,a_0\tilde{n}, \quad (\partial\tilde{n}/\partial\tau - 2i\alpha\zeta)\tilde{n} = -i\omega_0^2/\omega_p\,a_0^*\tilde{a}_1. \tag{124}$$

These equations have a self-similar solution of the same form $\tilde{a}_1 = \zeta A(\xi)$, $\tilde{n} = -2i\sqrt{\omega_0^3/\omega_p^3}\,N(\xi)$ as for the case without chirp:

$$\xi dA/d\xi + A = \Gamma_0 N, \quad (dN/d\xi - 2i\alpha N) = \Gamma_0 A. \tag{125}$$

Eliminating A yields

$$\xi d^2 N/d\xi^2 + (1 - 2i\alpha\xi)dN/d\xi - (\Gamma_0^2 + 2i\alpha)N = 0, \tag{126}$$

with solution

$$N(\xi) = N_0\,\mathrm{L}_{iq-1}(2i\alpha\xi), \tag{127}$$

where $q = \Gamma_0^2/(2\alpha)$, and $\mathrm{L}_\nu(z)$ is a Laguerre function. Using the second equation of (125) we find the corresponding probe:

$$A(\xi) = i\Gamma_0 N_0\,[\mathrm{L}_{iq}(2i\alpha\xi) - \mathrm{L}_{iq-1}(2i\alpha\xi)]/(2\alpha\xi). \tag{128}$$

As in the case without chirp, the result (shown in Fig. 22) can be interpreted as the Green's function of the system, or as approximate response to a short seed. It shows important features:

First, looking at the density modulations, we find that, far from both the seed and the plasma edge, their amplitude tends to a finite limit:

$$|N(\xi)| \to |N_0|\sqrt{(e^{2\pi|q|} - 1)/(2\pi|q|)} \quad \text{for } \alpha\xi \gg 1. \tag{129}$$

For $|q| > 1$, the square root is, approximately, $e^{\pi|q|}/\sqrt{2\pi|q|}$. The exponent can be interpreted as integral of the growth rate (98) over the time during which the detuning of pump and scattered wave, $\Delta/2 = \alpha t$, remains within the resonance bandwidth: $\int dt\, \Gamma(\Delta(t)) = \pi\Gamma_0^2/(2\alpha)$.

Second, as in the case of pump depletion, $A(\xi)$ has a maximum for small ξ, and the scattered wave, a_1, has a maximum close to $z = ct$, growing proportional to $\zeta \approx t$, with decreasing width proportional to $1/t$. Here, the linear (rather than exponential) growth is due to detuning. The suppression of the scattered wave behind the peak can be ascribed to dephasing: parts of the pump scattered at different times / positions oscillate at different frequencies, and thus interfere destructively. Remarkably, the chirp of the pump wave leads to superradiant scaling before pump depletion sets in.

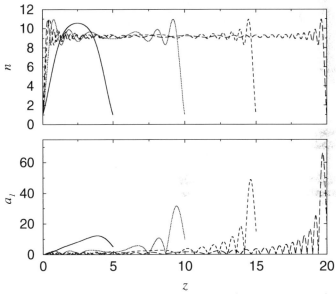

Figure 22. *Amplitudes of self-similar solutions for density n (top, scaled with N_0) and scattered wave a_1 (bottom, scaled with $\Gamma_0 N_0/\sqrt{\alpha}$); snapshots for $q = 1$ at $\sqrt{\alpha}\,t = 5, 10, 15, 20$; z-axis scaled with $\sqrt{\alpha}/c$*

11   Summary

We have described the parametric three-wave interaction underlying Raman scattering, using a cold fluid model for the plasma wave which couples incident and scattered wave. We have derived the dispersion relation for low-amplitude scattered waves, showing a region of instability and yielding the corresponding growth rate. Envelope equations allow to study depletion of the pump wave, or the case of a chirped pump pulse. Both these cases show superradiant scaling of the probe pulse. The corresponding contraction of the pulse duration should make stimulated Raman backscattering suitable for the amplification of short laser pulses.

References

Akhiezer, A.I. and Polovin, R.V. (1956) *Sov. Phys. JETP* **30**, 915.

Ayvazyan, V., et al., (2002) *Physical Review Letters*, **88**, 104802.

Ayvazyan, V., et al., (2006) *European Physical Journal D*, **37**, 297 – 303

Bingham, R. (2005) *In this volume.*

Bobroff, D. L. and Haus, H. A., et al. (1967) *J. Appl. Phys.*, **38** 390.

Bobrova, et al. (2002) *Physical Review E*, **65**.

Bonifacio, R., Casagrande, Pelegrini, C., (1987) *Opt. Comm.*, **61**, 55.

Bonifacio, R., McNeil, B. and Pierini, P. (1989) *Physical Review A*, **40**, 4467 – 4475.

Bonifacio, R., Piovella, N. and McNeil, B. (1991) *Physical Review A*, **44**, R3441 – R3444.

Bonifacio, R., Piovella, N. and Robb, G. (2005) *Nucl. Instr. & Meth. in Phys. Res. Sect. A-Accel. Spectr. Detect & Assoc. Equip.*, **543**, 645 – 652.

Bonifacio, R. et al. (2005a) *Nucl. Instr. & Meth. in Phys. Res. Sect. A-Accel. Spectr. Detect & Assoc. Equip.*, **546**, 634 – 638.

Bulanov, S.V., Pegoraro, F., Pukhov, A.M. and Sakharov, A.S. (1997) *Phys. Rev. Lett.* **78**, 4205.

Butler, A., Spence, D. and Hooker, S. (2002) *Physical Review Letters*, **89**.

Chen, P., Dawson, J.M., Huff, R.W. and Katsouleas, T. (1985) *Phys. Rev. Lett.* **54**, 693.

Chiou, T.C., Katsouleas, T. and Mori, W.B. (1996) *Phys. Plasmas* **3**, 1700.

Chiou, T. and Katsouleas, T. (1998) *Phys. Rev. Lett.*, **81**, 3411 – 3414.

Clark, T.R. and Milchberg, H.M. (2000) *Phys. Rev. E* **61**, 1954.

Colson, W. (2001) *Nucl. Instr. & Meth. in Phys. Res. Sect. A-Accel. Spectr. Detect & Assoc. Equip.*, **475**, 397 – 400.

Coverdale, C., Darrow, C., Decker, C., Mori, W., Tzeng, K., Marsh, K., Clayton, C. and Joshi, C. (1995) *Physical Review Letters*, **74**, 4659 – 4662.

Danson, C.N. et al. (1993) *Optics Comm.* **103**, 392.

Darrow, C.B. et al. (1992) *Phys. Rev. Lett.* **69**, 442.

Dorchies, F. et al. (1999) *Phys. Rev. Lett.* **82**, 4655.

Durfee III, C.G. and Milchberg, H.M. (1993) *Phys. Rev. Lett.* **71**, 2409.

Edwards, C.B. et al. (2003) *Proc. SPIE Int. Soc. Opt. Eng.* **4948**, 444.

Ehrlich, Y. et al. (1996) *Phys. Rev. Lett.* **77**, 4186.

Ersfeld, B. and Jaroszynski, D. (2005) *Phys. Rev. Lett.*, **95**, 165002.

Esarey, E. et al. (1996) *IEEE Trans. Plasma Sci.* **24**, 252.

Esarey. E. and Pilloff, M. (1995) *Phys. Plasmas* **2**, 1432.

Esarey. E. et al., (2002) *Phys. Rev. E* **65**, 056505.

Faure, J. et al. (2004) *Nature* (London) **431**, 541.

Forslund, D.W. et al. (1973) *Phys. Fluids* **18**, 1002.

Fritzler, S., et al., (2004) *Physical Review Letters*, **92**.

Geddes, C.G.R. et al. (2004) *Nature* (London) **431**, 538.

Ginzburg, V.L. and Frank, I.M., (1946) *JETP* **16**, 15.

Gordon, D.F. et al. (2003) *Phys. Rev. Lett.* **90**, 215001.

Gordon, D.F. et al. (2005) *Physical Review E*, **71**.

Hafizi, B. et al. (2000) *Phys. Rev. E* **62**, 4120.

Hogan, M.J. et al. (2005) *Phys. Rev Lett.* **95**, 054802.

Hubbard, R.F., Sprangle, P. and Hafizi, B. (2000) *IEEE Trans. Plasma Sci.* **28**, 1159.

Jackel, S. et al. (1995) *Opt. Lett.* **20**, 1086.

Jackson, J. D., (1975) *Classical Electrodynamics.* 2nd Edition, John Wiley and Sons Inc.

Jaroszynski, D., et al.(1993) *Phys. Rev. Lett.*, **71**, 3798 – 3802.

Jaroszynski, D. (1997) *Towards X-ray Free-Electron Lasers*, AIP Conf. Procs. **413**, 55 – 79.

Jaroszynski, D., et al.(1997a) *Physical Review Letters*, **78**, 1699 – 1702.

Jaroszynski, D., et al. (2000a) *Nucl. Inst. and Meth. in Phys. Res. Sect. A - Acc. Spect. Det. and Assoc. Equip.*, **445**, 317.

Jaroszynski, D. and Vieux, G. (2002) *Advanced Accelerator Concepts*, **647**, 902 – 913.

Jaroszynski, D., et al.(2000b) *Nucl. Instr. & Meth. in Phys. Res. Sect. A-Accel. Spectr. Detect & Assoc. Equip.*, **445**, 261 – 266.

Jaroszynski, D., et al. (2006) *Phil. Trans. R. Soc.*, **A 364**, 689.

Jelley, J. V., (1958) *Cerenkov Radiation and Its Applications.* Pergamon, London.

Joshi, C., (1990) *Physica Scripta*, **T30**, 90 – 94.

Katsouleas, T. et al. (1987) *Part. Accel.* **22**, 181.

Khachatryan, A., van goor, F., Boller, K., Reitsma, A. and Jaroszynski, D. (2004) *Physical Review Special Topics-Accelerators and Beams*, **7**.

Kostyukov, I. and Pukhov, A. (2004) *Phys. Plasmas* **11**, 5256.

Kruer, W. (1988) *The Physics of Laser Plasma Interaction.* Addison-Wesley, Reading, MA.

Krushelnick, K. *In this volume.*

LCLS Conceptual Design Report *www-ssrl.slac.stanford.edu/lcls/CDR.*

Landau, L. D., Liftshitz, E. M., and Pitaevskii, L. P., (1984) *Electrodynamics of Continuous Media.* Pergamon, New York.

Lawson, J.D. (1977) *The Physics of Charged Particle Beams.* Oxford University Press, London.

Lu, W. et al. (2006) *Phys. Rev. Lett.* **96**, 165002.

Leemans, W. P. et al. (2003) *Phys. Rev. Lett.* **91**, 074802.

Luchini, P. and Motz, H. (1990) *Undulators and Free-Electron Lasers.* Clarendon Press, Oxford

Luo, C., et al., (2003) *Science* **299**, 368Û371.

Malka, G. et al. (1997) *Phys. Rev. Lett.* **79**, 2053.

Malka, V., et al., (2002) **298**, 1596 – 1600.

Malkin, V., Shvets, G. and Fisch, N. (1999) *Phys. Rev. Lett.*, **82**, 4448.

Mangles, S., et al., (2004) *Nature*,**431**, 535 – 538.

Marshall, T.C.(1985) *Undulators and Free-Electron Lasers.* Clarendon Press, Oxford

Martins, S., et al. (2005) *IEEE Transactions on Plasma Science*, **33**, 558 – 559.

McKinstrie, C.J. and Bingham, R. (1992) *Phys. Fluids* **B4**, 2626.

McNeil, B., Robb, G. and Jaroszynski, D. (1999) *Optics Communications*, **163**, 203 – 207.

McNeil, B., Robb, G. and Jaroszynski, D. (2000) *Nucl. Instr. & Meth. in Phys. Res. Sect. A-Accel. Spectr. Detect & Assoc. Equip.*, **445**, 72 – 76.

Modena, A. et al. (1995) *Nature* (London) **377**, 606.

Nakajima, K., et al., (1995) *Physical Review Letters*, **75**, 984 – 984.

Ping, Y., Geltner, I. and Suckewer S., (2003). *Phys. Rev. E*, **67**, 016401.

Pukhov A. and Meyer-ter Vehn J., (2002) *Applied Physics B*, **74**, 355 – 361.

Raubenheimer, T. (1995) *Nucl. Instr. & Meth. in Phys. Res. Sect. A-Accel. Spectr. Detect & Assoc. Equip.*, **358**, 40 – 43.

Reitsma, A., Cairns, R., Bingham, R. and Jaroszynski, D. (2005) *Physical Review Letters*, **94**.

Reitsma, A., Trines, R. and Goloviznin, V. (2000) *IEEE Transactions on Plasma Science*, **28**, 1165 – 1169.

Rosenbluth, M.N. and Liu, C.S. (1972) *Phys. Rev. Lett.* **29**, 701.

Rosenzweig, J.B., Breizman, B., Katsouleas, T. and Su, J.J. (1991) *Phys. Rev. A* **44**, R6189.

Saldin, E., Schneidmiller, E. and Yurkov, M. (2002) *Nucl. Instr. & Meth. in Phys. Res. Sect. A-Accel. Spectr. Detect & Assoc. Equip.*, **483**, 516 – 520.

Saldin, E., Schneidmiller, E. and Yurkov, M. (2004) *Nucl. Instr. & Meth. in Phys. Res. Sect. A-Accel. Spectr. Detect & Assoc. Equip.*, **528**, 355 – 359.

Schwoerer H. et al. (2006) *Phys. Rev. Lett.* **99**, 014802.

Shvets, G., Wurtele, J. and Shadwick, B. (1997) *Phys. Plasmas*, **4**, 1872.

Shvets, G. et al. (1998) *Phys. Rev. Lett.*, **81**, 4879.

Smith, S.J. and Purcell, E.M., (1953) *Phys. Rev.* **92**, 1069.

Spence, D. and Hooker, S. (2001) *Physical Review E*, **6302**, art. no. – 015401.

Spence, D.J. and Hooker, S.M. (2001), *Phys. Rev. E* **63**, 015401.

Strickland, D. and Mourou, G. (1985) *Optics Comm.* **56**, 219.

Sun, G.-Z. et al. (1987) *Phys. Fluids* **30**, 526.

Tajima, T. and Dawson, J. (1979) *Phys. Rev. Lett.* **43**, 267.

Ta Phuoc K., et al., (2003) *Phys. Rev. Lett.* **91**, 195001.

Ta Phuoc K., et al., (2006) *Phys. Rev. Lett.* **97**, 225002.

Ter-Mikaelian, M.L.,(1972) *High-energy electromagnetic processes in condensed media.* Wiley-Interscience, New York

Rousse, A. et al. (2004) *Phys. Rev. Lett.* **93**, 135005.

Ting, A. et al. (1997) *Phys. Plasmas* **4**, 1889.

Tochitsky, S.Y. et al. (2004) *Phys. Rev. Lett.* **92**, 095004.

Trines, R., Goloviznin, V., Kamp, L. and Schep, T. (2001) *Physical Review E*, **63**, art. no. – 026406.

Tzeng, K.-C. et al. (1999) *Phys. Plasmas* **6**, 2105.

Umstadter, D. et al. (1996) *Science* **273**, 472.

Volfbeyn, P., Esarey, E. and Leemans, W.P. (1999) *Phys. Plasmas* **6**, 2269.

Wagner, R. Chen, S.-Y., Maksimchuk, A. and Umstadter, D. (1997) *Phys. Rev. Lett.* **78**, 3125.

Whittum, D.H., et al., (1990) *Physical Review Letters*, **64**, 2511.

Wiggins, S., et al., (2000) *Physical Review Letters*, **84**, 2393 – 2396.

Wilks, S., Katsouleas, T., Dawson, J., Chen, P. and Su, J. (1987) *IEEE Transactions on Plasma Science*, **15**, 210 – 217.

Zholents, A. (2003) *Proceedings of the 2003 Particle Accelerator Conference, VOLS 1-5*, 872 – 874.

Zewail, A. (1999) *Nobel Lecture 1999* Femtochemistry: atomic-scale dynamics of the chemical bond using ultrafast lasers.

5

High Intensity Laser-Plasma Interaction Experiments

Karl Krushelnick

Dept of Physics, Blackett Laboratory, Imperial College, London

1 Introduction

The technology of high power lasers has advanced significantly over the past decade. It is now possible to perform experiments with high energy "Petawatt" (10^{15} Watts) class laser systems at large laser facilities and similarly possible to perform high intensity experiments using ultra-short pulse laser systems (sub 50 fs) which have high repetition rates and which fit into a university scale laboratory (Perry and Mourou 1994). Both of these types of lasers are capable of producing unique states of matter which can have relativistic "temperatures" (Key et al. 1998), ultra-strong magnetic fields (Tatarakis et al 2002) and which can produce beams of energetic electrons (Mangles et al. 2004 - Faure et al 2004), ions (Clark et al. 2000) and gamma rays (Edwards et al. 2002). This has consequently led to a recent surge of interest in these systems for technological applications as well as for the examination of fundamental scientific issues.

In this chapter we will examine some of the unique phenomena that occur in ultra-high intensity interactions with both solid and gaseous density targets. We will use some recent experimental results from the high intensity laser facilities (Vulcan and Astra) at the Rutherford Appleton Laboratory in the UK as examples of such phenomena. In section 2 the production of electromagnetic radiation at very high order harmonics of the incident laser frequency will be examined, along with its use as a diagnostic of the extreme conditions of the laser produced plasma, – in particular as a method to measure magnetic fields within the plasma. In Section 3 progress on the use of intense lasers for generating relativistic electron beams electron is discussed – while in Section 4 recent work on ion acceleration from intense laser plasma interaction is reviewed. In Section 5 a discussion and an outlook for future applications is presented.

2 High harmonic emission

When a high power laser pulse is focused to very high intensities onto the surface of a solid target, a significant fraction of the electromagnetic radiation scattered from the interaction is composed of harmonics of the laser frequency which can extend into the soft x-ray region of the spectrum. Indeed the production of harmonics up to the 75th order (of 1.053 μm radiation) has been observed using focused laser intensities up to 10^{19} Wcm^{-2} with a conversion efficiency of 10^{-6} (Norreys et al. 1996).

The harmonic generation process can be understood as the result of an oscillating current of electrons which is dragged across the vacuum-solid interface by the electric field of the laser pulse or, equivalently, by a simple model based on the phase modulation experienced by the light upon reflection from the oscillating plasma-vacuum "boundary" (Lichters et al. 1996). The efficient generation of these harmonics is also of potential interest as a bright source of XUV radiation.

During high intensity laser plasma interaction experiments, measurements of the emission of high order harmonics can also be a powerful diagnostic of the physical processes and plasma conditions in the interaction region (Watts et al. 2002). High order harmonics are useful since they are generated at the same time and from the same location as the highest intensity interaction. There has also been significant progress on the theoretical understanding of the generation mechanism for these harmonics – so it is possible to infer a significant amount of information on the details of the interaction merely by measuring the properties of the harmonic radiation.

2.1   Basic experimental setup

As an example, measurements of high order harmonic emission were performed using the Vulcan laser system at the Rutherford Appleton Laboratory, which uses the technique of chirped pulse amplification to produce 0.65 - 1.0 ps pulses at 1.053 μm – in this case energies up to 70 J onto target. The rectangular laser beam, 20 × 11 cm^2, was focused by an off-axis parabolic mirror ($f/3$). Targets consisted of optically polished fused silica slabs set at 45° angle of incidence with the beam p-polarized. The energy on target and the pulse duration were monitored on a shot-to-shot basis. The intensity on target was up to $\sim 9 \times 10^{19}$ Wcm^{-2}. The laser pre-pulse was determined to be about 10^{-6} of the peak intensity using third order auto-correlation techniques (Danson et al. 1993). The sub-critical plasma produced by this prepulse in front of the target was measured by optical probing and was found to have a density scale-length which was typically less than 10 μm.

2.2   Plasma diagnostics using harmonics

2.2.1   Critical surface dynamics

The harmonic emission in the specularly reflected direction was observed by a slit-less flat-field XUV spectrometer. This consisted of a gold-coated cylindrical grazing-incidence collection mirror followed by a variable line-spaced grating, also at grazing-incidence. The acceptance angle of the mirror and grating was 1×10^{-6} sr. The dispersed radiation was split into two wavelength regions and detected separately so that harmonic emission from the 20th to 120th order could be detected simultaneously. The absolute spectrometer response was calculated using the measured mirror reflectivities, the grating calibration, the filter transmission and the calibrated MCP/CCD detector response.

A typical spectrum showing the 20th (526.5 Å) to 37th (284.6 Å) harmonics – corrected for the spectrometer response and background – is shown in Fig. 1a. The highest harmonic observed in this experiment was the 60th at 175.5 Å. The conversion efficiency varied from 4.6×10^{-6} at the 20th harmonic to $\sim 4.0 \times 10^{-8}$ at the 60th. A particularly noticeable feature of the harmonic spectrum is the amplitude modulation which occurs at every 2 to 4 harmonics over the entire range on both detectors. It should be noted that the modulations peak at different harmonics

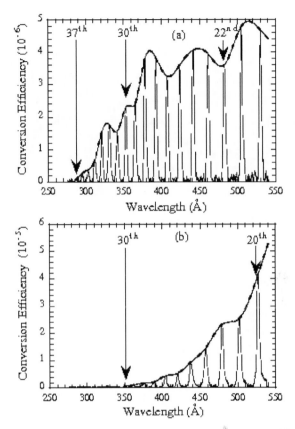

Figure 1. *Lineouts of typical harmonic spectra for (a) high intensity $I\lambda^2 \sim 10^{19}$ W/cm² and (b) low intensity $I\lambda^2 \sim 10^{18}$ W/cm².*

for different shots but could be consistently observed for high intensity shots. Figure 1b shows a typical harmonic spectrum at a lower intensity. This shows harmonics only up to the 30th at 351.0 Å. Comparing Fig. 1a and Fig. 1b it is clear that the modulations in the harmonic spectra are much reduced and that the structure is an intensity dependent effect.

Simulations using a fully relativistic one-dimensional Particle in Cell (PIC) code, LPIC (Richters et al. 1996), exhibit a similar modulation structure and suggest that this structure is intrinsic to the harmonic generation process and not related to internal propagation and reflection of harmonic light.

Figure 2 shows a simulated harmonic spectra of the reflected light for high and low $I\lambda^2$ similar to the experimental values in Fig. 1. At the lower value of $I\lambda^2$ (Fig. 2b) the conversion into harmonics is low with a smooth unmodulated roll-off. The noise in the spectrum shown is numerical and can be reduced when using a higher temporal resolution. The highest harmonic above noise is the 6th harmonic which compares to the 30th harmonic observed experimentally. At the higher value of $I\lambda^2$ (Fig. 2a) the harmonic emission is enhanced and the spectrum is modulated at every 4 to 5 harmonics. This is to be compared to a modulation of 2 to 4 harmonics observed experimentally. The highest harmonic above numerical noise in this case is the 55th.

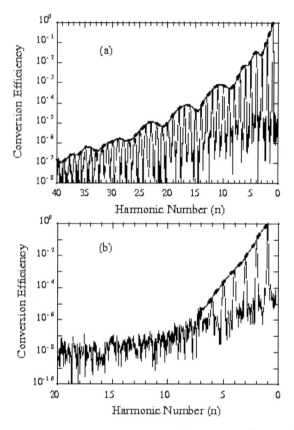

Figure 2. *The simulated harmonic spectra for (a) $I\lambda^2 = 1.23 \times 10^{19}$ W/cm^2 and (b) $I\lambda^2 = 1.37 \times 10^{18}$ W/cm^2. The conversion efficiency is given by $2|E(\omega)/E_L|^2$.*

The importance of such effects can be seen by taking the Fourier transform of the calculated harmonic spectrum and comparing this to the motion of the critical surface from the PIC simulations. The amplitude of the critical surface motion and the degree of non-linearity of the original oscillation is preserved.

Consequently, this demonstrates that it should be possible to use structures in the observed harmonic spectra to determine the highly non-linear dynamics of the critical surface in these interactions. However, it should be noted that for the experiments discussed, this could not be done using the observed spectra in Fig. 1 since only part of the spectrum was measured in this work (i.e., that in the soft X-ray/XUV region).

2.3   Magnetic field measurements

The generation of magnetic fields during high intensity laser solid target interactions is a subject which has been under investigation for many years and such fields have been predicted to be extraordinarily high (up to 1 GGauss) (Sudan 1993). In laser-plasma interactions, such spontaneous magnetic fields can be generated by non-parallel electron temperature and density gradients (Stamper et al. 1971), by the radiation pressure associated with the laser pulse itself (Sudan 1993), as well as by the current of fast electrons generated during the interaction (Mason et al. 1998).

Magnetic field measurements in laser produced plasmas have generally used external probe laser beams (at visible or near ultra-violet wavelengths) which transversely probe the laser-produced plasma using the Faraday effect. The main advantage of the use of an external probe beam is that the plasma is not disturbed by the measurement since the frequency of the probe laser is typically much greater than the characteristic plasma frequency. However, in practice refraction of the probe limits observations to regions of the plasma having an electron density less than about 10^{20} cm^{-3}. The magnitude of the magnetic field is of the order of 1-5 MG at these low densities (Borghesi et al. 1998) which is considerably less than the predicted magnetic fields generated at or near the critical surface.

However, very high field measurements are possible using polarimetry of the self-generated harmonics of the laser frequency and also using the fact that the plasma in the presence of a magnetic field is birefringent and/or optically active depending on the propagation direction of the electromagnetic wave (Tatarakis et al. 2002). The former produces the Cotton-Mouton effect (i.e., an induced ellipticity) and the latter produces the Faraday effect (i.e., rotation of the polarization vector). An important difference between these two effects at lower field strengths is that the Cotton Mouton effect scales like λ^3 while the Faraday effect scales like λ^2. Consequently, comparison of the polarization states of different harmonics can determine the relative importance of these effects. Therefore, measurements of the polarization parameters (i.e., the Stokes vectors) of the scattered harmonics can precisely determine the transition matrix which connects the initial and the final Stokes vectors of the e-m wave (harmonics) and hence also determine the magnetic field distribution. The formalism of the Stokes vector method (polarimetry) can be found in the review article by Segre (1999).

An important effect which simplifies the measurement of very large magnetic fields using polarimetry is the cutoff of the X-wave (i.e., the wave with electric field vector perpendicular to the B-field). When an electromagnetic wave propagates in a plasma with its k vector perpendicular to B (Cotton-Mouton effect), the extraordinary wave (X-wave), i.e., with an electric field vector perpendicular to the magnetic field, can experience cutoffs and resonances. Cutoffs occur when the index of refraction is equal to zero and resonances when the index approaches infinity. The X-wave is reflected when it encounters a cutoff and is absorbed in a resonance (at the upper hybrid frequency). The cutoffs of the various harmonics can exist only because of the presence of a magnetic field in the plasma (i.e., there is no other physical mechanism which can produce an X-wave cutoff). Consequently, they provide an unambiguous method for measuring the magnetic field strength in the overdense plasma. The ordinary wave (O-wave, with its electric field parallel to the external magnetic field) is unaffected by these cutoffs and resonances. Thus, observation of a harmonic cutoff in the x-polarized component means that the magnetic field in the plasma is greater than the cutoff magnetic field value. Note that the Stokes vector method and observations of the X-wave cutoffs are two separate methods for using polarimetry of the scattered harmonics to measure magnetic fields.

In the experiments, lower order harmonics (i.e., the 2nd, 3rd, 4th and 5th harmonics) generated at the critical surface as well as the 3/2 harmonic generated at the quarter-critical density surface were used as diagnostic tools to measure the self generated magnetic fields. Since these harmonics propagate out of the high density region isotropically and are generated at the critical surface with the same polarization as the incident laser beam, the study of the polarization properties can be used to infer the magnitude of magnetic fields through which they travel. Moreover, use of self-generated laser harmonics is convenient since they are produced at precisely the same time as the large magnetic fields are generated and propagate so that their k-vectors are perpendicular to the azimuthal magnetic fields in the plasma – which greatly simplifies data interpretation. All of the previous theoretical work imply that the largest magnetic fields are in the azimuthal direction about the laser axis of propagation – which is also what is observed in experiments.

In the Vulcan experiment at RAL the X-wave cutoffs and the Stokes parameters of the selected harmonics were measured using three polarimeters which use high dynamic range CCD arrays as detectors. Interference filters were placed at the entrance of the polarimeters in order to select the desired harmonic frequency. Each polarimeter had four channels consisting of one reference channel, two channels with polarizers set at 0° and 45° respectively and a channel with a polarizer at 45° followed by a $\lambda/4$ waveplate. The results of these measurements can completely determine the polarization state of the radiation emitted from the interaction. Calibration shots, i.e., shots without polarizers, were used to measure the efficiency of each channel.

2.3.1   Magnetic field from the X-wave cutoffs

When an electromagnetic wave propagates in a plasma with its k vector perpendicular to B, the extraordinary wave (X-wave), i.e., with an electric field vector perpendicular to the magnetic field, can experience cutoffs and resonances as shown in Figure 3. The X-wave is reflected when it encounters a cutoff and is absorbed in a resonance (at the upper hybrid frequency). The ordinary wave (with E parallel to B) is unaffected by these cutoffs and resonances. This means that if a field larger than the cutoff field exists in the plasma and if this field extends over a region larger than the location where that particular harmonic is generated – only the ordinary wave will be able to propagate to the detector and consequently only the s-polarized or O-wave can be observed. This is precisely what was measured in this experiment for the very highest intensity shots. Figure 4 presents the ratio of p-component (x-wave) to total emission for both the 3rd and 4th harmonic for a series of laser shots. At high intensities there are definite indications of X-wave cutoffs, implying the existence of a minimum magnetic field of 340 MegaGauss in the plasma. No cutoffs were observed for the 5th harmonic. This indicates that the peak magnetic field is below 450 MegaGauss and above 340 MegaGauss at intensities of 8×10^{19} W/cm^2.

2.3.2   Magnetic field measurements from Stokes vectors

From the X-wave cutoffs we can determine the magnetic field strength within a certain range. The lower value of the window is defined by the cutoff values of the highest order harmonic in which a cutoff is experimentally observed while the upper value is defined by the cutoff value of the lowest order harmonic which is not observed. On the other hand, the detailed measurement of the polarization state of the harmonics (Stokes parameters) can determine the transition matrix which connects the initial Stokes vector of the e-m wave (before the plasma) and the final Stokes vector (after the plasma) by which one can also estimate the magnetic field. However, for a given harmonic order this can only be done successfully for much lower magnetic fields than those required to observe a cutoff (Tatarakis et al. 2002; Wagner et al. 2004).

In general, the state of polarization of an electromagnetic wave can be described by the parameters of the Stokes vector **S** namely,

$$S = \begin{pmatrix} S_0 \\ S_1 \\ S_2 \\ S_3 \end{pmatrix} = I \begin{pmatrix} 1 \\ \cos(2\chi)\cos(2\psi) \\ \cos(2\chi)\sin(2\psi) \\ \sin(2\psi) \end{pmatrix}, \text{with } S_1^2 + S_2^2 + S_3^2 = S_0^2 = I^2 \tag{1}$$

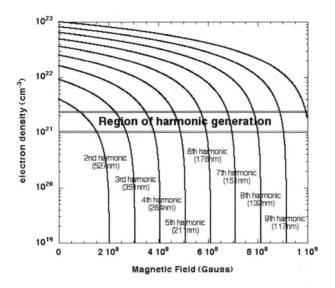

Figure 3. *Plot of cut-offs for various harmonics of 1.054 μm radiation (in terms of plasma electron density and magnetic field). Harmonics produced at high electron density will only exhibit a "cut-off" if a very large magnetic field exists in the plasma. This plot indicates that this technique may be suitable for measurements of field strengths up to 1 GigaGauss.*

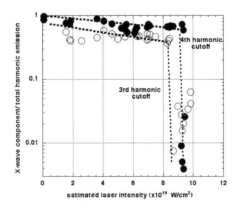

Figure 4. *X-wave/total harmonic emission of 3rd (open circles) and 4th harmonic (closed circles) for a series of laser shots. X-wave cutoffs for both harmonics are observable at intensities of greater than 8×10^{19} W/cm². No cutoffs were observed for the 5th harmonic and dotted lines are drawn as visual aids only.*

where I is the intensity of the radiation. For a Cartesian system with the z axis in the direction of propagation, the angle ψ defines the rotation of the ellipse, while the angle χ defines the degree of ellipticity of the wave ($\tan(\chi) = \pm b/a$, where a and b are the major and minor semi-axis of the ellipse, and $0 \leq \psi \leq \pi$ and $-\pi/4 \leq \chi \leq \pi/4$). As in any birefringent material, the change of the polarization state of the wave traveling through the plasma is described by

$\mathbf{S}_{out}(z) = \mathbf{M} \cdot \mathbf{S}_{in}(0)$, where $\mathbf{M}$ is the transition matrix which couples the initial (before the plasma) and the final Stokes vectors (after the plasma) of the wave. To calculate $\mathbf{M}$ one must solve the evolution equation, $\frac{d\mathbf{S}(z)}{dz} = \mathbf{\Omega} \times \mathbf{S}(z)$, where $\Omega = \frac{\omega}{c}(\eta_1 - \eta_2)$, and η_1, η_2 are the indices of refraction of the slow and fast characteristic waves as given by the Appleton-Hartree formula (Segre 1999). For the case where the $\mathbf{k}$ vector of the radiation is perpendicular to the magnetic field (i.e., $B_z = 0$), the η_1 and η_2 are the indices of refraction for the ordinary wave (O-wave) and the extra-ordinary wave (X-wave) respectively, and due to the different refractive indices, the wave becomes elliptically polarized when it propagates with its $\mathbf{k}$-vector perpendicular to the magnetic field.

The initial polarization of the harmonics was linear and was the same as that of the main interaction beam (p-polarized). This was verified experimentally from very low energy shots (less than 1 J). It is important to note that throughout these experiments there was also no observed change of the state of the polarization of the 3/2 harmonic (i.e., that the 3/2 harmonic was p-polarized for an incident p-polarized laser beam). This implies that the magnetic field which affects the polarization of the scattered harmonics is likely localized between the critical density surface and the quarter critical density surfaces where the 3/2 emission is produced. The measured magnetic field is therefore that produced in the very highest density regions. Indeed simulations suggest that the field is highly localized around the critical density surface.

Consequently, from these measurements we can estimate the magnitude of the magnetic field for a given plasma density scalelength. This was calculated from the fact that the field is assumed to decrease exponentially with a scalelength of 1 μm which is predicted by computational simulations of the interaction. At low intensities ($I \sim 10^{17}$ W/cm^2) measurements of the 5th, 4th, and 3rd harmonics of the laser give similar results ($B_{max} \sim 80$ MegaGauss) while as the intensity increases, the results diverge. Fields in excess of 300 MG are suggested by the measurements of the 5th harmonic.

Analysis was also performed to determine the relative importance of magnetic field components parallel to the propagation direction of the harmonics (i.e., the Faraday effect) and these measurements indicated that that this component is negligible – confirming that the magnetic field produced during the interaction is azimuthal "around" the laser focal spot.

2.3.3   XUV Measurements

To confirm these polarimetry measurements using the low-order harmonics a multichannel polarimeter for the higher order VUV harmonics was also constructed. This enables measurements at higher plasma densities and also for much larger magnetic fields and incident laser intensities. To perform magnetic-field measurements with harmonics of higher order than 5, a VUV-polarimeter is needed, as the wavelength of these harmonics is below 200 nm for Nd:Glass laser system such as Vulcan. The VUV-polarizers use a triple-mirror-configuration, which are partially polarizing due to the Fresnel-refection on the gold-coated mirror surfaces. Two VUV-polarisers were set-up orthogonally to measure the p- and the s-component of the harmonics. The polarized beams are then focused onto the slit of an Acton spectrometer, enabling the simultaneous measurement of many higher order harmonics. An open microchannel-plate (MCP) coupled to a charge-coupled-device (CCD) by a fiber-optic-bundle was used as the detector.

No cutoffs of the X-wave were observed for the emission of these harmonics in these experiments but the depolarization of the VUV harmonics was found to follow the λ^3 scaling as suggested by the Cotton-Mouton effect and in agreement with the measurements of lower order harmonics and suggest the existence of fields of 07 ± 0.1 GigaGauss.

3 Electron acceleration using high intensity lasers

A particularly exciting application of laser produced plasmas is the potential development of high field, "compact" electron accelerators (Mangles et al. 2004; Faure et al. 2004; Tajima and Dawson 1979). In recent experiments the use of both Petawatt class lasers and high repetition rate ultra-short pulse systems have been evaluated for relativistic electron acceleration applications. The acceleration mechanism for most previous experiments has been the production of relativistic plasma waves, which can trap and accelerate electrons. It has been found that for interactions in the "picosecond" pulse-duration regime the electron acceleration mechanism changes from relativistic plasma wave acceleration to direct laser acceleration as the intensity is increased. The limiting electron energy for plasma wave acceleration is due to the well known dephasing limit – however for direct acceleration mechanisms the ultimate limit is not clear. Although the relativistic electron bunches generated from these experiments are highly directional and contain high charge, they are emitted with an extremely large energy spread which makes many of the potentially important applications for electron beams unfeasible with these sources.

However, recently it has been shown that only in high power experiments using much shorter pulses (in the ten's of femtoseconds regime) is it possible to generate true "beams" of relativistic electrons which have low divergence and which have a relatively small energy spread ($< 5\%$) (Mangles et al. 2004; Faure et al. 2004).

This is an extremely important result since only if narrow energy bandwidths are achievable will the full range of applications become possible. The use of plasma acceleration consequently now offers the potential of significantly smaller and cheaper facilities for generating energetic electron beams, which, considered along with the current rapid developments in laser technology, could soon allow the construction of university laboratory sized accelerators for use in a wide range of experiments and applications. For example, table-top narrowband femto-second x-ray sources and free-electron lasers could become a reality – which may potentially lead to significant advances in both medicine and materials science. It may also be possible to use electron bunches generated in this way for injection into conventional rf accelerators or into subsequent plasma acceleration stages.

3.1 Self-modulated laser wakefield acceleration

High field electron acceleration techniques typically rely on electron plasma waves as the accelerating medium. In the laser wakefield accelerator concept (Tajima and Dawson 1979), these waves are excited immediately behind the laser pulse as it propagates through the plasma. Efficient energy transfer between the plasma wave and the electrons requires that both move at similar velocities and this is achieved through the use of low density plasmas (less than 10^{20} cm^{-3}), in which the phase velocity of the laser-excited plasma wave is equal to the laser pulse group velocity (which is close to the speed of light in vacuum). The longitudinal electric fields associated with the relativistic plasma waves are then able to accelerate relativistic particles injected externally, or even, for large amplitude waves, to trap particles from the plasma itself. Subsequently, particles can be boosted to high energy over very short distances by "surfing" on this electrostatic wave.

Acceleration schemes using lasers use the ponderomotive force of either a single very short pulse or a train of very short light pulses, each tailored to resonantly drive the relativistic plasma wave (resonance occurs when the laser pulse duration is about half of the electron plasma period, T_p).

Thus far, the schemes studied for producing wakefields are the laser wake field accelerator (LWFA) (Tajima and Dawson 1979), the laser beat wave accelerator (LBWA) (Everett et al.

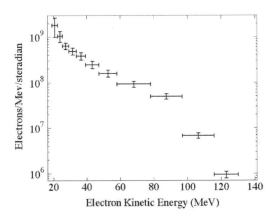

Figure 5. *Typical electron spectrum from self-modulated laser wake field from a 50 TW laser interaction with a helium gas jet target* $(n_e = 3 \times 10^{19}\ cm^{-3})$.

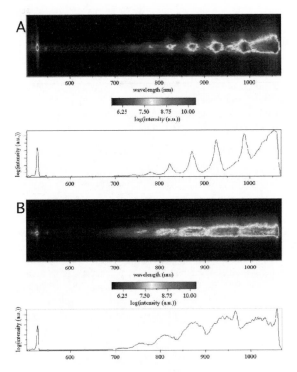

Figure 6. *Transmitted beam spectra with a)* $n_e = 4 \times 10^{18}\ cm^{-3}$ *and* $I = 2 \times 10^{19}\ Wcm^{-2}$ *and b)* $n_e = 7.5 \times 10^{18}\ cm^{-3}$ *and* $I = 1 \times 10^{19}\ Wcm^{-2}$. *The signal at 527 nm is due to the second harmonic. Stokes satellites (down shifted satellites) are not shown because the CCD detector used is not sensitive at those wavelengths.*

1994, Walton et al. 2002), the self modulated laser wake field accelerator (SM-LWFA) (Modena et al. 1995, Coverdale et al. 1995, Santala et al. 2001), and the forced laser wake field (F-LWFA) (Malka et al. 2002) accelerator. In the LWFA the laser pulse pushes electrons at its leading and trailing edges with optimal coupling when the resonance condition mentioned above is satisfied. Electrons injected at 3 MeV have been accelerated by this scheme up to 4.7 MeV (Aminaroff et al. 1998). In the LBWA, a train of short pulses (a beat frequency) is produced by co-propagating two laser pulses at slightly different wavelengths. Electric fields close to the GV/m level have been measured for this experimental configuration using CO_2 laser beams and injected electrons have been accelerated up to 30 MeV (Everett et al. 1994). In LBWA, only electrons injected externally into the wave have been accelerated. In contrast, in the SM-LWFA and F-LWFA and resonant LWFA schemes, the plasma wave amplitude becomes so large that electrons from the plasma itself are trapped by the wave and are boosted to very high energies (i.e., the plasma wave breaks so that no external electron source is needed).

The SM-LWFA regime uses long $(\tau_L >> T_p)$ laser pulses at intensities sufficient to strongly excite the self-modulation instability which is related to stimulated forward Raman scattering (Amiranoff et al. 1998). In this process, the laser pulse self-focuses and scatters upon encountering electron plasma waves. The beating of the scattered electromagnetic wave with the laser light amplifies the plasma waves, leading to instability and plasma wave growth. As a result, the initial laser pulse is strongly modulated and is gradually transformed into a train of very short pulses that naturally satisfy the resonance condition. The plasma can also act as a converging lens due to relativistic effects and can focus the laser beam, permitting interactions at higher intensity than in vacuum, and over a longer distance than the natural diffraction length.

An example of an electron spectrum from a SM-LWFA experiment is shown in Figure 5. This was produced from a laser plasma interaction experiment using the Vulcan laser system operating at 50 TW. This interaction was performed at a plasma density of $\sim 3 \times 10^{19}$ cm^{-3}, a laser intensity of $\sim 10^{19}$ Wcm^{-2} and a pulse length of about 1 psec. In this regime the peak electron energy observed was about 120 MeV — although because the acceleration mechanism is caused by an instability, experiments in this regime are characterized by considerable shot-to-shot fluctuations in the peak electron energy and in the total accelerated charge (Santala et al. 2001). Typical forward scattered laser spectra are shown in Figure 6. These show the characteristic scattered anti-Stokes sidebands (separated by the electron plasma frequency) which can be correlated to the production of the relativistic electron beam. The behavior of the sidebands on the Stokes (downshifted) side of the laser frequency is similar. As the plasma density increases the number of accelerated electrons can also dramatically increase – due to "wave-breaking" of the waves and self-trapping of electrons in the wake. This is evident in the forward scattered spectrum as a distinct broadening of the sidebands due to the transition from scattering from a large amplitude "single frequency" plasma wave to a scattering from a non-linear "broken" wave.

The divergence of the high energy electrons from interactions in the SM-LWFA typically increases with density and can reach values up to 15 degrees for electrons greater than 8 MeV.

3.2 Laser acceleration of electrons at intensities greater than 10^{20} W/cm^2

The recent development of Petawatt class lasers such as that at the Central Laser Facility at the Rutherford Appleton Laboratory in the UK has allowed experiments to be performed at much higher intensities than previously available. The intensity of a laser system is often described by the normalised vector potential of the laser field, $a_0 = eA/mc^2$ (A is the vector potential of the laser field and m is the electron mass). a_0 is also the normalised transverse momentum of the electron motion in the laser field. As a_0 approaches 1 the electron motion becomes relativistic. The previous experiments as described up until now were performed with a_0 between 1 and 5.

The Vulcan Petawatt facility allows experimental access to regimes where $a_0 >> 1$ (and in the experiment reported here $a_0 \approx 15$).

During this experiment the laser consistently produced 650 fs duration pulses delivering ~ 180 J on target. These pulses were focused with an $f/3$ off-axis parabolic mirror to produce a focal spot with an intensity FWHM (full width half maximum) of ~ 10 μm, thus generating intensities greater than 3×10^{20} Wcm^{-2}.

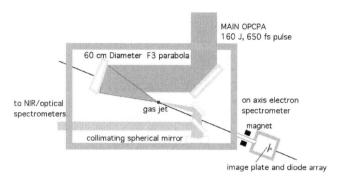

Figure 7. *Schematic of experimental setup for Petawatt electron acceleration experiments.*

The experimental setup is shown in Figure 7. The gas jet used has a 2 mm diameter supersonic nozzle and could produce plasma electron densities between 5×10^{18} and 2×10^{20} cm^{-3}. The density was controlled by varying the backing pressure of helium behind the valve of the gas jet and can be measured using the forward Raman scattering signal.

The electrons accelerated along the axis of laser propagation were measured using a high field magnetic spectrometer. The electrons exit the highly shielded main vacuum chamber through a small (25 mm) diameter tube to a secondary vacuum vessel – and this helps to reduce the level of background signal from low energy x-rays and scattered electrons. The entrance to the spectrometer is a 5 mm diameter hole that serves to collimate the electron beam to ensure sufficient energy resolution. The specially designed vacuum chamber allows the electron beam to pass between the pole pieces of an electromagnet that deflects the electrons off-axis. The correspondence between electron energy and deflection from the axis is determined by using a charged particle tracking code.

In these experiments, electrons are detected using an image plate, (Fuji BAS1800II) which is a re-usable film sensitive to ionizing radiation. Two sections of image plate are used, one to measure the electron signal, below axis, and another to measure the background signal above the axis. Since the background is due to x-rays from the interaction itself or from bremsstrahlung radiation emitted by the electrons as they pass through material before the detector plane it is assumed that the background is symmetric above and below the axis.

The relationship between image plate signal intensity and energy deposited in the plate is close to linear. The direct relationship between the number of electrons and the signal was calculated by placing a diode array directly behind the image plates. The ion implanted diodes used have an absolutely calibrated response to the number of electrons incident and by calculating the energy lost by the electrons as they travel through the plate before reaching the diodes it is possible to cross-calibrate the diode and image plate signals.

The image plate data exhibits a much larger dynamic range than the diodes and the image nature of the data allows much better noise discrimination. The resolution of the image plates, although not as high as x-ray film is significantly better than the diode array. The combination of the resolution and size of the image plates (each is 250 mm long) allows a reasonably broad energy range (for example 10 – 250 MeV) to be measured in a single shot.

The laser-plasma interaction was also diagnosed by measuring the transmitted spectrum of the laser. A portion of the transmitted beam was collimated and transported out of the vacuum chamber to a pair of near-infrared spectrometers; CCD cameras recorded the spectra on each shot at two different dispersions.

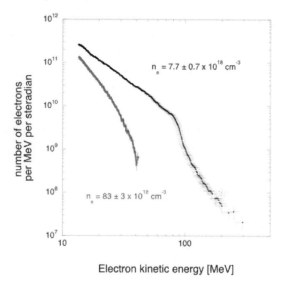

Figure 8. *Electron energy spectra at two densities from Petawatt interactions. The highest energy electrons are observed at an electron density of $n = (7.7 \pm 0.7) \times 10^{18} cm^{-3}$. At this density the spectrum is non-Maxwellian. For densities higher or lower than this the spectra are broadly similar and can be better described by an effective temperature.*

Figure 8 shows electron energy spectra obtained from shots with helium gas. These shots have been selected to show the trend observed as the density of the gas jet was varied. Electrons were accelerated to relativistic energies at all densities. At low densities ($n_e < 10^{19}$ cm^{-3}) the electron energy spectrum could be characterized by an effective temperature, that is the number of electrons N, with energy E was given by $N(E) \propto \exp(E/T_{eff})$. As the density was increased the maximum energy observed increased, along with T_{eff}. The spectrum also begins to take on a non-Maxwellian form. At $n_e = (0.7 \pm 0.1) \times 10^{19}$ cm^{-3} the acceleration is significantly enhanced, and the energy observed was up to about 300 MeV. The maximum is, in this case, limited by the noise level on the image plate. As the density was increased above 10^{19} cm^{-3} the acceleration was observed to be less effective and the spectrum regains its effective temperature form.

Figure 9 shows explicitly how the acceleration varied with electron density for both helium and deuterium shots. The characteristic electron energy in this case is given as that when the ratio of signal to noise the image plate is 10. This energy varies strongly with density exhibiting an apparent peak around 1×10^{19} cm^{-3}. The shots taken with deuterium gas produced similar spectra, and show the same variation with density, although the optimum density is shifted slightly to $(1.4 \pm 0.2) \times 10^{19}$ cm^{-3}.

Figure 9 also shows the relationship between the amount of forward Raman scattering (FRS)

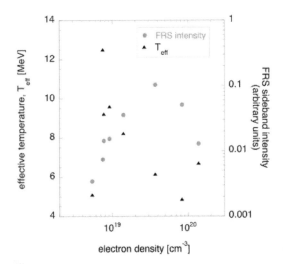

Figure 9. *Variation of observed electron "temperature" and amount of forward Raman scattered (FRS) light with plasma density in helium gas jet. Note the similar behavior in both data sets and the lack of correlation of FRS and acceleration.*

observed and the electron density. It is clear that the production of high energy electrons and FRS are uncorrelated. This is in marked difference to previous experiments (Modena et al. 1995; Coverdale et al. 1995; Santala et al. 2001) and indicates that the acceleration mechanism cannot be self-modulated laser wakefield acceleration (SM-LWFA) as in the experiments described in section 2.

A study by Gahn (Gahn et al. 1999) with 200 fs, 0.25 J pulses showed an enhanced acceleration at 2×10^{20} cm^{-3}. In this experiment there is strong evidence that electron acceleration is strongly correlated with channel formation and magnetic field generation. Simulations indicate that when the betatron motion of the electron in the self-generated magnetic field is resonant with the relativistic motion of the particle in the laser field electron can pick up energy directly from the laser pulse ("direct laser acceleration" (DLA)). This is similar to an inverse free-electron laser mechanism and can occur efficiently when the laser power becomes significantly greater than the critical power for self-focusing.

However the divergence of the beam suggested from these previous lower intensity experiments and simulations is large. With the much smaller divergence observed in the Vulcan Petawatt experiments the mechanism of direct laser acceleration seems to be more complex and may involve stochastic "de-phasing" processes (Mangles et al. 2005; Meyer-ter-Vehn and Sheng, 1999) and the combined effect of the laser electric field with the electrostatic field due to charge displacement during the intense laser plasma interaction.

A number of 2D-3V Particle in Cell (PIC) code simulations using the code OSIRIS (Hemker 2000) have been performed showing that a form of direct acceleration by the laser is the dominant acceleration mechanism. The PIC code simulations show the bunching of the accelerated electrons at twice the laser frequency and also that the electrons are accelerated within the laser pulse, the maximum energy coinciding with the maximum laser intensity.

The PIC code simulations have also indicated a possible cause of the observed density–

acceleration dependence. Simulations were performed at various densities simulating realistic laser pulses incident on a fully ionized plasma having a linear density ramp from vacuum up to bulk plasma density over ~ 600 μm (this is consistent with the type of nozzle used in these experiments). Each simulation was run for passage throughout the gas jet target. In high density runs $(1.4 \times 10^{20}$ cm$^{-3})$ the laser undergoes strong self focussing and filamentation. The filamented laser then undergoes a hosing type instability. This filamentation and hosing does not occur in the low density case.

If this hosing instability occurred at high densities in the experiment it may be responsible for reducing the effective intensity during the interaction as well as perhaps moving the electron beam off axis, such that the highest energy electron did not enter the electron spectrometer (in simulations the electron beam closely follows the laser pulse).

3.3 Mono-energetic beam production using ultra-short pulses

Recently electron acceleration experiments using the high power ultra-short pulse Titanium:Sapphire laser system at the Central Laser Facility of the Rutherford Appleton Laboratory (Astra) have been performed (Mangles et al. 2004). The laser pulses ($\lambda = 800$ nm, $E = 450$ mJ, $\tau = 40$ fsec) were focused with an $f/16.7$ off axis parabolic mirror onto the edge of a 2 mm long supersonic jet of helium gas to produce intensities on the order of 1.3×10^{18} Wcm^{-2}. The electron density (n_e) as a function of backing pressure on the gas-jet was determined by measuring the frequency shift ($\Delta\omega = \omega_{pe}$ where ω_{pe} is the electron plasma frequency) of satellite wavelengths generated by forward Raman scattering in the transmitted light. The plasma density was again observed to vary linearly with backing pressure within the range $n_e = 3 \times 10^{18}$ cm^{-3} – 5×10^{19} cm^{-3}. In this density range the wavelength of relativistic plasma waves produced (i.e., $\lambda_p = 2\pi\omega_{pe}/c$) can be less than the laser pulse length $c\tau_L$. For laser pulses which are less than the plasma wavelength, relativistic plasma waves can be generated "resonantly" in the wake of the pulse – while in the regime in which the laser pulse length is longer than the plasma wavelength, high intensity interactions are required to drive an instability in which the plasma waves are produced via "self-modulation" of the laser pulse envelope[10] at the plasma frequency. In the work described here it is likely that the plasma waves are driven so that they grow until "wave-breaking" occurs. This is a phenomenon which takes place at very large amplitudes such that the wave motion becomes so non-linear that wave energy is transferred directly into particle energy and the plasma wave loses coherence. Electrons which reach relativistic energies from wavebreaking of a plasma wave can be "injected" into an adjacent plasma wave where they can pick up even more energy (the "cold" wavebreaking electric-field amplitude for electron plasma waves is given by, $E = m_e c \omega_{pe}/e$). Note that in a regime between SM-LWFA and LWFA lies the F-LWFA (*forced* laser wakefield accelerator) where a short pulse is used which is only slightly longer than the resonant pulse length. This pulse is consequently focussed and compressed by the wave to produce large amplitude plasma waves which can break.

Note that in the experiment described here the electron energy spectrum was measured using an on-axis magnetic spectrometer similar to that in the Petawatt experiments. The electrons were also simultaneously measured with the high resolution image plate detectors well as using a much lower resolution array of diodes. The spectrometer magnet, image plates and diodes were set up to measure the spectrum over a wide energy range in a single shot.

Other diagnostics used in this case included the simultaneous measurement of the transmitted laser spectrum and transverse optical probing of the interaction with a frequency doubled laser probe beam. This was used to produce images of the plasma via shadowgraphy, and was independently timed so it could also be used to measure pre-pulse effects and plasma channel formation.

Electron acceleration was observed over a range of electron densities. With the plasma density below 7.3×10^{18} cm^{-3} no energetic electrons were observed (this corresponds to $c\tau_L < 2.5\lambda_p$). In this regime, the growth rate of the self-modulation instability is too low for a plasma wave to reach wavebreaking amplitudes. As the density was increased above 7×10^{18} cm^{-3}, very high energy electrons were suddenly produced with the most energetic electrons reaching up to 100 MeV. The output beam divergence was also measured and was found to be less than 5 degrees. However the most interesting aspect of these spectra is that, in this regime, the electron energies were exceptionally non-Maxwellian and, indeed, generally consisted of one or more narrow spiky features – each of which could have an energy bandwidth of less than 20%. This is in contrast to the energy spectra of previous laser acceleration experiments in which 100% energy spreads are observed. As the density was increased in our experiments, the peak energy of the observed electrons was observed to decrease and the spectra begin to assume a broad Maxwellian shape which was characteristic of previous experiments in the SM-LWFA regime.

The likely explanation for the difference observed in these spectra is due to the timing of the "injection" of electrons into the relativistic plasma wave. It appears that as the plasma wave reaches an amplitude which is just sufficient for wavebreaking only a few electrons are able to "fall" into the accelerating portion of the adjacent waves in the wake, and so these electrons all see an almost identical acceleration gradient. Since successive waves in the wakefield are of different amplitude (i.e., they have differing accelerating gradients) successive trapped bunches will be accelerated to different energies.

In addition, at low densities these electron bunches are not "dephased" since the propagation distance is only about 1 mm (or the length of the gas jet plasma). The dephasing distance is the length over which an electron outruns the plasma wave – and begins to be de-accelerated by the wave, and is given as $L_d = 2\pi x \omega^2 / \omega_{pe}^3$. For these experiments the conditions which showed the clearest, most reproducible evidence of these electron beams were those in which the de-phasing length, the gas jet length and the confocal parameter of the laser beam are all roughly 1 mm.

In contrast, at higher densities the de-phasing distance is much shorter than the interaction distance and so a "randomised" or quasi-Maxwellian distribution of electrons would emerge from the plasma.

In the final set of experiments the energy of the laser pulse was increased to about 500 mJ. These experiments showed that for the densities used to obtain "bumpy" spectra in the lower energy situations that very "mono-energetic" spectra could be observed. This is shown in Figure 10 in which 2 electron spectra are shown – from the same series of shots. The spectra are reasonably reproducible and the narrowest spectrum shows a beam at 52 MeV with an energy spread ($\Delta E / E$) of less than 5%.

This phenomenon is indeed what is observed when 2D particle-in-cell simulations of the interaction were performed using the code OSIRIS and it was found that for relatively low plasma densities as the plasma waves grow to an amplitude such that it they are observed to break – a group of electrons is injected into a particular phase position into the plasma wave and this group of electrons can be accelerated relatively uniformly. When the pulse is fully self-focused, some relativistic electrons have appeared, but at quite low energies. This is where wave-breaking occurs. As the laser pulse front begins to steepen, the wakefield amplitude grows and the electron energies increase until the pulse reaches its maximum peak intensity. At this point the electron energies are clearly "bunched" at a particular energy. After this time the average electron energies begin to drop and the distribution of electron energies is randomised – since the propagation distance is beyond the dephasing length for this interaction.

The energies are also diminished by the "erosion" of the pulse leading edge, coupled with the non-linear lengthening of the plasma wavelength in the wake which slows the electrons. This process continues so that the bunch again enters a region of accelerating field and the energies

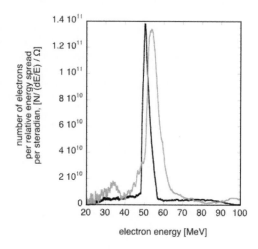

Figure 10. *Two measured electron spectrum with $E = 500$ mJ laser at a density of 2×10^{19} cm^{-3}. Shots are taken from the same shot series.*

increase once more. When the wake reaches its highest amplitude the trapped electrons are completely dephased with respect to the plasma wave.

It is clear from simulations that in our experiments the "bunches" of electrons are produced due to "wavebreaking" in the immediate vicinity of the laser pulse. These electrons are then accelerated through the entire length of the plasma – which is shorter than the dephasing distance. Consequently, the bunch of electrons can remain relatively mono-energetic after leaving the plasma. The requirements for this regime are that the plasma density has to be high enough so that wave-breaking is easily achieved for an interaction at a given density – but low enough so that the electron bunches produced are not de-phased before they leave the plasma.

To summarise, it has been demonstrated that ultra short pulse lasers can be used to produce relativistic bunches of electrons with energies up to 100 MeV. Accelerated electrons are not observed below a minimum density – but at densities slightly higher than this "mono-energetic" electron beams can be clearly observed in the spectrum. As the plasma density is increased still further such structures are randomised by the dephasing of the accelerated electrons with the plasma waves and the energy spread of 100% is observed.

The observation that laser produced plasmas alone can produce mono-energetic electron beams suggests that such sources hold great promise for future development of table top particle accelerators and that a wide range of applications may soon become possible.

4 Ion acceleration using short laser pulses

In this section, recent measurements of energetic ion emission from intense laser interactions with plasmas will be discussed (from both overdense and underdense interactions).

4.1   Observations of ions from the "front" of laser solid interactions

Some of the first plasma physics experiments to be performed using high powered lasers were energetic ion measurements (Gitomer et al. 1984). Such experiments have always been important for the elucidation of the highly transient, non-linear phenomena which occur when an intense laser pulse interacts with a plasma. The beginning of research into inertial confinement fusion (ICF), together with development of multi-kilojoule lasers initially brought about considerable interest in these experiments. Indeed studies of ion production from planar and spherically irradiated targets are important for diagnosing the coupling of laser radiation into hot dense plasmas as well as the efficiency of plasma compression of by lasers for ICF.

Results from experiments (Begay et al. 1982; Tan et al. 1984; Ehler et al. 1975) in the early 1980's using long wavelength CO_2 lasers indicated that ions with energies greater than 1 MeV could be produced. However, such work was undertaken using long pulse lasers (durations of ~ 1 ns) unlike the sub-picosecond duration laser pulses available today. Both theoretical and computational studies at the time (Gitomer et al. 1984) were able to model the plasma expansion from the target surface and the subsequent acceleration of ions in the plasma. The self-similar solution for the isothermal expansion of the laser produced plasma indicated the presence of a space-charge electric field which is able to accelerate the ions as the electrons expand into vacuum (sheath acceleration) (Wickens et al. 1978). This mechanism produces a maximum ion energy – i.e., a high-energy cut-off – which was similar to the experimentally observed ion energy distribution. Indeed, the maximum ion energy has also been shown experimentally to be related to the hot electron temperature agreeing with the sheath acceleration mechanism.

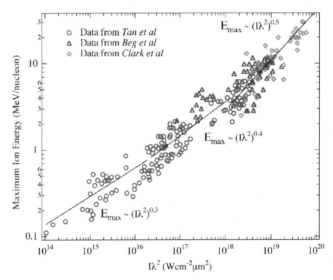

Figure 11. *Maximum ion energy against irradiance on target. In addition, heavy ions were measured using a Thomson parabola and CR-39 detector.*

Accumulated data from a number of published experiments over the past thirty years is shown in Fig. 11. Of interest is the maximum proton/ion energy for interactions above 10^{18} W/cm^2 μm^2 which is found to scale as $(I\lambda^2)^{0.5}$ i.e., when the oscillatory velocity of the electrons in the laser field becomes relativistic – which is similar to observed trend for the hot electron temperature in this regime. This suggests that the laser energy is mainly coupled to the plasma via ponderomotive $\mathbf{J} \times \mathbf{B}$ acceleration of electrons at these intensities.

In the short pulse regime ($\sim$ 1 psec) the first ion measurements made at the front of the target were found to be in general agreement with the earlier results from long pulse lasers in terms of the scaling of the maximum ion energy as a function of laser intensity (Fews et al. 1994). These ion spectra exhibited a sharp cut-off at energies up to 12 MeV.

Subsequent observations of the proton emission at the front of the target at high intensities indicated that there is significant structure in the proton spectrum as well as an angular emission profile which is non-uniform and which varies with energy. Two dominant components to the spatial distribution have been observed (Clark et al. 2000). At lower energy, a ring of ablated plasma is observed which contains low ($<$ 4 MeV/neucleon) energy ions/protons. This ring-like emission is formed at the target surface in the ablating plasma and is likely caused by self-generated magnetic fields in the plasma and the subsequent influence of this field on electron transport along the target surface. At higher energies, the proton emission has a qualitatively different spatial distribution and is much more diffuse. It is likely that the higher energy protons are accelerated by a combination of a "Coulomb explosion" generated by the laser during the interaction as well as the accelerating "sheath" fields produced by the hot electron population which escapes the plasma.

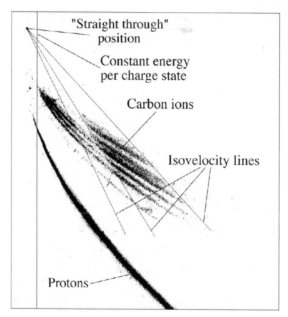

Figure 12. *Raw data from CR39/Thomson parabola ion spectrometer.*

The protons from such interactions originate from hydrocarbon/water vapor contaminants on the surface of the target (which exist within almost every vacuum system) as well as some from within the target material itself. In Figure 12 a scanned image of a piece of CR39, a nuclear track detector, is shown which has been used as the detection medium for a Thomson parabola ion spectrometer. This was produced from an interaction of the Vulcan 50 TW Nd:Glass laser with a 2 mm thick lead target at an intensity at about 2×10^{19} W/cm at RAL. The target was oriented at $45°$ to the incident direction of the laser. The ion spectrometer was positioned in front of the target and offset from the target normal by $15°$. This diagnostic enabled the heavier ions as well as the protons emitted from the target surface to be measured. The proton spectrum derived from this data (not shown) for this case exhibits an exponential fall-off in the number of protons up to 4 MeV followed by a flattening of the spectrum to the diagnostic limit of 11 MeV. From this spectrum - given the angular dispersion of the protons, it can be estimated that there are $\sim 10^{12}$ protons having an energy greater than 500 keV. The total energy in the proton "beam" emitted from the front surface of the target is about 5% of the total laser energy incident onto the target. By fitting an exponential $(\exp(-E/T))$ to the spectrum in the energy region less than 4 MeV, an ion temperature of 700 keV is obtained.

An interesting observation is that the proton spectrum often exhibits modulations, which are up to twice the background "continuum" level. These peaks in the spectrum are reproducible but vary in position from shot to shot and the energy "width" of the peaks is determined by the resolution of the ion spectrometers used.

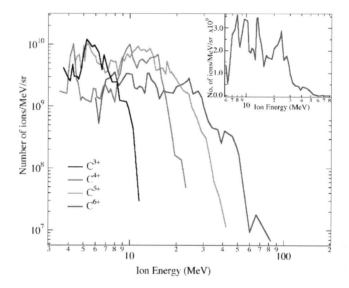

Figure 13. *Typical carbon spectrum from front surface interaction.*

The spectrum of the carbon ions can also be obtained from the piece of CR39 as shown in Figure 13. Ions with charge states up to C^{6+} are present and it can be seen that carbon ions having an energy up to 100 MeV are possible and that the maximum energy occurs for those ions with the highest ionization state. The carbon ions also have a modulated structure, similar to the protons but at a lower energy per nucleon. This is more apparent in the inset in Figure 13 where the C^{6+} spectrum is inset and plotted on a linear scale.

This can also be seen in the raw data shown in Figure 14. Along the carbon parabolas (ions are separated into different "parabolas" according to their charge/mass ratio in this type of ion spectrometer), distinct "band" like structures can be observed which indicate changes in the density of the number of pits along the parabolas. The bands occur between imaginary lines which pass through the "straight through" position. These lines indicate loci of constant velocity, which suggest that the carbon ions are accelerated in "velocity bunches" and acquire similar velocities despite having different charge states. It should be noted that the highest energy ions of each species lie along the line which indicates a constant energy per charge state – as if the highest energy ions are accelerated by falling through the same potential (i.e., before

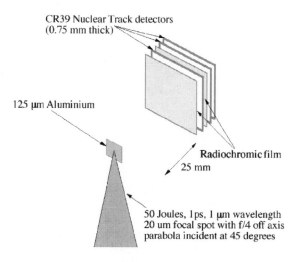

Figure 14.

the number of accelerated ions acts to reduce the accelerating fields in the sheath).

Lead spectra obtained during the same shot as the carbon and proton spectra were also obtained. From this data, Pd-like Pb^{36+} ions up to 220 ± 30 MeV and Kr-like Pb^{46+} ions up to 430 ± 40 MeV were measured with the Thomson parabola diagnostic. The quoted charge state represents an upper limit to the charge state which could be measured on the CR39. Ions of lower charge were also present, but individual parabolas could not be resolved on the detector. As in the case of the carbon ions, the ion species having the highest energy is the one with the highest charge state.

Targets can also be heated radiatively to more than 400 C using a heating element positioned near the target. The temperature was monitored using a thermocouple and was maintained at a constant level prior to the shot. The purpose of heating the target was to investigate the source of the protons in metallic targets (by removing the surface layers of contaminants through evaporation) and to observe the influence of removing the hydrogen on the acceleration of the heavy ions. The number of accelerated protons was substantially reduced by up to two orders of magnitude when compared to interactions with unheated gold or aluminum targets at a similar intensity of 2×10^{19} W/cm^2. The maximum carbon ion energy was increased by about 20 MeV and the population of these ions was pushed towards the high-energy end of the spectrum while the proton number and maximum energy were radically reduced. Indeed the reduction of the number of accelerated protons and the enhancement in the number and energy of heavier species has been confirmed by the use of nuclear activation diagnostics (Ledingham et al. 2000; Santala et al. 2000).

It is clear that very energetic ions can be generated at the front surface of the interaction of a high intensity laser with a solid target. It is likely that fast proton acceleration occurs near the critical surface by space charge acceleration from the hot electrons generated in this region and it is possible that a complex hot electron spectrum resulting from the laser plasma interaction would manifest itself as modulations in the proton and heavier ion spectrum. However the modulation features in the proton and heavier ion spectra which vary in position from shot to shot and the complex emission pattern of these ions make these ions more difficult to use for many applications.

4.2   Ions and protons from the rear of laser interactions with thin foil targets

The first measurements of ion acceleration in the "forward direction" i.e., in the direction of the laser propagation, were inferred from neutron measurements of beam-fusion neutrons from the interaction of intense laser pulses with deuterated targets (Norreys et al. 1998; Disdier et al. 1999). It was determined that deuterium ions with energies of at least several hundred keV could be accelerated into the target. These conclusions were obtained from measurements of the angular distribution of the neutrons from beam-fusion D-D interactions in the target. Modelling of these interactions has shown that the ponderomotive pressure of the laser can expel ions into the target to energies of a few MeV at these intensities. This process bores a hole into the overdense plasma – and indeed can generate an electrostatic shock, imparting even more energy to the ions (Silva et al. 2004).

A number of more recent measurements have directly recorded the production of proton and ion beams at the rear of thin solid targets irradiated by a high intensity laser beam. In Vulcan experiments at the Rutherford Appleton Laboratory protons with energies up to about 40 MeV were measured at the rear of the target. The laser wavelength was 1.054 μm, the pulselength was 0.9 – 1.2 ps and the incident energy on target was up to 100 J. The laser beam was focused onto the target surface using a *f.l.* = 225 mm on-axis parabolic mirror and was *p*-polarized incident at an angle of 45°. The peak intensity was more than 10^{19} W/cm^2.

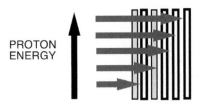

Figure 15. *Experimental setup for measurement of ion emission from rear surface of thin foil targets.*

The targets used were generally aluminium, ranging in thickness from 10 μm to 1 mm, cut into 5 mm by 5 mm squares. Behind each target, a CR39/RCF diagnostic stack was placed at a distance such that the entire beam of the particle emission could be captured by the diagnostic. This stack typically consisted of a piece of 110 μm thick RCF followed by 3 or more pieces of 0.75 mm thick CR39. The unique feature of this passive CR39/RCF stack diagnostic is that it is capable of simultaneously measuring the spatial distribution and the energy of the protons (and electrons) and was first used during these experiments (Clark et al. 2000). The diagnostic was placed about 25 mm behind the target as shown in Fig. 15. Quantitative measurements of the spatially integrated proton spectrum can be obtained using a similarly designed copper activation stack diagnostic (Santala et al. 2001).

Ions deposit energy in the CR39 as they pass through and damage the plastic. Most of the energy is deposited just as the proton stops at the Bragg peak. CR39 is subsequently "developed" in a NaOH solution that etches a pit at the surface of the plastic if a proton was stopped there. The particular advantage of this diagnostic is that knowledge of the range of protons in CR39 and radiochromic film enables the energy of protons which have produced etched pits to be determined as the protons pass through the stack (see Fig. 16).

A particularly interesting aspect of these first measurements was the observation that for some targets the protons were emitted in quasi-mono-energetic "ring" structures such that a narrow range of energies were emitted primarily at a particular angle and such that the ring radius decreased with increasing proton energy. The production of this ring structure has been

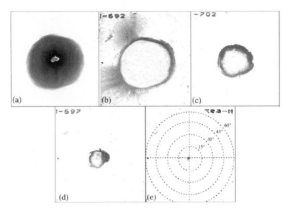

Figure 16. *Schematic of use of "sandwich" detectors for proton spectroscopy.*

confirmed through the use of other diagnostic techniques such as nuclear activation (Zepf et al. 2003, 2001).

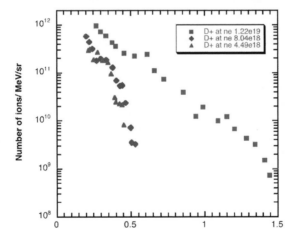

Figure 17. *CR-39 track detectors showing a ring-like structure and that the proton beam is emitted in a mono-energetic pattern that is energy dependent.*

The maximum divergence of this ring pattern was up to 60°. The ring pattern may be caused by deflection of the protons by magnetic fields within the target interior if one considers that the origin of these protons could be the front surface. Magnetic fields inside the target can be generated by the current of hot electrons that also propagate into the target ahead of the protons (Davies et al. 1997). The spatial distribution of the protons as measured by the CR39/RCF diagnostic is shown in Fig. 17. This data was taken from a shot with a laser intensity of 2×10^{19} W/cm² incident on a 125 μm thick aluminum target. Fig. 17a shows a scanned image of the front piece of radiochromic film. The signal on the film is contained within a well-defined radius from the central burn mark in the centre of the film. The angle subtended by the periphery of this signal covers a cone half angle of 30°. The center of the film is coincident with the rear target normal. Towards the center of the film, as the radius decreases, there is an abrupt change in optical density on the film at a cone half angle of 10°, while at the center, the film has been burned.

Figure 17 also shows the scanned images of the surfaces of the CR39 facing the target. The CR39 is only sensitive to protons and ions, however ions of species other than protons would be unlikely to pass though the first layer of RCF. The signal observed in these images is consequently predominantly due to protons. Observation of the size of the pits indicated that they were ~10 μm, which for these etching conditions would also confirm that the tracks were made by protons. Figure 17 indicates that most of the proton signal is situated on the ring. From the stopping range of the protons in the diagnostic, it can be estimated that these protons have an energy of 3 ± 1 MeV. The signal on the CR39 which forms the ring is saturated but an estimate of the number of protons can be derived by comparison of the CR39 signal with the outermost region of the RCF signal in Figure 17a. Clearly, there is a strong correlation between these two regions so it can be estimated that there are $\sim 10^{12}$ protons with energies in excess of 2 MeV. The contrast between the signal (protons/cm^2) contained in the ring and the signal inside the ring which is not saturated is about 10^4:1. Fig. 17c to Fig. 17e show the surfaces of the layers of CR39 further into the stack. As the energy of the protons increases, the radius of the ring pattern decreases, until at ~ 18 MeV, the proton signal is reduced to a "dot". In effect, as a function of angle from the rear target normal, a mono-energetic beam of protons has been measured which has rotational symmetry around the rear target normal. Clearly, this is an unusual observation, where it might have been intuitively expected that a broad spectrum of ions would be observed at each emission angle. If this was the case, then the signal on the CR39 would manifest itself as a "disk" of signal rather than a ring.

It is important to note that the actual signal levels in the interior of the rings shown here are not overexposed and indeed individual proton tracks can be observed and counted within the structure – although the signal level in the "ring" itself is saturated. Ring structure is, in fact, not observable when the proton signal was highest – i.e., for thin targets ($< 50 \mu$m) when in fact the emission pattern was observed to be uniform – but rather only occurred when the target was thicker ($50 - 200 \mu$m) such that the proton signal levels were reduced. The fact that the ring structure can be reproducibly observed only for "thicker" targets is important evidence that this structure may be due to relatively low fields within the target

Figure 18 shows data from another shot, with a similar intensity, when the laser was incident on a 10 μm aluminum target. This series of scanned CR39 images in Fig. 8b to Fig. 8d shows both sides of the 0.5 mm thick CR39. The darker region is the signal on the rear surface. In front of the first piece of CR39 a 250 μm polyethylene filter was positioned which covered half of the CR39 surface. This is equivalent to etching an intermediate surface on the CR39 between the front and back by staggering the distance of material which the protons have to traverse in the diagnostic. In this way, four proton energies can be derived from one piece of CR39.

The emission pattern of these protons is more uniform and less divergent than that from the thicker target discussed above. The number of protons produced also increases significantly as the target is made thinner until the energetic proton emission from the rear exceeds that observable at the "front" of the target. Indeed when thin ($< 50 \mu$m) Al foil targets are irradiated with 5×10^{19} Wcm^{-2} on target (using the Vulcan 100 TW facility) a uniform poly-energetic proton distribution is observed with no ring structure present.

There have been a large number of experiments which have investigated the acceleration of protons from the rear surface of very thin foil targets – and such beams seem to be emitted with unusual beam qualities (Najmudin et al. 2003). Several experiments have shown that

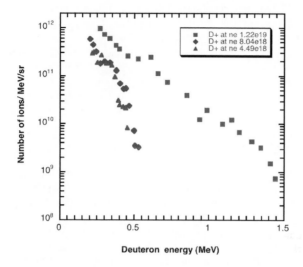

Figure 18. *Measured spectrum of deuterium ions resulting from intense laser interaction with deuterium gas jet targets.*

acceleration of these protons from the rear surface of the thin foil is the dominant mechanism in such experimental situations. This mechanism is the well known sheath acceleration which causes most of the observed acceleration of ions at the ablated plasma at the front. In this case the process occurs within a plasma created at the rear surface by the hot electrons which propagate through the target. Such acceleration can be well simulated using collisionless particle in cell codes and these codes suggest that the direction the proton emission can be affected by the shape of the rear surface and that it may be possible to "focus" this proton beam in order to deposit the energy into a small region of plasma for inertial confinement fusion applications (i.e., fast ignition and isochoric heating of plasma).

Indeed for targets composed of conducting materials the proton beam quality is much superior to that from insulating materials – which is indicative of improved hot electron flow through the material. This indicates that sheath acceleration from the rear surface may be dominant for those conditions, in agreement with the observations of Snavely (2000) who observed the highest accelerated proton energies to this point (58 MeV).

Depending on the laser intensity and the target thickness, simulations suggest that front surface (due to collisionless shock acceleration) and rear surface (due to sheath acceleration) mechanisms can become dominant in different experimental configurations. It is interesting to note that acceleration in the shock becomes dominant when the protons accelerated in this way reach the sheath acceleration region at the rear side of the target with a velocity greater than the ions already accelerated in the sheath region. Particle-in-cell modelling by Silva suggests that a signature for this effect is the formation of a plateau region between 15 - 30 MeV which can be clearly observed in data from various experiments. It is also the initial plasma temperature and target resistivity also plays a significant role in determining which process dominates.

One of the obstacles for applications of these beams is that the beams are generated with a broad energy spread – consequently methods to produce mono-energetic sources could be important for the development of these applications. Clearly the observation of energy resolved angular distribution in Figure 17 indicates that it is possible to generate mono- energetic proton beams rather easily with the insertion of a suitably designed filter (aperture).

These protons could then be used as an injector for subsequent acceleration in a conventional accelerator. Indeed, heavier ions could also be used for this purpose in the future. It is interesting to note that Cowan (Cowan et al. 2004) have shown that the emittance of protons from a laser-produced plasma source can be less than 0.004 π mm mrad, which is two orders of magnitude better than conventional injectors for accelerators. Heating the target also removes the proton layer from the rear of the target and enables the acceleration of heavier ion species (Hegelich et al. 2002).

4.3   Ions from laser interactions with underdense plasmas

Another potentially important source of energetic ions are those resulting from the interaction of intense laser beams with underdense plasmas. However, in this case the energetic ions are typically not measured along the direction of the laser beam but rather are produced primarily perpendicular to the direction of propagation.

The acceleration mechanism is also fundamentally different. The large ponderomotive force of the laser pulse acts to displace electrons in the plasma and set-up an electric field because of charge separation. Since the three dimensional shape of the focused laser pulse is typically longer than it is wide (i.e., "cigar" shaped) this implies that the electric field experienced by the ions in the focal volume during the passage of the pulse is primarily radial. Consequently, the pulselength of the incident laser will determine the duration of the acceleration. In experiments the radial acceleration of these ions is the principal observation (Krushelnick et al. 1999; Wei et al. 2004). The acceleration of ions using this technique is termed either as "Coulomb explosion" because of the space charge fields resulting from the charge displacement caused by the ponderomotive force of the laser pulse during the interaction or alternatively "ponderomotive shock acceleration".

Study of the ion dynamics is important as it can supply valuable information of the fundamental physics of the interaction of a high-intensity laser with underdense plasma, such as self-focusing and channelling due to relativistic and charge displacement effects. It is also directly related to the observations of anomalously high yields of neutrons resulting from hot channel formation (Tabak et al. 1994; Kodama et al. 2001; Roth et al. 2001). The maximum ion energy that can be produced by this process is roughly equal to the ponderomotive energy. Recently measurements of energetic ions accelerated during the interaction of a 0.25 PW laser at RAL with a gas-jet plasma at electron densities up to 1.4×10^{20} cm^3. For these investigations, the laser produced pulses with an energy up to 180 J in a duration of 0.5– 0.7 ps. The laser pulse was focused onto the edge of a supersonic gas jet (2 mm nozzle diameter) using an $f = 3$ off-axis parabolic mirror to a focal spot size of 10 μm in vacuum.

The energy spectrum of the ions at 100 degrees from the laser propagation direction was measured with a Thomson parabola ion spectrometer, positioned 80 cm from the interaction region. The ions were recorded on a 1 mm thick piece of CR39 nuclear track detector. The angular distribution of the ions was measured with a stack of several layers of radiochromic film (RCF) strips parallel to the direction of laser propagation placed at a distance of 6 cm radially from the interaction region. The angular distribution of ions in different energy ranges was determined through the use of aluminum filters of various thickness in front of the RCF. The spectra show two main characteristics: i) the maximum ion energy and the number of energetic ions are higher at high density; ii) at high density, a plateau is observed at high energy. In the high density case both He^{2+} and He^{1+} ions have been accelerated to high energy, the maximum energy for He^{2+} being 13.2 ± 1.0 MeV. The total number of He^{2+} ions with energy greater than 680 keV is $3 : 8 \times 10^{11}$. The maximum ion energy drops to 2.3 MeV and the total number of

He2 ions with energy greater than 590 ± 20 keV is reduced to 7.4×10^{10}. At even lower density, i.e., below 4×10^{18} cm^3, no energetic helium ions (greater than 100 keV =nucleon—the detector threshold) were observed. The He^{2+} ion spectrum exhibits a plateau in the energy range between 6–10 MeV and has a cutoff at the maximum ion energy recorded on the detector. The plateau structure is a characteristic of the acceleration due to laser driven laminar shock waves (Silva et al. 2004). An example of an ion emission spectrum due to an interaction with deuterium is shown in Fig. 18. These ions were measured using a Thomson parabola ion spectrometer positioned at 90 degrees from the direction of laser. The ion emission at low density (1.0×10^{19} cm^3) is predominantly at 90 degrees. For energies greater than 2 MeV, there is a narrow lobe with an angular spread of less than 4 degrees (FWHM). At higher plasma density, the angular spread is much greater. Even the high energy component ($E > 3.5$ MeV), which has a less broad angular emission, has an angular spread of 27 degrees, and is preferentially emitted in the forward direction. Studying the radial momentum of the ions as a function of radial distance shows that the ions are accelerated by the interaction of the collisionless shocks. Large populations of ions are seen to be accelerated to higher energies especially during the merging of the multiple shocks during simulations. This is shown most clearly by an abrupt increase in the radial momentum of the ions at radial displacements well beyond the laser beam radius and thus beyond the influence of its ponderomotive force. This acceleration process contributes to the formation of the plateau in the ion energy spectrum, as has been similarly observed in other simulations above critical density. The strong dependence of the ion acceleration on plasma density can be explained by the additional shock acceleration in high density, while such mechanisms do not occur at low density. In particular, ion acceleration due to the ponderomotive expulsion of plasma is found to be enhanced by collisionless shock acceleration.

5 Conclusions

In this chapter we have examined a variety of phenomena which occur in ultra-intense laser plasma interactions – i.e., the production of high harmonic radiation, large magnetic fields and energetic particle beam production. While this is not a conclusive list of interesting new physical phenomena made possible by recent developments in high power laser technology, these effects have a particular importance with regard to applications.

Clearly high order harmonic measurements can be used to determine a significant amount of information concerning these interactions. The experimentally observed modulation of the harmonic spectrum agrees with the PIC simulation and is due to the introduction of higher modes of oscillation of the critical surface $x(t)$. Observations of the modulational structure of harmonic emission from high intensity interactions may be important as a diagnostic tool of the behavior of the critical surface in future experiments, for example in "fast ignitor" type experiments (Tabak et al. 1994; Kodama et al. 2001; Roth et al. 2001). Measurements of the modulational structure of harmonic emission from several different angles can potentially provide 3-dimensional information on the critical surface dynamics under intense laser irradiation.

We have also shown that magnetic fields up to about 700 MegaGauss are generated during a relativistic intensity laser solid interaction. An innovative method using the propagation properties of the self generated harmonics has been used to measure these fields which is a confirmation of theoretical predictions made over the last ten years. Such diagnostic techniques will enable the collection of harmonic emission simultaneously from a wide range of angles which will consequently allow a highly detailed picture of these interactions to be obtained. For example this could lead to a method to generate a time-dependent "contour" plot of the magnetic fields surrounding the high intensity interaction region.

With regard to relativistic electron acceleration for low density plasmas the experiments on the Vulcan Petawatt facility has successfully accelerated electrons up a maximum energy up to about 300 MeV with low divergence. However the electron beams produced in this way have a very large energy spread making them not particularly useful for applications – except as an efficient source for MeV radiography.

On the other hand a reduction in the pulse length (and increase in the focal length) has shown that the SM-LWFA moves into a completely different regime in which relativistic "mono-energetic" electron beams can be produced. This seems to be the most attractive route for further scientific exploration as well as for the development of applications.

There are also many applications proposed for the recently observed proton and ion beams from thin foil targets. These include injectors for subsequent conventional accelerators; isotope production for positron emission tomography (Santala et al. 2001); fusion evaporation studies (McKenna et al. 2003), among others. Most of the potential applications require improvements in the laser technology or improvements in the understanding of the physics of ion acceleration in order to enhance make the application economically viable. Indeed for many applications there is a particular requirement for mono-energetic source and there are several researchers who have proposed methods and target configurations to generate such sources.

However the most significant application of this phenomenon occurred almost immediately after the discovery of these high quality ion beams – primarily because of the excellent beam quality of the proton beams emitted from the rear surface of these thin foils enables them to used very successfully as high resolution probes of electric fields with a second laser produced plasma. There have been a number of interesting observations using this technique including the measurement of solitons (Borghesi et al. 2001, 2002) in a laser produced plasma. This application is still under development since in fact the probing images are affected not only by electric fields in the target but also in some circumstances by plasma density variations and magnetic fields. Clearly, with so many potential applications, it is vital to understand in detail the dependence of the different acceleration mechanisms on the laser and target conditions.

References

F. Amiranoff, S. Baton, D. Bernard *et al.*, Phys. Rev. Lett. **81**, 995 (1998)

F. Begay *et al., Phys. Fluids,* **25**, 1675 (1982)

M. Borghesi, A. J. MacKinnon, A. R. Bell *et al., Phys. Rev. Lett.* **81**, 112 (1998)

M. Borghesi *et al., Plasma Physics and Controlled Fusion* **43** A267 (2001)

M. Borghesi *et al., Phys. Rev. Lett.* **88** 135002 (2002)

E. L. Clark, K. Krushelnick, M. Zepf, M. Tatarakis, F. N. Beg, A. Machacek, P. A. Norreys, M. I. K. Santala, I. Watts, A. E. Dangor, *Physical Review Letters,* **84**, 670 (2000)

E. L. Clark, K. Krushelnick, M. Zepf, F. N. Beg, A. Machacek, P. A. Norreys, M. I. K. Santala, M. Tatarakis, I. Watts and A. E. Dangor, *Physical Review Letters* **85**, 1654 (2000)

C. A. Coverdale, C. B. Darrow, C. D. Decker *et al., Phys. Rev. Lett.* **74**, 4659 (1995)

T. Cowan *et al., Physical Review Letters* **92**, 204801 (2004)

C. N. Danson *et al., Opt. Comm.* **103**, 392 (1993)

J. R. Davies *et al.*, Phys Rev E **56**, 7193 (1997)

L. Disdier *et al., Physical Review Letters* **82** 1454 (1999)

R. D. Edwards, *et al., Applied Physics Letters* **80**, 2129 (2002)

A. H. Ehler *et al., J. Appl. Phys* **46**, 2464 (1975)

M. Everett, A. Lal, D. Gordon *et al., Nature* **368**, 527 (1994)

J. Faure, Y. Glinec, A. Pukhov *et al. Nature* **431**, 541 (2004)

A. P. Fews *et al., Phys. Rev. Lett.* **73**,1801 (1994)

S. Fritzler *et al., Phys. Rev. Lett.* **89**, 165004 (2002)

C. Gahn, G. D. Tsakiris, A. Pukhov, J. Meyer-ter-Vehn, G. Pretzler, P. Thirolf, D. Habs, K. J. Witte, *Phys. Rev. Lett.* **83**, 4772 (1999)

C. G. R. Geddes, C. Toth, J. van Tilborg *et al., Nature* **431**, 538 (2004)

S. J. Gitomer *et al.*, *Phys. Fluids.* **29**, 2679 (1984)

M. Hegelich *et al.*, *Phys. Rev. Lett.* **89**, 085002 (2002)

R. Hemker, PhD dissertation UCLA (2000)

M. H. Key, M. D. Cable, T. E Cowan, K. G. Estabrook, B. A. Hammel, S. P. Hatchett, E. A. Henry, D. E. Hinkel, J. D. Kilkenny, J. A. Koch, W. L. Kruer, A. B. Langdon, B. F. Lasinski, R. W. Lee, B. J. MacGowan, A. MacKinnon, J. D. Moody, M. J. Moran, A. A. Offenberger, D. M. Pennington, M. D. Perry, T. J. Phillips, T. C. Sangster, M. S. Singh, M. A. Stoyer, M. Tabak, G. L. Tietbohl, M. Tsukamoto, K. Wharton, S. C. Wilks, *Physics of Plasmas*, **5**, 1966 (1998)

R. Kodama, *et al.*, *Nature* **412**, 798 (2001)

K. Krushelnick, E. L. Clark, Z. Najmudin et al., *Phys. Rev. Lett.* **83**, 737 (1999)

K. Krushelnick, I. Watts, M. Tatarakis, A. Gopal, U. Wagner, F. N. Beg, E. L. Clark, R. J. Clarke, A. E. Dangor, P. A. Norreys, M. S. Wei, and M. Zepf *Plasma Physics and Controlled Fusion* **44**, B233 (2002)

K. W. D. Ledingham *et al.*, *Physical Review Letters* **84**, 899 (2000)

R. Lichters, J. Meyer-ter-Vehn and A. Pukhov, *Phys. Plasmas* **3**, 3425 (1996)

V. Malka, S. Fritzler, E. Lefebvre, M.-M. Aleonard, F. Burgy, J.-P. Chambaret, J.-F. Chemin, K. Krushelnick, G. Malka, S. P. D. Mangles, Z. Najmudin, M. Pittman, J.-P. Rousseau, J.-N. Scheurer, B. Walton, and A.E. Dangor, *Science* **298**, 1596 (2002)

S. P. D. Mangles C. D. Murphy, Z. Najmudin, *et al.*, *Nature* **431**, 535 (2004)

S. P. D. Mangles, B. R. Walton, M. Tzoufras, et al., *Phys. Rev. Lett.* **94** 245001 (2005)

R. J. Mason and M. Tabak, *Phys. Rev. Lett.* **80**, 524 (1998)

P. McKenna *et al.*, *Phys. Rev. Lett.* **91**, 075006 (2003)

J. Meyer-ter-Vehn and Z. M. Sheng, *Phys. Plasmas* **6**, 641 (1999)

A. Modena, Z. Najmudin, A. E. Dangor, C. E. Clayton, K. A. Marsh, C. Joshi, V. Malka, C. B. Darrow, C. Danson, D. Neely, and F. N Walsh, *Nature* **377**, 606 (1995)

Z. Najmudin, K. Krushelnick, M. Tatarakis, E. L. Clark, C. N. Danson, V. Malka, D. Neely, M. I. K. Santala, and A. E. Dangor, *Physics of Plasmas* **10**, 438 (2003)

P. A. Norreys, M. Zepf, S.Moustaizis, A. P. Fews, J. Zhang, P. Lee, M. Bakarezos, C. N. Danson, A. Dyson, P. Gibbon, P. Loukakos, D. Neely, F. N. Walsh, J. S. Wark and A. E. Dangor, *Physical Review Letters* **76**, 1832 (1996)

P. A. Norreys *et al.*, *Plasma Physics and Controlled Fusion* **40** 175 (1998)

M. D. Perry and G. Mourou, *Science* **264**, 917 (1994)

M. Roth *et al.*, *Physical Review Letters* **86**, 436 (2001)

M. I. K. Santala, *et al.*, *Physical Review Letters* **84**, 1459 (2000)

M. Santala *et al.*, *Appl. Phys. Lett.* **78**, 19 (2001)

M. I. K. Santala, Z. Najmudin, K. Krushelnick, E. L. Clark, A. E. Dangor, V. Malka, J. Faure, R. Allott and R. J. Clarke, *Physical Review Letters* **86**, 1227 (2001)

S. E. Segre, *Plasma Physics and Controlled Fusion* **41** R57 (1999)

L. O. Silva *et al.* *Phys. Rev. Lett.* **92**, 015002 (2004)

R. A. Snavely, M. H. Key, S. P. Hatchett. T. E. Cowan, M. Roth, T. W. Phillips, M. A. Stoyer, E. A. Henry, T. C. Sangster, M. S. Singh, S. C. Wilks, A. J. Mackinnon, A. Offenberger, D. M. Pennington, K. Yasuike, A. B. Langdon, B. F. Lasinski, J. Johnson, M. D. Perry, E. M. Campbell, *Physical Review Letters*, **85**, 2954 (2000)

J. Stamper, K. Papadopoulos, R. N. Sudan, E. McLean, S. Dean and J. Dawson, *Physical Review Letters* **26** 1012 (1971)

R. Sudan, *Physical Review Letters* **70** 3075 (1993)

M. Tabak *et al.*, *Phys. Plas.* **1**, 1626 (1994)

T. Tajima and J. M. Dawson, *Phys. Rev. Lett.* **43**, 267 (1979)

T. Tan *et al.*, *Phys. Fluids* **27**, 296 (1984)

M. Tatarakis *et al.*, *Physics of Plasmas* **9**, 2244 (2002)

M. Tatarakis, I. Watts, F. N. Beg, E. L. Clark, A. E. Dangor, M. G. Haines, P. A. Norreys, M. Zepf and K. Krushelnick, *Nature* **415**, 280 (2002)

U. Wagner, M. Tatarakis, A. Gopal *et al.* *Phys. Rev. E* **70**, 026401 (2004)

B. Walton, Z. Najmudin, M. S. Wei *et al.*, *Optics Letters* **27**, 2203 (2002)

I. Watts *et al.*, *Physical Review Letters* 88, 155001 (2002)

M. S. Wei *et al.*, *Phys. Rev. Lett.* **93** 155003 (2004)

L. M. Wickens *et al.*, *Phys. Rev. Lett.* **41**, 243 (1978)

M. Zepf *et al., Physics of Plasmas* **8**, 2323 (2001)
M. Zepf *et al., Physical Review Letters* **90**, 064801 (2003)

6

Computational Challenges in Laser-Plasma Interactions

Ricardo Fonseca

DCTI, Instituto Superior de Ciências do Trabalho e da Empresa, Lisbon, Portugal
GoLP/CFP, Instituto Superior Técnico, Lisbon, Portugal

1 Introduction

The collective interaction of particles and fields in laser plasma interactions represent a complex and strongly nonlinear problem, with applications covering a wide range of physical scenarios. Traditionally the investigation of such systems has been done through theoretical and experimental work, developing analytical models to determine the behavior of our system, and carrying out sophisticated experiments to observe the system evolution. Notwithstanding the great success and power of this approach, the simultaneous interaction the extraordinary complexity and many degrees of freedom of laser plasma interactions render analytical modeling impractical. Also, many of the significant details of laser plasma interaction are extremely difficult to obtain experimentally, and other tools are required to further the understanding in this field.

With the advent of powerful computer systems, numerical simulation now plays a key role in physics, allowing for the testing of theoretical models and providing detailed information which is difficult to measure experimentally. Numerical modeling has matured to a level where it can be understood as separate discipline from theory and experiment, and has proved its value as and essential tools in developing theoretical models and understanding experimental results. Presently, it is clear that a detailed understanding of the physical mechanisms in laser-plasma interaction can only be achieved through the combined synergies of theory, experiment and simulation.

While analyzing these problems from a computational physics point of view, we start by developing a numerical model for our system. One than carries out a numerical experiment on a high-speed computer, allowing for our numerical system to evolve from some initial condition in accordance with the model used. This numerical experiment can give as much information regarding the details of the evolution of our system as one desires. The researcher can then use the results of each simulation to test the predictions based on simplified analytical models, or compare them with experimental results and infer details otherwise impossible to obtain experimentally. Finally, after properly benchmarked, these tools can be used to predict the outcome of future experiments, and help guide the experimental effort in this field.

2   Simulating plasmas

At the microscopic level plasmas are very simply physical systems: charged particles interacting through electromagnetic fields. However, even a very small spatial region has a very high number of particles, typically in the range of $\sim 10^{16} - 10^{18}$ particles/cm^3. Furthermore, there is a wide range of temporal and spacial scales, such as the plasma skin depth or the Rayleigh range that need to resolved, making this an outstanding computational challenge, that requires state of the art machines for detailed simulations.

State of the art computers can achieve performances on the order of $\sim 10^{14}$ floating point operations per second. This level of performance is achieved by splitting the workload between several (thousands) of CPUs, in what is known as massively parallel computing. These are extremely sophisticated and complex systems, that represent the pinnacle of computer science research in this field and that open the door to one-to-one detailed simulations of plasmas. Access to these systems is however limited and a popular alternative for university groups is a "do it yourself" approach pioneered by Sterling and co-workers in the mid 1990's, Sterling (1995). High performance computing machines that are able to run small/mid sized problems can be built using commodity computers connected by a standard network.

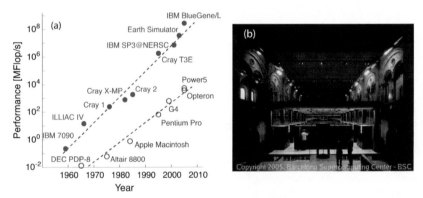

Figure 1. *(a) Evolution of computing power The red values show performance of large scale HPC machines, the blue values show performance of commodity computers; (b) Mare Nostrum computer at the Barcelona Supercomputing Center*

It should also be noted that running simulation is only half of the computational work. The data sets from current from current simulations are both large and complex. These sets can have up to five free parameters for field data: three spatial dimensions, time and the different components (i.e., E_x, E_y, and E_z). For particles, phase space has seven dimensions: three for space, three for momentum and one for time. Plots of y versus x are simply not enough and sophisticated graphics are needed to present so much data in a manner that is easily accessible and understandable. It is therefore crucial to also write the necessary tools to ensure that the results flow as quickly as possible from the simulation to the researcher, providing a tight integration between simulation codes and visualization infrastructure.

3 Numerical models

Based on the highly nonlinear and kinetic processes that occur during high-intensity particle and laser beam-plasma interactions, we must use a kinetic description of the plasma in our numerical model. Solving the plasma kinetic equations (e.g., Vlasov or Focker-Planck equations) numerically is not achievable in realistic geometries with nowadays computers, and presently the best choice is that of "particle simulations". This follows the pioneering work of Buneman and Dawson in the early 1960's (Buneman 1959, Dawson 1962) that showed we can accurately model the collective behavior of plasmas by following the evolution of a relatively small number (when compared to the number of particles in a real plasma) of simulation particles. This lead to the concept of a macro particle, that represents many plasma particles, and allows for the simulation of large spatial regions on the order of several tens of collisionless plasma skin depths.

As a first approach we might consider calculating the interaction between these particles directly, implementing a so called particle-particle algorithm. However, even using macro particles, realistic simulations would still lead to enormous requirements in computing power, requiring months of computing time in present day state of the art machines. The number of operations required for such algorithm will scale roughly as the square of the number of particles used, and a simulation using 10^8 particles and 10^4 time steps could take as much as 4 months in a 10 Tflop/s machine.

The solution to this problem comes from the fact that we are generally interested in situations where the relevant physics is dominated by the collective effects of the plasma, meaning that short-range collisions can be neglected. In this sense, rather than interacting particles directly we will do so through the electromagnetic fields, which we will sample on a grid. The field values are interpolated at particle positions to calculate forces acting on particles, and particle positions/velocities will be used to determine the electrical current density used to solve the field equations. This is a so called particle-mesh algorithm, which greatly reduces the computing power requirements. With this algorithm, the number of operations for a simulation with N_P particles and N_{cells} grid cells will scale as $\sim \alpha N_P + \beta(N_{cells})$, where α is an integer constant and β is a function of the number of cells. For a typical algorithm, the same simulation mentioned before, run on a 128^3 grid, would only take about 1 hour on the same machine. This extraordinary performance boost is obtained at the expense of spatial resolution, given that we are now limited by a grid thus limiting the shortest wavelengths that we can study.

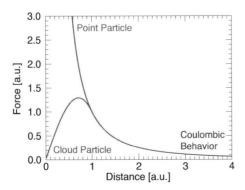

Figure 2. *Electrical force between two point particles and between two cloud particles.*

We must also consider the effects of our simulation model on short-range collisions in our plasma. As stated above, it is not possible to have as many particles in the simulation as in a real plasma, and macro particles are used represent many real particles. If we simply consider point macro particles the force between them will go to infinity as they cross, which means that forces between macro particles will be over estimated at short ranges, given that there are not enough particles to average out the effect of these short-range collisions. In this sense the collisional effects in our simulation will be much greater than in typical laboratory plasma, where the ratio between plasma and collision frequency is of the order of $\omega_p/\nu \sim 10^2 - 10^6$.

To address this issue we use finite sized particles in our simulation. Generally macro particles are considered to be clouds of real particles (sometimes described as the cloud-in-cell algorithm or CIC) that have a given shape, typically that of our simulation grid cell. As shown on figure 2 this means that forces between macro particles will go to zero for short distances, reducing the effect of overestimating short range collisions, while having a Coulombic behavior at large distances, thus retaining collective behavior effects. This compensates for the smaller number of particles used, and accurately simulates the conditions of interest.

It should be noted that this model is not collisionless and can be used to test collision operators. Furthermore, this algorithm can also be extended to handle strongly collisional plasmas, by implementing algorithm to treat short range collisions separately. For further reading see for example (Takizuja 1977).

4   Particle-in-Cell algorithm

Particle in cell codes usual share the same typical cycle, shown in figure 3 (Hockney 1981, Birdsall 1991). This cycle will start at $t = 0$ with some appropriate initial conditions for particles and and fields. The values of the electromagnetic field are interpolated at particle positions, and used to integrate the equations of motion for the particles. We then deposit the electrical current resulting from this motion on the simulation grid, and use it to integrate the field equations, closing the simulation cycle.

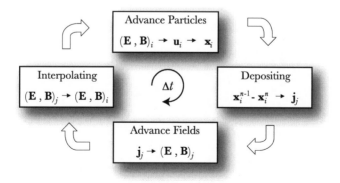

Figure 3. *Typical Particle-in-Cell simulation loop.*

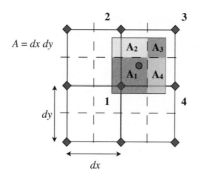

Figure 4. *Area Weighting scheme for field interpolation. The particle center is identified by the red circle.*

4.1 Particle pusher

Integrating the equations of motion for the particles is usually done through a leap-frog method, to increase accuracy in time. In this method positions and velocities are not time-centered, which in fact leads to second order accuracy in time. A detailed analysis of accuracy and stability for this algorithm can be found in (Birdsall 1991). If the positions are defined at time t^n and velocities are defined at time $t^{n+1/2}$ time differencing the motion equation we get:

$$\mathbf{x^{n+1}} = \mathbf{x^n} + \mathbf{v^{n+1/2}} \tag{1}$$

To advance particle velocities we need the forces acting on our particle, which in our situation is simply the Lorentz force. It should be noted however that we are dealing with relativistic particles, so we must use the following momentum equation:

$$\frac{d\mathbf{p}}{dt} = q\left(\mathbf{E} + \frac{\mathbf{v}}{c} \times \mathbf{B}\right) \tag{2}$$

To update particle momenta, we must take the field values from the grid and interpolate at the particle position, and also account for the shape of the particle that we have chosen. This algorithm must be consistent with the choice of current deposition, as described below, to ensure momentum conservation and avoid self-forces (i.e., dp/dt must be 0 for a single particle). For a particle shaped like the simulation cell this can easily be achieved through an area weighting scheme (figure 4). Assuming that that the field component required is defined at the lower left-hand corner of the simulation cell, this scheme can be interpreted as a bilinear interpolation of the four nearest values yielding:

$$E_{part} = \frac{A_1 E_1 + A_2 E_2 + A_3 E_3 + A_4 E_4}{A} \tag{3}$$

Where the E_k represent the field values in the corners of the cell, A_k the area of the particle falling closer to the cell corners, and A is the total cell area, as described in figure 4. Using the interpolated fields, we can now advance particle momenta in time. This is generally done using the algorithm proposed by (Boris 1970), that makes time centering of all quantities relatively simple. Since we are in the relativistic case we deal with the generalized velocity, $\mathbf{u} = \gamma\mathbf{v}$. As we will see $\mathbf{E}$ and $\mathbf{B}$ are usually defined at t^n, while $\mathbf{u}$ is defined at $t^{n+1/2}$. To deal with this issue the algorithm splits the time integration of particle velocities into several steps, separating the electric acceleration from the magnetic rotation. Starting with $\mathbf{u}^{-1/2}$ we first we add $1/2$ of the electric impulse, obtaining an estimate for $\mathbf{u}$ at time t^n:

$$\mathbf{u}' = \mathbf{u}^{n-\frac{1}{2}} + q\frac{dt}{2}\mathbf{E}^n \tag{4}$$

Using this value we get an estimate for the time centered Lorentz factor, γ^n, and perform half of the magnetic rotation:

$$\gamma' = \left(1 + u'^2\right)^{\frac{1}{2}}$$

$$\mathbf{u}'' = \mathbf{u}' + q\frac{dt}{2}\frac{\mathbf{u}'}{\gamma'} \times \mathbf{B}^n \tag{5}$$

Using this estimate for $\mathbf{u}$ we do a full magnetic rotation, and add the remaining electric impulse.

$$\mathbf{u}''' = \mathbf{u}' + q\,dt\,\frac{\mathbf{u}''}{\gamma'} \times \mathbf{B}^n$$

$$\mathbf{u}^{n+\frac{1}{2}} = \mathbf{u}''' + q\frac{dt}{2}\mathbf{E}^n \tag{6}$$

Finally, to advance particle positions in time we must get $\mathbf{v}^{\frac{1}{2}}$ from $\mathbf{u}^{\frac{1}{2}}$ and use it in the leap-frog equation above (1):

$$\gamma^{n+\frac{1}{2}} = \left[1 + \left(u^{n+\frac{1}{2}}\right)^2\right]^{\frac{1}{2}}$$

$$\mathbf{x}^{n+1} = \mathbf{x}^n + q\frac{\mathbf{u}^{n+\frac{1}{2}}}{\gamma^{n+\frac{1}{2}}}dt \tag{7}$$

It should be noted that in electromagnetic PIC codes particles should never move more than one cell per time step, since otherwise higher frequency waves would not be properly sampled by simulation particles. As we will see, the stability criteria for the field solver will actually enforce this, so that particle motion is not relevant for the choice of time-step.

4.2   Current deposition

To connect the effects of particle motion with the field equations, the algorithm must deposit the electrical current produced by the motion of the charged particles in the simulation on a grid. The current deposition scheme chosen must be consistent with the field interpolation scheme chosen for advancing the particles, and to insure that the algorithm is self-consistent we must also satisfy the continuity equation:

$$\frac{\partial \rho}{\partial t} = -\nabla \cdot \mathbf{j} \tag{8}$$

Where ρ is the charge density on the grid. There is also the added difficulty arising from the time centering required for current. Because of the leap-frog method chosen for the particle equations, particle positions and velocities are defined at different times, and the algorithm for the field equations require the electrical current to be define at $t^{n+1/2}$. We briefly describe

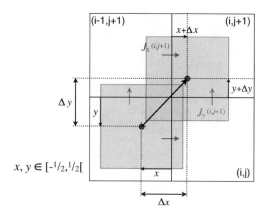

Figure 5. *Current deposition scheme. The particle center is identified by the red circle.*

here the algorithm proposed by (Villasenor 1992) that correctly solves this problem for particles shaped like the simulation cell. In this algorithm only the position of particles in the previous and current time step are required. Figure 5 shows the details of this algorithm in 2D.

For simplicity we define the particle position (x, y) as the distance to the lower left hand corner of cell (i,j). In one time step (time interval Δt), the charge will move from a position (x, y) to a position $(x + \Delta x, y + \Delta y)$. The horizontal component of the electrical current in cell (i, j), $j_{x(i,j)}$ can be calculated by determining the total amount of charge that crosses the cell boundary where this value is defined. For one time step the width of charge crossing this boundary is Δx. The initial height of charge is $1/2 - y$, and the final height of charge will be $1/2 - (y + \Delta y)$, which leads to an average height of $1/2 - y - \Delta y/2$. The total electrical current $j_{x(i,j)}$ will therefore be $\Delta x(1/2 - y - \Delta y/2)q/\Delta t$. For all components we get:

$$
\begin{aligned}
j_{x(i,j)} &= \Delta x \left(\tfrac{1}{2} - y - \tfrac{1}{2}\Delta y \right) q/\Delta t \\
j_{x(i,j+1)} &= \Delta x \left(\tfrac{1}{2} + y + \tfrac{1}{2}\Delta y \right) q/\Delta t \\
j_{y(i-1,j+1)} &= \Delta y \left(\tfrac{1}{2} - x - \tfrac{1}{2}\Delta x \right) q/\Delta t \\
j_{y(i,j+1)} &= \Delta y \left(\tfrac{1}{2} + x + \tfrac{1}{2}\Delta x \right) q/\Delta t
\end{aligned}
\tag{9}
$$

In situations where the particle crosses multiple cell boundaries, it is necessary to split motion into a series of smaller motions that cross only a single cell boundary (x and/or y), and apply the algorithm to these smaller motions.

4.3   Field solver

Using the updated values of current density, we can now integrate the field equations. As stated above we generally assume a well known initial condition for the fields (typically $E, B = 0$) and rewrite Maxwell's equations so that the time derivatives of **E** and **B** are in evidence. This leads to:

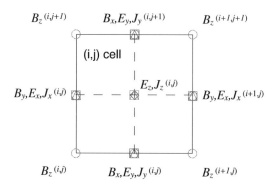

Figure 6. *Definition of field components on a simulation grid cell. Magnetic field components are marked with circles, electric field with squares, and current density with triangles.*

$$\frac{\partial \mathbf{E}}{\partial t} = 4\pi \mathbf{j} - c\nabla \times \mathbf{B}$$

$$\frac{\partial \mathbf{B}}{\partial t} = -c\nabla \times \mathbf{E} \tag{10}$$

Note that only the current density $\mathbf{j}$ is required for advancing the fields. The rotational operator is replaced by a finite-difference approximation on the mesh, and to achieve higher spatial accuracy we follow the technique devised by Yee (1966). To this end we define $\mathbf{E}$, $\mathbf{B}$ and $\mathbf{j}$ on shifted meshes as shown on figure 6. In 1D this can be viewed as defining $\mathbf{E}$ and $\mathbf{j}$ on points x_i, and $\mathbf{B}$ on points $x_{i+\frac{1}{2}}$). In the beginning of the time integration the magnetic and electric fields are centered at time t^n and the current density is centered at time $t^{n+1/2}$.

With the same reasoning we could also argue that it would be advantageous to use $\mathbf{E}$ and $\mathbf{B}$ centered at different time. However, due to the calculation of the Lorentz force (2) this is not usual, and second order accuracy in time is obtained by splitting the time integration in three steps: i) we first advance the magnetic field by half time step, $\mathbf{B}^n \to \mathbf{B}^{n+\frac{1}{2}}$, then ii) we advance the electric field by a full time step, $\mathbf{E}^n \to \mathbf{E}^{n+1}$, and finally iii) we advance the magnetic field by another half time step $\mathbf{B}^{n+\frac{1}{2}} \to \mathbf{B}^{n+1}$. The discretized equations can be easily derived, and as an example we show here the equation for the electric field in two dimensions:

$$E_{x(i,j)}^{n+1} = E_{x(i,j)}^n + \left(\frac{B_{z(i,j+1)}^{n+1/2} - B_{z(i,j)}^{n+1/2}}{\Delta y} - J_{x(i,j)}^{n+1/2} \right) \Delta t$$

$$E_{y(i,j)}^{n+1} = E_{y(i,j)}^n + \left(-\frac{B_{z(i,j+1)}^{n+1/2} - B_{z(i,j)}^{n+1/2}}{\Delta x} - J_{y(i,j)}^{n+1/2} \right) \Delta t \tag{11}$$

$$E_{z(i,j)}^{n+1} = E_{z(i,j)}^n + \left(\frac{B_{x(i,j+1)}^{n+1/2} - B_{x(i,j)}^{n+1/2}}{\Delta y} - \frac{B_{y(i+1,j)}^{n+1/2} - B_{y(i,j)}^{n+1/2}}{\Delta x} - J_{z(i,j)}^{n+1/2} \right) \Delta t$$

We can test the accuracy and stability of this method by seeing how it reproduces a plane electromagnetic wave in vacuum. If we use a test function in the form $(\mathbf{E}, \mathbf{B}) = (\mathbf{E}_0, \mathbf{B}_0)\, e^{-i(\omega t - \mathbf{k} \cdot \mathbf{x})}$ and replace it in the diffence equations in two dimensions (11) we obtain the following dispersion relation:

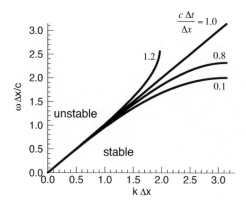

Figure 7. *Dispersion relation for a plane wave in 1D for several choices of time step.*

$$\left[\frac{\sin\left(\frac{\omega \Delta t}{2}\right)}{c\,\Delta t}\right]^2 = \left[\frac{\sin\left(\frac{k_x \Delta x}{2}\right)}{\Delta x}\right]^2 + \left[\frac{\sin\left(\frac{k_y \Delta y}{2}\right)}{\Delta y}\right]^2 \tag{12}$$

The wave frequency, ω, will only be real (no damping or growth of the wave) if the following equation is respected.

$$\frac{1}{c^2 \Delta t^2} > \frac{1}{\Delta x^2} + \frac{1}{\Delta y^2} \tag{13}$$

This is the so called Courant condition in 2D. Similar equations can easily be derived in 1D and 3D. Is the time step chosen is larger than Courant condition the discretized field equation will therefore become unstable. This will also limit the range of motion a particle can have in a single time-step. Equation (13) also ensures that an object moving at the speed of light, c, will never travel more than one cell, thus respecting the limitations imposed by the particle pusher.

However, one must not simply choose a time step which is much smaller than the Courant condition, as this will also lead to non physical results. Figure 7 shows the numerical solution of the dispersion relation that we obtain for 1D for several values of $c\Delta t/\Delta x$. As discussed above when $c\Delta t/\Delta x > 1$ the equations become unstable, but the solutions are only exact $(v_{ph} = c)$ when $c\Delta t/\Delta x$ is exactly 1.

For smaller values of Δt, the phase velocity of our waves will always be smaller than c, and for large wave numbers k, this velocity can drop as low as $0.637c$. Although this is not critical for most particle-in-cell simulations, in laser-plasma interactions where highly relativistic particles are expected, the velocity of these particles may exceed v_{ph}, leading to numerical Cherenkov radiation. It is therefore advised to use the largest possible time step, while maintaining simulation stability

4.4 Normalization

Particle-in-Cell simulations are usually done in normalized units. Besides being the usual units in theoretical plasma physics, these units provide a significant advantage in terms of simulations

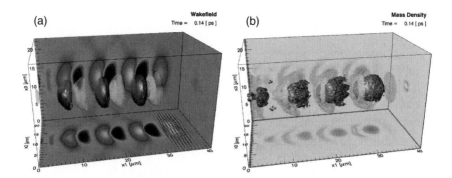

Figure 8. *Wakefield (a) and Plasma wave density (b) in the laser wakefield accelerator.*

since the multiplication by several constants (e.g., m_e, e and c) is avoided, thus leading to better performance from the simulation codes. Also, by expressing the simulation quantities in terms of the fundamental plasma quantities, the results are more general and not bound to some specific geometry we may have chosen.

One possibility, such as the one used in OSIRIS framework (Fonseca 2002) is normalizing charge to the electron charge, e, normalizing mass to the electron mass, m_e, normalizing speed to the speed of light in vacuum, c, and normalizing frequency to the electron plasma frequency or the laser frequency. If the electron plasma frequency is chosen then we get:

$$
\begin{aligned}
t' &= \omega_p t \\
\mathbf{x}' &= \frac{\omega_p}{c}\mathbf{x} \\
\mathbf{p}' &= \frac{\gamma v}{c}\mathbf{p}
\end{aligned}
\qquad
\begin{aligned}
\mathbf{E}' &= e\frac{\omega_p}{m_e c}\mathbf{E} \\
\mathbf{B}' &= e\frac{\omega_p}{m_e c}\mathbf{B}
\end{aligned}
\tag{14}
$$

For laser plasma interaction studies the laser intesity in terms of normalized vector potential, a_0, is not changed in simulation units.

5   Applications

As an example we show here results from a 3D simulation of the laser wakefield accelerator (LWFA), proposed by Tajima and Dawson in 1979 (Tajima 1979). In the LWFA a short ultra-high intensity laser pulse drives a plasma wave that can then be used to accelerate electrons. The wakefield is driven most efficiently when the laser pulse length $L = c\tau_L$ is approximately the plasma wavelength $\lambda_p = 2\pi c/\omega_p$. State-of-the art laser technology is already available for LWFA. Figures 8 (a) and (b) show the wakefield and the plasma density for a simulation of this accelerator. This simulation was done using the OSIRIS 2.0 framework (Fonseca 2002) in 3D for a 30 fs, 800 nm laser pulse, with an intensity $I = 10^19$ W/cm^2 propagating through a plasma with a density of $n_e = 1.38 \times 10^{19}$ cm^{-3}. The simulation was run for a total laser propagation length of 40 μm, using a $512 \times 160 \times 160$ grid, and a total of 1.05×10^8 particles.

Besides the information shown here, the simulation also provides detailed information on all plasma parameters, as well as time-resolved analysis of the evolution of our system, indicating that for these parameters, electrons with sufficient thermal energy are trapped by the plasma wave generated by the laser pulse and are continuously accelerated by gradients of the order of GeV/m.

6 Conclusions

Numerical modeling of laser plasma interactions present a formidable challenge to computational physicists. Present day state of the art machines already provide the necessary computing power for detailed realistic simulations on this field, putting the pressure on researchers to effectively take advantage of these outstanding tool. The particle in cell methods devised accurately model the physics involved, and can be extended to handle extra physics, providing great insight into the microphysics involved, and helping theoreticians and experimentalists alike. This is an exciting cutting edge discipline which has become a cornerstone of laser-plasma interaction studies.

Acknowledgments

The author would like to thank the organizers of the school for the kind invitation, V. Decyk who kindly allowed us the use of his BEPS1D code for the computational workshop, and also C. Ren and L. O. Silva for their invaluable help in the preparation of the computational lectures and this paper. Work partially supported by Fundação para a Ciência e Tecnologia, Portugal.

References

Birdsall, C. K., and Langdon, A. B. (1991), *Plasma Physics via Computer Simulations*, IoP Publishing, Bristol, UK.
Buneman, O. (1959), *Phys. Rev.* **115**, 503-517.
Dawson, J. M. (1962), *Phys. Fluids* **5**, 445-459.
Dawson, J. M. (1983), *Rev. Mod. Phys.* **55**, 403-447.
Dawson, J. M. (1995), *Phys. Plasmas* **2**, 2189-2199.
Fonseca, R. A. et al. (2002), *Lect. Notes Comp. Sci.* **2331**, 342-351.
Hockney, R. W., and Eastwood, J. W. (1981), *Computer Simulation Using Particles*, McGraw-Hill, New York.
Kruer, W. L. (1981), *The Physics of Laser Plasma Interactions*, McGraw-Hill, New York.
Lindmann, E. (1975), *J. Comp. Phys.* **18**, 66-78.
Sterling, T. et. al (1995), *Proceedings of the 24th International Conference on Parallel Processing*, I11-I24.
Tajima, T., and, Dawson, J. M. (1979), *Phys. Rev. Lett.* **43**, 267-270.
Takizuka, T., and Abe H., *J. Comp. Phys.* **25**, 205-219.
Vay, J. L. (2000), *J. Comp. Phys.* **165**, 511-521.
Villasenor, J., and, Buneman, O. (1992), *Comp. Phys. Comm.* **69**, 306-316.
Yee, K. S. (1966), *IEEE Trans. Ant. Prop.* **14**, 302-307.

7

Particle Acceleration in Plasmas

R. Bingham

Rutherford Appleton Laboratory, Chilton, Didcot, Oxon, U.K.

1   Introduction

Studies of charged particle acceleration processes remain one of the most important areas of research in laboratory, space and astrophysical plasmas. In this article, we present the underlying physics and the present status of high gradient and high energy plasma accelerators. We will focus on the acceleration of charged particles to relativistic energies by plasma waves that are created by intense laser and particle beams. The generation of relativistic plasma waves by intense lasers or electron beams in plasmas is important in the quest for producing ultra-high acceleration gradients for accelerators. With the development of compact short pulse high brightness lasers and electron positron beams, new areas of studies for laser/particle beam-matter interactions is opening up. A number of methods are being pursued vigorously to achieve ultra-high acceleration gradients. These include the plasma beat wave accelerator (PBWA) mechanism which uses conventional long pulse (~ 100 ps) modest intensity lasers ($I \sim 10^{14}$ W/cm^2 – 10^{16} W/cm^2), the laser wakefield accelerator (LWFA) which uses the new breed of compact high brightness lasers (< 1 ps) and intensities $> 10^{18}$ W/cm^2, self-modulated laser wakefield accelerator (SMLFA) concept which combines elements of stimulated Raman forward scattering (RFS), and electron acceleration by nonlinear plasma waves excited by relativistic electron and positron bunches the plasma wakefield accelerator. In the ultra-high intensity regime, laser/particle beam-plasma interactions are highly nonlinear and relativistic, leading to new phenomenon such as the plasma wakefield excitation for particle acceleration, relativistic self-focusing and guiding of laser beams, high-harmonic generation, acceleration of electrons, positrons, protons and photons. Fields greater than one GV/cm have been generated with monoenergetic electron beams accelerated to the 100 MeV range in millimetre distances recorded. Plasma wakefields driven by positron beams at the SLAC facility have accelerated the tail of the positron beam. In the near future, laser plasma accelerators will be producing GeV particles. Ion acceleration in solid laser-target interactions from foils and clusters irradiated by intense lasers is proving to be important ion accelerator for studies in fusion and in proton beam production. This opens up a whole area of uses from transmutation of elements for isotope production to pion production, leading perhaps to muon sources with applications in medicine and neutrino physics.

2   High energy plasma accelerators

Plasma is an attractive medium for particle acceleration (Dawson, 1989) because of the high electric field it can sustain with studies of acceleration processes remaining one of the most important areas of research in both laboratory space and astrophysical plasmas. The subject is also vast, making it almost impossible to cover all aspects. Our objective here is to concentrate mainly on the fundamental physics of particle acceleration by relativistic plasma waves that are generated, for example, by intense laser or particle beams. In a plasma-based accelerator (Tajima and Dawson, 1979), particles gain energy from a longitudinal plasma wave. To accelerate particles to relativistic energies, the plasma waves have to be sufficiently intense, with a phase speed close to the speed of light c in vacuum.

A number of laboratory schemes are used to generate intense plasma waves that can accelerate charged particles. The most successful are those based on large amplitude relativistic plasma waves generated by lasers, first proposed by Tajima and Dawson (Tajima and Dawson, 1979). Particle acceleration by relativistic electron plasma waves has been demonstrated in a number of experiments, the most recent (Modena et al., 1995; Gordon et al., 1998; Santala et al., 2001) producing more than hundred MeV electrons in distances of about 1 mm. The resulting accelerating fields as high as 1 GV/cm have been achieved in these experiments. Note that the maximum accelerating field in conventional high energy accelerators is of the order of 20 MV/m.

The rapid advance in laser technology on femtosecond pulse amplification in the mid 80's initiated the development of compact terawatt and petawatt laser systems with ultra-high intensities ($\geq 10^{18}$ W/cm^2), with modest energies (≤ 100 J) and short sub-picosecond pulses (≤ 1 ps). This new breed of lasers whether they be excimer, dye, gas or glass are commonly referred to as T^3 (tabletop-terawatt) lasers "T cubed". These lasers have now made it possible to study laser plasma interactions at ultra-high intensities or high brightness where the laser-electron interaction becomes highly nonlinear and relativistic resulting in a wide variety of interesting phenomena such as (i) plasma wakefield excitation, (ii) relativistic self-focusing and guiding of lasers in plasma channels, (iv) relativistic self-phase modulation, (v) photon acceleration, (vi) proton acceleration, (vi) harmonic generation, (vii)ultra - high magnetic field generation etc. In the past, several authors (Malka et al., 2002; Joshi et al., 1984; Wilson, 1990; Bingham, 1994; Joshi et al., 1995; Tajima, 1995; Umstadter et al., 1995; Esarey et al., 1996) described the essential physics of high energy plasma accelerators, as well as the progress at the time.

Short pulse high brightness lasers are not the only way to produce relativistic waves in plasmas. For a number of years some groups have been using the beat wave process to generate intense relativistic electron plasma waves in preformed plasmas. The beat wave process relies on using two long laser pulses (≤ 100 ps) of moderate intensity $I \simeq 10^{15}$ W/cm^2 co-linearly injected into a low density plasma such that the plasma frequency equals the difference frequency of the two laser beams. Under such conditions, large amplitude relativistic plasma waves are generated with a phase velocity equal to the group velocity of the laser beam which is close to the speed of light.

The generation of relativistic plasma waves using the laser wakefield or the beat wave mechanism was first proposed by Tajima and Dawson in 1979 (Tajima and Dawson, 1979) for producing an ultra-high gradient plasma accelerator. But it was not until 1994 that Clayton *et al.*,(Clayton et al., 1993) and Everett *et al.*, (Everett et al., 1994) demonstrated that high-energy injected electrons could be trapped and accelerated to significant energies by relativistic plasma waves. The main emphasis in the beat wave experiments was to demonstrate particle acceleration by the relativistic plasma wave. In these experiments, the maximum accelerating field gradient E was limited by the wave breaking, which occurs for a cold plasma when the plasma wave density perturbation δn equals the mean plasma number density n_0. Wavebreaking occurs when higher harmonics created due to nonlinear processes distort an initially sinusoidal wave into a lopsided triangular waveform with a steep leading edge, the gradient of which eventually becomes infinite. The profile becomes multi-valued and the wave turns over and "breaks" like an ocean wave on the beach, converting the wave energy into thermal energy of the particle. In reality, the maximum wave amplitude is less than this and is determined mainly by relativistic de-tuning due to the relativistic electron mass increase. A significant result was obtained from a series of wakefield experiments carried out in the UK, USA and France. Monoenergetic electron beams accelerated to the 100 Mev's energy range have been generated in multimetre length plasmas (Mangles et al., 2004; Geddes et al., 2004; Faure et al., 2004). In both of the proposed mechanisms for producing ultra-high accelerating electric fields, the electron quiver velocity in the plasma wave, defined as $v_{osc} = eE/m_e\omega_p$, is greater than the speed of light, where e is the magnitude of the electron charge, E is the electric field of the plasma wave, m_e is the electron rest mass, and ω_p is the electron plasma frequency. The electron quiver velocity in the field of the laser, defined as $v_{osc} = eE/m_e\omega_L$, in this case E is the laser electric field and ω_L is the laser frequency, for the beat wave case is not relativistic, whereas in the high brightness laser field it is relativistic. This makes a considerable difference to the physics of nonlinear interactions between the two processes. Alternatively, instead of laser beams, charged particle beams are also used for generating large amplitude plasma waves. Here, multi GeV electron and positron beams are used to further enhance the energy of accelerated particles (Katsouleas et al., 1999; Lee et al., 2001; Joshi et al., 2002; Blue et al., 2003) because of the incredibly strong electric fields that are induced in the plasma by space charge effects.

Conventional accelerators use radio frequency (RF) waves to accelerate particles, these devices are limited by the breakdown voltage of the waveguide structure to maximum accelerating fields of about 50MV/m. The accelerating gradients and focusing strength that have been demonstrated in plasma accelerators are orders of magnitude greater of order GV/m. The success of plasma accelerators is plotted in figure 1 known as the "Livingston curve" that charts the progress of particle physics accelerators over the decades. Since the early 90's significant progress has been made in both laser and particle beam driven plasma accelerators. Accelerating gradient although important is not the only parameter needed to make a successful accelerator, Luminosity and emittance are two others that have to match or better conventional accelerators. For example the next linear colliders being planned will have Luminosity in the region $10^{34} cm^{-2} s^{-1}$ which is beyond the present capability of plasma accelerators. Plasma accelerators are ideal at providing a compact, short pulses accelerator, they may also be useful at increasing the energy of conventional accelerators using the afterburner concept.

In this chapter, we review the underlying physics and the present status of plasma based charged particle accelerators which are of significant importance (Joshi et al., 2002; Blue et al., 2003; Mora et al., 2001; Bingham et al., 1987) in high energy physics and in medicine. Specifically, we shall concentrate on the generation of large amplitude electron plasma waves by the beat wave and laser wakefield mechanisms, and by particle beams. We will also touch on such topics as photon acceleration, relativistic self-focusing and plasma guiding, as well as the generation of high energy proton beams due to intense laser cluster/solid target interactions. The laser guiding process is of particular importance since intense non-diverging laser beams would be useful to deliver an extended interaction region for plasma wave accelerators which are limited at present to a Rayleigh length. On the other hand, powerful proton beams may be beneficial to medical radiation therapy and isotope production.

The lecture is organized in the following fashion. In Sec. 2 we present the physics of relativistic plasma wave acceleration. In Sec. 3 we discuss the plasma beat wave accelerator. The laser wakefield accelerator scheme is presented in Sec. 4. The model equations for laser wake field excitation are described in Sec. 5. Section 6 is devoted to the mechanism of self-modulated wakefields due to forward Raman scattering. Section 7 deals with the physics of photon acceleration in plasmas. Section 8 contains materials pertaining to relativistic self-focusing and optical guiding of intense laser beams. Section 9 presents the essential physics and observations of laser cluster and solid target interactions. Finally, conclusions and outlook are given in Sec. 10.

3 Relativistic plasma wave acceleration

Particle acceleration by relativistic plasma waves has gained a lot of interest lately due to the rapid advance in laser technology and the development of compact terawatt and petawatt laser systems with ultra-high intensities ($\geq 10^{18}$ Watts/cm^2), with modest energies (≤ 100 J) and short sub-picosecond pulses (≤ 1 ps). The strength of the electric field at the focus of these high-power short-pulse lasers $E_\perp$ is directly related to the laser intensity by $eE_\perp \approx 30\sqrt{I}$ GeV/cm. The electric field $E_\perp$ of a laser whose intensity I is 10^{18} W/cm^2 is 30 GV/cm at 10^{21} W/cm^2 the field is ~ 1 TV/cm. Direct use of the laser field for particle acceleration is not straightforward. Since the electric field of the laser is perpendicular to the propagation direction, the maximum energy gain is limited by the distance the particle moves across the wavefront before the electric field changes sign. However, the situation changes when one uses a plasma into which laser energy can be coupled.

Plasma as a medium for particle acceleration has a number of advantages. It has no electrical breakdown limit like conventional accelerating structures which are limited to a maximum field strength of less than 1 MV/cm. A plasma supports longitudinal plasma waves which oscillate at the plasma frequency $\omega_p \equiv \left(4\pi n_0 e^2/m_e\right)^{1/2}$, where n_0 and m_e are the electron number density and the mass, respectively. In these waves the plasma electrons oscillate back and forth at ω_p irrespective of the wavelength. Therefore, these waves can have arbitrary phase speed, v_{ph}; relativistic plasma waves have $v_{ph} \overset{<}{\sim} c$. The electric field E of relativistic plasma waves with an oscillatory density n_1, can be estimated from Poisson's equation, yielding $E = \epsilon\sqrt{n}$ V/cm, where n is the plasma number density in cm^{-3} and ϵ is the plasma wave amplitude or fractional density bunching n_1/n_0. For a plasma density of 10^{19} cm^{-3} accelerating gradients of 1 GV/cm are possible, which is more than thousand times larger than in conventional accelerators. This aspect of plasma accelerators is what makes them a very attractive alternative to conventional accelerators. The high accelerating gradients allows the possibility of building compact "table top" accelerators rather than multi-kilometer sized present day accelerators.

In their seminal paper on plasma based accelerators, Tajima and Dawson (1979) showed how intense short pulse lasers with a pulse length half the plasma wavelength could generate large amplitude relativistic longitudinal plasma waves. This scheme has become known as the Laser Wakefield Accelerator (LWFA). Alternative schemes to excite plasma waves using large laser pulses are: i) the Plasma Beat Wave Accelerator (Tajima and Dawson, 1979) (PBWA) where two long laser pulses with a frequency separation equal to ω_p beat together in a plasma to resonately excite the plasma wave; ii) the Raman forward scattering (RFS) instability where one long intense laser pulse is used, this is now called the self-modulated LWFA scheme (Krall et al., 1993). Alternatively instead of using lasers short relativistic particle beams cna also excite large amplitude relativistic longitudinal plasma waves.

In a beam driven plasma wakefield accelerator (PWFA) a large amplitude relativistic plasma waves is excited by a short, in comparison to the plasma wavelength, high charge relativistic beam. The Coulomb force of the beam's space charge expels plasma electrons, which rush back in after the beam has passed setting up a plasma oscillation (Chen et al., 1985). Both electrons or positrons can be used to excite the plasma wakefield, in the case of positrons electrons from the background plasma are pulled in by the bunch these electrons overshoot and set up the plasma oscillation. Tajima and Dawson (1979) showed that the maximum energy gain ΔW of a particle in a relativistic plasma waves with $v_{ph} \leq c$ is

$$\Delta W = 2\epsilon\gamma^2 m_e c^2, \tag{1}$$

where γ is the Lorentz factor associated with the phase velocity of the plasma wave $\gamma = \left(1 - v_{ph}^2/c^2\right)^{1/2}$. The phase velocity of the plasma wave is equal to the group velocity v_g of the laser in the plasma, viz. $v_{ph} = v_g = c\left(1 - \omega_p^2/\omega^2\right)^{1/2} \approx c\left(1 - \frac{\omega_p^2}{2\omega^2}\right)$, where ω is the laser frequency; therefore, $\gamma = \omega/\omega_p$ and the maximum energy gain is

$$\Delta W = 2\epsilon\frac{\omega^2}{\omega_p^2}m_e c^2. \tag{2}$$

It is clear that for given values of ω there is a trade-off to be considered in choosing ω_p. From the group velocity v_g we see that a low value of ω_p is required to minimize the phase slip of extremely relativistic electrons with respect to the wave while a high value of ω_p is necessary to maximize the accelerating field E. We want to maximize E by increasing ω_p but this minimizes the energy gain ΔW due to phase slip. As the electron accelerates it slips forward in phase and eventually outruns the useful part of the accelerating field. Knowing the plasma wavelength and the velocity difference between the plasma wave and particle $\Delta v \simeq c - v_g$, it is possible to determine the phase slip, $\delta = \frac{\omega_p^2}{\omega^2}\frac{Lk_p}{2}$ where L is the length of the acceleration stage. Clearly the maximum phase shift cannot exceed π. It can be shown that $5\pi/8$ is a preferable figure, since near zero and π the acceleration is small and inefficient. However, only half this range ie $5\pi/16$ is available. The reason for this is that the plasma is bounded in the traverse direction and consequently there is a radial filed in quadrature with the longitudinal field, which produces a strong defocusing force over the first half of the accelerating phase and a strong focusing force over the second half. This limits the acceptable range to only $5\pi/16$. To prevent phase slip the accelerator must be split into stages of length $\lambda_p\left(\omega^2/\omega_p^2\right)(5/16)$. The maximum energy gain occurs over a distance $L\left(=\Delta W/eE = 2\gamma^2 c/\omega_p\right)$, which is the limit of the de-phasing length. These three schemes are based on the generation of coherent large amplitude electron plasma waves travelling with a phase speed close to the speed of light.

4   Plasma beat wave accelerator

In the plasma beat wave accelerator (PBWA) a relativistic plasma wave is generated by the ponderomotive force of two lasers separated in frequency by the plasma frequency, such that the energy and momentum conservation relations are satisfied, viz. $\omega_1 - \omega_2 = \omega_p$ and $k_1 - k_2 = k_p$, where $(\omega_{1,2}, k_{1,2})$ are the frequencies and wavenumbers of the two lasers respectively, and k_p is the plasma wave wavenumber.

The beat pattern can be viewed as a series of short light pulses each $\pi c \omega_p$ long moving through the plasma at the group velocity of light which for $\omega_{1,2} \gg \omega_p$ is close to c. The plasma electrons feel the periodic ponderomotive force of these pulses. Since this frequency difference matches the natural oscillation frequency of the electron plasma wave, ω_p, the plasma responds resonantly to the ponderomotive force and large amplitude plasma waves would be build up.

If $\omega_p \ll \omega_{1,2}$ then the phase velocity of the plasma wave $v_{ph} = \omega_p/k_p = (\omega_1 - \omega_2)/(k_1 - k_2) = \Delta\omega/\Delta k$ equals the group velocity of the laser beams $v_g = c\left(1 - \omega_p^2/\omega_{1,2}^2\right)^{1/2}$ which is almost equal to c in an underdense plasma. Particles which are injected into the beat wave region with a velocity comparable to the phase velocity of the electron plasma wave, can gain more energy from the longitudinal electric field. Because ω_1 is close to ω_2 and much larger than ω_p, the Lorentz factor γ_p associated with the beat waves is

$$\gamma_p = \left(1 - \frac{v_{ph}^2}{c^2}\right)^{-1/2} = \frac{\omega_{1,2}}{\omega_p} \gg 1. \tag{3}$$

The beat wave process is related to stimulated Raman forward scattering (SRFS). Stimulated Raman scattering is the terminology used in plasma physics for the scattering of electromagnetic waves by longitudinal electron plasma waves. If the scattered electromagnetic wave propagates in the same direction as the incident electromagnetic wave we refer to this as forward scattering. Electron plasma waves are also sometimes referred to as Langmuir waves after E. Langmuir who was the first to discover them. The general equations describing the beat wave and SRFS are similar. It is sufficient to analyze the problem of plasma wave growth and saturation using the relativistic fluid equations for electrons, Maxwell's and Poisson's equations, namely

$$\frac{\partial n_e}{\partial t} + \nabla \cdot (n_e \mathbf{v}_e) = 0 \tag{4}$$

$$\left(\frac{\partial}{\partial t} + \mathbf{v}_e \cdot \nabla\right)(\gamma_e \mathbf{v}_e) = -\frac{e}{m_e}\left(\mathbf{E} + \frac{1}{c}\mathbf{v}_e \times \mathbf{B}\right) - \frac{3k_B T_e}{n_0 m_e}\nabla n_e, \tag{5}$$

$$\nabla \times \mathbf{E} = -\frac{1}{c}\frac{\partial \mathbf{B}}{\partial t}, \tag{6}$$

$$\nabla \times \mathbf{B} = \frac{4\pi}{c}\mathbf{J} + \frac{1}{c}\frac{\partial \mathbf{E}}{\partial t}, \tag{7}$$

and

$$\nabla \cdot \mathbf{E} = 4\pi e (n_e - n_0), \tag{8}$$

where n_e is the electron number density, $\mathbf{v}_e$ is the electron fluid velocity, $\gamma_e = \left(1 - v_e^2/c^2\right)^{-1/2}$ is the relativistic gamma factor, $\mathbf{E}$ and $\mathbf{B}$ are the electric and magnetic fields, respectively, k_B is the Boltzmann constant, and T_e is the electron temperature. Note that to study the stability of large amplitude plasma waves with respect to wave decay, and modulational instabilities due to the ponderomotive force of the plasma wave, the ion dynamics must also be included within the framework of a kinetic treatment (Bingham et al., 1987).

For the plasma beat wave study, one uses eqs. (4)-(8) with $v_e^2/c^2 \ll 1$ in the Lorentz factor and introduces slowly varying amplitudes to describe the nonlinear behavior of the laser field represented by

$$\mathbf{E}_{1,2} = \text{Re } \mathbf{E}_{1,2}'(x,t) \exp\left[i\left(k_{1,2}x - \omega_{1,2}t\right)\right], \tag{9}$$

and takes the plasma density perturbation as

$$\delta n_e = n_e - n_0 = \text{Re } \delta n_e'(x,t).\exp(ik_p x). \tag{10}$$

Notice note that we have not separated the linear time scale from the total time variation for the plasma density perturbation since this mode can be strongly nonlinear.

The equation for the plasma number density perturbation is found to be

$$\left(\frac{\partial^2}{\partial t^2} + \omega_p^2\right)\delta n_e = \frac{3}{8}\omega_p^2\left(\frac{\delta n_e}{n_0^2}\right)^2 \delta n_e - \frac{n_0}{2}\omega_{pe}^2\alpha_1\alpha_2 \exp(-i\delta t), \tag{11}$$

where $\alpha_j = eE_j/m_e\omega_j c$, $j = 1,2$, is the normalized quiver velocity in the field of each laser and $\delta = \omega_1 - \omega_2$ is the frequency mismatch. (Note that we have dropped the primes on δn_e and $\mathbf{E}_j$).

Rosenbluth and Liu (1972) solved eq. (11) in the limit of zero pump depletion, i.e., $\alpha_1 = \alpha_2 = $ constant, obtaining

$$\frac{\delta n_e(t)}{n_0} = \frac{\delta n_e(0)}{n_0} + \frac{1}{4}\alpha_1\alpha_2\omega_p t, \tag{12}$$

which shows that the plasma wave amplitude grows initially linearly with time. However, due to the second term in the right-hand side of eq. (12), which is a cubic nonlinearity in δn_e [cf the first term in the right-hand side of eq. (11)] and is due to the relativistic electron mass increase in the field of the Langmuir wave, the amplitude growth will slow down and saturation will eventually occur. Rosenbluth and Liu (1972) showed that the wave saturated at an amplitude level well before reaching the wave breaking limit of $\delta n_e/n_0 = 1$. We have

$$\frac{\delta n_{e\ max}}{n_0} = \left(\frac{16}{3}\alpha_1\alpha_2\right)^{1/3} \equiv \epsilon. \tag{13}$$

From equation (11) we see that the relativistic mass increase of the plasma electrons has the effect of reducing the natural frequency of oscillation. From the continuity equation one finds that the electron quiver velocity in the plasma wave is $v_{osc}/c = \delta n_e/n_0$ for $\omega_0/k_0 \simeq c$. The natural frequency of oscillation of the plasma wave is reduced, and is given by

$$\omega_p' = \omega_p\left(1 - \frac{3}{8}\frac{v_{osc}^2}{c^2}\right)^{1/2} \equiv \omega_p\left(1 - \frac{3}{8}\frac{\delta n_e}{n_0}\right)^{1/2}. \tag{14}$$

It was pointed out by Tang *et al* (1984) that by deliberately allowing for the relativistic mass variation effect (Sluijter and Montgomery, 1965) and having a denser plasma such that the plasma frequency was initially larger than the laser frequency difference the plasma wave would come into resonance as it grew, allowing a larger maximum saturation value to be attained. An increase of about 50% in the saturated wave amplitude can be achieved by this technique.

The longitudinal field amplitude of these relativistic plasma waves can be extremely large with a theoretical maximum obtained from Poisson's equation, and it is given by

$$E = \epsilon\sqrt{n_0} \text{ V/cm}, \tag{15}$$

where ϵ is the Rosenbluth and Liu (1972) saturation value defined by eq. (13). For the plasma densities of the order of 10^{19} cm^{-3} and saturated values of 30% obtained in present day experiments, the field strength can be of order 10^9 V/cm, which is close to the Coulomb field of a proton E_a at a distance of the order of a Bohr radius a_0, $E_a = 5 \times 10^9$ V/cm. This longitudinal field is capable of producing a GeV electron in a distance of 1cm.

An important consideration in the beat wave scheme is to have sufficiently intense lasers such that the time to reach saturation is short compared to the ion plasma period. When the timescale is longer than the latter, the ion dynamics become important and the electron plasma wave becomes modulationally unstable by coupling to low-frequency ion density perturbations (Aminaroff et al., 1995).

The growth of the plasma wave due to the beat wave mechanism is described by

$$\epsilon = \int_0^t \frac{\alpha_1\alpha_2\omega_p}{4}dt, \tag{16}$$

where $\alpha_{1,2} = eE_{1,2}/m_e\omega_{1,2}c$ is the normalized oscillatory velocity of an electron in the laser fields $E_{1,2}$. As the electron plasma wave grows its electric field amplitude given by eq. (1) becomes large enough that the velocity of an electron oscillating in this field becomes relativistic and the plasma frequency ω_p suffers a small red shift $\Delta\omega_p = -(3/16)\epsilon^2$ due to the relativistic increase in the electron mass. This red shift in the frequency causes the wave to saturate at

$$\epsilon_{sat} = \left(\frac{16}{3}\alpha_1\alpha_2\right)^{1/3}, \tag{17}$$

and the time for saturation is

$$\tau_{sat} = \frac{8}{\omega_p}\left(\frac{2}{3}\right)^{1/3}\left(\frac{1}{\alpha_1\alpha_2}\right)^{2/3}. \tag{18}$$

Other factors which can limit the interaction or acceleration length is diffraction of the laser beams or the pump depletion. Diffraction limits the depth of focus to the Rayleigh length which may be overcome by channelling of the laser (Butler et al., 2002; Dorchies et al., 1999). The pump depletion can be avoided by using more powerful lasers. By using intense short pulse lasers ion instabilities, such as the stimulated Brillouin and plasma modulational instabilities (Esarey et al., 1996), can be avoided. A number of experiments have been carried out which demonstrate that the theoretical estimates are in very good agreement with observations.

The experiments carried out at UCLA (Clayton et al., 1993; Everett et al., 1994) focused a two frequency carbon dioxide laser and injected a 2 MeV electron beam to the same point in a hydrogen plasma at a density of about 10^{16} cm^{-3}. The results showed that approximately 1% or 10^5 electrons of the randomly phased injected electrons are accelerated from 2 MeV to 30 MeV in the diffraction length of about 1 cm. This corresponds to a gradient of 0.03 GV/cm. The measured amplitude of the relativistic plasma waves is 30% of its cold wavebreaking limit, agreeing with the theoretical limit given by eq. (17). What is particularly significant about this experiment is that it demonstrated that the electrons were "trapped" by the wave potential. Only trapped electrons can gain the theoretical maximum amount of energy limited by de-phasing which occurs when the polarity of the electric field of the plasma wave seen by the accelerated electron changes sign.

A trapped electron, by definition, moves synchronously with the wave at the point of reflection in the wave potential. At this point, trapped electrons have a relativistic Lorentz factor $\gamma = \left(1 - v_{ph}^2/c^2\right)^{1/2}$. As the electron continues to gain energy it remains trapped (and eventually executes a closed orbit in the wave potential). Trapping also bunches electrons. In the UCLA experiment (Clayton et al., 1993), the plasma wave has a Lorentz factor of 33 which is synchronous with 16 MeV electrons. Therefore, all electrons observed above 16 MeV are trapped and move forward in the frame of the wave.

The experiment done at the Ecole Polytechnique (Aminaroff et al., 1995) also accelerated electrons but were limited to very small energy gains from 3 MeV to 3.7 MeV due to the saturation of the relativistic plasma wave by the modulational instability of the plasma wave coupling to the low frequency in acoustic mode. This instability is important for long pulses of the order of the ion plasma period ω_{pi}^{-1}, and it limits the wave amplitude to very small values. All beat wave experiments confirm earlier simulations (Mori et al., 1988) and theoretical work demonstrated the need to use short pulses to avoid competing instabilities.

The success of the experiments indicate that it should be possible to accelerate electrons to 1 GeV in a single stage laser plasma accelerator. In such an experiment an injected 10 MeV beam of electrons of 100 A could produce about 10^8 electrons at 1 GeV energies. The necessary laser power required is ~ 14 TW $\left(14 \times 10^{12}\text{W}\right)$ with a pulse duration of 2 ps, corresponding to laser energy of 28 Joules and wavelengths of 1.05 μm and 1.06 μm in a plasma with density 10^{17} cm^{-3} and interaction length $\simeq 3$ cm. From these parameters we find that the plasma wave will saturate at a value of $\delta n/n_0 \simeq 0.45$, resulting in a field gradient $E = 0.45\sqrt{n_0} \simeq 140$ MV/cm for $n_0 \simeq 10^{17}$ cm^{-3}. Assuming no self-focusing or laser guiding this accelerating field is constant over a Rayleigh length $R \simeq \pi\theta^2/\lambda \simeq 0.34$ cm, where θ is the spot size. This results in a maximum energy gain of

$$\Delta W = eE\pi R \simeq 150 \text{ MeV}. \tag{19}$$

By using a discharge channel or capillary channel (Dorchies et al., 1999; Butler et al., 2002) could increase the interaction length considerably. For an interaction length of about 1 cm a GeV is possible.

One of the problems to overcome is the ion instabilities due to the small ion plasma period, which is of order 15 psec. To avoid ion plasma instabilities, such as the ion modulational instability, the plasma density could be reduced. This has the effect of making the laser beams appear shorter, but it also reduces the maximum accelerating gradient.

5   Laser wakefield accelerator

In the Laser Wakefield Accelerator scheme, a short laser pulse, whose frequency is much larger than the plasma frequency, excites a wake of plasma oscillations (at ω_p) due to the ponderomotive force much like the wake of a motor boat. Since the plasma wave is not resonantly driven, as in the beat wave, the plasma density does not have to be of a high uniformity to produce large amplitude waves. As an intense pulse propagates through an underdense plasma, $\omega_0 \gg \omega_p$, where ω_0 is the laser frequency, the ponderomotive force associated with the laser envelope $F_{pond} \simeq -(m/2)\nabla v_{osc}^2$ expels electrons from the region of the laser pulse and excites electron plasma waves. These waves are generated as a result of being displaced by the leading edge of the laser pulse. If the laser pulse length ($c\tau_L$) is long compared to the electron plasma wavelength, then the energy in the plasma wave is re-absorbed by the trailing part of the laser pulse. However, if the pulse length is approximately equal to or shorter than the plasma wavelength, viz. $c\tau_L \simeq \lambda_p$, the ponderomotive force excites plasma waves or wakefields with a phase velocity equal to the laser group velocity, and are not re-absorbed. Thus, any pulse with a sharp rise or a sharp fall on a scale of c/ω_p will excite a wake. With the development of high brightness lasers the laser wakefield concept (Tajima and Dawson, 1979) has now become a reality. The focal intensities of such lasers are $\geq 10^{19}$ W/cm^2, with $v_{osc}/c \geq 1$, which is the strong nonlinear relativistic regime. Any analysis must, therefore, be in the strong nonlinear relativistic regime and a perturbation procedure is invalid.

The maximum wake electric field amplitude generated by a plane polarized laser pulse has been given by (Sprangle et al., 1988) in the one-dimensional limit as

$$E_{max} = 0.38(v_{osc}/c)^2 \left(1 + v_{osc}^2/2c^2\right)^{-1/2} \sqrt{n_0} \text{V/cm for } v_{osc}/c \sim 4, \text{ and } n_0 = 10^{18} \text{cm}^{-1},$$

$E_{max} \approx 2$GV/cm, and the time to reach this amplitude level is of the order of the laser pulse length. There is no growth phase as in the beat wave situation, which requires many plasma periods to reach its maximum amplitude.

To get larger wave amplitudes, it has been suggested by several groups (Johnson et al., 1994; Bonnaud et al., 1994; Dalla and Lontano, 1994) to use multiple pulses with varying time delay between pulses. Conclusive experiments (Malka et al., 2002; Nakajima et al., 1995; Nakajima et al., 1995; Marqués et al., 1996; Amiranoff et al., 1998; Siders et al., 1996; Bingham et al., 1992; Wilks et al., 1992; Mendonça, 2001; Antonsen et al., 2004; Tang et al., 1992; Decker et al., 1994; Coverdale et al., 1995; Modena et al., 1995) have been carried out to demonstrate the excitation of the plasma wakefield. The experiments use lasers that delivered energies up to 1J on target with a pulse length of 30 fs. producing a focused intensity of 3×10^{18} W/cm^2

corresponding to a normalized vector potential $a_o = eA/m_oc^2$ of 1.2. The density of plasma chosen was between $(2-6) \times 10^{19} cm^{-3}$ which corresponds to a period of 25 to 14 fs similar to the pulse length of the laser. The most impressive results so far have been carried out by groups in the UK, USA and France, (Mangles et al., 2004; Geddes et al., 2004; Faure et al., 2004). In these experiments monoenergetic electron beams are produced in millimetre sized plasmas. The energy of the beam is in the range 100 MeV corresponding to accelerating gradients of 1GV/cm. These set of experiments demonstrate the principle on monoenergetic beam formation using the wakefield scheme.

6 Equations describing laser wakefield

To understand the laser wakefield excitation mechanism, it is sufficient to use a model based on the one fluid, cold relativistic hydrodynamics and Maxwell equations together with a "quasi-static" approximation a set of two coupled nonlinear equations describing the self-consistent evolution in 1-D of the laser pulse vector potential envelope and the scalar potential of the excited wakefield. Accordingly, one starts with the relativistic electron momentum equation

$$\frac{\partial \mathbf{p}}{\partial t} + v_z \frac{\partial \mathbf{p}}{\partial z} = -e\left(\mathbf{E} + \frac{1}{c}\mathbf{v} \times \mathbf{B}\right), \tag{20}$$

where the relativistic momentum is

$$\mathbf{p} = m_e \gamma \mathbf{v}, \qquad \text{with} \qquad \gamma = \left(1 + p^2/m_e^2 c^2\right)^{1/2},$$

and $\mathbf{v}$ being the electron velocity. In eq. (20) we have assumed that all quantities only depend on z and t; z is the direction of the propagation of the (external) laser pump. The electromagnetic fields are given by

$$\mathbf{E} = -\frac{1}{c}\frac{\partial \mathbf{A}_\perp}{\partial z} - \hat{z}\frac{\partial \phi}{\partial z} \; ; \; \mathbf{B} = \nabla \times \mathbf{A}_\perp; \; \mathbf{A}_\perp = \hat{x}A_x + \hat{y}A_y, \tag{21}$$

where $\mathbf{A}_\perp$ is the vector potential of the electromagnetic pulse and ϕ the ambipolar potential due to charge separation in the plasma.

From eqs. (20) and (21) one finds that the perpendicular component of the electron momentum is

$$\frac{p_\perp}{m_e c} = \frac{e}{m_e c^2}\mathbf{A}_\perp \equiv \mathbf{a}(z,t). \tag{22}$$

Hence, we can write

$$\gamma = \left[1 + \left(\frac{p_\perp}{m_e c}\right)^2 + \left(\frac{p_z}{m_e c}\right)^2\right]^{1/2} \equiv \gamma_a \gamma_\|, \tag{23}$$

where

$$\gamma_a = \left(1 + \mathbf{a}^2\right)^{1/2}; \; \gamma_\| = \left(1 - \beta^2\right)^{-1/2}, \tag{24}$$

and $\beta = v_z/c$.

The equations derived from this model are now the longitudinal component of eq. (20), the electron continuity equation, Poisson's equation, and the wave equation for $\mathbf{a}(z,t)$, which are (Bingham et al., 1992), respectively,

$$\frac{1}{c}\frac{\partial}{\partial t}\left(\gamma_a\sqrt{\gamma_\|^2 - 1}\right) + \frac{\partial}{\partial z}\left(\gamma_a\gamma_\|\right) = \frac{\partial \phi}{\partial z}; \; \varphi \equiv \frac{e\phi}{m_e c^2}, \tag{25}$$

$$\frac{1}{c}\frac{\partial n}{\partial t} + \frac{\partial}{\partial z}\left(n\frac{\sqrt{\gamma_\parallel^2 - 1}}{\gamma_\parallel}\right) = 0, \tag{26}$$

$$\frac{\partial^2\varphi}{\partial z^2} = \frac{\omega_{p0}^2}{c^2}\left(\frac{n}{n_0} - 1\right), \tag{27}$$

and

$$c^2\frac{\partial^2\mathbf{a}}{\partial z^2} - \frac{\partial^2\mathbf{a}}{\partial t^2} = \omega_{p0}^2\frac{n}{n_0}\frac{\mathbf{a}}{\gamma_a\gamma_\parallel}. \tag{28}$$

Assuming a driving pulse of the form

$$\mathbf{a}(z,t) = \frac{1}{2}\mathbf{a}_0(\xi,\tau)\exp(-i\theta) + c.c., \tag{29}$$

where $\theta = \omega_0 t - k_0 z$, ω_0 and k_0 being the central frequency and the wavenumber, respectively, $\xi = z - v_g t$, $v_g = \partial\omega_0/\partial k_0$ is the group velocity, and τ is a slow time-scale such that

$$a_0^{-1}\frac{\partial^2 a_0}{\partial\tau^2} \ll \omega_0^2,$$

and accounting for changes in the pump due to the plasma reaction, we obtain from eq. (28)

$$\left[2\frac{\partial}{\partial\tau}\left(i\omega_0 a_0 + v_g\frac{\partial a_0}{\partial\xi}\right) + c^2\left(1 - \frac{v_g^2}{c^2}\right)\frac{\partial^2 a_0}{\partial\xi^2} + 2i\omega_0\left(\frac{c^2 k_0}{\omega_0} - v_g\right)\frac{\partial a_0}{\partial\xi}\right]\exp(-i\theta)$$

$$+c.c. = \left[c^2 k_0^2 - \omega_0^2 + \frac{n}{n_0}\omega_{p0}^2\gamma_a\gamma_\parallel\right]a_0\exp(-i\theta) + c.c., \tag{30}$$

where ω_{p0} is the plasma frequency of the unperturbed plasma, and c.c. stands for the complex conjugate. Equations (25), (26), (27) and (30) form the basic set for studying the laser wakefield excitation in the "envelope approximation".

In the weakly relativistic regime, the solution has the structure of a wake-field growing inside the electromagnetic pulse and oscillating behind the pulse with the maximum amplitude being reached inside the pulse. Using the quasi-static approximation the time derivative can be neglected in the electron fluid eqs. (25) and (26), yielding the following constants

$$\gamma_a\left(\gamma_\parallel - \beta_0\sqrt{\gamma_\parallel^2 - 1}\right) - \varphi = 1 \tag{31}$$

and

$$n\left(\beta_0\gamma_\parallel - \sqrt{\gamma_\parallel^2 - 1}\right) = n_0\beta_0\gamma_\parallel, \tag{32}$$

where $\beta_0 = v_g/c$. The constants of integration have been chosen in such a way that

$$n = n_0, \gamma_\parallel = 1, \varphi = 0, \tag{33}$$

for $\gamma_a = 1$ and $|a_0|^2 = 0$.

Using eqs. (29) and (32) the general system of eqs. (25)-(28) can be written as two coupled equations (Bingham et al., 1992) describing the evolution of the laser pulse envelope $\mathbf{a}_0$ and the scalar potential φ

$$\frac{\partial^2\varphi}{\partial\xi^2} = \frac{\omega_{p0}^2}{c^2}G, \tag{34}$$

and

$$2i\omega_0\frac{\partial a_0}{\partial\tau} + 2c\beta_0\frac{\partial^2 a_0}{\partial\tau\partial\xi} + \frac{c^2\omega_{p0}^2}{\omega_0^2}\frac{\partial^2 a_0}{\partial\xi^2} = -\omega_{p0}^2 H a_0, \tag{35}$$

where

$$G = \frac{\sqrt{\gamma_\parallel^2 - 1}}{\beta_0 \gamma_\parallel - \sqrt{\gamma_\parallel^2 - 1}} \quad \text{and} \quad H = 1 - \frac{\beta_0}{\gamma_a \left(\beta_0 \gamma_\parallel - \sqrt{\gamma_\parallel^2 - 1}\right)}.$$

The present set of nonlinear equations (34) and (35) are obtained by using a quasi-static approximation, which yields two integrals of the motion given by equations (31) and (32). The model is valid for electromagnetic pulses of arbitrary polarization and intensities $|a_0|^2 \geq 1$.

Equations (34) and (35) can be solved numerically in the stationary frame of the pulse. Equation (34) is solved with the initial conditions $\varphi = 0$ and $\partial\varphi/\partial\xi = 0$ by a simple predictor-corrector method. The envelope equation (35), which describes the evolution of the laser pulse in the presence of the wake potential, is written as two coupled equations for the real and imaginary parts of a_0, and they can be solved implicitly. Numerical solutions of eqs (34) and (35) show (Bingham et al., 1992) the evolution of the excited plasma wake-field potential φ and the electric field E_w as well as the envelope of the laser pulse. There is a significant distortion of the trailing edge of the laser pulse resulting in photon spikes. The distortion occurs where the wake potential has a minimum and the density has a maximum. The spike arises as a result of the photons interacting with the plasma density inhomogeneity with some photons being accelerated (decelerated) as they propagate down (up) the density gradient; this effect, predicted again by John Dawson and his group (Wilks et al., 1992) and by others (Mendonça, 2001), is called the photon accelerator. The distortion of the trailing edge increases with increasing ω_{p0}/ω_0. The longitudinal potential, viz. $e\phi/mc^2 > 1$ or $eE_z/m_e\omega_{p0}c > 1$, is significantly larger than fields obtained in the plasma beat-wave accelerator, which are limited by relativistic detuning, no such saturation exists in the laser wakefield accelerator. Furthermore, Reitsma et al (Reitsma et al., 2002) proposed a new regime of laser wakefield acceleration of an injected electron bunch with strong bunch wakefields. In particular, the transverse bunch wakefield induces a strong self-focusing that reduces the transverse emittance growth arising from misalignment errors.

7   Self-modulated laser wakefield accelerator

Self-Modulated LWFA is a hybrid scheme combining elements of stimulated Raman forward scattering (RFS) and the laser wakefield concept. Raman forward scattering describes the decay of a light wave at frequency ω_0 into light waves at frequency $\omega_0 \pm \omega_p$, and a plasma wave ω_p with $v_{ph} \simeq c$. Although Raman forward scattering generates relativistic plasma waves and was identified as the instability which generated MeV electrons in early laser plasma experiments, it was not considered a serious accelerator concept because the growth rate is too small for sufficient plasma wave amplitudes to be reached before the ion dynamics disrupt the process. However, coupled with the laser wakefield accelerator concept it becomes a viable contender. Short pulse lasers have been demonstrated in Refs. (Antonsen et al., 2004; Mori et al., 1994; Andreev et al., 1995) to self-modulate in a few Rayleigh lengths. This modulation forms a train of pulses with approximately $\pi c/\omega_p$ separation, which act as individual short pulses to drive the plasma wave. The process acts in a manner similar to a train of individual laser pulses. A comprehensive theoretical and simulation studies of RFS and self-modulation pulses in tapered plasma channels have been presented by Penano et al (Penano et al., 2002).

A number of groups e.g., (Modena et al., 1995; Gordon et al., 1998; Santala et al., 2001; Nakajima et al., 1995; Wagner et al., 1997) have recently reported experimental evidence for the acceleration of electrons by relativistic plasma waves generated by a modulated laser pulse. The most impressive results come from a group working with the Vulcan laser at RAL, UK.; this group, which consisted of research teams from Imperial College, UCLA, Ecole Polytechnique, and RAL, reported (Gordon et al., 1998; Santala et al., 2001; Modena et al., 1995; Malka et al., 2002) observations of electrons at energies as high as 120 MeV. The observations of energetic electrons was correlated with the simultaneous observation of $\omega_0 + n\omega_p$ radiation generated by Raman forward scattering. The experiments were carried out using a 25 TW laser with intensities $> 10^{18}$ W/cm^2 and pulse lengths < 1 ps in an underdense plasma $n_0 \sim 10^{19}$ cm^{-3}. The laser spectrum is strongly modulated by the interaction, showing sidebands at the plasma frequency (Gordon et al., 1998). Electrons with energies up to 100 MeV with an inferred minimum acceleration gradient of > 1.60 GV/cm from > 100 MeV electrons over a measured 600 μm interaction length have been observed (Gordon et al., 1998). Laser self-channelling of up to 12 Rayleigh lengths was also observed in the experiment. Such extensive self-channelling was only observed for electron energies up to 40 MeV; at the higher energy ~ 94 MeV the acceleration length was 7 times shorter (Gordon et al., 1998). Similar results of a modulated laser pulse at ω_p have been obtained by a separate Livermore experiment (Coverdale et al., 1995) using a 5 TW laser but only observed 2 MeV electrons. The difference in the energy is explained by a difference in the length of plasma over which acceleration takes place. In the RAL experiment the acceleration length is much larger. It is worth pointing out that in these experiments the accelerated electrons are not injected but are accelerated out of the background plasma. These experiments together with the short pulse wakefield have produced the largest - terrestrial acceleration gradients ever achieved - almost 200 GeV/m.

The Raman forward scattering instability is the decay of an electromagnetic wave (ω_0, k_0) into a forward propagating plasma wave (ω_p, k_p) and forward propagating electromagnetic waves the Stokes wave at $(\omega_0, -n\omega_p)$ and an anti-Stokes wave at $(\omega_0, +n\omega_p)$, where n is an integer; this is a four wave process. One of the earliest papers on forward Raman instabilities in connection with laser plasma accelerators was by Bingham (1983) and McKinstrie *et al* (1992) who discussed a purely temporal theory including a frequency mismatch. More recently a spatial temporal theory was developed in Refs. (Mori et al., 1994; Decker et al., 1996). In this theory the relativistic plasma wave grows from noise; calculating the noise level is non-trivial since there are various mechanisms responsible for generating the noise. For example, the fastest growing Raman backscatter and sidescatter instabilities cause local pump depletion forming a "notch" on the pulse envelope. The plasma wave associated with this notch acts as an effective noise source as seen in the simulations by Tzeng *et al* (1996). Simulations carried out by Tzeng *et al* (1996) are the first to use the exact experimental parameters and show significant growth within a Rayleigh length. In fact, there is also a remarkable agreement with the experimental results of (Modena et al., 1995; Gordon et al., 1998; Santala et al., 2001).

All these experiments rely heavily on extending the acceleration length which is normally limited to the diffraction length or Rayleigh length $L_R = (\omega_0/2c)\,\sigma_0^2$, where ω_0 is the laser frequency and σ_0 is the spot size. In present day experiments this is limited to a few mm, for example the RAL Vulcan CPA laser has a wavelength of 1μm, a spot size of 20μm resulting in a Rayleigh length $L_R \simeq 350\mu$m. To be a useful accelerator laser pulses must propagate relatively stably through uniform plasmas over distances much larger than the Rayleigh length. Relativistic self-focusing is possible if the laser power exceeds the critical power given by $P_c \approx 17\,\omega_0^2/\omega_p^2$ GW, which is easily satisfied for the present high power laser experiments. There are difficulties in relying on the laser pulse to form its own channel since the beam may break up due to various laser-plasma instabilities, such as Raman scattering and filamentation. Relativistic self-guiding over five Rayleigh lengths has recently been reported by (Chiron et al., 1996) using a 10TW laser in a plasma of density 5×10^{18} cm^{-3}. They also note that the effect disappears at larger powers and densities. Alternatively, plasma channels have been demonstrated (Durfee et al., 1995) to be very effective in channelling intense laser pulses over distances much larger than the Rayleigh length. In these experiments, a two laser pulse technique is used. The first pulse creates a breakdown spark in a gas target, and the expansion of the resulting hot plasma forms a channel which guides a second pulse injected into the channel. Pulses have been channelled up to 70 Rayleigh lengths (Durfee et al., 1995) corresponding to 2.2 cm in the particular experiment with about 75% of the energy in the injected pulse focal spot coupled into the guide.

In a preformed channel other instabilities may appear. For example, it has been shown by Shvets and Wurtele (Shvets et al., 1994) that a laser hose instability exists for parabolic channels. Plasma channels are not only important for laser plasma accelerators but have applications in high harmonic generation, for UV and soft x-ray lasers.

8   Particle beam driven plasma wakefield accelerators (PWFA)

The laser accelerator schemes are very effective at producing very high accelerating gradients over short plasma scales of the order of mms. Channelling coupled with guiding can increase this to several centimeters producing GeV's on a table top. Extending these table top accelerators needed for high energy physics requires meter scale plasma, alternatively high energies could be accomplished by staging hundreds of these relatively compact laser accelerators, each providing gains of up to 10 GeV. Staging however, is a formidable challenge requiring hundreds of lasers to be synchronized. Future lasers with intensities or order $10^{24}\omega/cm^2$ or greater created by optical parametric chirped pulse amplification may be powerful enough to propagate through meter long plasmas and produce ultra - high energy beams in the region 100 GeV - 1 TeV suitable for high energy physics experiments. Another approach is to use the existing high energy particle beams that can also creates wakefield in plasma (Chen et al., 1985; Rosenzweig et al., 1993).

A series of experiments at SLAC has demonstrated successfully the beam driven plasma wakefield accelerators (PWFA) (Blue et al., 2003). The SLAC beam is used in the experiments have intensities of the order of $10^{20}W/cm^2$. In a particle - beam driven plasma wakefield accelerator (PWFA) the Coulomb force of the charged particles space charge expels the plasma electrons if it is an electron beam and pulls them in if it is a positron beam. The displaced electrons snap back to restore charge neutrality and overshoot their original positions setting up a plasma wave that trails behind the beam. If the beam is about half a plasma wavelength long the plasma wave electric field amplitude in the linear theory where $eE_p/M\omega_p c \ll 1$ is given by (Joshi et al., 2002) $eE_p = 240(MeV/m)\left(\frac{N}{4\times10^{10}}\right)\left(\frac{0.6mm}{\sigma_z}\right)^2$ where σ_z is the bunch length, and N is the number of particles per bunch. The drivers of the plasma wakefield experiment in the SLAC experiment was the 28.5 GeV, 2.4ps long positron beam from the SLAC linac containing 1.2×10^{10} particles. The beam was focused at the entrance to a 1.4m long chamber filled with a lithium plasma of density $1.8 \times 10^{14} cm^{-3}$. The first two thirds of the positron beam set up by the relativistic plasma wakefields the last third of the beam was accelerated to higher energy. The main body of the beam decelerated at a rate corresponding to 49 MeV/m, while the back of the beam containing about 5×10^8 positrons in a 1 ps slice was accelerated by $79 \pm 15MeV$ in the 1.4m long plasma corresponding to an accelerating electric field of 56MeV/m. The results are in excellent agreeing with a 3-D PIC simulation (Blue et al., 2003) which predicted a peak energy gain of 78 MeV.

In the beam driven wakefield experiment the beam energy is transferred from a large number of particles in the core of the bunch to a fewer number of particles in the back of the same bunch. The wakefield thus acts like a transformer with a ratio of accelerating field to decelerating field of about 1.3 in the experiment. This is the first demonstration that plasma accelerators can be used to increase the energy of the high energy accelerator and leads the way for a possible energy doubler proposed by Lee et al (2001). In the energy doubler or plasma "afterburner" scheme a plasma wakefield accelerator using ten times shorter bunches can double the energy of a linear collider in short plasma sections of length 10m. By reducing the positron bunch by a factor 10 the accelerating gradient can be enhanced a hundred fold to 5 GV/m, which would convert the 50 GeV SLAC linear collider into a 100 GeV machine (Lee et al., 2001). To sustain the luminosity at the interaction point at the nominal level of the original beam without the plasma, the reduction in the number of higher energy particles is compensated by using plasma lenses to reduce the spot size. Higher density plasma lenses are added to the afterburner design just before the interaction point. This type of plasma lens exceeds conventional magnetic lenses by several orders of magnitude in focusing gradient and was successfully demonstrated (JST et al., 1992). Wakes generated by an electron or positron beam can also accelerate positrons and muons. These SLAC experiments have brought the concept of plasma accelerators to the forefront of high energy accelerator research.

9 Photon acceleration

Photon acceleration (Wilks et al., 1989; Mendonça, 2001) is intimately related to short pulse amplification of plasma waves and is described by a set of equations that are similar to those in wakefield accelerators. In the last section on the wakefield amplification we noted that it was much easier to increase the amplitude of the wakefield if we used a series of photon pulses. These pulses have to be spaced in a precise manner to give the plasma wave the optimum "kick". If the second or subsequent photon pulses are put in a different position, 1.5 plasma wavelengths behind the first pulse, the second pulse will produce a wake that is 180° out of phase with the wake produced by the first pulse. The superposition of the two wakes behind the second pulse results in a lowering of the amplitude of the plasma wave. In fact, almost complete cancellation can take place.

The second laser pulse has absorbed some or all of the energy stored in the wake created by the first pulse. From conservation of photon number density the increase in energy of the second pulse implies that its frequency has increased. The energy in the pulse is $W = N\hbar\omega$, where N is the total number of photons in the wave packet.

A simple and rigorous account of the frequency shift can be made by using the photon equations of motion (Mendonça and Silva, 1994)

$$\frac{d\mathbf{r}}{dt} = \frac{\partial \omega}{\partial \mathbf{k}} \quad , \quad \frac{d\mathbf{k}}{dt} = -\frac{\partial \omega}{\partial \mathbf{r}}, \tag{36}$$

where $\mathbf{r}$ and $\mathbf{k}$ are the photon position and the wavevector, respectively, and $\omega = (k^2 c^2 + \omega_p^2(\mathbf{r}, t))^{1/2}$ is the photon frequency.

A net frequency shift (or photon acceleration) occurs in non-stationary plasmas as well as in other time varying media. Let us take, as an example, the case of a laser (or photon) beam interacting with a relativistic ionization front with the velocity v_f. From eq. (36) one can readily obtain the net frequency shift of the laser beam

$$\Delta\omega = \frac{\omega_{pe}^2}{2\omega} \frac{\beta}{1 \pm \beta}, \tag{37}$$

where $\beta = v_f/c$, and the plus and minus signs pertain for co and counter-propagation of the photons with respect to the ionization front. This was confirmed by experiments (Dias et al., 1997). Quite recently, frequency shifts as large as 144 nanometers were observed for ionization fronts with a velocity of $0.99c$ (Lopes et al., 2002). Experiments designed to observe photon acceleration by laser produced wakefields are currently being prepared at RAL in collaboration with IST and Imperial College Groups.

10   Relativistic self-focusing and optical guiding

In the absence of optical guiding the interaction length L is limited by diffraction to $L = \pi R$, where $R = \pi \sigma^2 / \lambda_0$ is the Rayleigh length and σ is the focal spot radius. This limits the overall electron energy gain in a single stage to $E_{max} \pi R$. To increase the maximum electron energy gain per stage, it is necessary to increase the interaction length. Two approaches to keeping high energy laser beams collimated over a longer region of plasma are being developed. Relativistic self-focussing can overcome diffraction and the laser pulse can be optically guided by tailoring the plasma density profile forming a plasma channel (Monot et al., 1995; Chen et al., 1998; Sarkisov et al., 1999).

Relativistic self-focusing uses the nonlinear interaction of the laser pulse and plasma resulting in an intensity dependent refractive index to overcome diffraction (Schmidt and Horton, 1985). In regions where the laser intensity is highest the relativistic mass increase is greatest; this results in a reduction of the plasma frequency. The reduction is proportional to the laser intensity. Correspondingly, the phase speed of the laser pulse will decrease in regions of higher laser intensity. This has the effect of focusing a laser beam with a radial Gaussian profile. This results in the plane wavefront bending and focussing to a smaller spot size. Relativistic self-focusing has a critical laser power threshold, P_{cr}, given by

$$P \geq P_{cr} \simeq \frac{2m_e^2 c^5 \omega_0^2}{e^2 \omega_p^2} = 17 \frac{\omega_0^2}{\omega_p^2} \quad \text{GW}. \tag{38}$$

The laser must also have a pulse duration that is shorter than both the collisional and ion plasma periods to avoid the competing effects of thermal and ponderomotive self-focusing. The shape of the self-focused intense short pulses is also interesting. There is a finite time for the electrons to respond and the front of the pulse propagates unchanged. The trailing edge of the pulse compresses radially due to the nonlinear relativistic self-focusing. It is only the trailing edge of the pulse that is channelled. Several authors (Abramyan et al., 1992; Komashko et al., 1995; Cattani et al., 2001) have presented analytical studies of self-focusing and relativistic self-guiding of an ultrashort laser pulse in the plasma, taking into account ponderomotive force and relativistic mass variation nonlinearities. The ponderomotive force of superstrong electromagnetic fields expels electrons, thus producing "vacuum channels" that guide the radiation and stable channeling with power higher than the critical one can take place. Specifically, Cattani $et\ al$ (2001) presented a simple model to investigate multifilament structures of a circularly polarized laser beam in 2D planer geometry including electron cavitation that are created by the relativistic ponderomotive force. The governing equations describing filamentary structures are (Cattani et al., 2001)

$$\nabla^2 a_\perp + \left(1 - \frac{\alpha n}{\gamma_\perp}\right) a_\perp = 0, \tag{39}$$

and

$$\nabla^2 \varphi = \alpha(n - 1), \tag{40}$$

where $\varphi = \gamma_\perp - 1$ if and only if $n \neq 0$, $\gamma_\perp = \sqrt{1 + a_\perp^2}$. and $\alpha = N_0/(1 - k_z^2/k^2)$. Here, we have denoted $a_\perp = e|\mathbf{A}_\perp|/m_e c^2$, $N_0 = n_0/n_c$, $n_c = m_e \omega^2/4\pi e^2$, $\varphi = e\phi/m_e c^2$, and $\mathbf{r} = k\sqrt{1 - k_z^2/k^2}\mathbf{r}_\perp$, with $k_\perp \ll k_z$. Here $k_\perp$ is the transverse component of the laser wave number and k_z is the axial propagation constant. Equations (38) and (39) admits an exact solitonlike solution as well multifilament structures whose specific forms are presented by Cattani $et\ al$ (2001).

Another way of forming a guided laser is to create a preformed plasma with one long pulse laser, that heats and expands on a nanosecond timescale forming a plasma density channel

(Borisov et al., 1992; Durfee and Milchberg, 1993; Nikitin et al., 1999). Alternatively, long plasma channels could also be formed by high voltage capillary discharges which produce the same effect (Kaganovich et al., 1999; Spence and Hooker, 2001). Bulanov *et al* (1995) have presented the results of a two-dimensional particle-in-cell simulation of the nonlinear propagation of a short, relativistically intense laser pulse in an underdense plasma. They found that such a pulse can be focused in a plasma with strong magnification of its amplitude and channeling in a narrow channel shaped like a "bullet". Sprangle *et al* (1999) have studied the dynamics of short laser pulses propagating in plasma channels, taking finite-pulse-length as well as nonlinear focusing effects into account. They found that in plasma channels, pulses can undergo an envelope modulational which is damped in the front and initially grows in the back of the pulse. Finite-pulse-length effects also significantly increase the nonlinear focusing. A nonlinear theory for nonparaxial laser pulse propagation in plasma channels is presented by Esarey *et al* (2000) who, in the adiabatic limit, analyzed pulse energy conservation, nonlinear group velocity, damped betatron oscillations, self-steepening, self-phase modulation, and shock formation. In the non-adiabatic limit, the nonlinear coupling of FRS and the self-modulational instability leads to a reduced growth rate. The possibility of stable laser pulses in a tapered plasma channel for GeV electron acceleration has also been discussed (Sprangle et al., 2000), including wakefields and relativistic and nonparaxial effects.

Recently, it has been shown (Hoda et al., 1999) that relativistic self-focusing, in three-dimensions, of a linearly polarized laser pulse propagating in an underdense plasma with power above the threshold power P_{cr} is anisotropic and evolves differently in the plane in which the pulse electric field oscillates and in the plane in which the magnetic field oscillates. The effect of the pulse polarization and of the accelerated fast electrons on the propagation anomalies ("hosing" (Duda et al., 1999) and "snaking") of a high-intensity laser pulse has been investigated (Naumova et al., 2001) by means of fully electromagnetic relativistic two-dimensional particle-in-cell simulations. The results show that the hosing (Duda et al., 2000; Duda et al., 2002) occurs in the front of the pulse, and it corresponds to low-frequency oscillations of the electron current at the sharp front of the pulse and to a periodic change of the refractive index. The hosing of the pulse may be responsible for self-trapped electrons as well as for the excitation of "wake surface waves" (Macchi et al., 2001) at the walls of the self-focusing channel. Snaking occurs in the later evolution of long laser pulses almost independently of their polarization.

In the present laser experiments it is in principle feasible to reach 1 GeV range of energies without resorting to hybrid mechanisms. In the future, in order to achieve TeV energies substantial research must be carried out with all combinations as well as develop larger plasma channels and introduce staging. The present experiments will make it possible to judge whether a TeV machine is possible with desirable luminosities and emittances. Future high energy accelerators may combine conventional accelerators together with a plasma afterburner to increase the beam energy. The SLAC experiments suggest that the beam energy can be doubled if the pulse length is halved.

11 Laser cluster/target interactions

The interaction of intense short laser pulses with atomic clusters and solid targets has become an important area of research. Experiments show that the interaction of short (psec), intense $\left(\sim 10^{18} \mathrm{W/cm}^2\right)$ laser pulses with rare gas clusters is responsible for the production of highly energetic electrons and ions (Ditmire et al., 1998; Lezius et al., 1998; Kurshenlnick et al., 1999; Clark et al., 2000; Skobelov et al., 2002), x-ray emission in the keV range (McPherson et al., 1994), coherent high-harmonic generation (Donnelly et al., 1996), plasma waveguide formation (Ditmire et al., 1998b), and applications to nuclear fusion (Ditmire et al., 1999). The cluster expansion mechanism can be described using either a hydrodynamical model or a Coulombic repulsion model, depending on the laser intensity and pulse duration as well as on the cluster size and charge state. In the hydrodynamic regime, dominant for lower intensities $\left(< 10^{17}\text{W/cm}^2\right)$ and longer pulses, the electrons are held in the cluster via space-charge attraction from neighbouring ions and the cluster is heated to very high temperatures. Absorption of radiation is mainly due to above threshold ionisation and inverse bremsstrahlung. The pressure buildup inside the cluster core gives rise to a hydrodynamically expanding plasma, the electrons leaving first due to their greater mobility and creating a space-charge field which accelerates the ions, similar to what is observed during the expansion of a hot solid target plasma into vacuum (Wickens et al., 1979). Both the ionisation process and the thermal heating can easily be treated, providing good qualitative and quantitative agreements with experiments and ion energies up to 1 MeV (Ditmire et al., 1998b).

Recent advances in laser technology have made possible ultraintense $\left(> 10^{17}\text{W/cm}^2\right)$, ultrashort ($< 150$fs) laser pulses. To describe the interaction of such pulses with clusters, the hydrodynamic models are no longer valid, since for these powers and time scales Coulombic explosion of cluster ions is the dominant mechanism. In this regime, a sufficient number of electrons exit from the cluster core leaving behind a positively charged cluster which explodes due to electrostatic repulsion between the ions. This explosion is observed to take place on time scales too short for normal transport processes such as plasma heating and thermalization, determined by the sound transit time across the cluster, to occur. Also, it has been demonstrated by Lezius et al. (1998) that Coulomb explosions occur preferentially for smaller clusters and produce the most energetic ions.

Various studies have been carried out to investigate the Coulomb mechanism. Poth and Castleman 1998 proposed a molecular dynamic approach to model the temporal evolution of the cluster based on electrostatic repulsion within the cluster cores after the ionisation event. Ditmire (1998a) has carried out classical particle dynamics simulations to examine the ionization and subsequent explosion of small and medium-sized Ar clusters irradiated with high-intensity femtosecond laser pulses. Brewczyck et al (1998) have used a on-dimensional Fermi liquid model to describe Xe clusters in a strong laser field, where, within a few optical cycles, the field was large enough to evacuate a sufficient number of electrons from the cluster. The ions were found to be ejected layer by layer with very high velocities due to Coulombic repulsion and the energy of the hot electrons was transferred to the outermost cluster ions. Particle in cell simulations have been carried out by Eloy et al (2001) and the results demonstrate that with laser intensities of 10^{20} W/cm^2 ion energies with tens of MeV can be generated. These simulations also confirm the Coulombic explosion mechanism as being responsible for the energetic ions.

Recent experiments (Zweiback et al., 2000) on the interaction of intense ultrafast laser pulses with large van der Waals bonded clusters have shown that these clusters can explode with substantial kinetic energy. By driving explosions in deuterium clusters with a 35 fs laser pulse, Zweiback et al have accelerated ions to sufficient kinetic energy to produce DD nuclear fusion from the D(D,n)He^3 reaction. The fusion yield enhancement is due primarily to the greater ion energies produced in the Coulomb explosion of larger targets.

A recent simulation study (Chen et al., 2002) reported the emission of a hot electron jet from intense (10^{17} W/cm^2) femtosecond-laser-Ar cluster interactions. Channel betatron resonance (CBR) is believed to be the main accelerating mechanism for the generation of hot electron jets. Here, the ponderomotive force of linearly polarized intense pulse expels the background plasma electrons. The resulting space charge electric field accelerates the electrons. Energetic electrons in the laser fields, in turn, can make oscillations at the betatron frequency $\omega_\beta \approx \omega_p/(2\gamma)^{1/2}$.

When the latter coincides with the laser frequency seen by relativistically moving electrons, CBR occurs and results in energy transfer from laser light to electrons. Using three-dimensional Monte Carlo simulations, Hu and Starace (2002) demonstrated that interactions between an ultraintense laser short laser pulse (with an intensity of 8×10^{21} W/cm^2) and highly charged ions can produce GeV electrons. Here, the electron acceleration occurs due to the Lorentz force associated with intense electromagnetic fields.

On the other hand, several Petawatt experiments (Snavely et al., 2000; Hatchett et al., 2000) have demonstrated the generation of intense collimated proton beams from thin solid targets irradiated by laser pulses of one micron wavelength and 0.5-5 ps with intensities up to 3×10^{20} W cm^{-2}. Experimental observations reveal a large number of (3×10^{13}) of protons ejected off the back of thin (50-125 μm gold and plastic targets. The energies of the protons were reported to be in the range of 5-50 MeV, from a tiny spot approximately 400 microns in size. Furthermore, Maksimchuk *et al* (2000) observed a collimated beam of fast protons, with energies as high as 1.5 MeV and total number $\geq 10^9$, when a high-intensity ($\sim 3 \times 10^{18}$ W/cm^2) subpicosecond laser pulse is focused onto a thin foil target. Acceleration field gradients ~ 10 GeV/cm are inferred. Pukov (2001) carried out three-dimensional simulations of ion acceleration from a foil irradiated by a short-pulse laser at 10^{19} W/cm^2 intensity. He observed a ring-like angular distribution of the energetic ions resembling that detected in the experiment (Clark et al., 2000). Mackinnon *et al* (2002) reported the generation and enhancement of of multi-MeV proton and ion beams from solid targets (100 μm thick) irradiated by 100-fs laser pulses at intensities above 1 $\times 10^{20}$ W/c,$^{-2}$.

Various physical mechanisms for ion acceleration have been suggested (Mendonça et al., 2001). Physically, the ponderomotive force of a single laser pulse striking a thin slab of a target expels electrons which form cloud of negative charge around the back of the target. The space charge electric field of high energy electron cloud pulls positively charged ions from the back of this target which are rapidly accelerated to high energies. The ions are accelerated to extremely high energies over a short distance (almost 1 MeV/micron for protons)- order of magnitude higher than conventional ion accelerators. The ion acceleration may be determined by

$$\frac{d\mathbf{p}_i}{dt} = -m_e c^2 \nabla \gamma_L, \tag{41}$$

where $\mathbf{p}_i = m_i \gamma_i \mathbf{v}_i$ is the relativistic ion momentum, m_i is the ion mass, $\gamma_i = (1 - v_i^2/c^2)^{-1/2}$ is the ion gamma factor, and $\gamma_L = (1 + e^2|\mathbf{A}_\perp|^2/m_e^2 c^4)^{1/2} - 1$ is associated with the radiation pressure.

Other sources of the proton acceleration could be the resistance field created by electrons inside the target (Davies, 2002), the shock front that propagates through the solid material created by the laser pulse interacting with the overdense medium (Andreev et al., 2002), or the space charge field created by the electrons leaving the rear surface of the target. In principle, any type of high velocity ion can be generated simply by depositing atoms of the desired species onto the back of the target. One can also envision the possibility of creating an "ion lens" by shaping a concave section from target. Hence, the ejected ions would focus toward a point, further enhancing the brightness of the ion beam.

12 Conclusions and outlook

Plasma acceleration processes continue to be an area of active research. The initial studies of particle acceleration have provided fruitful for current drive schemes and laser accelerators. Particle acceleration in strongly turbulent plasmas is still in its infancy and requires a great deal more research. This area of research is important in astrophysical and space plasmas as well as in high energy physics.

The present and future laser experiments, however, are very far from the parameter range of interest to high energy physicists who require something like 10^{11} particles per pulse accelerated to TeV energies (for electrons) with a luminosity of 10^{-34} cm^{-2} s^{-1} for acceptable event rates to be achieved. The TeV energy range is > 1000 times greater than a single laser accelerating stage could provide at present, even if the interaction length can be extended by laser channelling there is still going to be the requirement of multiple staging, and more energetic lasers. For a TeV beam of 10^{11} particles per pulse and a transfer efficiency of 50% would require a total of 32 kJ of laser energy per pulse, for a 100 stage accelerator. This would require 100 lasers of about 300 J each with high repetition rates. Compared to the 56 J lasers in the proposed GeV accelerator and 1 Joule laser used in present day experiments (10^{10} particles per pulse would require 100$\times$ 30 J lasers). Beam driven plasma wakefields, which contain conventional accelerators with plasma accelerators may be the way forward in the near term while more intense lasers are developed using the optical parametric CPA process.

The work on plasma-based accelerators represents one area that is being explored by researchers in the advanced accelerator field. Other schemes being investigated at present for high-gradient acceleration are the inverse Cherenkov effect and the inverse free-electron laser effect. Still other researchers, realizing that the next collider will almost certainly be a linear electron-positron collider, are proposing a novel way of building such a device known as a two-beam accelerator, and there are many groups developing an entirely new type of electron lens using focusing by a plasma to increase the luminosity of future linear colliders (JST et al., 1992; Hirapetain et al., 1994). This plays on the fact that relativistic electron beams can be focused by a plasma if the collisionless skin depth c/ω_{pe} is larger than the beam radius.

Generally, when a relativistic electron beam enters a plasma, the plasma electrons move to neutralize the charge in the beam on the electron plasma period timescale. However, if the skin depth is larger than the beam radius, the axial return current flows in the plasma on the outside of the electron beam and the beam current is not fully neutralized, leading to the generation of an azimuthal magnetic field. Consequently, self-generated magnetic fields pinch or focus the beam in the radial direction. This type of plasma lenses (Su et al., 1990) exceeds conventional lenses by several orders of magnitude in focusing gradient.

Currently an experiment is underway at the Stanford Linear Accelerator Center (SLAC) to demonstrate wakefield excitation by an electron bunch. Clayton *et al* (2002) studied experimentally the transverse dynamics of a 28.5-GeV electron beam in a 1.4 m long underdense plasma (with $n_0 \sim 2 \times 10^{14}$ cm^{-3}). The transverse component of the wakefield excited by the short electron bunch focuses the bunch, which experiences multiple betatron oscillations. One is thus developing new technologies to reduce the size and cost of future elementary particle physics experiments based on colliding high energy beams, including even positron beams. Lee *et al* (2001) presented three-dimensional (3D) particle-in-cell simulations and physical models for plasma-wakefield excitation by positron beams. They found that the nonlinear wake of a positron bunch is smaller than that of an electron bunch, but it can be made comparable to the electron wake by employing a hollow plasma.

For laser plasma accelerators, the next milestone to be achieved is 100 MeV - 1 GeV energy levels with good beam quality. Electrons in this energy range are ideal as a driver for free-electron lasers (FEL). At higher energies, predicted of several GeV, it is possible to produce an x-ray FEL capable of biological investigations around the water window. Furthermore,a high intensity laser pulse interacting with a plasma can produce intense proton beams which can be used for treatment of oncological diseases (Bulanov et al., 2002). Yamagiwa *et al* (1999) have carried out two-dimensional particle-in- cell simulations to show that the longitudinal electric field induced by electron evacuation due to the intense light pressure can accelerate ions to several MeV in the direction of the laser propagation.

The work on plasma accelerators by high power lasers has applications in astrophysics in that one is able to explain energetic particles from compact high brightness objects. Raman scattering has already been discussed in the astrophysics literature, in particular the eclipsing pulsar radio sources (Thomson et al., 1994). Forward Raman scattering which has not been considered in this area will produce a relativistic plasma waves, which have the capability of accelerating particles to energies greater than TeV. For example, forward Raman scattering in a plasma of the density $10^7 - 10^8$ cm^{-3} is capable of generating plasma wave turbulence with mean electric field values of ≈ 1 V/cm. The quasilinear diffusion of particles in such fields over distances of $10^{15} - 10^{16}$ cm would result in such energetic particle, thus avoiding the need for the Fermi acceleration. It is envisaged that particle acceleration by relativistic plasma waves in Jets could sustain the acceleration process over hundreds of parsecs. Particles that lose energy through radiation losses are re-accelerated within the JET. The source of relativistic plasma waves are the relativistic particles themselves. Relativistic electrons generate the relativistic plasma wave in a manner similar to the plasma wakefield accelerator. Recently, Chen *et al* (2002) suggested that plasma wakefield mechanism could be responsible for ultrahigh-energy cosmic rays exceeding the Greisen-Zatsepin-Kuzmin (GZK) limit.

An exciting development for high powered lasers with ultra high intensities $> 10^{23}\omega/cm^2$ in the range is the study of Unruh radiation (Unruh, 1976; Davies, 1975), which requires very large acceleration to be detectable. This radiation has been likened to Hawking radiation (Hawking, 1974) from black holes. The weak decay of uniformly accelerated protons in the context of standard quantum field theory has been recently investigated by Vanzella and Matsas (2001). Experiments in this area can be carried out using high powered lasers in ionizing a gas rapidly, changing the refractive index within one cycle of the laser. Such rapid refractive index changes can produce a reference frame accelerating with a $\gtrsim 10^{20}$ g, where g is the acceleration on Earth.

These experiments are associated with vacuum energy which in the future could be harnessed to accelerate particle to very high energies greater than present day experiments. The ponderomotive force of a short laser pulse can generate ultra-bright attosecond electron bunches (Stupakov G V and Zolotorev, 2001) in vacuum. The laser-pulses profile can be tailored in such a manner that electrons are both focused and accelerated by the light pressure. In plasmas with the transverse density gradient $\partial n_0/\partial r$, a recent experiment (Takahashi et al., 2001) reveals the emission of second-harmonic due to the spatially asymmetric relativistically quivering electron motion produced by the ponderomotive force of an intense laser pulse. The power of the second-harmonic emission is (Takahashi et al., 2001)

$$P_{2\omega} \propto \frac{a_{\perp 0}^4}{1 + a_{\perp 0}^2/2} \left(\frac{n_0}{L_n} \right)^2,$$ (42)

where $a_{\perp 0} = e|\mathbf{A}_{\perp 0}|/m_e c^2$ corresponds to the intensity of the pump pulse and $L_n = n_0(\partial n_0/\partial r)^{-1/2}$ is the density gradient scalelength, which is largest at the beam edges.

Finally, we mention that short intense laser pulses can spontaneously create megagauss magnetic fields (Kim et al., 2002), which affect the dynamics of electrons in plasmas. Consequently, both the relativistic ponderomotive force and wakefields are influenced (Shukla, 1993; Shukla, 1999; Brodin and Lundberg, 1998) by the presence of these magnetic fields. The surfatron acceleration mechanism (Katsouleas and Dawson, 1983) can produce unlimited electron acceleration due to the cross wave-electric and external magnetic fields. In fact, Ucer and Shapiro (2001) and McClements *et al* (2001) have exploited the idea of surfatron mechanism for unlimited relativistic shock surfing acceleration of ions and acceleration of cosmic ray electrons by waves excited by ions reflected from supernova remnant (SNR) shocks, respectively. Furthermore, in a magnetized plasma we also have the possibility of radiation generation from the Cherenkov wake excited by an ultrashort intense laser pulses.

Recently, Yugami *et al* (2002) have presented a proof-of-principle experiment demonstrating the emission of radiation in the millimeter range (up to 200 GHz). The intensity of the radiation is proportional to the magnetic field strength. On the other hand, Yu *et al* (2002) have considered electron acceleration and high-harmonic (short-wavelength radiation) generation by an intense short linearly polarized laser pulse in an external magnetic field. They found that electrons can be strongly energized, and most of the energy gained by electrons is retained in its relativistic cyclotron motion, even though the pulse has disappeared. The energetic electrons, in turn, emit radiation at high harmonics of the cyclotron frequency. In view of the above mentioned recent works, it is suggestive that future research should concentrate on the formation of three-dimensional wakefields and subsequent particle acceleration as well as on the related phenomena (e.g., self-focusing and optical guiding, generation of small scale density and magnetic field perturbations (Gorbunov and Ramazashvili, 1998; Liseikina et al., 1999), etc.) by including the dc magnetic field and the background plasma nonuniformity. In closing, we stress that the generation of wakefields by relativistic electron or positron bunches, is certainly one of the potential candidates for high energy electron acceleration in existing PWFA experiments.

Acknowledgments

The author dedicates this paper to late John Dawson in the memory of his many pioneering and unmatched contributions in physics. Thanks to Alan Cairns, Warren Mori and Tom Katsouleas for a number of invaluable discussions.

References

Abramyan L A *et al* 1992 *Sov Phys JETP* **75** 978

Amiranoff F *et al* 1995 *Phys Rev Lett* **74** 5220

Amiranoff F *et al* 1998 *Phys Rev Lett* **81** 995

Andreev N E *et al* 1995 *Plasma Phys Rep* **21** 824

Andreev A A *et al* 2002 *Plasma Phys Controlled Fusion* **44** 1243

Antonsen T and Mora P 1992 *Phys Rev Lett* **9** 2004

Bingham R *et al* 1987 *Plasma Phys Control Fusion* **29** 1527

Bingham R *et al* 1992 *Plasma Phys and Controlled Fusion* **34** 557

Bingham R 1994 *Nature* **368** 496

Bingham R 1983 Rutherford Appleton Laboratory report RL83058

Blue B *et al* 2003 *Phys Rev Lett*

Bonnaud G *et al* 1994 *Phys Rev E* **50** R36

Borisov A B *et al* 1992 *Phys Rev Lett* **68** 2309

Brewczyk M *et al* 1998 *Phys Rev Lett* **80** 1857

Brodin G and Lundberg J 1998 *Phys Rev E* **57** 7041

Bulanov S V *et al* 1995 *Phys Rev Lett* **74** 710

Bulanov S V *et al* 2002 *Phys Lett A* **299** 240

Butler A *et al* 2002 *Phys Rev Lett* **89** 185003

Cattani F *et al* 2001 *Phys Rev E* **64** 016412

Chen *et al* 1985 *Phys Rev Lett* **54** 693

Chen S Y *et al* 1998 *Phys Rev Lett* **80** 2610

Chen P *et al* 2002 *Phys Rev Lett* **89** 161101

Chen L M *et al* 2002 *Phys Rev E* **66** 025402

Chiron A *et al* 1996 *Phys Plasmas* **3** 1373

Clayton C E *et al* 1993 *Phys Rev Lett* **70** 37

Clayton C E *et al* 2002 *Phys Rev Lett* **88** 154801

Clark E L *et al* 2000 *Phys Rev Lett* **84** 670

Coverdale C A *et al* 1995 *Phys Rev Lett* **74** 4659

Dalla S and Lontano M 1994 *Phys Rev E* **49** R1819

Davies P C W 1975 *J Phys A* **8** 609

Davies J R 2002 *Laser Part Beams* **20** 1

J M Dawson J M 1989 in *From Particles to Plasmas*, ed J W Van Dam (Reading, MA: Addition Wesley) p 131

Decker C *et al* 1994 *Phys Rev E* **50** 3338

Decker C D *et al* 1996 *Phys Plasmas* **3** 1360

Dias J M *et al* 1997 *Phys Rev Lett* **78** 4773

Ditmire T *et al* 1998 *Phys Rev A* **57** 369

Ditmire T 1998 *Phys Rev A* **57** R4094

Ditmire T *et al* 1998 *Opt Lett* **23** 322

Ditmire T *et al* 1999 *Nature* **398** 489

Donnelly T D *et al* 1996 *Phys Rev Lett* **76** 2472

Dorchies J R *et al* 1999 *Phys Rev Lett* bf 82 4655

Duda B J *et al* 1999 *Phys Rev Lett* **83** 1978

Duda B J *et al* 2000 *Phys Rev Lett* **83** 1978

Duda B J *et al* 2002 *Phys Rev Lett* **88** 125001

Durfee II C G and Milchberg H M 1993 *Phys Rev Lett* **71** 2409

Durfee II C G *et al* 1995 *Phys Rev E* **51** 2368

Eloy M *et al* 2001 *Phys Plasmas* **8** 1083

Esarey E *et al* 1996 *IEEE Trans Plasma Sci* **24** 252

Esarey E *et al* 2000 *Phys Rev Lett* **84** 3081

Everett M *et al* 1994 *Nature* **368** 527

Faure J *et al* 2004 *Nature* **431** 541

Geddes GCR *et al* 2004 *Nature* **431** 538

Gorbunov L M and Ramazashvili R R 1998 *JETP* **87** 461

Gordon D *et al* 1998 *Phys Rev Lett* **80** 2133

Hatchett S P *et al* 2000 *Phys Plasmas* **5** 4107

S W Hawking S W 1974 *Nature* **248** 30

Hirapetain G *et al* 1994 *Phys Rev Lett* **72** 2403

Hoda T *et al* 1999 *J Plasma Fusion Res* **75** 219

Hu S X and Starace A F 2002 *Phys Rev Lett* **88** 245003

Johnson D A *et al* 1994 *Physica Scripta* **T52** 77

Joshi C *et al* 2002 *Phys Plasmas* **9** 1845

Joshi C *et al* 1984 *Nature* **311** 525

Joshi C and Corkum P B 1995 *Phys. Today* **48** 36

JST Ng *et al* 2001 *Phys Rev Lett* **87** 24480

Kaganovich D *et al* *Phys Rev E* **59** 4769

Katsouleas T and Dawson J M 1983 *Phys Rev Lett* **51** 392

Katsouleas T *et al* 1999 *Comments Plasma Phys Controlled Fusion* **1** 99

Kim A *et al* 2002 *Phys Rev Lett* **89** 095003

Komashko A *et al* 1995 *JETP Lett* **62** 861

Krall J *et al Phys Rev E* **48** 2157

Kurshenlnick K *et al* 1999 *Phys Rev Lett* **83** 737

Lee S *et al* 2001 *Phys Rev E* **64** 045501

Lezius M *et al* 1998 *Phys Rev Lett* **80** 261

Liseikina T V *et al* 1999 *Phys Rev E* **60** 5991

Lopes N *et al* 2002 (submitted to Europhys Lett)

Macchi A *et al* 2001 *Phys Rev Lett* **87** 205004

Mackinnon A J *et al* 2002 *Phys Rev Lett* **88** 215007

Maksimuck A *et al* 2000 *Phys Rev Lett* **84** 4108

Malka V *et al* 2002 *Science* **298** 1596

Mangles SPD *et al* 2004 *Nature* **431** 535

Marqués J R *et al* 1996 *Phys Rev Lett* **76** 3566

McClements K G *et al* 2001 *Phys Rev Lett* **87** 255002

McKinstrie C J and Bingham R 1992 *Phys Fluids B* **4** 2626

McPherson A *et al* 1994 *Nature* **370** 631

Mendonça J T and Silva L O 1994 *Phys Rev E* **49** 3520

Mendonça J T 2001 *Theory of Photon Acceleration* (Bristol: Institute of Physics)

Mendonça J T *et al* 2001 *Meas Sci Technol* **12** 1801

Modena A *et al* 1995 *Nature* **377** 606

Monot P *et al* 1995 *Phys Rev Lett* **74** 2953

Mora P 2001 *Plasma Phys Control Fusion* **43** A31

Mori W B *et al* 1988 *Phys Rev Lett* **60** 1298

Mori W *et al* 1994 *Phys Rev Lett* **72** 1482

Mourou G A *et al* 1998 *Phys. Today* **51** 22 January 1998

Nakajima K *et al* 1995a in *Advanced Accelerator Concepts* eds Fontana W I and Schoessow P (New York: AIP Conf Proc No 335) p 145

Nakajima K *et al* 1995b *Phys Rev Lett* **74** 4428

Naumova N M *et al* 2001 *Phys Plasmas* **8** 4149

Nikitin S P *et al* 1999 *Phys Rev E* **59** 3839

Penano J R *et al* 2002 *Phys Rev E* **66** 036402

Poth L and Castleman Jr A W 1998 *J Phys Chem A* **102** 4975

Pukov A 2001 *Phys Rev Lett* **86** 3562

Reitsma A J M *et al* 2002 *Phys Rev Lett* **88** 014802

Rosenbluth M and Liu C S 1972 *Phys Rev Lett* **29** 707

Rosenzweig J *et al* 1993 *Phys Rev A* **44** R6189

Santala M *et al* 2001 *Phys Rev Lett* **86** 1227

Sarkisov G S *et al* 1999 *Phys Rev E* **59** 7042

Schmidt G and Horton W 1985 *Comments Plasma Phys Controlled Fusion* **9** 85

Shukla P K 1993 *Phys Fluids* **B5** 3088

Shukla P K 1999 *Phys Plasmas* **6** 1363

Shvets J and Wurtele J S 1994 *Phys Rev Lett* **73** 3540

Siders C W *et al* 1996 *Phys Rev Lett* **76** 3570

Skobelov I Yu *et al* 2002 *JETP* **94** 73

Sluijter F W and Montgomery D 1965 *Phys Fluids* **8** 551

Snavely R A *et al* 2000 *Phys Rev Lett* **85** 2945

Spence D J and Hooker S 2001 *Phys Rev E* **63** 015401

Sprangle P *et al* 1988 *App Phy Lett* **53** 2146

Sprangle P, Hafizi B and Serafim P 1999 *Phys Rev Lett* **82** 1173

Sprangle P *et al* 2000 *Phys Rev Lett* **85** 5110

Stupakov G V and Zolotorev M S 2001 *Phys Rev Lett* **86** 5274

Su J J *et al* 1990 *Phys Rev A* **41** 3321

Tajima T and Dawson J M 1979 *Phys Rev Lett* **43** 267

Tajima T 1995 *Laser and Particle Beams* **3** 351

Takahashi E *et al* 2001 *Phys Rev E* **65** 016402

Tang C M *et al* 1984 *App Phys Lett* **45** 375

Sprangle P *et al* 1992 *Phys Rev Lett* **69** 2200

Thomson C *et al* 1994 *Ap J* **422** 304

Tzeng K C *et al* 1996 *Phys Rev Lett* **76** 3332

Ucer D and Shapiro V D 2001 *Phys Rev Lett* **87** 075001

Umstadter D *et al* 1996 *Science* **273** 472

Unruh W G 1976 *Phys Rev D* **14** 870

Vanzella D A T and Matsas G E 2001 *Phys Rev Lett* **87** 151301

Wagner R *et al* 1997 *Phys Rev Lett* **80** 3125

Wickens L W and Allen J E 1979 *J Plasma Phys* **22** 167

Wilks S C *et al* 1989 *Phys Rev Lett* **62** 2600

Wilks S C *et al* 1992 in *Nonlinear and Relativistic effects; Research Trends in Physics* ed Stefan V (New York: AIP Conf Proc No 402)

Wilson E J N 1990 *Physica Scripta* **T30** 69

Yamagiwa M *et al* 1999 *Phys Rev E* **60** 5987

Yu W *et al* 2002 *Phys Rev E* **66** 036406

Yugami N *et al* 2002 *Phys Rev Lett* **89** 065003

Zweiback J *et al* 2000 *Phys Rev Lett* **84** 2634

8

Laboratory Astrophysics Using High Energy Density Photon and Electron Beams

R. Bingham

Rutherford Appleton Laboratory, Chilton, Didcot, Oxon, U.K.

1   Introduction

Modern high-power lasers and electron beams are unique tools that are able to deliver pulses that have enormous energy densities to target. During the laser pulse, matter is heated to temperatures of millions of degrees kelvin and attains pressures that are equivalent to millions of atmospheres. These conditions are equivalent to those at the centre of stars. They allow measurements of plasma conditions that are of astrophysical interest, and to test the complex models of these processes with unprecedented precision. Laser-plasma results have been applied to the study of such diverse environments as active galactic nuclei (Levinson and Blandford 1995) and the Earth's bow shock (Bell et al. 1988). More recent applications include the hydrodynamics of supernovae (Remington et al. 2000) supernova remnants (SNRs) (Woolsey et al. 2001), the collision of galactic clouds (Perry et al. 2000). Intense electron beams such as the SLAC linear collider can perform experiments to test the validity of measurements of the highest energy cosmic rays and investigate the physics of e^+e^- plasmas, relativistic jets as well as possible acceleration mechanisms. Such experiments are made possible by ensuring that certain key dimensionless parameters in the plasma have values similar to those of the space and astrophysical plasmas of interest (Ryutov et al. 2001, Ryutov et al. 1999). This scaling was an important development from the point of view of designing experiments. The disparity in spatial and temporal scales makes it impossible to model astrophysical phenomena exactly. However with scaling certain dimension can be found to be similar making experiments meaningful. Another important reason for doing laboratory scaled experiments is to test numerical simulation models.

The construction of petaWatt lasers and in the future exaWatt lasers offers a unique possibility of extending these observations to a new class of objects and to answer some problems of great astrophysical interest. For example, the physics at play in the neutron-star atmosphere requires radiation pressure that is roughly the same order of magnitude, or considerably larger in the case of high luminosity X-ray pulsars, as the particle pressure, each of which are less than the confining magnetic field pressure. The "accelerating" interface between the radiation fluid and the particle fluid in the very strong gravitational field is hydro-dynamically unstable. Neutron stars are the endpoint of most core-collapse supernovae. Modelling of the atmospheres surrounding neutron stars are difficult, due to the extreme conditions encountered. Up until now, there has been no conceivable way for testing the overall radiative-MHD modelling in the bulk.

At present laser beams can deliver $10^{21} W/cm^2$ on target this will be extended in the near future to $10^{23} W/cm^2$ using upgrades of the current petawatt class of lasers to 100 petaWatt. Particle physics beams such as the SLAC electron beam can deliver $10^{21} W/cm^2$ on target and therefore compliment the laser in certain regimes. Intense particle beams are important in the study of relativistic jets e^+e^- plasmas and some acceleration models. Early work in laboratory astrophysics included the study of blast waves (Grun et al. 1991), equations of state and hydrodynamic processes in supernovae experiments (Takabe 1993, Drake 1999, Remington et al. 1999) including Rayleigh Taylor and Richtmyer-Meshkov Instability. These instabilities are particularly important in the understanding of type II supernovae, particularly during the breakout from the surface of the star by the blast wave.

High power laser experiments can also provide precise and detailed information on the equation-of-state of both hot and warm dense matter. The shock Hugoniot of warm, strongly correlated plasmas is an area that is still largely unexplored but that has enormous implications for Jupiter-like planetary structure calculations. Recent experiments have looked at the behaviour of shock compressed water using the Vulcan laser with the aim of validating assumptions used in modelling gaseous planets. However, little is known of the shock behaviour of mixed materials such as methanes, alcohols etc that are also of direct relevance to these conditions.

2   Generating neutron star atmospheres

The use of ultra-intense laser pulses from a new generation of petaWatt lasers to generate for the first time, the conditions appropriate to the atmospheres of magnetised neutron stars in an earth based laboratory. The development of plasma under such conditions is a breakthrough in our ability to study highly dynamical phenomena such as photon bubble instabilities, thought to be present in the low altitude atmosphere above the surface of magnetised neutron stars.

In a series of papers, theoretical studies (Hsu et al. 1997, Burnard 1991, Jernigan et al. 2000) have investigated such effects by solving the self-consistent multi-dimensional time dependent equations of magneto-radiation-hydrodynamics governing the accretion of matter onto the polar caps. These calculations have shown that the settling mound on the surface of a neutron star does indeed develop a new form of "turbulence" in which photon bubbles form in the medium and onto the polar caps. The numerical calculations show clear evidence for a substantial coalescence of the photon bubbles to become relatively large, rising, optically thin pockets within the settling mound filled with hot ($T \sim 10 keV$) radiation, embedded in optically thick, settling plasma. The discovery of such photon bubble instabilities in the accretion mound of X-ray pulsars has important consequences for probing the physics of the accretion column of a neutron star. Low-mass x-ray binaries exhibit similar phenomena.

A major advance in our understanding of photon bubble instabilities and related dynamical phenomena on the surface of neutron stars could be made if the conditions present in the low altitude of the neutron star atmosphere could be duplicated in an earth based laboratory. To duplicate such conditions in the laboratory we require radiation temperatures of the order of 1 keV at densities of order $10^{-3} gcm^{-3}$ to generate radiation pressures that would substantially exceed thermal pressures and have radiation pressure dominate in a super-Eddington flow. A stability analysis shows that for these temperatures and optical depths of order 1000, typical in super-Eddington flow, magnetic field strengths as low as 0.1 Giga-Gauss would be required to give rise to the photon bubble instability. These field strengths would also be required to prevent sideways adiabatic expansion of photon bubbles and confine the plasma to flow in one direction. Finally, the gas would also have to have an extremely high acceleration to simulate the effective gravity of a neutron star that is about $10^{14} cm/s^2$.

Recent advances in the development of ultra-bright, short pulse lasers may make this a possibility. There is now ample and detailed theoretical modelling which predicts that field strengths of order 10^9 Gauss are generated in petaWatt lasers. Detailed 2-D Particle in cell simulations predict that 1.5 Giga fields are generated with petaWatt lasers. The magnetic field generated in ultra-intense interactions is about 1/3 of the oscillating B field of the light wave for intensities up to $10^{22} W cm^{-2}$, in reasonable agreement with PIC simulations. Undoubtedly, PIC simulations require careful evaluation (with suitable selection of diagnostic particles). Recently, experiments demonstrated two rather unique methods of measuring the B field in these interactions. The first is by charged particle (i.e. proton) deflection (Clark et al. 2000) and the second by depolarisation of harmonics via the Cotton-Mouton effect (Tatarakis et al. 2000) where a plane EM wave with B field perpendicular to the k-vector becomes elliptically polarized, the degree of ellipticity depends upon the strength of the field. The magnetic fields they measured are in agreement with the theoretical modelling that has been done. The densities required pose no special difficulties. The acceleration required is also not a problem: the radiation pressure of the light wave from a petaWatt laser is so large that it expels all the plasma from the focal region and bores a hole at $10^{21} W cm^{-2}$ creating a high vacuum. A formidable challenge will be the generation of 1 keV radiation temperatures.

3   Cosmic ray acceleration at energetic shocks.

Collisionless plasma shocks occur throughout the Universe, heating the interstellar medium and accelerating charged particles. Energetic collisionless plasma shocks are associated with supernovae - the end state of large stars - and observed 100's to 1000's of years later as a remnant. These stellar remnants, particularly at an early age are believed to accelerate nuclei to $10^{15} - 10^{16}$ eV via a strongly non-linear process of self-generated magnetic field, and self-generated magnetic turbulence. Furthermore, the cosmic ray particle population contains negligible mass yet carries a significant fraction of the shock energy. Thus, the cosmic ray particles are dynamically significant and also relativistic resulting in a compressive plasma equation of state and strong shock modification. This further influences the acceleration of charged particles.

This problem can be addressed through very large-scale plasma simulation, computational techniques are not feasible, but laboratory methods are both possible and experimentally very exciting. Using lasers such as Vulcan, Astra and Gemini and future upgrades the interaction of supersonic magnetised plasmas with relativistic electron beams is possible. The essential ingredients are high Mach number collisionless plasma shocks and energetic particle injection, with key parameters that are scaled to the astrophysical event.

The observation of ultrahigh energy cosmic ray (UHECR) events exceeding the Greisen-Zatsepin-Kuzmin (GZK) cutoff (Greisen et al. 1966, Zatsepin and Kuzmin 1966), protons with energy 5×10^{19} eV and above cannot propagate further than $50 Mps$ due to their strong interaction with the cosmic microwave background, present a formidable theoretical challenge. Even if the GZK cutoff can be circumvented the existing models namely the Fermi mechanism as well as its variants such as the diffusive shock acceleration is not effective in reaching ultra high energies (Achterberg 2000), this has led to several authors (Chen et al. 2002) to propose the plasma wakefield accelerator concept. Large amplitude electrostatic relativistic plasma waves or "wakefields" can be excited by powerful lasers or particle beams (Bingham these proceedings). The particles are accelerated by the electrostatic wave. Alfven waves have been proposed as an alternative to the laser in astrophysics in generating the wakefield. The ponderomotive force of the Alfven wave is responsible for generating the electromagnetic relativistic plasma wave. An experiment demonstrating this process has been proposed by Pisin Chen (2002) and his collaborators using an electron positron plasma. Recent infrared observations (Jester et al. 2006, Uchiyamah et al. 2006) of 3C273 by the Spitzer Space Telescope are providing evidence that high energy protons with energies of $10^{16} - 10^{19}$ eV are produced in the jets of 3C273, backing up the model of (Chen et al. 2002). These results will lead the way to new developments in ultrahigh particle acceleration.

For much slower shock waves such as those found in the interaction of old supernovae remnants with interstellar clouds x-ray emission is observed at the interface. These x-rays are produced by lower energy keV electrons through the process of bremsstrahlung and line radiation. The free energy is the reflected ion beam from the shock surface forming a two stream type instability.

The formation of opposing ion streams moving perpendicular to the magnetic field leads to various instabilities. The model we consider for electron energization and X-ray emission is one based on collisionless coupling of the shock wave and the ambient medium through various instabilities driven by counterstreaming ion beams. Specific instabilities resulting in strong wave activity are the modified two stream instability, Buneman and ion acoustic instabilities giving rise to both energetic electrons and ions. The modified two stream instability will lead to significant electron energization forming a high energy tail on the electron distribution function. The instability produces waves around the lower-hybrid frequency range. The special role played by lower-hybrid waves in plasma astrophysics is due to the fact that these waves can be in simultaneous resonance with both the relatively slow ions and the fast electrons. Indeed, by virtue of the fact that the wave frequency is much larger than the ion cyclotron frequency, the criterion for resonance interaction between the ions and the waves is identical to the usual Cerenkov resonance criterion, $\omega = \mathbf{k}.\mathbf{v}_i.\omega$ is the wave frequency, $\mathbf{k}$ is the wavenumber and $\mathbf{v}_i$ is the ion velocity, and the effect of the magnetic field on the ions reduces to simple phase mixing of their distribution functions. At the same time, since the lower-hybrid wave frequency is much smaller than the electron cyclotron frequency, the magnetic field suppresses the transverse motions of the electrons, and the criterion for resonance between the electrons and waves is identical to the criterion for longitudinal Cerenkov resonance with electrons, $\omega = k_{\parallel} v_{\parallel e}, k_{\parallel} \left(v_{\parallel e} \right)$ is the component of the wavevector (electron velocity) in the direction of the magnetic field which is in the z direction. When $k_{\parallel} <| k |$, the resonance conditions are simultaneously satisfied for $v_e \gg v_i$. It is precisely the ions reflected from the shock front component that is the main free energy reservoir in many plasma astrophysics problems or the flow of plasma through neutral gas in the anomalous ionization problem. The excitation of lower-hybrid waves by these ions and the subsequent absorption of the waves by the electrons is thus a mechanism for transmission of the energy in the shock to the electrons and for the electron heating needed in anomalous ionization or the creation of accelerated electron flows in shock wave fronts. These accelerated electrons are responsible for bremsstrahlung and line radiation.

4   Electron-positron plasmas

It is interesting to note that the source region of radio emission in polar-cap models of pulsars is thought to be a relativistic, strongly magnetised, one-dimensional electron-positron plasma. We may soon be able to generate such electron-positron plasmas in the laboratory. For example, positron production as high as 0.1% of the initial electron density is predicted by computer modelling and by theory and occurs during double-sided irradiation of thin high Z foils with ultra-intense laser pulses (Liang et al. 1998). The principle is this: electrons are accelerated to multi-MeV energies within an optical cycle either by Brunel-type resonance absorption or the $\mathbf{j} \times \mathbf{B}$ force of the light wave. The electrons penetrate the target but are then decelerated in the foil by the enormous ponderomotive pressure of the opposite directed laser beam. The electrons are confined in the focal area by the very large self-generated magnetic fields.

The deceleration causes copious numbers of γ-rays to be generated. These, in turn, create the electron-positron pairs via three-body process. Calculations suggest that 10^{11} or higher electron-positron pairs can be generated for each laser shot. Clearly some of the positrons will be annihilate within the target material, but the rapid, collisionless expansion of the plasma after the laser shot will allow a significant number to escape into the vacuum. This generation process requires the irradiation of high Z materials, and the planned experiments will concentrate on optimising the production rates using these targets. However, theory predicts that the production mechanism will saturate when intensities on target rise above $10^{20} W\,cm^{-2}$ due to the increased ponderomotive energy of the electrons and the increased difficulty of stopping them inside the solid density plasma. It has been proposed that the pulse duration is increased to ~ 10 psec to compensate for this saturation process and maintain the intensity at $\sim 10^{20} W\,cm^{-2}$ for optimum pair production.

The electron-positron plasma in a pulsar atmosphere also exists within a background of low Z ions, a rather different environment to those described above. Scaling of the pair production process to lighter element targets closely matches the pulsar atmosphere composition. In particular, if the focal area is increased to maintain optimal intensity on target, rather than the pulse duration, then it may be possible to compensate the Z scaling reduction of the pair production with an increased interaction volume. The generation of electron-positron plasma with a low Z background ion source will then allow a rich array of pulsar relevant physics to be studied. For example, in a relativistic electron-positron fireball scenario, electromagnetic instabilities, such as the Weibel instability, produce large magnetic field structures and subsequent synchrotron radiation. Time resolved studies would reveal the presence of low frequency radiation that may be initially trapped and then released later. Preliminary 3-D PIC simulations demonstrate the role of the Weibel instability plays in interpenetrating electron-positron plasmas and has been proposed as a possible x-ray burst model (Medvedev et al. 2004).

5 Physics of type II supernovae

In the subatomic world few particles are as elusive as the neutrino. Yet after photons they are the most abundant of the observed particles in the Universe, released from the fusion-burning core of stars, supernovae explosions, gamma ray bursts and left over from the Big Bang. Neutrinos play an important role in the Universe, for example reviving the stalled shock in core collapse. In addition, the Sudbury Neutrino Observatory (SNO) has settled the most pressing issue in neutrino physics - that of the neutrino mass. The mass resulting from the neutrinos will be sufficient to influence the structure of galaxies. It may also provide a solution to the baryon-antibaryon asymmetry in the Universe.

The interaction of the neutrino flux from core collapse supernovae can produce energetic particles through streaming instabilities analogous to those generated in laser-plasmas interactions. By using the refractive index concept for neutrinos, it is also possible to demonstrate that neutrino oscillations in matter are analogous to electromagnetic mode conversion in plasmas.

The telltale sign that the neutrinos have mass is flavour oscillation - neutrinos appear to oscillate from one flavour to another as they travel. For this to happen there must be a mass difference between the two flavours. Current estimations of this mass difference come from the Super-Kamiokande experiment in Japan, where atmospheric neutrino observations show that some of the muon neutrinos are turning into another flavour. The results imply at least one neutrino has a mass of 0.05 eV; neutrinos in a galaxy with this mass can influence the galactic structure. Conclusive evidence from SNO of neutrino oscillations has solved the long-standing solar neutrino problem. Nevertheless, further measurements with new facilities are required to fully understand the phenomenon of neutrino oscillations.

Type II supernovae require an understanding of different areas of science including particle physics, general relativity, hydrodynamics and plasma physics. Some aspects of supernovae explosions can be studied in scaled laboratory experiments. Of most use here is the high energy density physics experiments used in the quest for inertial fusion. Type II supernovae are giant stars more than 10 solar masses that have come to the end of their fusion burning cycle. The core is composed mostly of iron the most stable nuclei. With no energy being generated from fusion, the core begins to implode under the force of gravity. At the centre of the core the protons and electrons are forced together by gravity forming neutrons and the core ends up as a proto neutron star, liberating 10^{58} neutrinos, at the same time gravity still causes the outer layers to implode. The formation of the proto neutron star forms a dense wall that incoming material bounces off creating an outward propagative pressure wave that quickly turns into a shock wave. This shock wave stalls by losing energy in dissociating the incoming nuclei. The only way to revive the shock wave is for the liberated neutrinos to heat the electrons behind the shock to about 500 keV. This is the first problem in trying to make an explosion out of an implosion. A core collapse supernovae releases most of its energy as a burst of neutrinos, 99% of the released energy is carried away by the neutrinos with almost 1% being carried away by gravitational waves and kinetic energy of the exploding mantle. A small percentage of this kinetic energy ends up as the flare of light, brighter than the entire galaxy. Under such extreme conditions the intensity of neutrinos radiating out from the neutrino sphere is approximately $10^{34}W/cm^2$ falling to about $10^{30}W/cm^2$ close to the stalled shock wave. At these intensities enhanced absorption of neutrinos by the plasma is possible through collective effects (Bingham et al. 2004, Silva and Bingham 2006). The neutrinos drive up plasma waves by a streaming instability these plasma waves heat the electrons to the required temperature of 500 keV at which point the star explodes. The neutrinos effectively drive a wakefield on their way through the star. As the temperature rises the instability switches off, this occurs at around 500 keV just sufficient to restart the explosion. A relativistic kinetic model (Bingham et al. 2004), describing electrons and neutrinos in lepton plasmas is required to solve this problem. Such a model is similar to e-beam and photon beams interacting with plasmas. The outward propagating shock wave leads to a second problem in supernovae dynamics namely one associated with the blast wave as it propagates outward through different stellar interfaces. Structures will grow at the interface through the Richtmyer-Meshkov instability and later through the Rayleigh-Taylor instability (Drake 2006). These instabilities have important and observable effects associated with the appearance of heavy elements in the photosphere of the supernovae. Early appearance of heavy elements indicates non-spherical explosions. Experiments on the interface stabilities have been carried out by a number of researchers using scaled experiments with high power lasers such as Nova (Drake 2006).

Conclusions

I have presented only a few topics in the field of laboratory astrophysics using intense photon or particle beams. It was not possible in the lecture to cover more ground. I hope that I have given a flavour of the many different astrophysical problems that can be investigated experimentally as well as by simulations. An important book has just appeared on the subject by Paul Drake (2006). I would recommend it to anyone interested in laboratory astrophysics.

Acknowledgments

The author would like to thank the Centre for Fundamental Physics (CfFP) for support.

References

Achterberg, A. (2000)in proceedings of the IAU Symposium edited by PCH Martens and S Truruta, *Astron Society of the Pacific, San Francisco* **Vol 95**
Bell, AR. et al. (1988) *Phys. Rev. A*, **38**, 1363
Bingham, R. (2006) *These proceedings*
Bingham, R. et al. (2004), *Plasma Phys Contr F* **46**, B327
Burnard, DJ. et al. 1991, *Astrophys J.* **367** 575
Chen, P. et al, (2002) *Phys. Rev. Lett.* **89**, 161101
Clark, EL. et al, (2000), *Phys Rev Lett.* **84**, 670
Drake, RP. (1999) *J. Geophys. Res.* **104**, **(A7)**, 14505
Drake, RP. (2006) *High Energy Density Physics, Springer, New York*
Griesen, K. (1966) *Phys. Rev. Lett.* **16**, 748
Grun, J. et al. (1991) *Phys. Rev. Lett.* **66**, 2738Ũ2741
Hsu, JJL. et al. (1997), *Astrophys J. Part 1* **478** 663
Jernigan, JG. et al. (2000), *Astrophys J. Part 1* **530** 874
Jester et al. (2006), *Astrophys. J. arxiv.org/also astro-ph/0605529*
Liang, EP. et al. (1998), *Phys. Rev. Lett.* **81**, 4887
Levinson, A, and Blandford. R, (1995), *Mon. Not. R. Astron. Soc.* **274**, 717
Medvedev. M, et al. (2004) *Astrophys. J*, **618**, L75
Perry. TS, et al. (2000), *Astrophys J.Suppl S* **127**, 437
Remington, BA. (1999), *Science*, **284**, 5419, 1488-1493
Remington, BA. et al. (2000), *Phys Plasmas* **7**, 1641
Ryutov, D. et al. (2001), *Phys Plasmas* **8**, 1804
Ryutov, D. et al. (1999), *Astrophys. J.* **518**, 821
Silva, L O. and Bingham, R. (2006), *J Cosmology and Astroparticle Physics* **5**, 11
Takabe, HJ. (1993) *Plasma Fusion Res.* **69**, 1285
Tatarakis, M. et al. (2002) *Nature* **415** (6869)
Uchiyama, U. et al. *Astrophys. J. arxiv.org/also astro-ph/0605530*
Woolsey, N. et al. (2001), *Phys Plasmas* **8**, 2439
Zatsepin, GT. and Kuymin, VA. (1966) *Sov.Phys. JETP* **4** 78
Wolfenstein, L. (1983), *Phys. Rev. Lett.* **51**, 1945

9
Acceleration of Photons and Quasi-Particles

J. T. Mendonça

CFP, Instituto Superior Técnico, Lisboa, Portugal

1 Abstract

We explore a new view on wave propagation in time-varying plasmas, where resonant interactions of monochromatic waves with photons and quasi-particles is considered. A distribution of quasi-particles, is used to describe a short laser beam or a generic turbulent state. Resonant interactions occur when the phase velocity of the large scale wave (for instance, a wakefield) is equal to the group velocity of the short wavelength, or quasi-particles, associated with the turbulent spectrum. It is shown that quasi-particle Landau damping can take place, as well as quasi-particle beam instabilities, thus establishing a direct link between the short and the large wavelength perturbations of the medium. Quasi-particle trapping is also discussed.

2 Introduction

Plasmas are usually in a turbulent state. This is valid for natural, e.g. astrophysical or space plasmas, and for laboratory plasmas. It is therefore important to understand how monochromatic waves and large scale structures can develop, propagate or be damped inside the turbulent medium. Here we present a global approach, where broadband turbulence is described as a gas of quasi-particles. A large variety of different types of quasi-particles can be consider within the same formalism. First of all the photons, which behave in the plasma as dressed particles in the sense of field theory, can be used to describe the particle-like behavior of electromagnetic radiation. In the case of broad-band radiation, associated for instance with a short laser pulse, the field phase is not relevant for a large variety of physical situations, and a photon number distribution can be used to describe the evolution of the electromagnetic field (Mendonça, 2001). Other quasi-particles can similarly be considered, such as plasmons, phonons, driftons (for drift wave trubulence) (Mendonça et al., 2003), chargeons (for dust-charge fluctuations) (Mendonça et al., 2001), etc. In the resulting new picture of a turbulent plasma, a major role will be played by the acceleration of the individual photons and quasi-particles by the background plasma perturbations.

Such an approach is valid as long as we can identify two distinct space and time scale ranges. The short scale will be associated with the internal oscillations of the quasi-particles, and is determined by the typical values of their frequency and wavelength. In the language of Quantum Mechanics, this would be identified with the particle energy and the de Broglie wavelength. And a long scale will be associated with individual monochromatic waves or large scale structures, which will act on the individual quasi-particles as a mean field background. Our approach would then be similar to a statistical theory of classical quasi-particles interacting with a mean field, but where the particles represent a simplified view of the short wavelength turbulence.

In such a picture, the main physical processes are those leading to a direct coupling between large scale structures (the wave) and short scale structures of the medium (the turbulence). This coupling establishes a direct channel for energy transfer between structures of very different sizes (Mendonça et al., 2003). The energy transfer from short to large scales can be due to kinetic or to fluid instabilities of the quasi-particle gas. The opposite case of an energy transfer from large to the short scales is due to wave damping, and can associated with the resonant interaction between a long wavelength wave and the quasi-particles. This extends the concept of a non-collisional Landau damping from the real particles (electrons and ions, in the original concept) to the quasi-particle distributions.

The universality of the resonant processes involving quasi-particles should be stressed. Several particular examples can already be found in the literature, in particular: i) photon acceleration by relativistic ionization fronts (Dias et al., 1997), or by laser wakefields (Dias et al., 1998); ii) electron plasma waves propagating in a photon gas (Bingham et al., 1997); this can be relevant to laser-plasma interactions and to astrophysical phenomena; iii) ion acoustic waves in a plasmon gas, where the plasmons describe the electron plasma wave turbulence (Vedenov et al., 1967; Mendonça and Bingham, 2002); this process recently became important for the interpretation of Peta-Watt laser experiments relevant to fast ignition (Mendonça et al., 2005; Norreys et al., 2005); iv) dust lattice waves in a turbulent plasma sheath (Shukla et al., 2000); these waves can become parametrically unstable due to the turbulent fluctuations of the plasma sheath where dust plasma crystals are formed, and can eventually lead to melting and sublimation of the crystalline structure. The background quasi-particles here are plasma phonons; v) zonal flows in what one could call a drifton gas (Mendonça et al., 2003; Trines et al., 2005; Lebedev et al., 1995; Smolyakov et al., 2000; Lashmore-Davies et al., 2001) (which describes a large spectrum of drift wave turbulence); this can be a relevant process for anomalous transport in magnetic fusion plasmas. Similar processes could also be identified in fluid dynamics, involving zonal flows and Rossby waves in the oceans or in the atmosphere of the planets (Busse et al., 1994).

But the universality of the physical picture described here is not exhausted by these examples, and a much larger number of relevant situations can be added. As an extreme case, we could consider the excitation of gravitational waves by a photon gas (Mendonça, 2002), where the fluid is replaced by the vacuum metric field and the photons can be considered as quasi-particles in the sense of short scale electromagnetic wavepackets.

3   Motion of quasi-particles

Let us start with the simplest case of photons moving in a time-dependent isotropic plasma. In the geometric optics approximation we describe the photon motion with the aid of the well known ray tracing equations. These ray equations can be seen as the equations of motion of individual wave-packets, or quasi-particles, and they can be written in the following canonical form:

$$\frac{d\vec{r}}{dt} = \frac{\partial \omega}{\partial \vec{k}} \quad , \quad \frac{d\vec{k}}{dt} = -\frac{\partial \omega}{\partial \vec{r}} \tag{1}$$

where $\vec{r}$ and $\vec{k}$ represent the (mean) position and (mean) wave-vector of the considered wavepacket, and can be associated with the position and momentum of the photon, seen here as a classical quasi-particle of the electromagnetic field. The frequency (or energy) ω plays the role of the Hamiltonian, as determined by

$$\omega(\vec{r}, \vec{k}, t) = \sqrt{k^2 c^2 + \omega_p^2(\vec{r}, t)} \tag{2}$$

where c is the velocity of light in vacuum and ω_p the local plasma frequency of the non-stationary medium. It can be seen from this Hamiltonian approach that photon acceleration will take place in the medium, because of the time dependence of the plasma frequency. Such an effect was experimentally observed in microwaves and optics, using relativistic ionization fronts (Dias et al., 1997). A straightforward generalization of the photon ray description to other quasi-partcles would simply involve the use of a different dispersion relation. For instance, in the case of plasmons, we would have to replace c by the electron thermal velocity $S_e = \sqrt{3T_e/m_e}$.

If we want to use a more consistent physical description of a given turbulent state, instead of a single particle trajectory we will have to use an ensemble of trajectories corresponding to a gas of quasi-particles. The plasma medium will then be made of a collection of particles (such as electrons and ions), plus a population of quasi-particles (such as photons, plasmons or phonons). This results in a generalized kinetic view of plasma physics, which can be called photon kinetics or wave kinetics. In order to understand, in generic terms how a wave kinetic theory can be establish, we will discuss first the general dispersion relation of electrostatic waves in a plasma, and next the kinetic equations for quasi-particles.

4 Electrostatic waves

For simplicity, we will restrict our discussion to electrostatic waves in non-magnetized plasmas. Notice that these waves will be the large scale structures or the present theoretical approach. The general f dispersion relation of electrostatic waves with frequency ω and wavevector $\vec{k}$, evolving in a turbulent plasma is

$$\epsilon(\omega, \vec{k}) + \chi_{qp}(\omega, \vec{k}) = 0, \tag{3}$$

A quasi-particle susceptibility χ_{qp} is added here to the usual dielectric function $\epsilon(\omega, \vec{k}) = 1 + \sum_\alpha \chi_\alpha$, where the sum is taken over the different particle species (electrons, ions, eventually dust particles). Such a dispersion relation can be derived from the equation of the propagation for the scalar potential Φ (or, in alternative, for the density perturbation $\tilde{n}$). For instance, for an electron plasma wave propagating in a plasma with a photon background, this equation takes the form (Bingham et al., 1997)

$$\left(\frac{\partial^2}{\partial t^2} + \omega_p^2 - S_e^2 \nabla^2 \right) \Phi = \int \alpha_k N(\vec{k}) \frac{d\vec{k}}{(2\pi)^3} \tag{4}$$

where $N(\vec{k})$ is the photon occupation number and α_k a momentum dependent parameter. In a more general situation, for unspecified electrostatic waves and arbitrary quasi-particles, we can write the potential equation as (Mendonça et al., 2003)

$$L(\vec{r}, t, N(\vec{k}'))\Phi = 0, \tag{5}$$

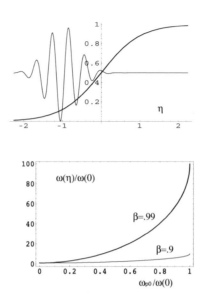

Figure 1. *(a) Electromagnetic wavepacket (photon) moving across a relativistic ionization front, with $\eta = x - \beta ct$. The short wavelength scale of the transverse field justify the use of the quasi-particle description, valid in the geometric optics approximation. (b) Maximum frequency shift (or photon acceleration), in co-propagation, for two different values of the front velocity.*

where L is a space-time differential operator which, will depend on the quasi-particle occupation number $N(\vec{k}')$, for different values $\vec{k}'$ of the quasi-particle momentum. Notice that this is already a nonlinear dispersion relation, because $N(\vec{k}')$ is also generally dependent on the the the potential perturbation Φ. For harmonic perturbations of the form $\exp i(\vec{k} \cdot \vec{r} - \omega t)$, for both Φ and the perturbed occupation number $\tilde{N}(\vec{k}')$ to be specified later, we can obtain

$$\epsilon(\omega, \vec{k})\Phi = \int g(\vec{k}, \vec{k}')\tilde{N}(\vec{k}')d\vec{k}'. \tag{6}$$

Obviously, $\epsilon(\omega, \vec{k}) = 0$ will be the dispersion relation of the slow wave, in the absence of turbulence. But this will be modified by the existence of the quasi-particle gas. In order to proceed further, we meed to estimate the quantity $\tilde{N}(\vec{k}')$ with Φ, as discussed next.

5   Wave kinetic equation

A wave kinetic equation can be built for the turbulent plasma if we identify some invariant quantity N, in the form of an integral over the sixth dimensional (position and momentum) phase space $(\vec{r}, \vec{k}')$, as given by

$$N = \int N(\vec{k}', \vec{r}, t)d\vec{r}d\vec{k}'. \tag{7}$$

In most situations, the quantity appearing inside the integral is nothing but the energy density divided by the frequency, and takes the obvious physical meaning of a quasi-particle number density or wave action: $N(\vec{k}', \vec{r}, t) = W(\vec{k}', \vec{r}, t)/\hbar\omega'$, where we can take $\hbar = 1$, and where ω' is the energy of the quasi-particles with momentum $\vec{k}'$. We can thus identify this frequency with the function $\omega' = \omega'(\vec{k}')$, as determined by the linear dispersion relation of the short wavelength turbulence modes. This is valid for the electromagnetic turbulence (where the photons in a plasma can be seen as an extreme example of quasi-particles and more appropriately described as dressed particles in the usual sense of field theory), as well as for the electrostatic turbulence of the electron plasma or the ion acoustic types (plasmons and phonons). This is also valid in the case of the pseudo-three-dimensional drift wave turbulence, which can be called a drifton gas, as described by a modified Hasegawa-Mima equation (Smolyakov et al., 2000; Lashmore-Davies et al., 2001).

A consistent definition of the quasi-particle number density, where the relation between these two possible definitions is clarified, can be given in terms of the Wigner function for the electrostatic or the electromagnetic field (Tsintsadze and Mendonça, 1998; Hall et al., 2002). Once identified the quantity describing the density distribution $N(\vec{k}', \vec{r}, t)$ in the quasi-particle phase space, we can establish the corresponding Liouville's theorem that states its total time invariance

$$\frac{d}{dt} N(\vec{k}') = \left(\frac{\partial}{\partial t} + \vec{v}' \cdot \frac{\partial}{\partial \vec{r}} + \vec{F}' \cdot \frac{\partial}{\partial \vec{k}'} \right) N(\vec{k}') = 0, \tag{8}$$

where, for simplicity, we use: $N(\vec{k}') = N(\vec{k}', \vec{r}, t)$. Here, $\vec{v}' = \partial\omega'/\partial\vec{k}'$ is the quasi-particle velocity (or equivalently, the group velocity of the turbulence wavepackets) and $\vec{F}' = d\vec{k}'/dt$ is the force acting on these quasi-particles due to large scale perturbations of the medium. This force term includes refraction effects (which maintain the value of ω'), as well as quasi-particle acceleration (implying the variation of the energy, or frequency, ω'). Quasi-particle kinetic equations similar to equation (8) have been widely used in the past, but the importance of the force term has only recently been recognized. It will be shown in this paper that this term plays and essential role in plasma turbulence.

We now consider $N(\vec{k}') = N_0(\vec{k}') + \tilde{N}(\vec{k}')$, and linearize the kinetic wave equation around the unperturbed state $N_0(\vec{k}')$, by assuming a slow perturbation with the frequency ω and the wavevector $\vec{k}$. The result is

$$\tilde{N}(\vec{k}') = -i \frac{\vec{F}' \cdot \partial N_0/\partial \vec{k}'}{(\omega - \vec{k} \cdot \vec{v}')}. \tag{9}$$

Noting that we can write the equivalent force as

$$\vec{F}' = -\frac{\partial \omega'}{\partial \vec{r}} = -i\vec{k} f'(\vec{k}')\Phi, \tag{10}$$

one can relate the density perturbation $\tilde{N}(\vec{k}')$ with Φ. Thus, returning to eq. (6), we obtain

$$\epsilon(\omega, \vec{k}) = -\int f(\vec{k}, \vec{k}') \frac{\vec{k} \cdot \partial N_0/\partial \vec{k}'}{(\omega - \vec{k} \cdot \vec{v}')} d\vec{k}', \tag{11}$$

where we have used $f(\vec{k}, \vec{k}') = f'(\vec{k}')g(\vec{k}, \vec{k}')$. Comparing this equation with (3), we see that the integral in the right-hand side is nothing but the quasi-particle susceptibility χ_{qp}.

6 Resonant contributions

In order to identify the resonant and non-resonant contributions to χ_{qp}, we consider the parallel and perpendicular motions of the turbulence quasi-particles with respect to the direction of the propagation of the slow wave

$$\vec{v}' = u\frac{\vec{k}}{k} + \vec{v}'_\perp \quad , \quad \vec{k}' = p\frac{\vec{k}}{k} + \vec{k}'_\perp. \tag{12}$$

Notice that the parallel velocity is a function of the parallel and the perpendicular momenta: $u = u(p, \vec{k}'_\perp)$. Equation (11) shows that resonant interactions of the wave with the quasi-particles can occur when the parallel velocity equals the phase velocity of the slow wave. We can then write

$$\chi_{qp} = \int f(\vec{k}, \vec{k}'_\perp, p)\frac{\partial N_0/\partial p}{(\omega/k - u)}d\vec{k}'_\perp dp. \tag{13}$$

The function $f(\vec{k}, \vec{k}'_\perp, p)$ and the equilibrium distribution N_0 are, in general, continuous and single valued functions, and this integral only has one pole at $p = p_0$, determined by the resonance condition $\omega/k = u$. Developing u around its resonance value $u_0 = u(p_0, \vec{k}') = \omega/k$, and introducing a parallel function $G(p)$, defined by

$$G(p) = \int \frac{f(\vec{k}, \vec{k}'_\perp, p)}{(\partial u/\partial p)_0}N_0(\vec{k}'_\perp, p)d\vec{k}'_\perp, \tag{14}$$

we can finally write the quasi-particle susceptibility in the form $\chi_{qp} = \chi_r + i\chi_{im}$, where the real and imaginary parts are determined by

$$\chi_r = -P\int \frac{\partial G(p)/dp}{(p - p_0)}dp \quad , \quad \chi_{im} = -\pi\left(\frac{\partial G}{\partial p}\right)_0, \tag{15}$$

and $P\int$ represents the principal part of the integral. This shows that the resonant and non-resonant contributions to the quasi-particle susceptibility have distinctive properties. The non-resonant real part leads to a small correction of the linear dispersion relation. However, the resonant imaginary part can lead to the wave damping or growth, according to the sign of the derivative of the parallel function $G(p)$ at $p = p_0$. This qualitatively important effect can thus be identified with quasi-particle Landau damping.

7   Beam instabilities

Let us first consider the simple and physically relevant case of a Gaussian beam of quasi-particles, described by

$$G(p) = G_0 \exp\left(-\frac{(p - \bar{p})^2}{2\sigma^2}\right), \tag{16}$$

where G_0 represents the beam intensity and σ is the spectral width. The maximum value for the the resonant part of quasi-particle susceptibility corresponds to $p = \bar{p} \pm \sigma$, and it is equal to

$$\chi_{max} = \chi_{im}(\bar{p} \pm \sigma) = \mp\frac{\pi}{e\sigma}G_0. \tag{17}$$

We see that it decreases with an increasing spectral width, and it is proportional to the intensity of the beam: $|\chi_{max}| \propto G_0$. This means that the resulting kinetic instabilities will typically have a growth rate proportional to G_0.

Another important case corresponds to the mono-energetic particle beam with a negligible spectral width, $\sigma \sim 0$, such that (for one-dimensional problems), it can be represented by

$$N(\vec{k}') = N_0\delta(\vec{k}'_\perp)\delta(p - \bar{p}). \tag{18}$$

In order to study this case, we can write χ_{qp} in the form

$$\chi_{qp} = f(k, 0, \bar{p}) \int \frac{\partial N / \partial p}{(\omega / k - u)} dp = -f(k, 0, \bar{p}) \int \frac{N_0 \delta(p - \bar{p})}{(\omega / k - u)^2} dp, \tag{19}$$

and finally obtain

$$\chi_{qp} = -\frac{f(k, 0, \bar{p}) k^2 N_0}{(\omega - k u_0)^2} = -\frac{\Omega_{qp}^2}{(\omega - k u_0)^2}. \tag{20}$$

We see that this takes the familiar form of the susceptibility of electron or ion beams with velocity u_0 and density N_0. The frequency Ω_{qp} plays the role of a quasi-particle plasma frequency, proportional to the square-root of the beam density. Here, again, the contribution of this term to the total wave dispersion relation will become relevant for nearly resonant conditions, such that $\omega \simeq k u_0$, We can then use $\omega = k u_0 + \eta$, with $|\eta| \ll k u_0$. The dispersion relation (11) will take the form

$$\epsilon(\eta, k) = \frac{f(k, 0, \bar{p}) k^2 N_0}{\eta^2}. \tag{21}$$

For solutions such that $\text{Im}(\eta) > 0$, we will have a hydrodynamic type of beam instability, with growth rates that vary with the beam density, typically between $N_0^{1/2}$ and $N_0^{3/2}$. Thus, there appear much stronger than the kinetic beam instabilities associated with the inverse process of Landau damping (Bingham et al., 1997; Mendonça and Bingham, 2002; Smolyakov et al., 2000; Mendonça, 2002; Silva et al., 2000).

8 Quasi-particle trapping

We can also push forward the quasi-particle concept and return to the motion of individual quasi-particles. Such a motion is determined by the characteristics of the kinetic wave equation (8):

$$\frac{d\vec{r}}{dt} = \vec{v}' \quad , \quad \frac{d\vec{k}'}{dt} = \vec{F}' = -f'(\vec{k}') \nabla \Phi(\vec{r}, t), \tag{22}$$

where $\vec{r}$ is the position of the quasi-particles (average position of individual short wavelength wavepackets). These characteristic equations coincide with the ray equations (1). Obviously, the force $\vec{F}'$ is modulated due to the existence of the monochromatic slow wave. If we assume the propagation of this slow wave along the direction Ox, as defined by $\Phi(\vec{r}, t) = \Phi_0 \cos(kx - \omega t)$, we can describe the parallel motion of the quasi-particles by

$$\frac{dx}{dt} = \left(u - \frac{\omega}{k} \right) \quad , \quad \frac{dp}{dt} = -k f'(p) \Phi_0 \sin(kx - \omega t). \tag{23}$$

The perpendicular motion is trivially determined by $\vec{k}'_\perp = \text{constant}$. These equations show the existence of an elliptic fixed point at

$$u(p) = \frac{\omega}{k} \quad , \quad x = \frac{\pi}{k}. \tag{24}$$

This means that, for quasi-particles such that $u(p) \simeq \omega / k$, we will have trapped oscillations at the bottom of the slow wave potential, with small amplitudes $\tilde{x} = x - \pi/k$ around the fixed point. From eqs. (24) we can then derive

$$\frac{d^2 \tilde{x}}{dt^2} = -k^2 f'(p) \frac{\partial u}{\partial p} \Phi_0 \tilde{x}, \tag{25}$$

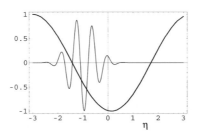

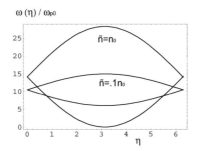

Figure 2. *(a) Electromagnetic wavepacket (photon) trapped by a relativistic electron plasma wave (wakefield). (b) Maximum range of photon frequency shifts, for trapped photon motion near the separatrix, for a wakefield gamma factor $\gamma = 10$, and for two distinct values of the wakefield amplitude.*

which is the equation for the harmonic oscillator with a frequency

$$\omega_b = k\sqrt{f'(p)(\partial u/\partial p)\Phi_0},\tag{26}$$

which is basically the bounce frequency for deeply trapped oscillations of quasi-particles in the slow wave potential Φ. The similarities with the electron bounce frequency are striking. Kinetic effects leading to the appearance of the Kruer mode in a quasi-particle gas is then conceivable.

9   Plasmons and driftons

We illustrate the above general formalism by considering two specific examples . The first one corresponds to the ion acoustic waves moving in a plasmon gas. More specifically, we assume an isotropic plasma with a broadband electron plasma wave turbulence (Mendonça and Bingham, 2002). In this case, we have the usual dielectric function

$$\epsilon(\omega, \vec{k}) = 1 - \frac{k^2 v_s^2}{\omega^2(1 + k^2\lambda_D^2)},\tag{27}$$

where v_s is the ion acoustic velocity and λ_D the electron Debye length. The plasmon gas can be characterized by the following expressions for the velocity and force

$$\vec{v}' = 3v_{the}^2 \frac{\vec{k}'}{\omega'} \quad , \quad \vec{F}' = -\frac{1}{2\omega'}\frac{e^2}{\epsilon_0 m_e}\frac{\partial \tilde{n}_e}{\partial \vec{r}}.\tag{28}$$

Here, $\tilde{n}_e$ is the perturbed electron plasma number density which, for the ion acoustic waves can be related with the potential perturbation by $\tilde{n}_e = \epsilon_0 \Phi/(e\lambda_D^2)$. This leads to a force, given by eq. (10) with

$$f'(\vec{k}') = \frac{1}{2\omega'} \frac{e}{m_e \lambda_D^2},$$ (29)

where $\omega' \simeq \omega_{pe}$. On the other hand, the calculation of χ_{qp} leads to

$$f(\vec{k}, \vec{k}') = \frac{k^2 \omega_{pe}^2}{\omega^2 n_0 m_i (1 + k^2 \lambda_D^2)^2 \omega'^2}.$$ (30)

This completely characterizes the problem. For example, the bounce frequency ω_b of a trapped plasmon in the potential well of the ion acoustic waves will be given by

$$\omega_b \simeq k \frac{\omega_{pe}}{\omega'} \left(\frac{e\Phi_0}{2m_e} \right)^{1/2}.$$ (31)

Comparing this with the electron bounce frequency $\omega_{be} = (e\Phi_0/m_e)^{1/2}$, we conclude that the plasmon behaves as a particle with the electron charge and an effective mass equal to $(2m_e/k^2)$, where we have assumed $\omega' = \omega_{pe}$.

Our second example will be that of a zonal flow in a drifton gas (or, a broad band drift wave turbulence). In this case, we have a pseudo-three-dimensional model in the plan (r, θ) perpendicular to the toroidal magnetic field B_0, which can be described by a modified Hasegawa-Mima equation (Smolyakov et al., 2000; Lashmore-Davies et al., 2001). In this case, we have $\epsilon(\omega, \vec{k}_\perp) = 1$, which means that there is no linear dispersion relation for the zonal flows. However, the quasi-particle susceptibility χ_{qp} associated with the drifton gas allows for the existence of a nonlinear dispersion relation determined by the function

$$f(\vec{k}, \vec{k}_\perp') = -\frac{k^2 v_s^2}{\omega} \frac{k_\theta'^2 \rho_s^2 k_r'}{(1 + k'^2 \rho_s^2)},$$ (32)

where $\rho_s = (v_s/\omega_{ci})$ is the ion acoustic Larmor radius. The electron drift wavepackets, or driftons in our quasi-particle description, can be characterized by the dispersion relation

$$\omega' = k_\theta' \left(V_0 + \frac{V_*}{1 + k_\perp'^2 \rho_s^2} \right),$$ (33)

where $V_0 = ck\Phi_0/B_0$, and V_* is the electron diamagnetic drift velocity. This means that the force acting on the driftons is

$$\vec{F}' = -\frac{\partial \omega'}{\partial \vec{r}_\perp} = -f'(\vec{k}_\perp') \frac{\partial}{\partial \vec{r}_\perp} \Phi,$$ (34)

where $\Phi(\vec{r}_\perp, t)$ is the slow potential perturbation associated with the zonal flow, and

$$f'(\vec{k}_\perp) = \frac{ck}{B_0} k_\theta'.$$ (35)

Here again, the functions $f(\vec{k}, \vec{k}_\perp')$ and $f'(\vec{k}_\perp)$ will completely characterize our problem and allow us to study the various kinetic and hydrodynamic beam instabilities, as well as to establish the values of the bounce frequency and the quasi-particle diffusion.

10   Conclusions

The use of the quasi-paricle concept, the definition of a quasi-particle susceptibility, and the existence of resonant interactions between quasi-particles and slow waves, leads to a general view of the plasma turbulence where long wavelength perturbations can be Landau damped by the turbulent medium. Examples of this type of general behavior are the cases of the photon Landau damping of relativistic electron plasma waves, the plasmon Landau damping of the ion acoustic waves, the phonon Landau damping of dust lattice waves in dusty plasmas or in ordinary liquids, or the drifton Landau damping of zonal flows in magnetic fusion plasmas. The existence of this wide variety of examples results from the universality of the resonant coupling between large scale and small scale structures in fluids and plasmas, and it can be seen as different manifestations of anomalous viscosity associated with the turbulent state.

On the other hand, for appropriate turbulent spectra (or quasi-particle distributions), resonant damping can be replaced by resonant amplification, and consequently lead to instabilities. This could be described as an anomalous negative viscosity. The excitation of large scale structures by small scale turbulence in fluids has been sometimes described as in terms of a negative viscosity. But here we prefer to relate it to the existence of quasi-particle distributions. The concept of quasi-particle is indeed one of the basic concepts of the field theory and its importance to plasma physics is stressed here. The resonant interactions between these particles and wave perturbations are efficient channels for the energy exchange between small and large scale events in a fluid without the need for energy cascading. Their equations of motion describe the energy exchange of these quasi-particles with the medium, and show that quasi-particle acceleration and trapping by waves and by moving large scale perturbations of the medium can eventually take place. Photon acceleration by electron plasma waves and by relativistic ionization fronts (Dias et al., 1997) are nothing but a particular and somewhat spectacular example of this much larger concept.

Furthermore, beams of nearly mono-energetic quasi-particles, with distributions of the type: $N(\vec{k}') = N_0 \delta(\vec{k}' - \vec{k}'_0)$, can excite waves as they move through the medium, in the same way as electron beams or laser pulses (photon beams) can excite electron plasma waves. This can be seen as quasi-particle beam instabilities, and several examples have been studied (Bingham et al., 1997; Mendonça and Bingham, 2002; Smolyakov et al., 2000). Finally, mention should be made to the derivation of quasi-linear equations describing diffusion in quasi-particle phase space, which can answer the question of the global energy transfer between the large and small scales of plasma perturbations, for given initial conditions. Particular examples are the expressions for the photon diffusion coefficient when interacting with an electron plasma wave spectrum (Bingham et al., 1997), or the plasmon diffusion coefficient in the plasmon wavenumber space, expressed in terms of the phonon distribution (Mendonça and Bingham, 2002). Similar expressions could be derived for different kinds of drift wave turbulence interacting with a spectrum of shear flows or zonal flows, as shown by the above general formalism.

References

R. Bingham, J. T. Mendonça and J. M. Dawson, *Phys. Rev. Lett.* **78**, 247 (1997).

F. H. Busse, *Chaos* **4**, 123 (1994).

J. M. Dias *et al. Phys. Rev. Lett.* **78**, 4773 (1997).

J.M. Dias, L.O. Silva and J.T. Mendonça, *Phys. Rev. ST: Accelerators and Beams*, **1**, 031301 (1998).

B. Hall, M. Lisak, D. Anderson *et al.*, *Phys. Rev. E*, **65**, 035602(R) (2002).

B.B. Kadomtsev, *Plasma Turbulence*, Academic Press, N.Y. (1965).

C. N. Lashmore-Davies, D. R. McCarthy and A. Thyagaraja, *Phys. Plasmas* **8**, 5121 (2001).

V. B. Lebedev et al., *Phys. Plasmas* **2**, 4421 (1995).

J.T. Mendonça, *Theory of Photon Acceleration*, Institute of Physics, Bristol (2001).

J.T. Mendonça, N.N. Rao and A. Guerreiro, *Europhys. Lett.*, **54**, 741(2001).

J. T. Mendonça, *J. Phys. A: Math. and Gen.* **34**, 9677(2001); Plasma Phys. Control. Fusion **44**, B225 (2002).

J. T. Mendonça and R. Bingham, *Phys. Plasmas* **9**, 2604 (2002).

J. T. Mendonça, R. Bingham and P.K. Shukla, *Phys. Rev. E* **68**, 016406 (2003).

J.T. Mendonça et al., *Phys. Rev. Lett.*, **94**, 245002 (2005).

P. Norreys et al., *Plasma Phys. Contr. Fusion*, **47**, L49 (2005).

P. K. Shukla, *Phys. Rev. Lett.* **84**, 5328 (2000).

L.O. Silva et al., *IEEE Trans. Plasma Sci.*, **28**, 1202 (2000).

A. I. Smolyakov, P. H. Diamond and V. I. Shevchenko, *Phys. Plasmas* **7**, 1349 (2000).

R. Trines et al., *Phys. Rev. Lett.*, **94**, 165002 (2005).

N. L. Tsintsadze and J. T. Mendonça, *Phys. Plasmas* **5**, 3609 (1998); Phys. Rev. E **62**, 4276 (2000).

A. A. Vedenov, A. V. Gordeev and L. I. Rudakov, *Plasma Phys.* **9**, 719 (1967).

10
Relativistic Phenomena in Plasma Solitons

Francesco Pegoraro

Physics Department, University of Pisa, Pisa, Italy [1]

1 Introduction

A relativistic plasma interacting with an ultrashort, ultraintense laser pulse exhibits new phenomena where the nonlinearity of the relativistic particle kinematics and the nonlinearity of the magnetic part of the Lorentz force become dominant (see e.g., Bulanov S.V. et al. 2001a). Electromagnetic (e.m.) solitons and vortices are part of this complex nonlinear interaction (Pegoraro et al. 2001) and represent the basic ingredients of the long time electron behaviour in the wake of the laser pulse (Naumova et al. 2001a). Electromagnetic solitons are found to occur in regimes where the energy of a laser pulse propagating in an underdense plasma is depleted, resulting in the downshift of the pulse frequency (Bulanov S.V. et al. 1995). This provides a mechanism of e.m. energy trapping in the form of slowly propagating "subcycle-solitons" (Bulanov S.V. et al. 1999).

Solitons have been known for a long time as a special kind of "wave" that occurs in media where nonlinearity (the dependence of the propagation velocity on the wave amplitude) and dispersion (the dependence of the propagation velocity on the wave frequency) play equally important roles (see e.g., Whitham 1974, Zakharov et al. 1984). In these media small amplitude waves evolve in the form of wave packets that spread as they propagate, while finite amplitude waves can develop into coherently propagating structures due to their nonlinear self-interaction. The physics of solitons in different nonlinear media is an open and fast evolving field and their

[1]In collaboration with:

M. Borghesi, Department of Pure and Applied Physics, The Queen's University of Belfast, UK;

S.S. Bulanov, Institute of Theoretical and Experimental Physics, 117218 Moscow, Russia;

S.V. Bulanov & T.Zh. Esirkepov, Advanced Photon Research Centre, JAERI, Kizu, Japan;

F. Califano, Physics Department, University of Pisa, Pisa, Italy;

D. Farina & M. Lontano, Istituto di Fisica del Plasma, CNR, Milano, Italy;

T.V. Liseikina & V.A. Vshivkov, Institute of Computational Technologies SD RAS, Novosibirsk, Russia;

N.M. Naumova, CUOS, University of Michigan; Ann Arbor, Michigan, USA ;

K. Nishihara, Institute of Laser Engineering, Osaka University, Osaka, Japan;

H. Ruhl & Y. Sentoku, Nevada Terawatt Facility, Reno, NV USA.

intriguing properties have lead scientists to create new fields in physics and in mathematics (Degasperis 1998): indeed solitons have been investigated intensively starting from the 60's, but up to now very few results are known about three-dimensional solitary waves.

2   Relativistic plasma solitons

In a plasma interacting with an ultra-high intensity e.m. wave, wave dispersion appears due to finite electron inertia while nonlinearity is due to the relativistic dependence of the electron mass on its energy and to the electron density redistribution under the ponderomotive pressure of the high intensity e.m. fields. The interplay of the relativistic nonlinearity and of the plasma dispersion results in the appearance of different types of coherent nonlinear structures such as self-focusing channels that guide the laser pulse and relativistic electron vortices associated with quasistatic magnetic fields.

Analytical, numerical and experimental techniques must be used to investigate these nonlinear structures. Indeed relativistic plasma solitons were first found in computer simulations (Bulanov S.V. et al. 1999, see also the results presented by Langdon and Lasinski 1983) and only later detected in laboratory experiments (Borghesi et al. 2002). In addition, when investigating these nonlinear effects, one has to account for the fact that only low-dimensional models can be solved analytically because the complexity of the interaction between the e.m. wave and the plasma, due to the high dimensionality of the problem, to the lack of symmetry and to the importance of nonlinear and kinetic effects, makes analytical methods unable to provide a detailed description. On the other hand, powerful methods for investigating the laser-plasma interaction have become available through the advent of modern supercomputers and to the developments of applied mathematics. Supercomputers can now perform fully three-dimensional simulations of the interaction of high intensity laser pulses with plasmas which allow us to obtain detailed information about the nonlinear structures that are generated by a strong e.m. field.

Relativistic plasma solitons are self-trapped and spatially confined nonlinear e.m. waves that propagate in a plasma without diffractive spreading. They are formed during the interaction of ultra-intense laser pulses with plasmas and are generated as the result of the frequency downshift due to pulse energy depletion (Bulanov S.V. et al. 1995). Self-trapping appears because a high amplitude e.m. wave modifies the local refractive index through the relativistic increase of the electron mass and the redistribution of the electron density under the pondermotive pressure of the radiation.

Different relativistic soliton branches have been discussed in the literature (Tsintsadze and Tskhakaya 1977, Schamel et al. 1977, Kozlov et al. 1979, Kaw et al. 1992, Sudan et al. 1997, Mofiz et al. 1985, Rouhani et al. 2002): of special interest here is the branch corresponding to non-propagating solitons, which are called "subcycle solitons" because the e.m. fields inside them have a spatial structure that corresponds to a single half-cycle oscillation (Esirkepov et al. 1998, Bulanov et al. 1999, Bulanov et al. 2001b, Lontano et al. 2003a). The ion dynamics affects these soliton branches differently: on times long on the electron dynamical times when ions can no longer be assumed to remain at rest, subcycle solitons evolve into quasineutral slowly expanding "post-soliton" structures (Naumova et al. 2001b) characterized by a slowly growing hole in the plasma density. The quasistationary nature of the charge-separation electric field associated to these post-soliton structures makes their experimental detection possible (Borghesi et al. 2002).

2.1   1-D analytical results

We shall first recall some analytical results that have been discussed in the case of 1-D relativistic solitons in plasmas. In 2-D and 3-D configurations no such exact solutions are available due to the complexity of the nonlinear plasma response to multi-dimensional e.m. fields and, in 3-D, of the fields' topological structure.

We start by introducing the well known equations (see e.g., Bulanov S.V. et al., 2001b) that describe finite amplitude circularly polarized waves in a cold collisionless unbounded relativistic plasma with immobile ions. In the Coulomb gauge div $A = 0$, Maxwell's equations give

$$\Delta A - \frac{1}{c^2}\partial_{tt}A - \frac{1}{c}\nabla\partial_t\varphi - \frac{4\pi e n_e}{m_e c^2 \gamma}(P + \frac{e}{c}A) = 0, \tag{1}$$

$$n_e = n_i(x) + \frac{1}{4\pi e}\Delta\varphi, \tag{2}$$

where $P \equiv p - eA/c$ is the canonical electron momentum, the relativistic Lorentz factor is $\gamma = [1 + (P + eA/c)^2/(m_e c)^2]^{1/2}$ and $n_i(x)$ is the density of the fixed ion background. We model the electron response using the hydrodynamic equations of a cold electron fluid. Then the electron momentum equation can be put in the form

$$\partial_t P = \nabla(e\varphi - m_e c^2 \gamma) + \frac{1}{\gamma}(P + \frac{e}{c}A) \times (\nabla\times P), \tag{3}$$

while the continuity equation is automatically implied by Eqs.(1,2). We consider 1-D solutions where $\partial_y = \partial_z = 0$. Then the Coulomb gauge gives $A_x = 0$. We can choose initial conditions such that the conserved y and z components of the canonical momentum vanish ($P_y = P_z = 0$).

Assuming the e.m. wave to be circularly polarized, we introduce the new independent variables $X = x - v_s t$ and $\tau = t$ and look for solutions of the form:

$$A_\perp = A_y + iA_z = A(X)\exp[i\omega((1 - \beta_s^2)\tau - v_s X/c^2)], \qquad p_x/m_e c = \beta_s b(X), \tag{4}$$

so that $|A_\perp|$ is independent of τ. Inserting Eq.(4) into Eqs.(1-3) and assuming the ion density to be homogeneous, we obtain the following coupled system of coupled ordinary differential equations

$$\left(\gamma - \beta_s^2 b\right)'' = \frac{\omega_{pe}^2 b}{(\gamma - b)c^2}, \qquad a'' + \frac{\omega^2}{c^2}a = \frac{\omega_{pe}^2 \gamma_s^2}{(\gamma - b)c^2}a, \tag{5}$$

where $\gamma = \left(1 + a^2 + \beta_s^2 b^2\right)^{1/2}$, $\gamma_s = \left(1 - \beta_s^2\right)^{-1/2}$, $\beta_s = v_s/c$, $a = eA/m_e c^2$, and a prime denotes a differentiation with respect to the variable X. This system of equations corresponds to the Hamiltonian motion of a "particle" in a two-dimensional potential field (Kozlov et al., 1979, Kaw et al., 1992).

Setting $b = 0$ in Eqs.(5) we obtain a purely transverse nonlinear e.m. wave with $A(X)$ constant and frequency ω given by (see Akhiezer and Polovin 1956)

$$\omega^2/\gamma_s^2 = \omega_{pe}^2/(1 + a^2)^{1/2}, \tag{6}$$

which is equivalent to $\omega^2 = k^2 c^2 + \omega_{pe}^2/(1 + a^2)^{1/2}$ with wavenumber $k = v_s \omega/c^2$. In this case v_s is the group velocity of the wave and the wave phase velocity is $\omega/k = c^2/v_s$. Localized solutions are obtained by imposing the boundary conditions $a(\infty) = b(\infty) = 0$ (Marburger and Tooper 1975). In this latter case the system of Eqs.(5) describes one-dimensional relativistic e.m. solitons propagating through a cold collisionless plasma. For a small but finite amplitude, an

isolated "envelope soliton" solution is described by the well known hyperbolic secant expression
(Tsintsadze and Tskhakaya 1977)

$$a = \frac{2[1 - (\omega/\omega_{pe}\gamma_s)^2]^{1/2} \exp[i\omega((1 - \beta_s^2)\tau - v_s X/c^2)]}{\cosh\left[k_p^2 X \left(1 - (\omega/\omega_{pe}\gamma_s)^2\right)^{1/2}\right]}, \tag{7}$$

with frequency $\omega \approx \omega_{pe}\gamma_s(1 - a_m^2/8)$, and amplitude $a_m = a(0,0)$.

An exact solution can be found in the limit of a soliton with zero propagation velocity
$\beta_s = 0$ (Kurki-Suonio et al. 1982, Esirkepov et al. 1998). In the case $\beta_s = 0$, p_x vanishes,
$\gamma = (1 + a^2)^{1/2}$, $\gamma_s = 1$, and Eqs.(5) reduce to

$$a'' + k_p^2[(\omega/\omega_{pe})^2 - (1 + k_p^2\gamma'')/\gamma]a = 0, \tag{8}$$

from which we obtain a soliton solution of the form

$$a(X,\tau) = \frac{2[1 - (\omega/\omega_{pe})^2]^{1/2} \cosh\left[k_p^2 X \left(1 - (\omega/\omega_{pe})^2\right)^{1/2}\right] \exp(i\omega\tau)}{\cosh^2\left[k_p^2 X \left(1 - (\omega/\omega_{pe})^2\right)^{1/2}\right] + 1 - (\omega/\omega_{pe})^2}, \tag{9}$$

where $k_p = \omega_{pe}/c$. The soliton frequency ω depends on the soliton amplitude a_m as

$$a_m = 2\omega_{pe}(\omega_{pe}^2 - \omega^2)^{1/2}/\omega^2. \tag{10}$$

This single-hump soliton solution is stable if the electron density inside it does not vanish. This
imposes the constraints $a_m < \sqrt{3}$ and $1 > (\omega/\omega_{pe}) > \sqrt{2/3}$.

The solution given by Eq.(9) can been extended to the case of a soliton in a uniform magnetised plasma. In this case, the maximum allowed soliton amplitude depends on the magnetic
field intensity and on its sign. For negative values of Ω/ω, with $\Omega = eB/m_ec$, solitons exist with
amplitudes appreciably larger and frequencies lower than in the unmagnetised case (Farina et
al. 2000).

Ion motion has important effect on the propagating envelope (multi-humped) solitons and
single cycle solitons. It can be shown (in the framework of two fluid, electron and ion, cold
equations) that no solution can be found for propagation velocities v_s smaller than a critical
value $v_{s,cr}$ (Farina and Bulanov S.V. 2001a, 2001b, 2005). The non propagating solution given
in Eq. (9) is not continuously connected to those with $v_s \neq 0$ and thus its structure will change
on the ion dynamical time.

Thermal and kinetic effects are important at low propagation speeds. The relativistic hydrodynamic equations can be extended to a hot multi-component plasma, e.g., by assuming an
adiabatic closure for each species (Lontano et al. 2001). In this framework, 1-D relativistic e.m.
solitons in warm electron-positron plasmas have been studied (see in addition Berezhiani et al.
(2002)). An isothermal hydrodynamic model has been formulated (Lontano et al. 2002) starting
from the exact solution of the relativistic Vlasov equation for circularly-polarized e.m. radiation
in a 1-D multi-component plasma. Relativistic solitons in an electron-positron plasma and in a
quasi-neutral electron-ion plasma have been investigated and solutions with extremely high field
intensities have been obtained (Lontano et al. 2002, 2003b).

2.2 Analytical models and numerical results for higher-dimension relativistic subcycle solitons

In two and three dimensions exact solutions for relativistic e.m. solitons of the type described
above are not available and it is worth noting that the nonlinear wave evolution in 3-D regimes

differs drastically from the 1-D and 2-D cases, as shown, e.g., by the problem of wave collapse (Zakharov 1972).

Numerical simulations with Particle in Cell (PIC) codes both in 2-D (Bulanov S.V. et al. 1999) and in 3-D (Esirkepov et al. 2002) indicate that long-lasting coherent soliton-like structures, where e.m. energy from the laser pulse is self trapped, do occur. Indeed, relativistic solitons are now routinely observed in two dimensional simulations (Sentoku et al. 1999, Mima et al. 2001, Naumova et al 2001b). The solitons found in these simulations consist of slowly

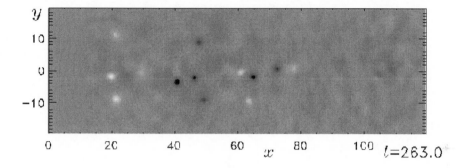

Figure 1. *Subcycle solitons in 2D soliton simulations: the oscillatory, out of the plane, component of the electric field of the multiple s-polarized subcycle solitons left behind in the plasma by an s-polarized short pulse. Bright and dark correspond to opposite oscillation phases (Naumova et al. 2001a). Here, and in the following, distances are given in units of the pulse wavelength and time in units of the pulse oscillation period.*

or non propagating electron density cavities inside which an e.m. field is trapped and oscillates coherently with a frequency below the unperturbed plasma frequency and a spatial structure corresponding to half a cycle.

In the 2-D case, the analogues of Eq.(8) can be obtained in the limit of weak nonlinearity for linearly polarized s- and p- solitons. In s-polarized solitons the z-component of the electric field and the azimuthal component of the magnetic field oscillate in time, while the electron density distribution remains constant. Assuming for the vector potential a time dependence of the form $a(r) \exp\left[-i(\omega_{\text{pe}} - \delta\omega)t\right]$, we find

$$r^{-1}\partial_r r \partial_r a - k_p^2[2\delta\omega/\omega_{\text{pe}} - |a|^2]a = 0. \tag{11}$$

This equation describes localized 2-D solitons with frequency shift $\delta\omega$ and radius r_0 which depend on the soliton amplitude as $\delta\omega \sim a^2$ and $r_0 \sim 1/a$. This scaling agrees with the dependence of the frequency on the soliton amplitude found above in the case of a planar circularly polarized 1-D soliton. In p-polarized solitons the electric field is azimuthal and the magnetic field is directed along z and for the amplitude $a_\varphi(r)$ we obtain

$$r^{-1}\partial_r r \partial_r a_\varphi - r^{-2}a_\varphi - k_p^2(2\delta\omega/\omega_{\text{pe}} - |a|^2)a_\varphi = 0. \tag{12}$$

The typical spatial structure and field orientation of the s- and p- solitons obtained by numerical integration of Eqs.(11,12) are discussed in Bulanov S.V. et al. (1999, 2001b). The information provided by these analytical models is found to be in agreement with the results of the numerical simulations of 2-D solitons, as illustrated in the figures (not shown here) presented in these references.

Here we briefly recall two important results that have been displayed by these simulations. Collisions between two 2-D s-solitons have been observed and shown to lead to the synchronization and eventual merging of two s-solitons, differently from the results obtained for circularly polarized 1-D solitons, where each soliton maintains its amplitude and propagation speed while their phases change. This indicates that in 2-D some of the characteristic properties of solitons are not recovered.

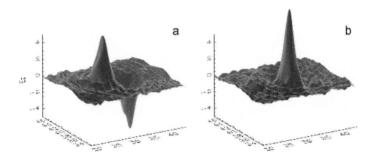

Figure 2. *Two 2-D subcycle solitons become synchronized and merge into a single, more intense and thus narrower soliton (Bulanov S.V. 2001a).*

In an inhomogeneous plasma the propagation of the subcycle solitons is strongly affected by the inhomogeneity of the medium. A simple model of the soliton motion derived by referring to the propagation of a wave packet in a non-uniform dispersive medium shows that the wave packet is accelerated towards the low density side of the plasma and, if a density channel is formed e.g., by the laser pulse, the soliton oscillates in the transverse direction. As a result, the soliton moves towards the plasma vacuum interface (Sentoku et al. 1999), as shown in Figure 3.

This effect can be exploited in order to extract and detect the solitons. The trapping of the e.m. energy becomes weaker as the solitons move towards regions of lower density until their local oscillation frequency becomes larger than the ambient plasma frequency when they burst into low frequency e.m. radiation. These radiation bursts might be detected and used as a soliton diagnostic.

Three dimensional simulations of laser induced subcycle relativistic e.m. solitons have been performed (Esirkepov et al. 2002), using the REMP - Relativistic Electro-Magnetic Particle-mesh code based on the Particle-in-Cell method and density decomposition scheme (Esirkepov 2001).

We could expect that, knowing the properties of the various types of two dimensional solitons, one could have easily guessed the topological structure of a three dimensional soliton. However, in generalizing the 2-D results to a fully 3-D configuration a new problem arises from the vector nature of the fields i.e., from the topological constraints that are encountered when trying to confine the e.m. fields inside a finite 3-D domain. In view of the difficulty of presenting an analytical model of a relativistic 3-D e.m. soliton, we describe explicitly the numerical results obtained in (Esirkepov et al. 2002).

The pulse considered in these simulations has a dimensionless amplitude $a = eE_z/(m_e\omega c) = 1$, corresponding to the peak intensity $I = 1.38 \cdot 10^{18} \text{W}/\text{cm}^2$ for a $\lambda = 1\mu\text{m}$ laser pulse. The pulse propagates in a plasma with density $(\omega/\omega_{pe})^2 = 7.7$ along the x-axis and is linearly polarized

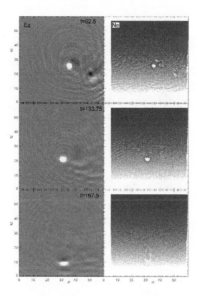

Figure 3. *Evolution in an inhomogeneous plasma with a density gradient along the y axis of a soliton generated by a laser pulse propagating along x. The oscillatory electric field out of the plane E_z and the electron density are shown at different normalized times (Sentoku et al. 1999).*

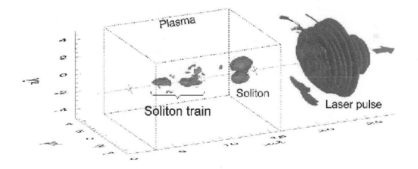

Figure 4. *Isosurface of the electromagnetic energy density of the laser pulse and of the 3-D soliton in the pulse wake (Esirkepov et al. 2002).*

along z, with a Gaussian envelope with FWHM size $8\lambda \times 5\lambda \times 5\lambda$. Its focal plane is placed in front of the plasma slab at the distance of 3λ. Ions and electrons have the same absolute charge, and the mass ratio is $m_i/m_e = 1836$. The length of the plasma slab is 13λ. The boundary conditions are periodic along the y- and z-axes and absorbing along the x-axis for both the EM radiation and the quasiparticles.

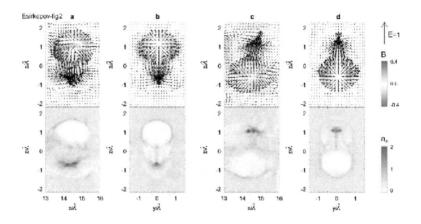

Figure 5. *3-D soliton structure. In the upper row the arrows represent the electric field in the x-z and y-z planes, and the background shading indicates the magnetic field perpendicular to the planes. In the bottom row, the electron density is shown (Esirkepov et al. 2002).*

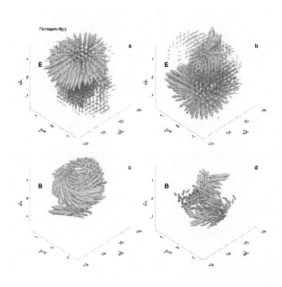

Figure 6. *Three dimensional structure of the electric (a),(b) and magnetic (c),(d) fields in the soliton (Esirkepov et al. 2002).*

These simulations are computationally expensive: the simulation box has $660 \times 400 \times 400$ grid points with a mesh size of 0.05λ. The total number of quasiparticles is $426 \cdot 10^6$. The simulations were performed on 16 processors of the NEC SX-5 vector supercomputer in Cybermedia Center, Osaka University.

The soliton found in these simulation is shown in Figures. 4,5,6, and consists of oscillating electrostatic and electromagnetic fields confined in a prolate cavity in the electron density. The cavity size is approximately $2\lambda \times 2\lambda \times 3\lambda$ and is is azimuthally symmetric. The electric field inside the cavity is so large that the quivering distance of electrons in the z-direction is of the order of the cavity size, which results in pulsations of the cavity.

The soliton has an electrostatic component and an inductive poloidal component, while the magnetic field is toroidal so that its structure resembles the lowest eigenmode of a cavity resonator with a deformable wall.

In the equatorial plane the structure of the 3-D soliton is similar to that of a 2-D s-soliton, while that in the perpendicular planes is similar to a 2-D p-soliton. The electrostatic and electromagnetic components are of the same order of magnitude.

The electromagnetic field trapped in the oscillating cavity pulsates at a frequency Ω_S, smaller than the surrounding unperturbed Langmuir frequency, $\Omega_S \approx 0.87\omega_{pe}$.

The density of cavity walls is $2-3\ n_{cr}$ where n_{cr} is the critical density for the laser radiation. Therefore the e.m. energy of the fields oscillating at fundamental (soliton) frequency can not be radiated away, and, in addition, the soliton oscillation does not resonate with plasma waves. However, a high frequency radiative decay process occurs due to the nonlinear oscillation of the electrons in the z-direction. As a result, a linearly polarized 3-D soliton emits antenna-like radiation. In the case of a circularly polarized 3-D soliton, the azimuthal nonlinear oscillation of the electrons causes the emission of an e.m. wave with a spiral spatial pattern, similar to coherent synchrotron radiation, as was shown in (Esirkepov et al. 2004). In both cases the radiated e.m. wave has a frequency much higher than the plasma frequency and thus of the oscillating fields inside the soliton, $\sim \gamma^3 \Omega_S$, where γ is the effective Lorentz-factor of the electron density responsible for emission. Because of the characteristic radiation power and polarization, this decay process is important for plasma diagnostics and provides a channel for the transformation of a part of the laser pulse energy into high-frequency radiation.

2.3   Subcycle soliton evolution

Besides supporting the analytical results and models on the soliton structure in configurations with different dimensionality, numerical simulations allow us to identify the mechanism of sub-cycle soliton generation and, on longer time scales, to account for the effect of the ion dynamics. As mentioned above, the results obtained in the 1-D limit, which show that no solution corresponding to non-propagating solitons is found when mobile ions are considered, indicate that subcycle solitons are expected to be significantly affected by the ion dynamics.

Adiabatic pulse energy depletion and soliton generation

The physical mechanism that produces sub-cycle solitons, as seen in the PIC simulations, is different from the standard process where the nonlinear steepening of the wave is counterbalanced by the effect of dispersion. In the present case the dispersion effects come into play because of the frequency downshift of the laser pulse.

Interacting with the underdense plasma, the laser pulse loses its energy as it generates electrostatic and magnetostatic wake fields behind it. Since the frequency of these fields is much lower than the carrier frequency of the laser pulse the laser-plasma interaction can be taken to be adiabatic: thus the ratio between the e.m. energy density and frequency is adiabatically conserved and the decrease of the laser pulse energy is accompanied by the downshift of its carrier frequency (Bulanov et al. 1992) . When the downshifted value of the carrier frequency approaches the value of the local plasma frequency, the process stops to be adiabatic.

Nevertheless, PIC simulations clearly show that the pulse frequency decreases further and becomes smaller than the local non-relativistic plasma frequency. This provides the mechanism for the e.m. wave energy trapping into a non-propagating (zero group velocity) soliton.

The effectiveness of this mechanism can be verified by correlating the location of the soliton formation in the PIC simulations with the pulse depletion length. Because of the general nature of this depletion process, we can expect that relativistic subcycle solitons represent an important channel of conversion of the energy of propagating relativistically intense e.m. radiation into trapped radiation and eventually into plasma energy (see also the results presented in Tushentsov et al. (2001)).

Soliton generation during Brillouin backscattering

An additional mechanism of soliton generation has been recently pointed out (Weber et al. 2005a, 2005b) for lower intensity pulses. In 1X-2V PIC numerical simulations of the interaction of a laser-pulse with $I \, \lambda^2 \; = \; 10^{16} \; W \; \mu m^2/cm^2$) and a slab of underdense plasma ($n_e = 0.3n_{cr}$), a saturation regime of the stimulated backward Brillouin scattering (SBBS) has been observed.

This regime has been attributed to the creation of several deep quasineutral plasma cavities, where part of the radiation at a down-shifted frequency is trapped. The envisaged process of frequency downshift consists of two steps involving the parametric decay into a transverse mode at the local plasma frequency and into a longitudinal kinetic plasma mode at the frequency $0.8\omega_{pe}$ and the subsequent slow adiabatic frequency down-shift down to 0.25ω.

During this process a plasma cavity is formed and the trapped radiation redistributes itself according to the lowest order eigenstate, an half-cycle soliton (Weber et al. 2005c). The quasi-stationary half-cycle solitons can be described (Lontano et al. 2005) in term of the kinetic model developed by Lontano et al. (2003b), showing the role of the plasma temperature in determining the soliton amplitude.

Soliton transformation into slowly expanding post-solitons

The soliton formation time is much shorter than the ion response time so that ions can be assumed to be at rest during this phase. For times longer than $(m_i/m_e)^{1/2}$ oscillation periods of the e.m. fields inside the soliton, the ponderomotive pressure created by these fields starts to dig a hole in the ion density and the parameters of the soliton change. Although this implies that, strictly speaking, subcycle solitons cease to behave as solitons on the ion timescale, numerical simulations show that, independently of the dimensionality of the configuration, a low frequency e.m. wave packet remains in the plasma and is well confined inside a slowly expanding plasma cavity (Naumova et al. 2001b, Esirkepov et al. 2002). This nonlinear e.m. wave packet has been defined as a "post-soliton".

We can describe the scenario of the post-soliton formation as follows. In the first phase, for times shorter than the ion dynamical time, the ponderomotive pressure of the e.m. fields inside the soliton is balanced by the force due to the electric field which appears due to charge separation. The ponderomotive pressure displaces the electrons outwards so that, on ion times, the Coulomb repulsion in the electrically non neutral ion core pushes the ions away.

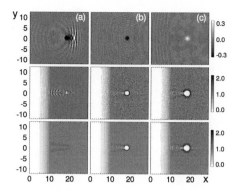

Figure 7. *Post soliton formation by the interaction of an s-polarized laser pulse, with dimensionless amplitude a = 1, with an underdense plasma $n_0 = 0.36\,n_{cr}$. The z-component of the electric field (first row), the electron density (second row) and the ion density (third row) in the x, y plane at t = 30, t = 70, t = 120. Dimensionless units are used (Naumova et al. 2001b).*

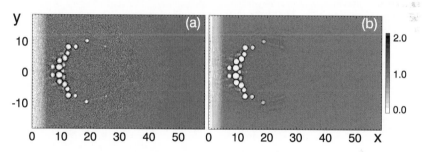

Figure 8. *Post soliton necklace. Plasma and pulse parameters are as in Figure (7) but the pulse is wider (Naumova et al. 2001b).*

The typical ion kinetic energy corresponds to the electrostatic potential energy which is of the order of $m_e c^2 a$, with a the dimensionless amplitude of the oscillating e.m. fields inside the soliton. The ion expansion in the radial direction leads to the formation of a hole in the ion density. This plasma cavity forms a resonator for the trapped e.m. field. As this cavity expands, the amplitude and the frequency of the e.m. field decrease. Since the radius R of the cavity increases very slowly compared to the period of the e.m. field oscillations we can exploit the adiabatic invariant given by the ratio between the energy and the frequency of the e.m. field $\int E^2 dV/\omega_s = const$.

As a simple analytical model to describe the e.m. field inside the post-soliton, we can use the well known electric- or magnetic-dipole oscillations inside a spherical resonator where the frequency is proportional to the inverse of the cavity radius R. From $\int E^2 dV/\omega_s = const$ we obtain that $E^2 \sim R^{-4}$. Under the action of the e.m. pressure $\langle E^2 \rangle /8\pi$ the walls of the cavity expand. In the "snow plough" approximation (Zel'dovich and Raizer 1967) all the mass of the plasma pushed by the e.m. pressure, $M = 4\pi n_0 m_i R^3/3$ where n_0 is the unperturbed plasma density and R is the shell radius, is located inside a thin shell. Then from Newton second law, with $4\pi R^2 \langle E^2 \rangle = 4\pi \langle E_0^2 \rangle (R_0^4/R^2)$, where E_0 is the amplitude of the electric field inside the soliton and $R_0 \approx c/\omega_{pe}$ is the soliton radius, we obtain

$$\frac{d}{dt}\left(R^3\frac{dR}{dt}\right) = \frac{3R_0^4\left\langle E_0^2\right\rangle}{8\pi n_0 m_i R^2}, \tag{13}$$

and, defining the expansion timescale $\tau = (6\pi R_0^2 n_0 m_i / \left\langle E_0^2\right\rangle)^{1/2}$,

$$\frac{R}{R_0} = \left[\frac{(2\tau^2)^{2/3} + (2\tau^2 + t^2 + t\sqrt{4\tau^2 + t^2})^{2/3}}{(2\tau^2)^{1/3}(2\tau^2 + t^2 + t\sqrt{4\tau^2 + t^2})^{1/3}} - 1\right]^{1/2}. \tag{14}$$

Asymptotically, when $t \to \infty$, the post-soliton radius increases as $R \approx R_0(2t/\tau)^{1/3}$, the amplitude of the e.m. field and its frequency decrease as $E \sim t^{-2/3}$ and $\omega_s \sim t^{-1/3}$ (in the case of a cylindrical cavity, we would find $R \sim t^{2/5}$, for $t \to \infty$).

2-D numerical PIC simulations, see Figure (7), show the formation of a single post-soliton and its expansion (Naumova et al. 2001b). In the case of wider laser pulses a cloud of solitary waves is produced in the pulse wake. At later times this soliton cloud evolves into post-solitons as shown in Figure (7) for an s-polarized pulse with amplitude $a = 1$, width 30λ and length 15λ in an underdense plasma ($n/n_{cr} = 0.64$) at $t = 70$. In the distribution of the electron and ion densities we see a "necklace" of the post-solitons.

As these post-solitons expand, they coalesce into large regular structures (post-soliton bubbles) that continue to grow and remain stable against non-spherical (non-circular) deformations, as shown in Figure 9. This geometrical stability is discussed in Bulanov S.V. and Pegoraro (2002), where the stabilizing effect of the mass accretion on the onset the Rayleigh-Taylor instability is analyzed. The evolution of a 3-D soliton into a post-soliton is also evident in the 3-D simulations presented in Esirkepov et al. (2002).

3   Experimental detection of relativistic solitons

The identification of the nonlinear structures, such as self-focussing channels, current filaments, magnetic vortices and solitons and the investigation of their time evolution represent a difficult experimental challenge. In the case of solitons, a key ingredient in their identification may be represented by the bursts of low frequency radiation that they are expected to emit when they exit an inhomogeneous plasma from the low density side, as seen in 2-D and 3-D simulations. No positive soliton identification based on an unambiguous detection of such bursts is as yet available. On the contrary, the transformation of the non-propagating subcycle solitons into quasi-neutral, long lived, post solitons makes it possible to infer the occurrence of solitons in the plasma, through the deflection of the proton beam in the proton imaging technique (Borghesi, et al. 2001) .

3.1   Post-soliton detection by proton imaging techniques

The first experimental observation of a post-soliton bubble was reported by Borghesi et al. (2002). The experiment was carried out at the Rutherford Appleton Laboratory employing the VULCAN Nd:glass laser operating in the Chirped Pulse Amplification mode (CPA). The VULCAN CPA pulse was split in two separate 1 ps, 1 μm, 20 J pulses (CPA_1 and CPA_2) which were focussed onto separate targets in a 10-15 μm FWHM focal spot giving an average intensity of about 10^{19} W/cm^2.

The CPA_1 pulse was used as the main interaction pulse and focused into a preformed plasma. The plasmas were produced by exploding thin plastic foils (0.3 μm thick) with two 1 ns, 0.527 μm

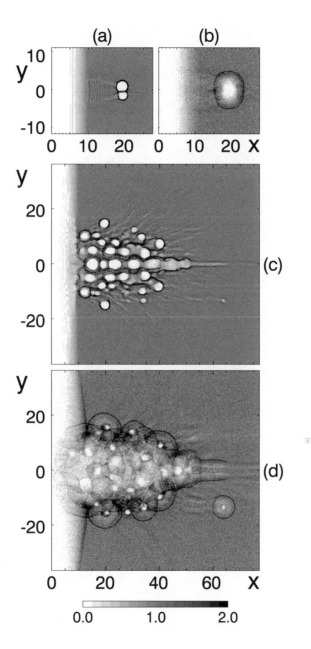

Figure 9. *Post soliton coalescence and expansion. The merging of two post-solitons, as described by 2-D PIC simulations, is illustrated in frames a,b. For a wider and more intense pulse with a = 3 a cloud of solitons is formed after the laser pulse has filamented, as shown in frame c. In As a result of the post-soliton merging, a big hole appears in the plasma density, as shown in frame d at t = 550 (Borghesi et al. 2002).*

laser pulses at a total irradiance of about 5×10^{14} W/cm^2. The delay between plasma formation and interaction was typically 1 ns. The CPA_2 pulse was focussed onto a 3 μm Al foil in order to produce a beam of multi-MeV protons, which were used as a transverse particle probe of the interaction region.

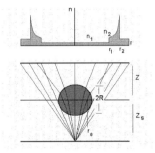

Figure 10. *Proton imaging geometry: proton deflection by the electric field in a post-soliton of radius R at distance Z_s from the proton source (Borghesi, et al. 2002).*

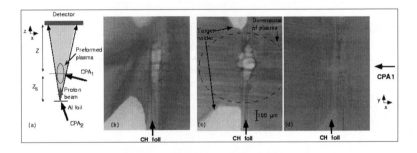

Figure 11. *Proton images of the plasma obtained with $6-7$ MeV protons and recorded, in different shots, at various delays after the interaction (Borghesi et al. 2002).*

As the proton source is small, the proton probe could be used in a point-projection imaging arrangement with the magnification determined by $1 + Z/Z_s$, where $Z = 2cm$ and $Z_s = 2mm$ are respectively the object-to detector and object-to source distances. Bubble-like structures in proton images of laser-produced plasmas were interpreted as the macroscopic remnants of solitons formed during the interaction with the ultraintense pulse (Borghesi et al. 2002).

The observed structures correspond to localized depletion regions in the cross section of a transverse-propagating proton probe beam. The bubbles appear as the protons are deflected away by the localized electric field at the edges of the cavitated areas: 5-6 bubble-like structures are clearly visible at the centre of the plasma corresponding to delays of 45 ps after the CPA_1 interaction. In particular a large structure with radius of approximately 50 μm is seen at the centre of the picture.

In the single particle approximation, which applies when the Debye shielding of the electrons accompanying the probe proton beam occurs on a scalelength bigger than the bubble size, it is possible to find both the size of the bubble and the value of the quasistatic electric field inside from the observed proton image of the bubble. In the case of protons with $\approx 6\ MeV$ energy the effect of the oscillating e.m. field is averaged out during the proton transit time. We take the quasistatic electric field $E_{\parallel}$ inside the post-soliton due to charge separation to be of the order of $E_{\parallel} \approx E_0(R_0/R)^4$ and to be directed normally to its walls and to vanish outside the cavity.

Furthermore, we assume that the post-soliton bubble has a spherical shape and that the $E_{\parallel}$ has a linear spatial dependence $E_{\parallel} = E_{\parallel 0}(r/R)$. Then, the transverse momentum gained from $E_{\parallel}$ by a proton crossing with velocity v_z and impact parameter r_0 a bubble of radius R, is given for $r_0 < R$ by $\Delta p_{\perp} \approx (2eE_{\parallel 0}r_0/v_z)\sqrt{1 - r_0^2/R^2}$ and $\Delta p_{\perp} = 0$ for $r_0 > R$.

The resulting proton density distribution at the distance Z from the bubble located at a distance $Z_s < Z$ from the point-like proton source is approximately constant inside the radius $r_1 = R(1 + Z/Z_s)$, i.e. inside the geometrical projection of the bubble on the image plane, has a jump at r_1 and tends to infinity at the caustic at r_2.

For relatively small deflections, $r_2 = r_1 + \delta r$ with $\delta r \approx (r_1/2)(eE_{\parallel 0}Z_s/m_p v_z^2)^2$. From the structure of the central bubble we can estimate $\delta r \approx 0.3r_1$ with $r_1 \approx 530\mu m$. For $m_p v_z^2/2 = 6\ MeV$, with $Z = 2cm$ and $Z_s = 2mm$, we obtain $E_{\parallel 0} \approx 4 \cdot 10^7\ V/cm$, $R \approx 50\mu$.

4   Attosecond pulse generation

The interaction of a low-frequency relativistic soliton in a plasma with the electron density modulations of a wake Langmuir wave can be exploited in order to produce a train of single-cycle high-intensity e.m. pulses. Methods for obtaining attosecond pulses have been recently presented in the so-called λ^3 regime of laser pulses with wavelength limited foci and single cycle pulse duration (Naumova, et al., 2004a, 2004b, 2005), or in the so-called sliding mirror regime, where under appropriate experimental conditions electrons in a thin plasma layer can move only along the plasma surface (Pirozhkov et al. 2005).

At very large intensities new nonlinear phenomena appear which arise from nonlinear quantum electrodynamics effects such as pair creation (Bulanov S.S. 2004, Bulanov S.S. et al. 2005a). In order to reach such regimes, an alternative method of generating ultra-short e.m. pulses was proposed by Isanin et al. 2005. This method exploits the interaction between a relativistic soliton and the density modulations of a Langmuir wakefield in a plasma. The pulse energy enhancement and the pulse compression arise because the electron density modulations in the wake wave act as parabolic relativistic mirrors for the relativistic soliton produced by a different laser pulse in the plasma. These results are based on the realization that nonlinear plasma processes can be harnessed ("relativistic engineering") in order to concentrate the e.m. radiation in space and in time and to produce e.m. pulses of unprecedented high intensity or short duration that can be used to explore ultra-high energy density effects in plasmas (Bulanov S.V. et al. 2003).

One-dimensional (1-D) analytical calculations show that the frequency of the reflected pulse is up-shifted according to Einstein's relationship for a relativistic reflecting mirror at normal incidence $\omega'' = \omega(1 + \beta_{ph})/(1 - \beta_{ph})$, by a factor $2\gamma_{ph}^2$, and that its intensity is proportional to γ_{ph}^3, where γ_{ph} is the Lorentz factor corresponding to the phase velocity v_{ph} of the wake wave. For the frequencies of interest for the e.m. fields of a soliton trapped in the plasma, the reflection coefficient is taken to scale as $\gamma_{ph}^{-1/2}$.

Recent 2-D numerical PIC simulations of the interaction of a breaking wake plasma wave with a soliton confirm the viability of this scheme of soliton reflection. In these simulations, the wake wave and the soliton are induced by two laser pulses that propagate in the plasma perpendicularly to each other: the soliton which is left in the wake of the first laser pulse interacts with the wake wave produced by the second pulse. As a result of this interaction, we observe the formation of a train of single cycle e.m. pulses that carry part of the energy of the soliton. The plasma and pulse parameters used in these simulations can be achieved in present day experiments.

4.1   1-D Analytic results

When an intense short laser pulse interacts with a plasma, it induces a wakefield consisting of nonlinear Langmuir waves with a phase velocity $v_{ph} = \beta_{ph}c$ equal to the group velocity of the laser pulse. If the laser pulse propagates in a low density plasma, the latter is close to the speed of light in vacuum.

The nonlinearity of the strong wakefield leads to the steepening of its profile and to the formation of sharply localized maxima, "spikes", in the electron density. At wavebreak the electron density in the spikes tends to infinity, $n_e(X) = (n_0\beta_{ph})/(\beta_{ph} - \beta_u(X))$ with $X = x - v_{ph}t$, but remains integrable. Here n_0 is the ion density and the ratio β_u between the speed of the electrons and the speed of light varies from $-\beta_{ph}$ to β_{ph}. Close to the wave breaking conditions, we can write $n_e(X) \sim n_0[1 + \lambda_p\delta(X)]/2$, where λ_p is the wavelength in the wavebreaking regime. This density spike partially reflects a counterpropagating e.m. wave.

The properties of the reflected pulse can be derived by performing a Lorentz transformation to the reference frame where the wake plasma wave is at rest. In this frame, the reflection coefficient of a plasma foil is given, according to the derivation by Bulanov S.V. et al. 2003, by

$$\rho(\omega') = -q/(q - i\omega'), \tag{15}$$

where $q = 2\omega_{pe}(2\gamma_{ph})^{1/2}$ and ω' is the soliton frequency in this moving frame.

The form and amplitude of the reflected pulse in the moving frame are then obtained by an inverse Fourier transform followed by the inverse Lorentz transformation of the vector potential: the frequency of the reflected pulse is up-shifted by $2\gamma_{ph}^2$, its length is compressed by $2\gamma_{ph}^2$ while the amplitudes of the electric and magnetic fields are increased by $\gamma_{ph}^{3/2}$.

The Lorentz factor γ_{ph} of the wakefield generated by a laser pulse in plasma is $\gamma_{ph} \approx \omega_d/\omega_{pe}$, where ω_d is the frequency of the laser pulse that generates the wake plasma wave. The frequency up-shift is $2\gamma_{ph}^2\omega \approx 2\gamma_{ph}^2\omega_{pe} \approx 2\gamma_{ph}\omega_d$. Thus, for a $1\mu m$ wavelength laser pulse, corresponding to the critical plasma density $n_{cr} \approx 10^{21}cm^{-3}$, the factor $2\gamma_{ph}$ required to generate an attosecond reflected pulse is of order 10^3, i.e. the density of the plasma must be of the order of $4 \times 10^{15}cm^{-3}$. Denoting by a_0 the dimensionless amplitude of the soliton (defined in terms of the soliton frequency in the laboratory frame), the intensity of the e.m. field is given by $I_0 \approx (a_0/\gamma_{ph})^2 10^{18}W/cm^2$. Then, the intensity of the reflected e.m. pulse is $I_{ref} \approx \gamma_{ph}^3 I_0 = a_0^2\gamma_{ph}10^{18}W/cm^2$. The paraboloidal shape of breaking wake plasma wave focuses the reflected e.m. pulse, further increasing its amplitude. In the case of the soliton, the enhanced scaling of the reflected wave intensity, $I_{ref} \approx \gamma_{ph}^5 I_0$, leads to $\approx a_0^2\gamma_{ph}^3 10^{18}W/cm^2$.

4.2   2-D Particle-in-Cell simulation results

Two-dimensional PIC simulations of two laser pulses that propagate in a plasma along perpendicular directions show the soliton reflection process (Bulanov S.S. et al. 2005b). The first pulse is initialized at $t = 0$ and $y = -18$ and propagates along the y-direction. Its wavelength is $\lambda_1 = 2\lambda_2$ and its dimensionless amplitude is $a_1 = 0.5$, which corresponds to the peak intensity $(0.25/4) \times 10^{18}$ W/cm$^2 \times (1\mu m/\lambda_2^2)$.

It is linearly polarized along the z-axis, has a Gaussian shape in both directions, and its longitudinal half-width is 4 while its transverse half-width is 6. The second pulse, chosen shorter

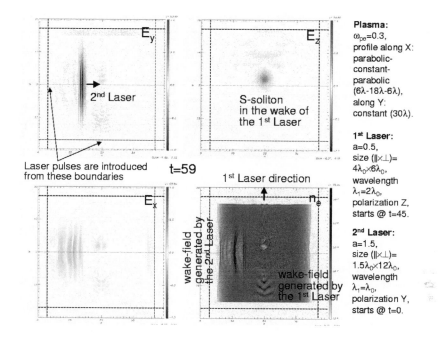

Figure 12. *Interaction geometry between the soliton and the wakefield.*

and wider than the first, generates the wakefield. It is initialized at $t = 45$ and $x = 2$ and propagates along the x-axis. Its wavelength is λ_2 and its dimensionless amplitude is $a_2 = 1.5$, which corresponds to the peak intensity 2.25×10^{18} W/cm$^2 \times (1\mu m/\lambda_2^2)$.

The pulse shape is Gaussian in both directions, its longitudinal half-width is 1.5 and its transverse half-width is 12. This pulse is linearly polarized along the y-axis so as to distinguish it from the reflected part of the soliton produced by the first pulse.

These pulses propagate in a plasma with electron density $n_e = 0.09 n_{cr}$, which corresponds to a Langmuir frequency $\omega_{pe} = 0.3\omega_2$. The plasma is localized in the region $5 < x < 35$ and $-15 < y < 15$. The plasma density profile along x axis is parabolic-constant-parabolic (6, 18, 6), and is constant along y. The parabolic density profile near the plasma-vacuum boundaries is chosen so as to make the laser pulse interaction with the plasma smoother than in the case of a sharp plasma-vacuum interface.

In Figure 12 we illustrate the plasma conditions at $t = 59$, before the interaction between the soliton generated in the wake of the first pulse and the wakefield induced by the second pulse.

In Figure 13 the dynamics of the soliton-wakefield interaction is shown as a set of pairs of figures corresponding to increasing time. As it propagates towards the soliton each electron density maximum in the wake acts as a fast propagating semitransparent parabolic mirror that partially reflects the e.m fields of the soliton. When the wakefield falls on the soliton, part of the soliton is reflected ($t = 74$) in the direction of wakefield propagation. The process is repeated when the subsequent maxima of the electron density interact with the soliton.

Thus a set of short e.m. pulses is formed ($t = 78$, $t = 82$). The frequency of the fields in the

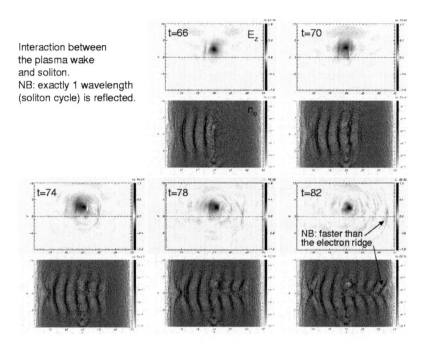

Figure 13. *Train of single cycle reflected pulses*

reflected single-cycle e.m. pulses is up-shifted and their longitudinal size is much smaller than the size of the soliton.

According to the results of simulations the soliton amplitude is approximately equal to 0.6, while the first laser pulse amplitude is equal to 0.5. The reflected pulse length is 6 times smaller than the soliton size. The reflected pulse amplitude is approximately equal to 0.3, but due to the increased frequency its intensity is 3.9×10^{18} W/cm^2, which is several times larger than expected from the 1-D analysis. This high intensity value is due to the additional intensification which appears because of the parabolic profile of wake wave that focuses the reflected pulse.

Conclusions

Analytical and numerical investigations show that in the complex physics of the interaction of high intensity ultrashort laser pulses with plasmas, fundamental physical mechanisms can be identified that form the basic blocks of the nonlinear physics of continuous media such as vortices and solitons. The basic features of the soliton formation and dynamics have been described. Solitons evolve into long-lived quasineutral coherent structures (post solitons). The static electric field in these structures can be detected and investigated with the use of proton imaging. A train of single cycle e.m. pulses with up-shifted frequency and increased intensity can be produced with presently available lasers by exploiting the interaction between a relativistic plasma soliton and the density modulations of a Langmuir plasma wave.

Acknowledgments

We are pleased to acknowledge the support the supercomputing center of Cineca (Bologna).

References

Akhiezer, A.I., and Polovin, R.V., (1956), *Sov. Phys. JETP*, **30**, 915.
Berezhiani, V.I., et al., (2002), *Phys. Rev. E*, **65**, 047402.
Borghesi, M., et al., (2001), *Plasma Phys. Control. Fus.*, **43**, A267.
Borghesi, M., et al., (2002), *Phys. Rev. Lett.*, **88**, 135002.
Bulanov, S.S., (2004), *Phys. Rev. E*, **69**, 036408.
Bulanov, S.S., et al., (2005a), *Phys. Rev. E*, **71**, 016404.
Bulanov, S.S., et al., (2005b), in preparation.
Bulanov, S.V., et al., (1992), *Phys. Fluids B*, **4**, 1935.
Bulanov, S.V., et al., (1995), *Plasma Phys. Rep.*, **21**, 550.
Bulanov, S.V., et al. (1999), *Phys. Rev. Lett.*, **82**, 3440.
Bulanov, S.V., et al., (2001a), *"Reviews of Plasma Physics"*, ed. V.D. Shafranov,
    Kluwer Academic, Consultants Bureau, New York **22**, 227.
Bulanov, S.V., et al., (2001b), *Physica D*, **152-153**, 682.
Bulanov, S.V., Pegoraro, F., (2002), *Phys. Rev. E*, **65**, 066405.
Bulanov, S.V., et al., (2003), *Phys. Rev. Lett.*, **91**, 085001.
Degasperis, A., (1998), *Am. J. Phys.*, **66**, 486.
Esirkepov, T.Zh., et al., (1998), *JETP Letters*, **68**, 36.
Esirkepov, T.Zh., (2001), *Comput. Phys. Comm.*, **135**, 144.
Esirkepov, T.Zh., et al., (2002), *Phys. Rev. Lett.*, **89**, 275002.
Esirkepov, T.Zh., et al., (2004), *Phys. Rev. Lett.*, **92**,255001.
Farina, D., et al., (2000), *Phys. Rev. E*, **62**, 4146.
Farina, D., and Bulanov, S.V., (2001a), *Phys. Rev. Lett.*, **86**, 5289.
Farina, D., and Bulanov, S.V., (2001b), *Plasma Phys. Rep.*, bf 27, 641.
Farina, D., and Bulanov, S.V., (2005), *Plasma Phys. Contr. Fus.*, **47**, A73.
Isanin, A.V., et al., (2005), *Phys. Lett. A*, **337**, 107.
Kaw, P.K., et al., (1992), *Phys. Rev. Lett.*, **68**, 3172.
Kozlov, V.A,. et al., (1979), *Sov. Phys. JETP*, **76**, 148.
Kurki-Suonio, T., et al., (1982), *Phys. Rev. A*, **40**, 3230.
Langdon, A.B., and Lasinski, B.F., (1983), *Phys. Fluids*, **26**, 582.
Lontano, M., et al., (2001), *Phys. Plasmas*, **8**, 5113.
Lontano, M., et al., (2002), *Phys. Plasmas*, **9**, 5113.
Lontano, M., et al., (2003a), *Laser Part. Beams*, **21**, 541.
Lontano, M., et al., (2003b), *Phys. Plasmas*, **10**, 639.
Lontano, M., et al., (2005), *Laser Part. Beams*, in press.
Marburger, J.H., and Tooper, R.F., (1975), *Phys. Rev. Lett.*, **35**, 1001.
Mima, K., et al., (2001), *Phys. Plasmas*, **8**, 2349.
Mofiz, U.A., et al., (1985), *Phys. Fluids*, **28**, 826.
Naumova, N.M., et al., (2001a), *Phys. Plasmas*, **8**, 4149.
Naumova, N.M., et al., (2001b), *Phys. Rev. Lett.*, **87**, 185004.
Naumova, N.M., et al., (2004a), *Phys. Rev. Lett.*, **92**, 063902.
Naumova, N.M., et al., (2004b), *Phys. Rev. Lett.*, **93**, 195003.
Naumova, N.M., et al., (2005), *Phys. Plasmas*, **12**, 056707.
Pegoraro, F., et al. (2001), in *Atoms, Solids And Plasmas In Super-Intense Laser Field*,
    ed. D. Batani et al., Kluwer Academic/Plenum Publishers.
Pirozhkov, A.S., et al., (2005), *Phys. Lett. A*, in press.
Rouhani, M.R., et al., (2002), *Phys. Rev. E*, **65**, 066406.
Sentoku, Y., et al., (1999), *Phys. Rev. Lett.*, **83**, 3434.
Schamel, H., et al., (1997), *Phys. Fluids*, **20**, 1286.
Sudan, R.N., et al., (1997), *Phys. Plasmas* **4**, 1489 .
Tsintsadze, N.L., and Tskhakaya, D.D., (1977), *Sov. Phys. JETP*, **45**, 252.

Tushentsov, M., et al., (2001), *Phys. Rev. Lett.*, **87**, A265002.
Weber, S., et al., (2005a), *Phys. Rev. Lett.*, **94**, 055005.
Weber, S., et al., (2005b), *Phys. Plasmas*, **12**, 043101.
Weber, S., et al., (2005c), *Phys. Plasmas*, in press.
Whitham, G.B., (1974), in *Linear and Nonlinear Waves*, Wiley, New York.
Zakharov, V.E., (1972), *Sov. Phys. JETP*, **35**, 908.
Zakharov, V.E., et al., (1984), in *Theory of Solitons*, Plenum Press, New York.
Zel'dovich, Ya.B., Raizer, Yu.P., (1967), in *Physics of Shock Waves and High-Temperature Hydrodynamic Phenomena*, Academic Press, New York.

11

Interference Stabilization of Atoms in a Strong Laser Field

M.V. Fedorov

General Physics Institute, Russian Academy of Science, Moscow, Russia

1 Introduction

The key subject of these lectures is Interference Stabilization (IS) of atoms in a strong laser field. Physics of the phenomenon, its various manifestations, as well as many related subtopics will be discussed. Generalizations to the case of two-colour interference stabilization will be briefly described. As it will be shown, owing to the achievable in this case very high degree of stabilization, the phenomenon appears to be very promising for making progress in such an important field as propagation of laser pulses in a gas medium. Since the first work on IS (Fedorov and Movsesian, 1988), there have been a lot of publications on this subject. I am not going to mention here all the relevant papers.

Many references, as well as a rather detailed description of IS, can be found in my book "Atomic and Free Electrons in Strong Light Field" (Fedorov, 1997), which I recommend as the main textbook on the subject I am going to discuss. Also I would recommend two review papers on the strong-field stabilization and, in particular on IS: Gavrila, 2002 and Popov et al, 2003.

2 Definitions of stabilization

Let an atom be interacting with a laser field characterized by its electric field strength

$$\vec{\varepsilon}(t) = \vec{\varepsilon}_0(t) \cos(\omega t), \tag{1}$$

where $\vec{\varepsilon}_0(t)$ is the field strength amplitude or pulse envelope slowly depending on time t, and ω is the laser frequency. The Hamiltonian of an atomic electron in the field (1) can be taken in the form

$$H = \frac{\widehat{\vec{p}}^{\,2}}{2m} + U(\vec{r}) - \vec{d} \cdot \vec{\varepsilon}(t), \tag{2}$$

where $\widehat{p\vec{p}} = -i\hbar\partial/\partial\vec{r}$ is the electron momentum operator, $U(\vec{r})$ is the atomic potential, and $\vec{d} = -|e|\,\vec{r}$ is the dipole moment, $e < 0$ is the electron charge.

The equation to be solved is the Schrödinger equation for the electron wave function $\Psi(\vec{r}, t)$

$$i\hbar \frac{\partial\Psi}{\partial t} = H\Psi. \tag{3}$$

Let $\psi_n(\vec{r})$ and $\psi_E(\vec{r})$ be eigenfunctions of the unperturbed atomic Hamiltonian

$$H_0 = \frac{\widehat{p}^{\,2}}{2m} + U(\vec{r}), \tag{4}$$

corresponding to negative $E_n < 0$ and positive $E > 0$ energies

$$H_0\psi_n = E_n\psi_n, \quad H_0\psi_E = E\psi_E. \tag{5}$$

If initially an atomic electron is in one of the bound states, i.e., $\Psi_{in} = \psi_n \exp(-iE_n t/\hbar)$ and if, by solving the Schrödinger equation (3), we can find the electron wave function Ψ_f after the laser pulse has gone, projections of Ψ_f upon the field-free states of the continuous spectrum determines the ionization probability density

$$\frac{dw_i}{dE} = |\langle\psi_E|\Psi_f\rangle|^2. \tag{6}$$

Integrating over E, the probability density gives the total probability of ionization

$$w_i = \int dE\,\frac{dw_i}{dE} = \int dE\,|\langle\psi_E|\Psi_f\rangle|^2. \tag{7}$$

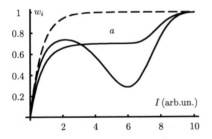
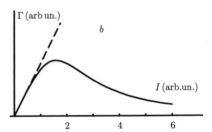

Figure 1. (a) *Probability of ionization per pulse: the dashed curve is the FGR-with-saturation ionization probability* (13) *and solid curves - different types of stabilization regimes;* (b) *The rate of ionization: the dashed line* -Γ_{FGR} (12), *the solid curve - the stabilization regime.*

If the laser field is not strong, its influence on the atomic electron can be described in the lowest order of perturbation theory. If the light frequency ω is large enough for providing one-photon transitions to the continuum from the initially populated bound state, $\hbar\omega > |E_n|$, the weak-field ionization can be described in the first-order of non-stationary perturbation theory, which gives

$$w_i = \frac{1}{4\hbar^2} \int dE \left| \int dt \exp\left[\frac{i}{\hbar}(E - E_n - \hbar\omega)t\right] \vec{d}_{E,n}\cdot\vec{\varepsilon}(t) \right|^2$$

$$= \frac{1}{4\hbar^2} \int dE \int dt \int dt' \exp\left[\frac{i}{\hbar}(E - E_n - \hbar\omega)(t - t')\right] \left(\vec{d}_{E,n}\cdot\vec{\varepsilon}(t)\right)\left(\vec{d}_{nE}\cdot\vec{\varepsilon}(t')\right), \tag{8}$$

i.e., the probability of ionization per pulse is determined by the Fourier transform of the pulse envelope. In Eq. (8) $\vec{d}_{E,n}$ is the dipole bound-free matrix element

$$\vec{d}_{E,n} = \langle \psi_E | \vec{d} | \psi_n \rangle = e \int \vec{r} \, d\vec{r} \, \psi_E^*(\vec{r}) \, \psi_n(\vec{r}). \tag{9}$$

For a hydrogen atom, as well as for any atoms with energies E far from autoionzing states, the matrix element (9) is a very smooth function of E compared to the exponential factor in Eq. (8) (the width of the latter is of the order of h/τ, where τ is the pulse duration). For this reason, the function $\vec{d}_{E,n}$ in Eq. (8) can be approximated by a constant. This is the so called "flat continuum approximation".

In this approximation the integral over energy E in Eq. (8) can be approximated by the δ-function $2\pi\hbar\delta(t-t')$ to give

$$w_i = \frac{\pi}{2\hbar} \int dt \, \left| \vec{d}_{E,n} \cdot \vec{\varepsilon}_0(t) \right|^2. \tag{10}$$

In the case of a rectangular pulse $\varepsilon_0(t) = const.$, Eq.(8) gives

$$w_i = \Gamma\tau, \tag{11}$$

where Γ is the well known Fermi Golden Rule (FGR) rate of ionization

$$\Gamma = \frac{\pi}{2\hbar} \left| \vec{d}_{E,n} \cdot \vec{\varepsilon}_0 \right|^2. \tag{12}$$

One of the applicability conditions of Eq. (11) is $\Gamma\tau \ll 1$. Generaliztion to the case of longer laser pulses is rather well known (for a derivation see, e.g., the book by Fedorov, 1997, section 1.6 b)

$$w_i = 1 - \exp(-\Gamma\tau). \tag{13}$$

As the coefficient Γ in front of τ in this expression is still the FGR ionization rate, Eq. (13) can be referred to as describing the Fermi Golden Rule with saturation regime of ionization.

All the derived equations for the probability of ionization per pulse (8), (10), (11), and (13) describe monotonously growing functions of the laser pulse peak intensity $I = c\varepsilon_{0,\,max}^2/8\pi$. The last of these equations, (13), describes both growth and saturation of the ionization probability at the level $w_i = 1$, which corresponds to total ionization of atoms. In contrast to such a behavior, by definition, stabilization of atoms corresponds to one of the regimes characterized by the curves of Figure 1 a: under stabilization conditions the probability of ionization per pulse $w_i(I)$ either saturates at a level smaller than one or even becomes a falling function of the peak laser intensity. Of course, such regimes can have only limited validity ranges, and at very large intensities stabilization inevitably turns into destabilization, where the ionization probability $w_i(I)$ becomes fast growing again and saturates only at the complete ionization level $w_i = 1$.

An alternative definition of stabilization is illustrated by the curves of Figure 1 b. This definition is valid for rectangular pulses or pulses having a well pronounced plateau region where the pulse envelope is constant, $\varepsilon_0(t) = const$. In this case the rate of ionization is well defined. In a weak field it is given by the FGR formula (12), according to which $\Gamma \propto I$. In the stabilization regime $\Gamma(I)$ becomes a falling function as is shown qualitatively in Figure 1 b.

Note that the given definitions do not explain yet when stabilization regimes can occur and what is the underlying physics. Answering these questions is a subject of many theoretical and some experimental works.

In these lectures I will concentrate only on one physical mechanism of stabilization, known as Interference Stabilization (IS) of Rydberg atoms. In the case of IS, both definitions illustrated by Figures 1 a and 1 b are valid at proper conditions.

In principle some generalizations, which I hope to discuss briefly in the last lecture, extend the validity region of IS to not too highly excited atomic levels, which means that the mechanism can work not only in Rydberg atoms. But the main bulk of investigations of IS was carried out just for Rydberg atoms. So, before going further to the analysis of IS, let us discuss briefly the physics of Rydberg atoms.

3   Rydberg atoms

A Rydberg atom is an atom in which one of this valence electrons is excited to a high (Rydberg) level. Energies of such states are similar to those of a hydrogen atom, $E_n = -1/2n^2$, where $n \gg 1$ and atomic system of units ($\hbar = m_e = |e| = 1$) is henceforth used. This means that such quantitues as energy, frequency, length, time, field strength, and intensity are measured in units of their characteristic atomic values

$$E_{at} = \frac{me^4}{\hbar^2} \approx 27.2\,\text{eV}, \quad \omega_{at} = \frac{me^4}{\hbar^3} \approx 4.1 \times 10^{16}\text{sec}^{-1},$$

$$t_{at} = \frac{2\pi}{\omega_{at}} = \frac{2\pi\hbar^3}{me^4} \approx 1.5 \times 10^{-16}\,\text{sec}, \quad r_{at} = \frac{\hbar^2}{me^2} = 0.53 \times 10^{-8}\,\text{cm},$$

$$\varepsilon_{at} = \frac{|e|}{r_{at}^2} = \frac{m^2|e|^5}{\hbar^4} \approx 5 \times 10^9\,\text{V/cm}, \quad I_{at} = \frac{c\varepsilon_{at}^2}{8\pi} = \frac{cm^4e^{10}}{8\pi\hbar^8} \approx 3 \times 10^{16}\,\text{W/cm}^2. \tag{14}$$

As n is very large, spacing between neighboring Rydberg levels is approximately equal to

$$E_{n+1} - E_n \approx \frac{1}{n^3} \equiv \omega_K, \tag{15}$$

and the right-hand side of Eq. (15) can be interpreted as a classical Kepler frequency ω_K of a particle motion in the Coulomb field for the particle's total energy equal to E_n. The corresponding classical Kepler period of motion is

$$t_K = \frac{2\pi}{\omega_K} = 2\pi n^3. \tag{16}$$

Note that though these classical characteristics can be associated with any Rydberg state with a given n, the wave functions of such states are not similar at all to a classical localized particle. The function $\psi_n(r)$ is stationary, i.e. it does not describe any motion. The size of a region where $\psi_n(r) \neq 0$ is very large: it can be found from the condition $E_n = -1/r$, which gives $r_{\max} = 2n^2 \gg 1$. To get localized and moving quantum states, one has to construct a wave packet, i.e. a superposition of many wave functions $\psi_n(r)$. This can be and has been done experimentally (see references in the book by Fedorov, 1997, Ch. 6). One can excite either one selected Rydberg level or a superposition of Rydberg states depending on the pulse duration τ and its relation with the Kepler period t_K (16). If $\tau \gg t_K$, the exciting laser pulse is spectrally narrow and one can excite a selected Rydnerg level E_n. If oppositely, $\tau \ll t_K$, a spectrally wide laser pulse excited coherently many Rydberg levels and forms a localized wave packet.

As well as in a Hydrogen atom, Rydberg levels in any atom are degenerate with respect to the electron angular momentum l, which varies from zero to $l_{\max} = n - 1$. In a pure form IS exists for states with small l ($l \ll n$).

4 Quasiclassical dipole matrix elements

In the definition (9) of the dipole matrix element $\vec{d}_{E,n}$, in the case $n \gg 1$, the functions $\psi_n(\vec{r})$ and $\psi_E(\vec{r})$ can be approximated by Coulomb functions. Owing to this, in principle, the integrals on the right-hand side of Eq. (9) can be calculated from the Gordon formulas (Landau and Lishitz, 1977, Appendix f). However, these formulas are very cumbersome and inconvenient for estimates or analytical calculations. Fortunately, for Rydberg states and states of the continuum close to the ionization threshold ($E \ll 1$) if appears possible to get rather simple expressions for the dipole matrix elements (Delone et al., 1983, 1989, 1994; Fedorov, 1997, Ch. 1.4 c).

The derivation is based on the use of the WKB or quasiclassical expressions for the wave functions ψ_n and ψ_E. Such expressions for the radial parts of these functions are known to be given by

$$\chi_n(r) = \frac{N_n}{\sqrt{p_n(r)}} \exp\left[i \int_0^r dr'\, p_n(r')\right] + c.c., \quad \text{and}$$

$$\chi_E(r) = \frac{N_E}{\sqrt{p_E(r)}} \exp\left[i \int_0^r dr'\, p_E(r')\right] + c.c., \tag{17}$$

where $N_n = (2\pi n^3)^{-1/2}$ and $N_E = (2\pi)^{-1/2}$ are the normalization constants and $p_{n,E}$ denotes the classical radial electron momentum at $l = 0$ and energies E_n and E:

$$p_n(r) = \left[2\left(E_n + \frac{1}{r}\right)\right]^{1/2}, \quad p_E(r) = \left[2\left(E + \frac{1}{r}\right)\right]^{1/2}. \tag{18}$$

The integral in Eq. (9) contains angular and radial parts. The angular part is rather simple and gives only a numerical factor of the order of one. The radial part is determined by the integral of the form

$$J = \int_0^\infty r\, dr\, \chi_n(r)\, \chi_E(r). \tag{19}$$

Let us assume now that the main contribution to this integral is given by distances r such that $1 \ll r \ll r_{\max} = 2n^2$ and that $E \sim |E_n|$. Then $E, |E_n| \ll 1/r$ and the expressions in Eqs. (18) can be expanded in powers of E and E_n:

$$p_n(r) \approx \sqrt{\frac{2}{r}} + E_n \sqrt{\frac{r}{2}}, \quad p_E(r) \approx \sqrt{\frac{2}{r}} + E \sqrt{\frac{r}{2}}. \tag{20}$$

As $r \gg 1$, the functions $\exp\left[\pm i \int_0^r dr'\, p_n(r')\right]$ and $\exp\left[\pm i \int_0^r dr'\, p_E(r')\right]$ are very fast oscillating in their dependence on r. If the oscillations in χ_n and χ_E do not compensate each other, contribution of such terms to the integral (19) is very small. Only the terms containing a difference of $p_n(r')$ and $p_E(r')$ can be retained to reduce Eq. (19) to the form

$$J = N_n N_E \int_0^\infty dr\, \frac{r}{\sqrt{p_n(r)p_E(r)}} \exp\left[i \int_0^r dr'\, (p_E(r') - p_n(r'))\right] + c.c.$$

$$\approx \frac{1}{\pi (2n)^{3/2}} \int_0^\infty dr\, r^{3/2} \exp\left[i \frac{\sqrt{2}}{3}\, (E - E_n)\, r^{3/2}\right] + c.c.$$

$$= \frac{2^{2/3}}{\pi 3^{1/3}} \Gamma\left(\frac{2}{3}\right) \cos\left(\frac{\pi}{6}\right) \frac{1}{n^{3/2}(E - E_n)^{5/3}}, \tag{21}$$

where now $\Gamma(x)$ is the gamma-function

$$\Gamma(x) = \int_0^\infty dt\, t^{x-1}\, e^{-t}. \tag{22}$$

If the initial- and final-state energies E_n and E are related approximately by the energy conservation rule for one-photon absorption, $E - E_n \approx \omega$, we get finally the following quasiclassical estimate for the bound-free dipole matrix element

$$d_{n,\,E} \sim \frac{1}{n^{3/2}\omega^{5/3}}\,, \tag{23}$$

where for the simplicity all the numerical coefficients of the order of one are dropped. Substituted into Eq. (12), this expression for the dipole matrix element gives the following quasiclassical estimate for the FGR ionization rate

$$\Gamma \sim \frac{\varepsilon_0^2}{\omega^{10/3}} = \frac{I}{\omega^{10/3}}, \tag{24}$$

where I is the laser intensity and, as usual, all quantities are in atomic units.

Note that directly from the form of Eq. (21) we can estimate the electron-ion distance r_0 which gives the main contribution to the integral over r:

$$r_0 \sim \frac{1}{(E - E_n)^{2/3}} \sim \frac{1}{\omega^{2/3}}. \tag{25}$$

At $r > r_0$ the function under the symbol of integral in Eq. (21) oscillates very quickly and this cancels a contribution of large-r regions to the integral. If, roughly, $|E_n| \sim \omega$, we easily find that the above-assumed condition $1 \ll r_0 \ll r_{\max} = 2n^2$ is satisfied at $\omega \ll 1$. This is the applicability condition of the quasiclassical approach: both bound and continuous spectrum states under consideration must be close to the ionization threshold (compared to the atomic unit of energy $E_{at} \approx 27\,\text{eV}$).

5   Qualitative explanation of the IS physics

Now we can give at last a description of the IS phenomenon. Let us begin from a very qualitative picture. Let us assume that initially an atom is excited to some Rydberg level E_n with $n \gg 1$ and, e.g., $l = 0$. Let such an atom be ionized by a laser field with the electric field strength determined by Eq. (1) and frequency ω exceeding the binding energy $|E_n|$, $\omega > |E_n|$.

If the field is weak enough for the lowest order perturbation theory to be applicable, the only process that can occur is a direct one-photon bound-free transition, $E_n \to E$. However, in a stronger field many additional processes can accompany and compete with the direct photoionization.

One group of such competing processes is that of Raman-type two-photon transitions between neighboring Rydberg levels $E_n \to E \to E_{n'}$ (Figure 2). These transitions can result in a coherent re-population of Rydberg levels. Transitions form repopulated Rydberg levels to the continuum $E_n \to E$ and $E_{n'} \to E'$ can interfere with each other. As it appears, the interference is destructive and it suppresses photoionization, thus giving rise to IS.

This is a very brief and very qualitative explanation of the phenomenon of interference stabilization. Though very simple, it is sufficient for qualitative determination of the main condition when the IS phenomenon can occur. The solid curve at the upper part of Figure 2 describes the

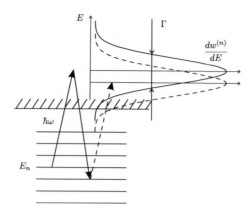

Figure 2. *Direct photoionization from a Rydberg level E_n, Λ-type Raman transitions to Rydberg levels, and interfering transitions to the continuum (solid and dashed lines with arrows at the top)*

probability density $dw^{(n)}/dE$ of photoionization from an isolated level E_n in its dependence on the photoelectron energy E. The curve is centered at $E_n + \omega$ and its width Γ coincides with the FGR rate of ionization (ionization width) of the level E_n (12).

Transitions to the continuum from other levels $E_{n'}$ result in similar curves but centered at different positions, $E_{n'} + \omega$ (the dashed curve in Figure 2). If the neighboring curves do not overlap, there is no interference and no stabilization. If they do overlap, transitions from neighboring Rydberg levels can interfere with each other and IS can occur. Hence, the threshold condition for IS is the condition that the FGR ionization width (12) exceeds spacing between neighboring Rydberg levels:

$$\Gamma \geq E_n - E_{n-1}. \tag{26}$$

By using Eqs. (15) and (24) for $E_{n+1} - E_n$ and Γ we can find immediately that the condition of Eq. (26) takes the form

$$\varepsilon_0 \geq \omega^{5/3} \quad \text{or} \quad I = \varepsilon_0^2 \geq \omega^{10/3}. \tag{27}$$

6 Two-level model

In a general form the IS problem involves interaction of infinitely many Rydberg levels with all states of the continuum. In addition, each Rydberg level and each state of the continuum are degenerate with respect to angular momentum, and re-population of all these states has to be taken into account.

Several approaches were worked out to simplify the problem. The simplest one is a model of two nondegenerate close levels E_1 and E_2 (with the indexes "1" and "2" not related to the principal quantum number of a real atom) interacting with a nondegenerate continuum.

In this model the total electron wave function in a laser field has a form of superposition

$$\Psi = C_1(t)\,\psi_1\,e^{-iE_1 t} + C_2(t)\,\psi_2\,e^{-iE_2 t} + \int dE\,C_E(t)\,\psi_E\,e^{-iEt}, \tag{28}$$

where ψ_1, ψ_2, and ψ_E are field-free atomic wave functions. Equations for the expansion coefficients $C_1(t)$, $C_2(t)$, and $C_E(t)$ can be obtained directly from the Schrödinger equation in the so called Rotating Wave Approximation (RWA). RWA means that $\cos(\omega t) = \frac{1}{2}\left(e^{i\omega t} + e^{-i\omega t}\right)$ in the interaction energy is replaced by only one term of this sum, such that it provides an approximate energy conservation rule $E \approx E_{1,2} + \omega$ and compensates large differences $\pm(E - E_n)$.

In this approximation the equations to be solved are given by

$$i \dot{C}_\alpha(t) = -\frac{1}{2} \int dE\, e^{i(E_\alpha - E + \omega)t}\left(\vec{\varepsilon}_0(t) \cdot \vec{d}_{\alpha,E}\right) C_E(t), \quad \alpha = 1,\, 2,$$

$$i \dot{C}_E(t) = -\frac{1}{2} \sum_{\alpha=1,2} e^{i(E - E_\alpha - \omega)t}\left(\vec{\varepsilon}_0(t) \cdot \vec{d}_{E,\alpha}\right) C_\alpha(t). \tag{29}$$

The last equation can be formally integrated

$$C_E(t) = \frac{i}{2} \int_{-\infty}^{t} dt' \sum_{\alpha=1,2} e^{i(E - E_\alpha - \omega)t'}\left(\vec{\varepsilon}_0(t') \cdot \vec{d}_{E,\alpha}\right) C_\alpha(t') \tag{30}$$

and substituted to the equations for $C_\alpha(t)$ (the first line of Eqs. (29)) to reduce the problem to a set of two integro-differential equations

$$i \dot{C}_\alpha(t) = -\frac{i}{4} \int dE\, e^{i(E_\alpha - E + \omega)t}\left(\vec{\varepsilon}_0(t) \cdot \vec{d}_{\alpha,E}\right) \int_{-\infty}^{t} dt' \sum_{\beta=1,2} e^{i(E - E_\beta - \omega)t'}\left(\vec{\varepsilon}_0(t') \cdot \vec{d}_{E,\beta}\right) C_\beta(t'). \tag{31}$$

This set of equations is absolutely equivalent to the original one (29). To make them simpler, we can use the same flat-continuum approximation that was used above for the derivation of the first-order ionization probability per pulse (10). In this approximation the dipole matrix elements $\vec{d}_{\alpha,E}$ and $\vec{d}_{E,\beta}$ are substituted by some constants and taken off from the integrals. After this the integral over energy E is calculated to give the δ-function $\int dE\, \exp[iE(t' - t)] = 2\pi\delta(t' - t)$. Then the integral over t' is easily taken too to give

$$i \dot{C}_\alpha(t) = -\frac{i}{4}\left(\vec{\varepsilon}_0(t) \cdot \vec{d}_{\alpha,E}\right) \sum_{\beta=1,2} e^{i(E_\alpha - E_\beta)t}\left(\vec{\varepsilon}_0(t') \cdot \vec{d}_{E,\beta}\right) C_\beta(t). \tag{32}$$

In the case of almost constant pulse envelope $\varepsilon_0(t) \approx const.$, the substitution $C_\alpha = e^{iE_\alpha t} B_\alpha(t)$ reduces Eqs. (32) to the form of equations with constant coefficients

$$i \dot{B}_\alpha(t) - E_\alpha\, B_\alpha(t) = -\frac{i}{2} \sum_\beta \Gamma_{\alpha\beta}\, B_\beta, \tag{33}$$

where $\Gamma_{\alpha\beta}$ are components of the tensor of ionization widths determined as a direct generalization of the FGR ionization rate (12):

$$\Gamma_{\alpha\beta} = \frac{\pi}{2}\left(\vec{\varepsilon}_0 \cdot \vec{d}_{\alpha,E}\right) \times \left.\left(\vec{\varepsilon}_0 \cdot \vec{d}_{E,\beta}\right)\right|_{E = E_\alpha + \hbar\omega}. \tag{34}$$

The described transition from Eqs. (29) to (34) is known as the adiabatic elimination of the continuum (Fedorov, 1997, Ch. 8.1).

The last but important assumptions simplifying further the derived equations concerns the components of the ionization width tensor (34). Let us assume that all of them are almost equal to each other $\Gamma_{\alpha\beta} \approx \Gamma$, where $\Gamma \propto I = \varepsilon_0^2$ is just the FGR ionization rate (12). Under this assumption Eqs. (33) take the form

$$i \dot{B}_1(t) - E_1\, B_1(t) = i \dot{B}_2(t) - E_2\, B_2(t) = -\frac{i}{2} \Gamma\left(B_1(t) + B_2(t)\right). \tag{35}$$

7 Quasienergies

As the coefficients in Eqs. (33) are constant, these equations have stationary solutions

$$B_\alpha(t) = \exp(-i\gamma t)\, b_\alpha, \tag{36}$$

where b_α are constants, and γ is a quasienergy. Before calculating quasienergies for the system under consideration let us discuss briefly general features of this concept (Fedorov, 1997, Ch. 1.6).

If the Hamiltonian of a system depends periodically on time t with a period T, $H(t+T) = H(t)$, the Scrödinger equation has solutions of the form

$$\psi(t) = e^{-i\gamma t}\, u_\gamma(t), \tag{37}$$

where $u_\gamma(t)$ are periodical functions of time with the same period as the Hamiltonian, $u_\gamma(t+T) = u_\gamma(t)$. Eq. (37) makes the contents of the well known Floquet theorem.

The constants γ and functions $u_\gamma(t)$ are called quasienergies and quasienergy wave functions. There is a complete analogy of these concepts with those of quasimomentum and Bloch functions in crystals with a potential energy periodically depending on coordinates. If the Hamiltonian is Hermitian and if the problem is solved exactly, all quasienergies are real. But, if the continuum is taken into account approximately in the framework of the adiabatic elimination procedure, this can give rise to a non-Hermitian effective Hamiltonian of the remaining bound-state subsystem and, consequently, to complex quasienergies. This is just the case under present consideration.

Note, however, that many authors use a different approach giving rise to complex quasienergies. This is a class of so called quasistationary theories, in which the quasienergy wave functions diverge at infinitely long distances and obey the "expanding spherical wave" (Gamov) boundary conditions (see, e.g., the review paper by Manakov et al, 1986).

In our case, by substituting solutions of the form (37) to Eqs. (35), we get a set of two algebraic equations for the constants b_α

$$\left(\gamma - E_1 + \frac{i}{2}\Gamma\right) b_1 + \frac{i}{2}\Gamma b_2 = 0,$$

$$\frac{i}{2}\Gamma b_1 + \left(\gamma - E_2 + \frac{i}{2}\Gamma\right) b_2 = 0. \tag{38}$$

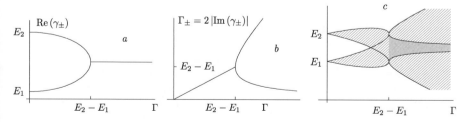

Figure 3. (a) *Real parts of quasienergies* $\gamma_\pm$, (b) *width of quasienergy levels* $\Gamma_\pm$, *and* (c) *quasenergy zones in dependence on the parameter* $\Gamma \propto \varepsilon_0^2$

The compatibility condition of these two equations gives the second-order equation for γ

$$(\gamma - E_1)(\gamma - E_2) + \frac{i}{2}\Gamma\left(2\gamma - E_1 - E_2\right) = 0, \tag{39}$$

solutions of which are given by

$$\gamma_\pm = \tfrac{1}{2} \left\{ E_1 + E_2 - i\Gamma \pm \sqrt{(E_2 - E_1)^2 - \Gamma^2} \right\}. \tag{40}$$

This equation shows that drastic changes in the solutions occur when the interaction constant Γ (which is proportional to ε_0^2) becomes larger than the level spacing $E_2 - E_1$. The point $\Gamma = E_2 - E_1$ is the branching point, below which (at $\Gamma < E_2 - E_1$) the root square in Eq. (40) is real, whereas above the branching point (at $\Gamma > E_2 - E_1$) it becomes imaginary.

The real parts of quasienergies $\gamma_\pm$ (40) are shown in Figure 3 a in their dependence on the parameter Γ. With a growing laser field strength, the level "centers of mass" are seen to come closer and closer to each other to collapse finally, at $\Gamma = E_2 - E_1$, into a single value equal to $\tfrac{1}{2} (E_2 - E_1)$.

The curves of Figure 3 b describe the dependence on Γ of the widths of quasienergy levels determined by imaginary parts of quasienergies, $\Gamma_\pm = 2 \left| \mathrm{Im} \, (\gamma_\pm) \right|$. In contrast to real parts of quasienergies, in the region $\Gamma \leq E_2 - E_1$ the widths $\Gamma_\pm$ of the quasienergy levels γ_+ and γ_- are equal to each other and to the weak-field FGR ionization width Γ (12), and they grow monotonously till $\Gamma = E_2 - E_1$. In the region $\Gamma > E_2 - E_1$ two branches of the functions $\Gamma_\pm(\Gamma)$ split: $\Gamma_-(\Gamma)$ continues growing, and even faster than in the region $\Gamma < E_2 - E_1$, whereas $\Gamma_+(\Gamma)$ appears to be a falling function. In the asymptotic case $\Gamma \gg E_2 - E_1$ Eq. (40) yields:

$$\Gamma_- \approx 2\Gamma \quad \text{and} \quad \Gamma_+ \approx \frac{(E_2 - E_1)^2}{2\Gamma} \propto \frac{1}{\varepsilon_0^2}. \tag{41}$$

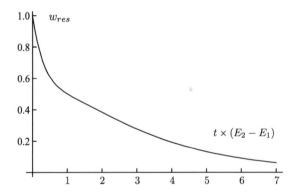

Figure 4. *Residual probability of finding an atom in bound states ψ_1 and ψ_2 vs. the pulse duration t.*

At last, the picture of Figure 3 c describes the broadened quasienergy levels in the form of quasienergy zones. "Gravity centers" of these zones are given by $\mathrm{Re} \, (\gamma_\pm)$ whereas widths of zones are equal to $\Gamma_\pm$. As it's seen from this picture, with a growing field strength ε_0, at first the zones broaden, then, at $\Gamma \sim E_2 - E_1$, they overlap completely and form something like a quasicontinuum, and in a strong field ($\Gamma > E_2 - E_1$) one of two quasienergy zones narrows (the darker shadowed zone in Figure 3 c), and its width (Γ_+) corresponds to a falling branch of the curves in Fig 3 b. The "gravity centers" of both quasienergy levels in the strong-field region appear to be localized exactly at the middle between the field-free energies E_1 and E_2.

Formation of a narrowing quasienergy level means that one of the two atomic eigenstates in a strong field has an increasing life-time and, hence, the population accumulated at this level appears resistant against ionization by a strong laser field, and the atom is stabilized. This conclusion is easily checked by a direct solution of the initial-value problem for a rectangle pulse shape (Fedorov, 1997, Ch. 8.1). Not reproducing here the formulas which are rather cumbersome, we show the result of calculations in Figure 4. In this picture the residual probability to find an atom in one of two bound states is plotted in its dependence on the pulse duration t (in units of $1/(E_2 - E_1)$) at $\Gamma = 1.5 \, (E_2 - E_1)$. One can clearly see manifestations of a bi-exponential decay: at first the function $w_{res}(t)$ falls rather quickly and then much slower, with the decay rates, correspondingly, Γ_- and Γ_+ (41). By plotting a series of such curves for different values of Γ, one can check that with a growing Γ the long-living wing of the curve in Figure 4 becomes even longer, i.e. the degree of stabilization increases.

8 Multilevel system of Rydberg levels

Of course, the described two-level model is a rather imperfect imitation of a real Rydberg atom. But, nevertheless, the main features of this model find their reflection in more complicated and realistic theoretical descriptions.

The first generalization that was considered (Fedorov and Movsesan, 1988; Fedorov, 1997, Ch. 8) was a model of an infinitely large amount of nondegenerate Rydbrg levels and a nondegenerate continuum. In such a model many steps of a solutions are very similar to those described above for the two-level model. In particular, the expansion of the wave function in a series of the field-free eigenfunctions differs from that of Eq. (28) by larger number of terms corresponding to bound states

$$\Psi = \sum_n C_n(t) \, \psi_n \, e^{-iE_n t} + \int dE \, C_E(t) \, \psi_E \, e^{-iEt} \tag{42}$$

and, besides, now the energies of the field-free levels can be specified as $E_n = -1/n^2$.

Equations for the expansion coefficients $C_n(t)$ look identically to Eqs. (29) with the only substitution $\alpha \to n$. Also all the approximations and assumptions are in this case the same as as in the two-level model. They are: RWA, adiabatic elimination of the continuum, and the assumption about equal components of the ionization-width tensor $\Gamma_{n,n'}$.

These simplifications result in equations for the functions $B_n(t) = e^{-iE_n t} C_n(t)$ similar to those of Eqs. (35)

$$i\dot{B}_n(t) - E_n \, B_1(t) = -\frac{i}{2} \Gamma \sum_{n'} B_{n'}(t) \tag{43}$$

As previously, these equations have stationary (quasienergy) solutions $B_n(t) = e^{-i\gamma t} b_n$, and algebraic equations for the constants b_n are similar to those of Eqs. (38)

$$(\gamma - E_n) \, b_n = -\frac{i}{2} \Gamma \sum_{n'} b_{n'}. \tag{44}$$

As previously, quasienergies γ are determnined by the compatibility condition for Eqs. (44).

In the case of infinitely many Rydberg levels the corresponding equation can be found easily with the help of the following trick. Let us divide the left- and right-hand sides of Eqs. (44) by $\gamma - E_n$ and sum the resulting equation over n. Then we get a single homogeneous equation for $\sum_n b_n$, which has nonzero solutions only if the coefficient in front of this expression equals zero, which gives

$$1 + \frac{i}{2} \sum_n \frac{\Gamma}{\gamma - E_n} = 0. \tag{45}$$

Evidently, this equation is more complicated than Eq. (39) for the two-level model. But, nevertheless, its solutions can be found rather easily in two limiting cases of small and large values of Γ (compared to the spacing between neighboring Rydberg levels (15)), $\Gamma \ll n^{-3}$ and $\Gamma \gg n^{-3}$.

The first case corresponds to the weak-field limit, in which quasienergies must be close to the field-free atomic energies. In this case one can retain only one term in the sum over n in Eq. (45) to get $\gamma_n \approx E_n - \frac{i}{2}\Gamma$.

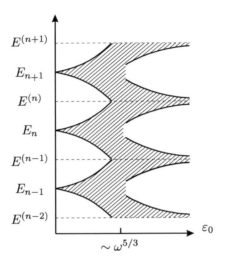

Figure 5. *Quasienergy Rydberg zones (levels) in a laser field in dependence on the field strength amplitude ε_0.*

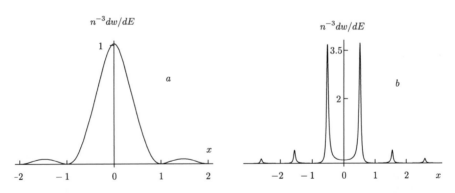

Figure 6. *Photoionization probability density vs. the dimensionless electron energy x, $x = (E - E_n - \omega)n^3$, (a) $\Gamma n^3 = 2\pi/10$ and (b) $\Gamma n^3 = 2\pi$; E_n and n are the energy and principal quantum number of the initially populated Rydberg level*

The second case, $\Gamma \gg n^{-3}$, is the strong-field limit. In this case, as Γ is very large, Eq. (45) can be satisfied only if the sum over n is anomalously small, and this happens when the quasienergy γ is close to the middle between some neighboring Rydberg levels, i.e., to $E^{(n)} = \frac{1}{2}(E_n + E_{n+1})$.

By substituting the summation index $n \to n'$ in the sum on the right-hand side of Eq. (45) and by expanding the the function $1/(\gamma - E_{n'})$ in powers of $\delta\gamma \equiv \gamma - E^{(n)}$, in the first order in

$\delta\gamma$ we reduce Eq. (45) to the form

$$1 = \frac{i}{2}\,\Gamma\,\delta\gamma \sum_{n'} \frac{1}{\left(E^{(n)} - E_{n'}\right)^2} = \frac{i}{2}\,\Gamma\,n^6\,\delta\gamma \sum_{n'} \frac{1}{\left(n' - n + \frac{1}{2}\right)^2} = \frac{i}{2}\,\Gamma\,n^6\,\pi^2\,\delta\gamma, \qquad (46)$$

which gives

$$\gamma_n \approx E^{(n)} - \frac{2\,i}{\pi^2\,n^6\,\Gamma}\,. \qquad (47)$$

So, all the strong-field quasienergies, or quasienergy zones, are centered at $E^{(n)}$ and with a growing field strength they narrow $\propto 1/\varepsilon_0^2$. All this is very similar to the two-level model except that in the case of infinitely many Rydberg levels all quasienergy levels narrow, whereas in the two-level model one of the two quasienergy levels remains wide. The broadening and then narrowing Rydberg qusienergy zones are shown in Figure 5.

The described reconstruction of the Rydberg quasienergy spectrum finds its reflection in the energy distribution of photoelctrons shown in Figure 6 for two cases, of weak (a) and strong (b) fields. A striking difference between the weak- and strong-field cases is evident. In a weak field the photoelectron spectrum has a Lorenz-like shape centered exactly at initial-state energy plus ω. In contrast, in a strong field the spectrum gets a multipeak structure and the peaks are located at energies equal to real parts of quasienergies $\mathrm{Re}\,(\gamma_\pm) = E^{(n\pm1)}$ plus ω. In a weak field the photoelectron peak broadens with a growing field strength, whereas in a strong field the peaks narrow. This is one of many effects accompanying and related to strong-field IS.

9 Degeneracy of Rydberg levels

To make our model closer to a realistic atom we can consider now a system of Rydberg levels and continua degenerate with respect to the electron angular momentum l. This degeneracy adds many new transitions that need to be taken into account. They are shown schematically in Figure 7.

In principle, in addition to Raman-type transitions between Rydberg levels with a given l, the atomic population can migrate to states with higher values of the angular momentum. Also, sub-states of the continuous spectrum with higher l can play the role of additional decay channels, and the question is "how do all these additional transitions affect IS?"

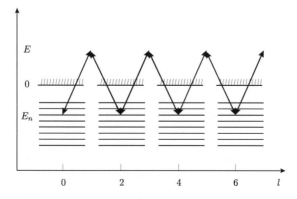

Figure 7. *Raman type transitions in the manyfold of degenerate Rydberg levels.*

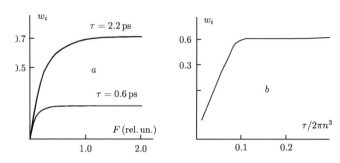

Figure 8. (a) *Ionization probability w_i vs. fluence F (in relative units) at two given values of the pulse duration τ and (b) w_i vs. τ at a given value of fluence F (Fedorov et al., 1996).*

Some answers are described in the paper by Fedorov et al. (1996) (see also he book by Fedorov, 1997, Ch. 8). Both RWA and continuum adiabatic elimination approximations were used. However, the approximation of identical components of ionization width tensor cannot be used anymore. The dependence of $\Gamma_{nl,\,n'l'}$ on n and n' remains smooth and can be ignored. But the dependence on the angular momentum quantum numbers l and l' cannot be considered as smooth and must be taken into account exactly.

For this reason the arising system of many coupled equations for the probability amplitudes $C_{nl}(t)$ cannot be solved analytically. Its numerical solutions were found and some of them are shown in two pictures of Figure 8.

In the first of these two figures the probability of ionization is plotted in its dependence on fluence F at two given values of the pulse duration τ. By definition, fluence is the product of the peak pulse intensity by the pulse duration, $F = I_{\max} \times \tau$. In Figure 8 *a* the fluence is evaluated in relative units around $F \sim 7\,\mathrm{J/cm^2}$.

The curves of Figure 8 *a* show that with growing fluence, the ionization probability saturates at a level smaller than one. Moreover, at a shorter pulse duration and at a given fluence, the ionization probability is seen to be smaller than for a longer pulse duration. As at a given fluence a shorter pulse duration corresponds to a higher intensity, the curves of Figure 8 *a* show that the ionization yield decreases with growing pulse intensity. This is a clear indication of strong-field atomic stabilization.

In the second picture of Figure 8 the ionization probability is plotted in its dependence on the pulse duration τ at a given fluence F. The pulse duration is evaluated in units of the Kepler period t_K for the initial Rydberg state ψ_n (16).

The plateau of the curve in Figure 8 *b* corresponds to the first-order perturbation theory, in which the FGR ionization rate (12) is time-independent and proportional to intensity, whereas the total probability of ionization per pulse is a given constant: $w_i \sim \Gamma\tau \propto I \times \tau = F = const$. On the other hand, the fall of the curve $w_i(\tau)$ at small values of the pulse duration is a manifestation of strong-field stabilization, because at a given fluence a decreasing pulse duration corresponds to a growing pulse intensity and, in accordance with the curve of Fig. 8 *b*, to a falling ionization yield.

Many other interesting results were also obtained in the framework of the approach we discuss here (Fedorov and Fedorov, 1998). One of them, shown in Figure 9 concerns time evolution of populations under the IS conditions.

The two curves of Figure 9 *a* show the total residual probability to find an atom in bound states at a current time t. The upper and lower curves correspond to weak- and strong-field

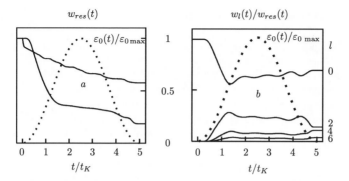

Figure 9. (a) *Time dependence of the total residual probability to find an atom in bound states in weak and strong fields (correspondingly, the upper and lower curves) and (b) relative partial probabilities to find an atom in bound states with given values of the electron angular momentum l; the dotted lines show the pulse envelope $\varepsilon_0(t)$.*

cases. The dotted line is the pulse envelope $\varepsilon_0(t)$ normalized by its maximal value.

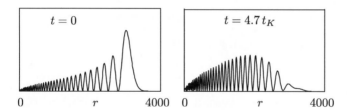

Figure 10. *A structure of the oscillating electron wave packet formed via Raman-type transitions.*

The strong-field curve of this picture shows that ionization of an atom takes place mainly in the region of a front wing of the pulse, where the field is not yet too strong. In the central part of the pulse the ionization process is strongly suppresses, which explains why on the whole stabilization by a strong laser pulse occurs.

The behavior of $w_{res}(t)$ characterized by the lower curve in Figure 9 a can be interpreted as the dynamical stabilization, i.e. stabilization occurring in some time domain of a laser pulse, where the field reaches its maximal value.

A series of curves in Figure 9 b describes the relative probabilities to find an atom at a time t in bound states with arbitrary n but given l. As it is seen, in a strong field excitation of higher-l states appears to be significantly suppressed. This means that there is not only stabilization with respect to ionization but also strong-field stabilization of population migration to bound states with a high angular momentum.

Finally, Figure 10 describe two snapshots of the evolving electron wave packet created owing to Raman-type transitions during the process of ionization in the IS regime. The wave packet is seen to have some degree of localization, and it's center of mass oscillates with a period equal to the classical Kepler period (16), $t_K = 2\pi n^3$, where n is the principal quantum number of the initially populated Rydberg level. In the case shown in Figure 10 the pulse duration equals to $5\, t_K$.

10 A fully quasiclassical approach

One of the approximations that has been used but not mentioned till now is the assumption that all free-free transitions can be ignored. In reality this is not evident. Moreover, free-free transitions can give rise to such a well known phenomenon as Above-Threshold Ionization (ATI) (see. e.g., the book by Fedorov, 1997, Ch. 7, and references therein).

In the case of one-photon ionization from Rydberg levels ATI means that an atomic electron can absorb more than one photon and form in the photoelectron additional peaks at energies around $E_n + (1 + s)\omega$, $s = 0,\ 1,\ 2\ldots$). In principle, ATI was included into the theory of IS in the framework of the Essential State Model (ESM) by Deng and Eberly (1984, 1985).

The result was rather simple: all the formulas and results in the above-described theory of IS ignoring ATI remain valid if the FGR ionization rate Γ is replaced by the renormalized interaction constant

$$\widetilde{\Gamma} = \frac{2\Gamma}{1 + \sqrt{1 + 2\pi\Gamma n^3}}. \tag{48}$$

Though very nice, this result can leave a feeling of dissatisfaction because it can seem to be too simple. Besides, ESM is based on a rather strong assumption, validity of which cannot be proved rigorously.

This is the so called "pole approximation", according to which whenever in equations singular expressions like $\left.\frac{1}{E-E'-\omega-i\varepsilon}\right|_{\varepsilon\to 0}$ are met, they are treated as the δ-functions: $i\pi\delta(E - E' - \omega)$ with dropped principal value parts $P\frac{1}{E-E'-\omega}$. Construction of a theory of ATI free from this assumption appears to be rather difficult. One example, where this was done (under some other restrictions), is the fully quasiclassical theory of IS (Fedorov, 1997, Ch. 8, p. 4; Fedorov and Tikhonova, 1998; Ivanov et al., 1998).

Fully quasiclassical means that the quasiclassical approximation is used for solving the Schrödinger equation of a Rydberg electron in a strong laser field rather than for calculating dipole matrix elements as it has been done above in Section 2. In principle, in three dimensions this is a rather difficult problem. To simplify it, it was assumed that for low angular momentum states angular motion of an electron can be rather slow compared to the radial motion. For this reason, in the lowest order in slow motions, the angular part of the electron kinetic energy operator can be dropped in the Schrödinger equation.

In this approximation the angle θ between the electron radius-vector and the field polarization appears to be just a parameter in the radial Schrödinger equation, and the problem is reduced to the on-dimensional one. When solutions are found they are averaged over the angle θ to take into account that initially the electron radius-vector has no preferred orientation.

The radial equation to be solved is

$$i\frac{\partial\chi(r,t;\theta)}{\partial t} = \left\{-\frac{1}{2}\frac{\partial^2}{\partial r^2} - \frac{1}{r} + \varepsilon_0(t)\sin(\omega t)\,r\,\cos(\theta)\right\}\chi(r,t;\theta). \tag{49}$$

Its field-free solution is given by the first equation of Eqs. (17). In the presence of a laser field it is found in the form

$$\chi(r) = \frac{N_n}{\sqrt{p_n(r)}}\left\{\exp\left[i\int_0^r dr'\,p_n(r') + iS_+(r,t;\theta)\right] + \exp\left[-i\int_0^r dr'\,p_n(r') + iS_-(r,t;\theta)\right]\right\}, \tag{50}$$

where $S_\pm(r,t;\theta)$ are unknown functions. In the quasiclassical approximation equations for them can be reduced to a rather simple form

$$\frac{\partial S_\pm}{\partial t} \pm \frac{\partial S_\pm}{\partial \tau} = \varepsilon_0(t)\sin(\omega t)\,r\,\cos(\theta), \tag{51}$$

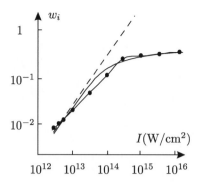

Figure 11. *Probability of ionization by a short laser pulse: analytical (the solid line) and exact numerical (dots) solutions (Tikhonova et al., 1999), the dashed line is the lowest order perturbation theory.*

where $\tau \equiv \tau(r)$ is the time it takes for a classical electron to cross a distance from $r' = 0$ to $r' = r$

$$\tau(r) = \int_0^r \frac{dr'}{p_n(r')}. \tag{52}$$

Solutions of Eqs. (51) are given by

$$S_\pm(r, t; \theta) = \cos(\theta) \, \mathrm{Im} \left\{ \int_{-\infty}^t dt' \, e^{-i\omega t'} \, \varepsilon_0(t') \, r_{cl}(t - t' \mp \tau(r)) \right\}, \tag{53}$$

where $r_{cl}(t)$ is the solution of the Newton equation for radial motion of an electron in the Coulomb field or, in other words, the solution of equation $\tau(r) = t$.

Many interesting results can be and have been deduced from the described strong-field quasiclassical solutions of the Schrödinger equation (50), (53). Some of them are structure and evolution of electron wave packets in the continuum, structure of ATI peaks in the photoelectron spectrum, etc. Not dwelling upon all of them let us reproduce here only one final analytical formula obtained for the total probability of ionization per pulse

$$w_i = \frac{1}{4\pi n_0^3} \int_{-\infty}^{\infty} dt \left\{ 1 - \int_0^1 dx \, J_0^2 \left(2^{2/3} 3^{1/6} \Gamma \left(\frac{2}{3} \right) \frac{\varepsilon_0(t)}{\omega^{5/3}} x \right) \right\}, \tag{54}$$

where J_0 is the zero-order Bessel function, $\Gamma(2/3)$ is the gamma-function, and $x = \cos(\theta)$. Note that Eq. (54) is derived for an arbitrary shape of the pulse envelope $\varepsilon_0(t)$ and all the ATI processes are completely taken into account. But, on the other hand, the validity of this result is restricted by the condition of relatively short pulse duration, $\tau < t_K$.

For a trapezoidal pulse with a short duration the dependence of the ionization probability (54) on the peak laser intensity I is plotted in Figure 11 by the solid line. This result is compared with the exact numerical solution of the 3D Schrödinger equation carried out by Tikhonova et al. (1999) and shown in Figure 11 by dots. It is clear that the analytical and exact numerical solutions of Eq. (54) agree very well with each other. Both analytical and numerical solutions describe saturation of the ionization probability at a level smaller than one, which corresponds to stabilization of an atom in a strong light field.

11   Experiment

There have been several experiments where strong-field stabilization of atoms has been observed or attempts to observe it were made. Not dwelling upon all of them let us describe here briefly only one of these works (Hoogenraad et al., 1994) in which the existence of stabilization was proved quite unambiguously and which is most closely related to the theory of IS.

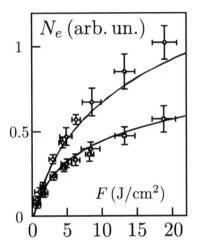

Figure 12. *Experimentally measured electron yield N_e vs the fluence F at $\tau = 2.7$ ps (the upper curve) and $\tau = 0.6$ ps (the lower curve).*

Suppression of photoionization was observed by Hoogenraad et al. (1994) in Ba atoms excited preliminary to the Rydberg state $6s27d$ and then ionized by a strong laser pulse. In the first series of measurements the electron yield due to one-photon ionization was measured in its dependence on the pulse duration varied in the interval 2.7-0.25 ps at a given value of fluence $F = 7.8$ J/cm^2. The experimental curve (not shown here) looks very similar to the later obtained theoretical one (Figure 8 a). As said above, the fall of this curve at small values of τ (and, hence, high values of intensity I) can be considered as a direct evidence of strong-field ionization suppression or atomic stabilization.

The second experimental result by Hoogenraad et al. (1994) shown in Figure 12 concerns the electron yield measured in its dependence on the laser fluence at two values of the pulse duration, $\tau = 2.7$ and 0.6 ps. Note that these values of the pulse duration were taken in the calculations by Fedorov et al. (1996) to reproduce most closely the experimental conditions of the work by Hoogenraad et al. (1994).

The curves of Figure 12 are similar qualitatively to the initial stages of the theoretical curves of Figure 8 b. Moreover, the dependence $w_i(F)$ calculated with the help of Eq. (54) under experimental conditions of the work by Hooegenraad et al. (1994) for $\tau = 0.6$ ps appears to be practically indistinguishable from the lower curve in Figure 12. All this indicates clearly that the atomic stabilization was observed in the work by Hoogenraad et al. 1994) and the observed phenomenon in nothing else but interference stabilization.

12 Two-color interference stabilization

As an extension of the effect of interference stabilization in a strong-field single-frequency laser field described above, let us consider here the phenomenon of two-colour interference stabilization (Fedorov and Poluektov, 2004). This will be a direct generalization of the simplest two-level model of Section 5.

In a single-frequency field there are two conditions under which the two level model can be considered as rigorously applicable: (i) the levels must be close to each other and sufficiently far from any other atomic levels and (ii) bound-free dipole matrix elements of transitions to the same states of the continuum from these levels must be approximately equal to each other. Both of these two conditions hardly can be satisfied in any realistic atoms.

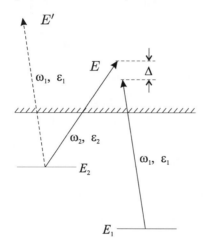

Figure 13. *A scheme of two atomic levels under the conditions of a Raman-type resonance in a two-colour filed* (55).

By extending the consideration to the case of a two-colour laser field and by manipulating with laser frequencies and intensities, we can easily satisfy the two mentioned conditions or their analogs.

So, let us take the field strength of a laser field in the form

$$\varepsilon(t) = \varepsilon_1(t)\cos(\omega_1 t) + \varepsilon_2(t)\cos(\omega_2 t) \tag{55}$$

and let us consider a scheme of transitions shown in Figure 13. In Eq. (55) and Figure 13 $\varepsilon_1(t)$, ω_1 and $\varepsilon_2(t)$, ω_2 are the pulse envelopes and frequencies of the two fields under consideration.

The frequencies are assumed to satisfy approximately the condition of the Raman-type resonance transitions for two arbitrary selected atomic levels E_1 and E_2. In other words, the resonance detuning

$$\Delta = E_2 + \omega_2 - E_1 - \omega_1 \tag{56}$$

is assumed to be small compared to ω_1 and ω_2 and to $|E_n - E_{1,2}|$, where E_n are energies of atomic levels different from E_1 and E_2. It is assumed that $\omega_1 > |E_1| > \omega_2 > |E_2|$, i.e., an electron at the level E_1 can make a transition to the continuum by absorbing one photon ω_1 but not ω_2. From the level E_2 transitions to the continuum can occur both via absorption of photons ω_1 and ω_2.

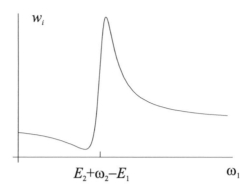

Figure 14. *The Fano curve describing spectral dependence of the ionization probability on the probe-field frequency ω_1 (in the presence of a strong pump field).*

A scheme of transitions of Figure 13 has been widely discussed in literature (see references in the papr by Fedorov and Poluektov, 2004) in connection with theoretical and experimental investigation of the phenomenon known as the Light-Induced Continuum Structures (LICS). This phenomenon is associated usually with the case when only one of the two fields (ε_2) is strong, whereas the other one (ε_1) is weak and is considered as a probe field. The probe field ionizes an atom, and the main subject of investigations is the spectral dependence of the ionization yield on the probe-field frequency ω_1. The final result is a Fano-type curve $w_i(\omega_1)$ (Figure 14) with parameters depending on the strong-field amplitude and frequency.

This formulation of the problem differs from that on the strong-field stabilization, and the latter consists of the following. Let the field (ε_1) be strong enough to ionize an atom completely during the pulse duration. Let then the second field (ε_2) be added and let us look for conditions under which this addition suppresses ionization rather than accelerate it. Such a suppression can arise only owing to interference of direct transitions to the continuum from the level E_1 and transitions via the level E_2. Hence, if the effect takes place, this is a clearly understandable manifestation of IS in a rather simple scheme of levels and transitions.

In a complete analogy with the consideration of Section 5 we can present the electron wave function in the form of a superposition of discrete states (ψ_1 and ψ_2) and continuum. But now, owing to a possibility of a transition $E_1 \rightarrow E'$ (see Figure 13) we must take into account two groups of wave functions in the continuum, ψ_E and ψ'_E. We can use again the RWA and adiabatic elimination of the continua to reduce the set of equations for the probability amplitudes $C_1(t)$ and $C_2(t)$ to find an atom at the levels E_1 and E_2 into the form

$$i\dot{C}_1 + \frac{1}{4}\left\{\alpha_1(\omega_1)\varepsilon_{10}^2(t) + \alpha_1(\omega_2)\varepsilon_{20}^2(t)\right\}C_1 = -\frac{1}{4}\varepsilon_{10}(t)\varepsilon_{20}(t)\alpha_{12}\,C_2,$$

$$i\dot{C}_2 - \left(\Delta - \frac{1}{4}\left\{\alpha_2(\omega_1)\varepsilon_{10}^2(t) + \alpha_2(\omega_2)\varepsilon_{20}^2(t)\right\}\right)C_2 = -\frac{1}{4}\varepsilon_{10}(t)\varepsilon_{20}(t)\alpha_{21}\,C_1, \tag{57}$$

where $\alpha_i(\omega_j)$ and α_{12} are the complex dynamical polarizabilities of the levels E_i, $i, j = 1, 2$ and the cross-polarizability

$$\alpha_i(\omega) = \int dE\, |d_{i\,E}|^2 \left(\frac{1}{E - E_i - \omega - i\delta} + \frac{1}{E - E_i + \omega}\right)_{\delta \to 0} \tag{58}$$

and

$$\alpha_{21} = \int dE\, d_{2\,E}d_{E\,1}\left(\frac{1}{E - E_1 - \omega_1 - i\delta} + \frac{1}{E - E_1 + \omega_2}\right)_{\delta \to 0} \approx \alpha_{12}; \tag{59}$$

α'_i, α'_{12} and α''_i, α''_{12} are their real and imaginary parts, and integrations over E in Eqs. (58), (59) include in themselves summations over discrete intermediate levels E_n different from E_1 and E_2.

Written in the form (57), equations for the probability amplitudes $C_1(t)$ and $C_2(t)$ are very convenient for demonstrating a nice scaling effect. If both pulse envelopes $\varepsilon_{10}(t)$ and $\varepsilon_{20}(t)$ depend on time t only via the ratio t/τ, where τ is the pulse duration common for both high- and low-frequency pulses, then Eqs. (57) are invariant with respect to the following scaling transformation:

$$\Delta \to \lambda\Delta, \; \varepsilon^2_{1,20} \to \lambda\varepsilon^2_{1,20}, \; \tau \to \tau/\lambda, \; t \to t/\lambda. \tag{60}$$

with an arbitrary λ. Owing to this scaling effect, it is convenient to use relative quantities such as the ratio of laser intensities $x = I_2/I_1$, relative detuning $\delta = \Delta/I_1$, relative pulse duration $\theta = \tau \times I_1$, where in all such definitions all quantities are in atomic units. By finding complex quasienergies of the system, $\gamma_\pm$, we can define widths of the quasienergy levels $\Gamma_\pm = -2\gamma_\pm$ and relative widths $g_\pm = \Gamma_\pm/I_1$, etc.

Not dwelling upon any further details (see the paper by Fedorov and Poluektov, 2004), we reproduce here only some most important results. A typical dependence of widths $g_\pm$ on the resonance detuning δ at a given intensity ratio x is shown in Figure 15.

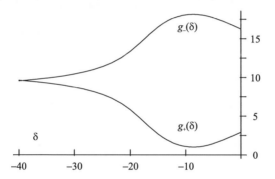

Figure 15. *The normalized widths $g_\pm$ of quasienergy levels vs. the normalized detuning δ at a given value of the intensity ratio $x = I_2/I_1$*

Specifically, this and all other results shown below were received for $1s2s$ (E_1) and $1s4s$ (E_2) states of a He atom and laser frequencies equal to $\omega_1 = 8.44\,\text{eV}$ (the 2nd harmonics of a dye laser) and $\omega_2 = 1.17\,\text{eV}$ (Nd:YAG laser).

One of the quasienergy widths in Figure 15 ($g_+(\delta)$) has a clear minimum. As usual, a small width of a quasienergy level is related to a long living part of the atomic population and to stabilization of an atom. For this reason the detuning δ where the function $g_+(\delta)$ has a minimum, is interpreted as an optimal detuning δ_{opt}. An explicit analytical expression for δ_{opt} is given by

$$\delta_{opt}(x) = \frac{1}{4}\left\{\alpha'_2(\omega_1) - \alpha'_1(\omega_1) - \frac{\alpha'_{12}}{\alpha''_{12}}[\alpha''_2(\omega_1) - \alpha''_1(\omega_1)]\right\}$$

$$+\frac{x}{4}\left\{\alpha'_2(\omega_2) - \alpha'_1(\omega_2) - \frac{\alpha'_{12}}{\alpha''_{12}}[\alpha''_2(\omega_2) - \alpha''_1(\omega_2)]\right\}, \tag{61}$$

If we change now the intensity ratio x and if we want to always remain at the minimum of the curve $g_+(\delta)$, we have to change the detuning synchronously with x in accordance with Eq. 61. The question is how does $[g_+(\delta)]_{\min} \equiv g_+[\delta_{opt}(x)]$ change with a changing intensity ratio x? The answer is given by the calculated dependence shown in Figure 16.

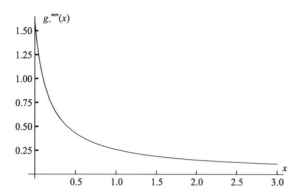

Figure 16. *The width of the narrower quasienergy level minimized with respect to the detuning δ vs. the intensity ratio x.*

According to this result, in the framework of the two-level model under consideration, the width of the narrower quasienergy level optimized over detuning at all values of the intensity ratio x, falls unrestrictedly approaching zero at very large x. Limitations of this narrowing can be related only to the model applicability restrictions.

In any case, a possibility to get a very strong narrowing of the quasienergy level γ_+ can be considered as an indication that the achievable degree of stabilization in a two-color laser field can be very large. This assumption is confirmed by direct calculations of the ionization probability. Some results of such calculations are shown in Figure 17.

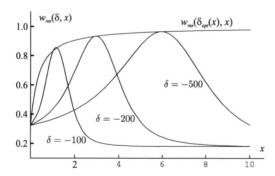

Figure 17. *The residual probability of finding an atom in bound states after the pulses have gone vs. the intensity ratio x at three given values of the detuning δ and at $\delta = \delta_{opt}(x)$ (61).*

In this picture three resonance-like curves correspond to three different but given values of the detuning δ. The resonance-like behavior arises because at a given δ, variations of x bring the system to the optimal conditions and then take it off the optimum. The envelope of these curves corresponds to the detuning tuned to the optimum at all values of x. The calculations are performed for coinciding rectangular envelopes $\varepsilon_1(t)$ and $\varepsilon_2(t)$ and $\theta = 0.1$.

The picture of Figure 18 shows the stabilization effect in a slightly different formulation. Instead of investigation of the dependence $w_{res}(x)$, we assume now that the intensity ratio is constant. Specifically, in the calculations corresponding to Figure 18 we take $x = 3$. A varying parameter in these calculations is the laser intensity I_1 which changes synchronously with I_2 so

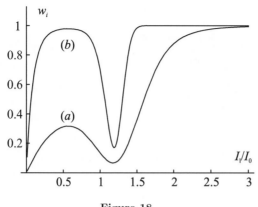

Figure 18.

that always $I_2 = 3I_1$. The quantity plotted in Figure 18 is the total probability of ionization, $w_i = 1 - w_{res}$. The intensity I_1 is normalized by an arbitrary constant intensity I_0. Appearance of such an arbitrary constant is a consequence of the above-mentioned scaling effect. The detuning and pulse duration are taken to be related to the constant I_0 but independent of the variable I_1. Specifically, in Figure 18 $\Delta = -300\,I_0$ and $\tau = 0.1/I_0$ for the curve (a) and $1/I_0$ for the curve (b) (as usual, here Δ, τ, and I_0 are in atomic units).

The curves of Figure 18 are typical for the stabilization phenomenon. The curve (a) consists of three parts: perturbation theory limit (a growing function $w_i(I/I_0)$ at small intensities), a falling function $w_i(I/I_0)$ in the stabilization region, and a growing function $w_i(I/I_0)$ at large values of intensity - destabilization region and quick complete ionization ($w_i = 1$). The case (b) has an additional part: complete ionization ($w_i = 1$) between the perturbation theory and stabilization regions. This means that at the parameters corresponding to the curve (b), the synchronous increase of the laser intensities I_1 and I_2 at first increases the probability of ionization until it reaches the saturation value ($w_i = 1$). But then, an even further growth of the laser intensities gives rise to a very unexpected result: the probability of ionization falls and, in the specific case of the curve (b) in Figure 18 it falls to the level as low as 0.2, which indicates a very high degree of stabilization. Note that, for example, at $I_0 = 3 \times 10^{10}\,\mathrm{W/cm^2}$ the maximal degree of stabilization (i.e., the minimum of the curves in Figure 18) is achieved at $I_1 \approx 4 \times 10^{10}\,\mathrm{W/cm^2}$ and $I_2 \approx 1.2 \times 10^{11}\,\mathrm{W/cm^2}$. Under these conditions the pulse duration, corresponding to the the curve (b) in Figure 18 is $\tau \approx 3$ps.

13 Conclusion

So, the given above overview of strong-field interference stabilization shows that the phenomenon is rather well studied theoretically and, to some extent, experimentally. As shown above, the two-colour scheme can provide significantly higher achievable degree of stabilization. In principle, this result can open a rather important area of practical applications for the pulse propagation problem. Indeed, let a strong laser pulse propagate in an atomic gas medium and let the length of its propagation be limited by ionization of atoms. Then, if we send to the same medium two pulses with appropriately chosen parameters, two-colour stabilization effect can decrease ionization losses rather significantly. This can result in a rather well pronounced increase of the propagation length of two pulses. Verification of this hypothesis requires further theoretical work and experimental investigation.

References

Delone, N.B. et al. (1983), J. Phys. B **16**, 2369 (1989), **22**, 2941; (1994), **27**, 4403.

Deng Z. and Eberly, J.H. (1984), Phys. Rev. Lett. **53**, 1810; (1985) JOSA B **2**, 486.

Fedorov, M.V. and Movsesian, A.M. (1988), *J. Phys. B* **21**, L155.

Fedorov, M.V. et al. (1996), J. Phys. B **29**, 2907.

Fedorov, M.V. (1997), *Atomic and Free Electrons in a Strong Light Field* (Singapore: World Scientific).

Fedorov M.V. and Fedorov S.M. (1998), Optics Express **3**, 271.

Fedorov, M.V. and Tikhonova, O.V. (1998), Phys. Rev. A **58**, 1322.

Gavrila, M. (2002), *J. Phys. B* bf 35, R147.

Hoogenraad. J. et al. (1994), Phys. Rev. A **50**, 4133.

Ivanov, M.Yu. et al. (1998), Phys. Rev. A **58**, R793.

Landau, L.D. and Lifshitz, E.M. (1977), *Quantum Mechanics* (New York: Pergamon Press).

Manakov, N.L. et al. (1986), Physics Reports **141**, 319.

Popov, A.M. et al. (2003), *J. Phys. B* **36**, R125.

Tikhonova, O.V. et al. (1999), Phys. Rev. A **60**, R749.

12
Entanglement and Wave Packet Structures in Photoionization and Other Decay Processes of Bipartite Systems

M.V. Fedorov

General Physics Institute, Russian Academy of Science, Moscow, Russia

A structure of wave packets arising in the process of a bipartite system decay is investigated and its connection with entanglement of quantum states is discussed. As long as the effect of wave packet spreading can be ignored, the ratio R of the single-particle to coincidence wave packet widths is shown to coincide with the degree of entanglement determined by the Schmidt correlation number K. Mathematically single-particle and coincidence schemes correspond to absolute and conditional probability densities. Coordinate and momentum wave-packets for each particle are compared. A series of reciprocity and identity relations between the corresponding wave packet widths is found to occur. Uncertainty relations following from these relations are found. The coordinate and momentum uncertainty products for each particle are shown to be related directly to the degree of entanglement K: the single-particle and coincidence uncertainty products are shown to be, correspondingly, $\sim K \geq 1$ and $\sim 1/K \leq 1$. A series of specific examples are considered.

1 Introduction

Very rapid breakup of physical systems has interesting physical consequences, and these are becoming more easily accessible in the atomic and molecular domain as laser-atom and laser-molecule interactions enter the sub-femtosecond time domain. Fundamental quantum issues are engaged by the creation of quantum-entangled states of matter, and breakup processes can be a rich source of entanglement for detailed study, both theoretically and experimentally. Two-party entanglement is already complex enough to pose interesting questions, and a number of them arise in the entanglement between the outgoing fragments in photo-ionization and photo-dissociation under very intense laser irradiation.

The key conditions which must be provided in both strong and weak fields and short and long pulses are the need to maintain entanglement during the break-up interaction and providing a relatively weak subsequent interaction of the daughter particles after the break-up. In all the processes considered in this paper these requirements are satisfied although for electron-positron pair production we have to assume application of superstrong-fields, at the current limit of accessibility. In all cases, by measuring wave packet parameters of the separating free particles, we can draw conclusions about the degree of entanglement that had been accumulated by the particles in their parent compound system and during the decay process. Very interesting questions about the role of a weak residual interaction between particles and about decoherence that can be caused by this interaction have not yet been investigated well enough, and we hope to return to these problems elsewhere.

The concept of entanglement, known since Schrödinger (1935), has become a central topic in contemporary quantum optics and quantum information. Roughly, entanglement of quantum states is the non-factorization of multi-particle wave functions. Entanglement is widely described and studied in the case of quantum systems characterized by a limited number of discrete variables, and measures of entanglement have been reviewed recently in this context (Plenio and Virmani, 2005). However, systems of particles characterized by continuous variables also attract the growing attention of researchers.

Examples include parametric down-conversion (PDC) (Rubin (1996), Burlakov et al. (1997), Law et al. (2000), Monken et al. (1998) Walborn et al. (2003), Law and Eberly (2004), Howell et al. (2004), D'Angelo et al. (2004), Chan et al. (2004)), entanglement and wave-packet structure analysis in photoionization and photodissociation (Fedorov et al. (2004)), atom-photon entanglement in spontaneous emission (Chan et al. (2003), Fedorov et al. (2005)), electron-positron entanglement in strong-field pair production (Krekora et al. (2005), Fedorov al. (2006)), entanglement of field frequencies (Huang and Eberly (1993), Law et al. (2000), Grice et al. (2001)), energy-time variables (Kwiat et al. (1993), Brendel et al. (1999)), momenta (Barity (1990)), and angular momenta (Arnaut and Barbosa (2000), Franke-Arnold et al. (2002), Mair et al (2001), Oemrawsingh et al (2005)).

All of these follow the careful consideration of the entanglement of the dimensionless coordinates and/or momenta of photon mode creation and destruction operators by Reid and Drummond (1988), Reid (1989) and Ou et al. (1992) in the context of observability of the Einstein-Podolsky-Rosen "paradox" (1935). Further discussions can be found on related topics as well (Freyberger et al. (1999), Opartny and Kurizki (2001), Opartny et al. (2003), Rau et al. (2003), Yonezawa et al. (2004)). A review of experimental characterizations has been given by Bowen et al., (2004).

There are many reasons why investigation of entanglement in systems with continuous variables is interesting. One of them is to establish links between such differing areas of physics as quantum optics, strong-field physics, wave packet physics, quantum electrodynamics, atom optics, etc. Another point concerns using methods well known in one of these fields for determining (in principle, even measuring) parameters used in a different field.

A fruitful element of such an analysis is the realization of close connections between coincidence measurements and conditional probability densities in quantum mechanics, as well as between single-particle measurements and unconditional probability densities (Fedorov et al., 2004, 2005). We can also point to the possibility to go beyond the question whether entanglement exists or not and address the question how much entanglement may be present.

Previously we determined that a specific degree of pure state entanglement in an EPR-like scenario is related directly to the ratio of single-particle and coincidence wave packet widths, which opens a way of experimentally measuring the degree of pure state entanglement in bipartite systems. On the theoretical, or conceptual level, our consideration shows that the modern and very natural interpretation and specification of the famous predictions by Einstein, Podolsky and Rosen (1935) is provided by using the concepts of conditional and unconditional probability densities of bipartite states Fedorov et al. (2005).

These general results were obtained for several different physical processes, such as photoionization of atoms and photodissociation of molecules (Fedorov et al., 2004), PDC (Rau et al. (2003), Law and Eberly (2004), Howell et al. (2004), D'Angelo et al. (2004), Chan al. (2004)), spontaneous emission of a photon as well as Raman and Rayleigh scattering (Chan et al. (2002, 2003), Fedorov et al. (2005), and multiphoton pair production (Fedorov et al., 2006).

In this lecture we will give an overview and generalization of the earlier results. In the next section we will consider a model of a double-Gaussian bipartite wave function of the most general form. Then, in Sections 3-6 we will consider specification of this general consideration for the above-indicated specific processes. We will show that the model of general double-Gaussian bipartite wave functions can be successfully used for unification of all the specific cases we consider though they differ from each other both in their physics and in the ranges of involved field and particle parameters.

2 General consideration

2.1 Definitions: entanglement, coincidence and single-particle measurements, conditional and unconditional probabilities

By definition, a bipartite wave function $\Psi(x_1, x_2)$ is entangled if it cannot be factorized, i.e., reduced to a product of two single particle functions

$$\Psi(x_1, x_2) \neq \psi_1(x_1) \times \psi_2(x_2). \tag{1}$$

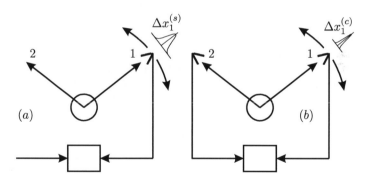

Figure 1. *(a) Single-particle and (b) coincidence schemes of measurements.*

Bipartite wave functions can be used to define both conditional and unconditional probability densities. The conditional probability density of finding the particle "1" in a vicinity of a point x_1 under the condition that the particle "2" has some given coordinate x_2 is given by

$$\frac{dw^{(c)}}{dx_1} = \frac{|\Psi(x_1, x_2)|^2}{\int dx_1 |\Psi(x_1, x_2)|^2}. \tag{2}$$

The unconditional probability density of finding the particle "1" in the vicinity of a point x_1 has the form

$$\frac{dw^{(u)}}{dx_1} = \int dx_2 |\Psi(x_1, x_2)|^2. \tag{3}$$

Experimentally, conditional and unconditional probability densities can be observed with the help of, correspondingly, coincidence-scheme and single-particle measurements, which is illustrated by the picture of Figure 1.

In the single particle scheme numbers of particles "1" are registered at various positions of the corresponding detector whereas the particle "2" can take arbitrary positions. All signals of the detector registering particles "1" are taken into account. As a result a single particle wave-packet of this particle is reconstructed and its single-particle wight $\Delta x_1^{(s)}$ is measured.

In contrast in the coincidence scheme one uses two detectors registering particles "1" and "2", and only coinciding (joint) signals from both detectors are recorded. If a position of the detector registering particles "2" is kept constant and the detector "1" is scanned, one finds a coincidence-scheme wave packet of the particle "1" and its coincidence width $\Delta x_1^{(c)}$. For these wave packets and widths we the following couple of identities between theoretical and experimental concepts:

$$conditional \equiv coincidence$$

$$unconditional \equiv single - particle$$

The second of these equalities assumes, in particular, that wave packets measured via single-particle measurements are identical to the unconditional probability density distributions of Eq. (3) : $dw^{(s)}/dx_{1,2} \equiv dw^{(u)}/dx_{1,2}$.

2.2 Double-Gaussian bipartite wave function

Let us consider a bipartite double-Gaussian wave function of the most general form:

$$\Psi(x_1, x_2) = \exp\left\{ -\frac{(\alpha x_1 + \beta x_2 + c_1)^2}{2a^2} \right\} \times \exp\left\{ -\frac{(\gamma x_1 + \delta x_2 + c_2)^2}{2b^2} \right\}, \qquad (4)$$

where x_1 and x_2 are coordinates of two particles, α, β, δ, and γ are arbitrary constants. As we are interested here only in wave packet widths, the normalizing factor is dropped in Eq. (4) and in all similar equations below. We regard the two exponential factors on the right-hand side of Eq. (4) and the parameters a and b as initial-state proto-functions and their widths, which are determined by the specific processes responsible for formation of such wave functions. Parameters c_1 and c_2 in Eq. (4) can describe, for example, a drift of the corresponding proto-function wave packets. The ratio of the protofunction's widths is the control parameter η:

$$\eta = \frac{b}{a}, \qquad (5)$$

which we have shown (Fedorov et al., 2004, 2005) to be the key in determining the degree of entanglement.

The number of independent parameters characterizing the wave function of Eq. (4) can be diminished by a redefinition of the protofunction's widths, e.g., as a, $\beta \rightarrow \alpha a$, $\alpha\beta$, and $b, \delta \rightarrow \gamma b$, $\gamma\delta$. But the form of Eq. (4) is more convenient, because often each of the parameters a, b, α, β, γ, and δ has its own physical sense. Note that a previous examination of double-Gaussian separability by Wodkiewicz and Englert (2002) differs substantially from ours. It is based on an operator density matrix formalism and is focused on bipartite systems governed by a harmonic oscillator Hamiltonian, and thus not capable of breakup.

The wave function of Eq. (4) is factorized in the variables ξ_1 and ξ_2, where

$$\xi_1 = \alpha x_1 + \beta x_2, \quad \text{and} \quad \xi_2 = \gamma x_1 + \delta x_2, \qquad (6)$$

and, hence, there is no entanglement in these variables. But in the variables x_1 and x_2 the wave function of Eq. (4) does not generally factorize and does describe an entangled quantum-mechanical state. The Jacobian of transformation from the variables $\{\xi_1, \xi_2\}$ to $\{x_1, x_2\}$ is given by

$$D \equiv \frac{D(\xi_1, \xi_2)}{D(x_1, x_2)} = \alpha\delta - \beta\gamma. \tag{7}$$

As we will see soon, the Jacobian D is crucially important for some processes to be considered.

2.3 Double-Gaussian conditional and unconditional wave packet widths

Let us find first the conditional width of the wave packet $|\Psi(x_1, x_2)|^2$, as given in (4), with respect to coordinate x_1. Conditional means in this case that the x_1 wave packet width must be found for a fixed value of the second particle's coordinate x_2. That is, we arrange that the detector registering particles "1" is scanned while the position of the detector registering particles "2" is held fixed.

By regrouping terms under the symbols of the exponent in Eq. (4) we get

$$|\Psi(x_1, x_2)|^2 = \exp\left\{-\frac{\alpha^2 b^2 + \gamma^2 a^2}{a^2 b^2} x_1^2 + A x_1 + B\right\}, \tag{8}$$

where A and B do not depend on x_1 and, at a given value of x_2, are constant. From Eq. (8) we immediately find the conditional width of the first particle's coordinate wave packet

$$\Delta x_1^{(c)} = \frac{ab}{\sqrt{\alpha^2 b^2 + \gamma^2 a^2}}. \tag{9}$$

To find the single-particle, or unconditional, width of first particle's wave packet, we have to integrate $|\Psi(x_1, x_2)|^2$ over x_2, which is easily done to give

$$\int dx_2 |\Psi(x_1, x_2)|^2 \propto \exp\left\{-\frac{D^2 x_1^2}{\beta^2 b^2 + \delta^2 a^2}\right\}, \tag{10}$$

where D is the $\xi - x$ transformation Jacobian of Eq. (7). The single-particle wave packet width determined by Eq. (10) is given by

$$\Delta x_1^{(s)} = \frac{\sqrt{\beta^2 b^2 + \delta^2 a^2}}{|D|}. \tag{11}$$

As shown by Fedorov et al. (2004, 2005), as long as there is no wave packet spreading, the ratio of the single-particle and conditional wave packet widths can be considered as a measure of entanglement of a bipartite quantum state, and this ratio is given by

$$R = \frac{\Delta x_1^{(s)}}{\Delta x_1^{(c)}} = \frac{\sqrt{\alpha^2 b^2 + \gamma^2 a^2}\sqrt{\beta^2 b^2 + \delta^2 a^2}}{ab|\alpha\delta - \beta\gamma|} = \frac{\sqrt{\alpha^2 \eta^2 + \gamma^2}\sqrt{\beta^2 \eta^2 + \delta^2}}{\eta|D|}. \tag{12}$$

Note that explicit expressions for the wave packet widths of Eqs. (9) and (11) can be reversed to find form measured values of $\Delta x_{1,2}^{(s,c)}$ the protofunction widths a and/or b, which are often unknown.

2.4   The Schmidt number

Another characteristic of entanglement is the correlation Schmidt number K introduced by Grobe, et al., 1994 (see also Nielsen and Chuang, 2000). It plays an important role as a degree of entanglement and is sometimes called a linear (rather than logarithmic) entropy. In the case of qubit pairs it is directly related to the concurrence measure of entanglement (Wootters, 1998). Its fundamental significance arises from its connection with the bi-orthogonal Schmidt modes that uniquely correlate the two particles (Nielsen and Chuang (2000), Ekert and Knight (1995), Eberly (2006)). For continuous entanglement the Schmidt number is defined by Grobe et al (1994) as the inverse purity, i.e., the inverse of the trace of the squared reduced density matrix (of either particle) of a pure bipartite state. This definition is in our case equivalent to the following one:

$$K^{-1} = N^{-2} \int dx_1 dx_2 dx'_1 dx'_2 \Psi(x_1,\, x_2) \Psi^*(x'_1,\, x_2) \Psi^*(x_1,\, x'_2) \Psi(x'_1,\, x'_2), \tag{13}$$

where

$$N = \int \int dx_1 dx_2 |\Psi(x_1,\, x_2)|^2 \,. \tag{14}$$

This definition of Schmidt number is defined for continuous entanglement, in contrast to the one given by Nielsen and Chuang (2000), which is not. For the double-Gaussian wave function of Eq. (4) all integrals in Eqs. (13) and (14) are easily calculated. We find the interesting identity:

$$K \equiv R = \frac{\sqrt{\alpha^2 \eta^2 + \gamma^2} \sqrt{\beta^2 \eta^2 + \delta^2}}{\eta |D|} \,. \tag{15}$$

That is, in the case of non-spreading double-Gaussian wave packets the Schmidt number K and the single-to-conditional width ratio parameter R coincide, as was already noted in special cases considered by Fedorov et al. (2004, 2005), and similar to a relation observed for large η by Chan et al. (2002, 2003). We focus on the identity of the K and R parameters because measuring wave packet widths is feasible, at least in principle, and this opens a way of measuring experimentally the degree of entanglement characterized by the Schmidt number.

Note that the $K - R$ equality can be violated by spreading of proto-wave packets. Spreading makes the wave-packet width ratio depend on time, $R \to R(t)$, whereas the Schmidt number K remains constant. In this case the identity of the K and R parameters (15) occurs only at the initial stage of evolution when spreading is not seen yet.

Eq. (15) describes a curve having a minimum at $\eta_0 = \sqrt{|\gamma \delta / \alpha \beta|}$, and the minimal value of the Schmidt number $K(\eta)$ (or $R(\eta)$) is given by

$$K_{\min} = K(\eta_0) = \frac{|\alpha \delta| + |\beta \gamma|}{|D|} \,. \tag{16}$$

In principle, this equation does not necessarily support a rather usual assumption that $K_{\min} = 1$. Moreover, if the Jacobian D is small, in accordance with Eq. (16) the minimal value of the Schmidt number can even be large. This is not the case for many earlier processes overviewed below where $K_{\min} \approx 1$. But here is at least one exception. This is the electron-positron pair production to be considered in Section 6. For this process, there are special conditions when D is small and $K_{\min}$ is large. The physical origin of this result will be discussed.

2.5 Double-Gaussian wave packets in the momentum representation

A double Fourier transformation of the coordinate bipartite wave function of Eq. (4) gives its momentum representation

$$\widetilde{\Psi}(k_1,\, k_2) = \int dx_1 dx_2 e^{-i(k_1 x_1 + k_2 x_2)} \Psi(x_1,\, x_2)$$

$$\propto \exp\left\{-\frac{a^2(\delta k_1 - \gamma k_2)^2}{2D^2}\right\} \exp\left\{-\frac{b^2(\beta k_1 - \alpha k_2)^2}{2D^2}\right\}. \tag{17}$$

From here we get immediately the momentum conditional wave packet widths. For example, for the first particle its wave packet's width in the momentum representation at a given value of the second particle's momentum k_2 is given by

$$\Delta k_1^{(c)} = \frac{|D|}{\sqrt{\delta^2 a^2 + \beta^2 b^2}}. \tag{18}$$

As previously, to get the unconditional (single-particle) wave packet width $\Delta k_1^{(s)}$ we integrate $|\widetilde{\Psi}(k_1,\, k_2)|$ over k_2 to get

$$\int dk_2 |\widetilde{\Psi}(k_1,\, k_2)|^2 \propto \exp\left\{-\frac{a^2 b^2 k_1^2}{\gamma^2 a^2 + \alpha^2 b^2}\right\}, \tag{19}$$

which gives

$$\Delta k_1^{(s)} = \frac{\sqrt{\gamma^2 a^2 + \alpha^2 b^2}}{ab}. \tag{20}$$

2.6 Identity, reciprocity, and uncertainty relations

By comparing Eqs. (18), (20) with Eqs. (9), (11), we find that there are the following identity relations between the coordinate and momentum conditional and unconditional wave packet widths:

$$\Delta x_1^{(c)} = \frac{1}{\Delta k_1^{(s)}}, \quad \Delta x_1^{(s)} = \frac{1}{\Delta k_1^{(c)}}. \tag{21}$$

Note that these equalities are different from those usually associated with the Heisenberg uncertainty relations. Indeed, for example the first equality of Eqs. (21) connects the conditional coordinate uncertainty with the unconditional momentum uncertainty, i.e., coordinate and momentum uncertainties found under different rather than the same conditions.

Also, by comparing the same equations for the wave packet widths [(18), (20), (9), and (11)] with Eq. (15) for the parameters K and R, we find the following uncertainty relations

$$\Delta x_1^{(s)} \times \Delta k_1^{(s)} = K \geq 1 \quad \text{and} \quad \Delta x_1^{(c)} \times \Delta k_1^{(c)} = \frac{1}{K} \leq 1. \tag{22}$$

The first of these two equations is related directly to the well known Heisenberg uncertainty inequality. The second inequality of Eq. (22) could be called the "Einstein uncertainty relation" because it belongs to the EPR-type relations (Einstein et al., 1935) although their interpretation in terms of conditional uncertainties was given only by Fedorov et al. (2005). Experimental achievement of an uncertainty product smaller than one was reported in the work by Howell et al. (2004). An alternative experimental verification of the EPR predictions was given by D'Angelo et al. (2004).

The choice of the particle "1" for our analysis is of course arbitrary. All the same can be done with the other particle with the help of substitutions $\alpha \rightleftharpoons \beta$ and $\gamma \rightleftharpoons \delta$. These substitutions do not change $|D|$ and K. As for the wave packet widths for particle "2", we can avoid writing them explicitly by presenting instead the following additional identity relations (which follow from explicit expressions for all the widths)

$$\Delta x_2^{(c,s)} = \frac{ab}{|D|} \Delta k_1^{(c,s)} \quad \text{and} \quad \Delta k_2^{(c,s)} = \frac{|D|}{ab} \Delta x_1^{(c,s)}. \tag{23}$$

These relations show that the information that can be obtained from measurements of coincidence and single-particle coordinate and momentum uncertainties for, e.g., particle "1" is sufficient for calculating the same parameter for the particle "2" without any measurements on it.

Eqs. (21)-(23) were derived in the paper by Fedorov et al. (2005) for a special kind of Gaussian protofunctions, arising in the problem of spontaneous emission of a photon by an atom. The derivation presented here suggests wide generality of these relations. Deviations of protofunctions from the Gaussian case can change these results quantitatively but not qualitatively. This has been demonstrated in a calculation made for PDC by Law and Eberly (2004), as we show below in Sec. V.

3   Photoionization of atoms and photodissociation of molecules

In the case of photoionization of an atom, the particles whose quantum states can be entangled are the photoelectron and the residual ion. The variables on which the bipartite function depends are the electron and ion position vectors $\vec{r}_e$ and $\vec{r}_i$.

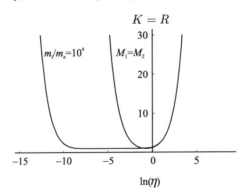

Figure 2. *The dependence of K on $\ln(\eta)$ for photoionization (wide curve) and photodissociation (narrow curve), as taken from the paper by Fedorov et al. (2004). Qualitatively the same wide curve characterizes the dependence $K(\eta)$ in the cases of spontaneous emission (Fedorov et al. (2005) and Section 4 below) and electron-positron pair production in the ultra-relativistic limit considered by Fedorov et al. (2006) (see also Section 6 of this lecture) and the narrow curve also characterizes $K(\eta)$ in the case of PDC (see the paper by Law and Eberly, 2004) and Section 5 below).*

In this case the protofunctions, the product of which gives a total bipartite wave function, have clear physical origins: these are (i) the relative electron-ion motion wave function $\Psi_{rel} \equiv \Psi_1$ depending on $|\vec{r}_e - \vec{r}_i|$ and (ii) the center-of-mass wave function $\Psi_{c.m.} \equiv \Psi_2$ depending on the electron-ion center of mass position vector $(m_e \vec{r}_e + m_i \vec{r}_i)/M$, where m_e, m_i, and $M = m_e + m_i$ are the electron and ion masses and their sum.

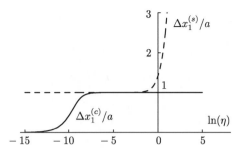

Figure 3. *Ratios of coincidence (solid) and single-particle (dashed) wave packet widths to a, the initial relative-motion width, for the electron, the picture from the paper by Fedorov et al. (2005) rescaled from atom-photon to ion-electron entanglement.*

If an atom is ionized by a linearly polarized field, photoionization is most efficient in the direction of polarization. If we assume that electron and ion detectors are installed on a line crossing the initial atomic center-of-mass position and parallel to the field's polarization, we come to a one-dimensional situation with electron and ion position vectors $\vec{r}_e$ and $\vec{r}_i$ replaced by their projections on the polarization direction $x_e \equiv x_1$ and $x_i \equiv x_2$.

If we assume that initially an atom is put into a trap and cooled, its center-of-mass wave function is Gaussian. As for the relative-motion wave function, it can be Gaussian too if the ionizing field is not too strong and its pulse envelope has a Gaussian form. In other cases the approximation of the relative-motion wave function by a Gaussian one can be considered as a model.

In any case, we come to the total wave function of the form

$$\Psi(x_1, x_2) = \exp\left\{-\frac{(|x_1 - x_2| - \overline{v}\,t)^2}{2a^2}\right\} \exp\left\{-\frac{(m_e x_1 + m_i x_2)^2}{2b^2 M^2}\right\}, \qquad (24)$$

where a and b are the widths of, respectively, the relative and translational motion, and $\overline{v}$ is the average velocity of the photoelectron with respect to the ion. Eq. (24) is written down in the "no-spreading" approximation, i.e., valid for a time t shorter than the spreading time of both relative and translational motion wave packets.

The protofunction widths are given by: $a = \overline{v}/\Gamma_i$, where Γ_i is the Fermi-Golden-Rule ionization rate, and $b = \Delta r_{at}$ - the width of the atomic wave function for an atom put, e.g., into a trap (and cooled to be in the Gaussian ground state of a parabolic trap potential).

By assuming that the electron and ion detectors are installed so that $x_1 > 0$ and $x_2 < 0$, we get $|x_1 - x_2| = x_1 - x_2$, and this reduces Eq. (24) to the form of Eq. (4) with

$$\alpha = 1, \quad \beta = -1, \quad \gamma = \frac{m_e}{M} \ll 1, \quad \text{and} \quad \delta = \frac{m_i}{M} = 1 - \frac{m_e}{M} \approx 1. \qquad (25)$$

In such a case the Jacobian D of Eq. (7) approximately equals one, $D \approx 1$, and expression (15) for the Schmidt number is reduced to

$$K = \frac{\sqrt{M^2\eta^2 + m_e^2}\sqrt{M^2\eta^2 + m_i^2}}{M^2\eta}. \qquad (26)$$

The dependence $K(\eta)$ for photoionization is described by the wide curve in Figure 2.

The narrow curve corresponds to photodissociation of molecules. In this case the Schmidt number is determined by the same equation as Eq. (26) but with the electron and ion masses m_e

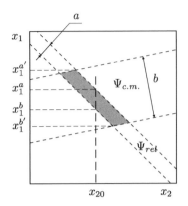

Figure 4. $x_e - x_i$ map $(x_e = x_1$ and $x_i = x_2)$.

and m_i substituted by masses of molecular fragments M_1 and M_2. The narrow curve of Figure 2 corresponds to $M_1 = M_2$ (e.g., photodissociaton of any diatomic molecule with two identical atoms).

The curves of Figure 2 show that entanglement is large in two asymptotic cases, $\eta \ll 1$ ($b \ll a$) and $\eta \gg 1$ ($b \gg a$), which means that one of the proto- wave packets (either that of translational or relative motion) must be much narrower than the other one. At intermediate values of η, $K \approx 1$, i.e. there is no entanglement. Figure 2 shows that the intermediate "no-entanglement" region is large for atoms and small for molecules. In this sense a molecule is a much more appropriate object than an atom for production of highly entangled states.

Specifically, coincidence and single-particle measurements give information about wave packet width. The high-entanglement conditions correspond to anomalously small conditional width or anomalously large single-particle width. These predictions follow directly from Eqs. (9), (11), (18), (20) and are illustrated by the curves of Figure 3.

In this picture ratios of the coincidence and single-particle electron wave packet widths to a, the initial electron relative motion packet width, are described by solid and dashed lines. When both ratios equal 1 we have the "normal", or "no-entanglement" case. Lower and higher wings of the curves describe the entanglement-induced anomalous narrowing and broadening of wave packets.

Similar results occur for the ion wave packets. The curve of Figure 2 for an atom follows from the results described by the curves of Figure 3 and the definition of the parameter R given by Eq. (12). Also, owing to the relations of Eqs. (21) and (22), the same curves describe normalized widths of wave packets in the momentum representation. Note that the entanglement-induced narrowing of wave packets shown in Figure 3, and observable in principle in coincidence measurements, has many common features with the effects known in atom optics as "measurement-induced localization of atoms". In various contexts these effects have been discussed by many authors - see the review by M. Freyberger et al. (1999) and references therein.

An origin of the entanglement-related wave packet narrowing and the conditions of its appearance can be illustrated also by the diagram characterizing localization regions of the proto-functions Ψ_{rel} and $\Psi_{c.m.}$ (Figure 4). Boundaries of these regions are indicated by the slanting dashed lines. The overlapping region is the region where the total electron-ion wave function is not small. x_{20} is some given value of the ion detector coordinate in the coincidence scheme measurements. These measurements register only electrons with coordinates x_1 at the vertical line starting from x_{20} and located in the shaded area, i.e, $x_1^b < x_1 < x_1^a$.

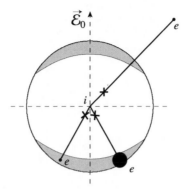

Figure 5. $x_e - x_i$ map ($x_e = x_1$ and $x_i = x_2$).

The difference x_1^a and x_1^b is just the coincidence width of the electron wave packet, $\Delta x_1^{(c)} = x_1^a - x_1^b$. On the other hand, in the single-particle scheme of measurements the whole shaded area gives contributions to the electron detector signals to be registered, and the single particle width of the electron wave packet is given by $\Delta x_1^{(s)} = x_1^{a'} - x_1^{b'}$. So, evidently, $\Delta x_1^{(s)} > \Delta x_1^{(c)}$, and the degree of entanglement is large, $R = \Delta x_1^{(s)}/\Delta x_1^{(c)} = K > 1$.

Note that in this case the necessary condition of high entanglement if a large aspect ratio of overlapping region in the $x_1 - x_2$ map. This is a general rule for many other phenomena. But there is at least one exception to be discussed below in Section 6 on pair production.

An alternative view of the same wave-packet propagation and entanglement pictures is given by Figure 5), which can be considered as a cross section of a three-dimensional picture. Here crosses denote original atomic center-of-mass positions, $\vec{\varepsilon}_0$ is the field-strength amplitude of a laser field that ionizes the atom, i indicates a position of the ion detector. The moon-like shaded regions and black dots describe localization regions of the corresponding relative-motion and center-of mass protofunction.

The total electron-ion wave function is not negligibly small only if these two regions overlap. This determines where the electron detector must be installed to register any signals. In particular, the upper black spot indicates wrong positions and two lower black spots - good positions of the electron detector. Among two lower black spots a narrower one corresponds to a higher degree of entanglement.

Note finally that for massive particles the coordinate picture of wave packet evolution is complicated by wave packet spreading effect, which makes wave-packet widths and their ratio depend on time $R_{\text{coord}} = \Delta x_1^{(s)}/\Delta x_1^{(c)} \to R_{\text{coord}}(t)$. On the other hand in the momentum representation the time evolution of a bipartite wave function adds to $\tilde{\Psi}(k_1, k_2)$ of Eq. (17) only a time-dependent phase factor which does not affect the momentum wave packet widths and their ratio and leaves them constant: $R_{\text{mom}} = \Delta k_1^{(s)}/\Delta k_1^{(c)} = \text{const}$.

The same can be said about the Schmidt number K: for non-interacting particles K does not depend on time both for massive and mass-less particles. Hence, spreading destroys the above described identity relations between the coordinate and momentum wave packet widths. Also for spreading wave packets the coordinate width-ratio parameter R_{coord} cannot be used directly to evaluate the degree of entanglement. But all other previously described relations remain valid for the momentum picture.

In particular, for any particles, both with and without wave packet spreading, the momentum width-ratio parameter equals the Schmidt number $R_{\mathrm{mom}} = \Delta k_1^{(s)}/\Delta k_1^{(c)} \equiv K$. This means that measurements of the coincidence and single-particle momentum wave-packet widths can be used for measuring the degree of entanglement independently of whether wave packet spreading occurs or not in the coordinate picture.

4 Spontaneous emission of a photon by an atom

If an initially excited atom emits a photon, the subsequent photon quantum state can be entangled with the atomic translation-motion state. The entangled variables in this case are the photon and atomic center-of-mass position vectors $\vec{r}_{ph}$ and $\vec{r}_{at}$. The two protofunctions in this case describe the entanglement-free photon and the atomic center-of-mass motion, the arguments of which however are entangled, i.e., are given by superpositions of $\vec{r}_{ph}$ and $\vec{r}_{at}$. Rigorously, an entanglement-free photon wave has an exponential tail with a sharp front edge. For simplicity it can be modelled crudely by a Gaussian function to reduce the total wave function to the form of Eq. (4).

Entanglement-related effects in such a case appear qualitatively similar to those described above for photoionization. But there are some important differences. A photon does not have a mass. For this reason a ratio like m_e/M does not have any sense for an atom-photon pair. Recall that photoionization could be described in the dipole approximation, and the injected momentum from the ionizing laser field played no role in the electron-ion recoil.

Similarly, in the case of spontaneous emission any momentum associated with the preparation of the atom in its excited state is ignored. Only the momentum of the emitted photon is responsible for the atom's recoil. But for this reason, in the investigation of atom-photon recoil entanglement, we cannot use the dipole approximation for describing the atom-photon interaction, as shown by Rzazewski and Zakowicz (1992).

With these remarks taken into account, we find that for spontaneous emission the parameters of a double-Gaussian wave function are given by (Fedorov et al., 2005)

$$\alpha = 1, \quad \beta = -1, \quad \gamma = \frac{v_{rec}}{c} \ll 1, \quad \text{and} \quad \delta = 1 - \frac{v_{rec}}{c} \approx 1, \tag{27}$$

where $v_{rec} = \hbar\omega/Mc$ is the atomic recoil velocity, ω is the mean frequency of the emitted photon, and $\hbar\omega/c$ is its momentum. By comparing Eqs. (27) and (25), we see that the only difference between them is the replacement of the mass ratio m_e/M by the velocity ratio v_{rec}/c. From this comparison we deduce that, as said above, qualitatively the atom-photon entanglement picture for spontaneous emission is identical to the electron-ion entanglement in photoionization.

Quantitatively, the parameter v_{rec}/c is much smaller than m_e/M: typically $v_{rec}/c \sim 10^{-8}$ whereas $m_e/M \sim 10^{-4}$. This means that for spontaneous emission high-entanglement regions correspond to even much more extreme values of the control parameter η than in the case of photoionization. For this reason the problem of experimental creation of high-entanglement atom-photon states can be more difficult than that of electron-ion and, particularly, molecular-fragment high-entanglement states. The control parameter η is determined by the same Eq. (5) as previously with b being the width of the atomic translational-motion wave function and a being determined by the radiative width γ of the excited atomic level: $a = c/\gamma$.

5   Parametric down-conversion (PDC)

PDC has been one of the most widely used methods for generating discrete two-particle entanglement (in the polarization states), but only recently has discussion of PDC entanglement in continuum states attracted wide attention. In this case photon wave packet structures are mostly discussed in the momentum representation (see, e.g., the works by Rubin (1996), Burlakov et al. (1997), Monken et al. (1998), Wallborn et al. (2003), Law and Eberly (2004), Howell (2004) and references therein). Here we will give an overview of the most important cases by emphasizing mainly specific conditions under which PDC biphoton wave packets can be reduced to the general forms discussed in the Section 2, both in the coordinate and momentum representations.

In PDC a pump photon (frequency ω_p) splits in a nonlinear crystal into two other photons (frequencies ω_1 and ω_2). In the stationary case the energy conservation law is exactly satisfied and gives $\omega_1 + \omega_2 = \omega_p$. In the so called "degenerate" case we will consider here frequencies of emitted photons are equal, $\omega_1 = \omega_2 \equiv \omega = \omega_p/2$.

Let us consider here (for simplicity) only the plane geometry in which both the pump wave vector $\vec{k}_p$ and the wave vectors $\vec{k}_{1,2}$ of emitted photons have zero y-components, and both a crystal and a laser beam are unlimited in the y-direction. Let the pump be an extraordinary wave propagating along the z-axis and the pump be converted into two ordinary waves "1" and "2" ($e \rightarrow o+o$ or type-I synchronism in terms of definitions by Rubin, 1996). Then absolute values of the pump and the emitted-photon wave vectors are given by $k_p = n \times 2\omega/c$ and $k_{1,2} = \widetilde{n} \times \omega/c$ where n and $\widetilde{n}$ are the extraordinary and ordinary wave refractive indices, correspondingly, at frequencies 2ω and ω. The relation between n and $\widetilde{n}$ is controlled by orientation of the crystal with respect to the laser-beam (z-) axis. In the case of orientation providing collinear phase-matching, $\widetilde{n} = n$.

Burlakov et al. (1997) show that the biphoton wave function in the momentum representation can be presented in the form

$$\widetilde{\Psi}(k_{1x}, k_{2x}) = \chi \int dx\, dz\, E_p^*(x, z)\, f_x(x)\, f_z(z)\, \exp\left(ix\Delta_x + iz\Delta_z\right), \tag{28}$$

where Δ_x and Δ_z are the phase-matching detunings

$$\Delta_x = k_{1x} + k_{2x}, \quad \Delta_z = k_{1z} + k_{2z} - k_p = \sqrt{k_1^2 - k_{1x}^2} + \sqrt{k_2^2 - k_{2x}^2} - k_p, \tag{29}$$

$f_x(x)$ and $f_z(z)$ in Eq. (28) are the functions restricting the maximal possible interaction region by the crystal width (d) and length (L)

$$f_x(x) = \begin{cases} 1, & -d/2 < x < d/2 \\ 0, & |x| > d/2 \end{cases}, \quad f_z(z) = \begin{cases} 1, & -L/2 < z < L/2 \\ 0, & |z| > L/2 \end{cases}, \tag{30}$$

and χ is the nonlinear susceptibility. The position-dependent slow amplitude of the pump $E_p^*(x, z)$ can be expressed in the following way via its "momentum" representation

$$E_p^*(x, z) = \int dk_{px}\, \widetilde{E}_p^*(k_{px}) \exp\left[-ixk_{px} - iz\left(\sqrt{k_p^2 - k_{px}^2} - k_p\right)\right] \tag{31}$$

to reduce Eq. (28) to an alternative equally well known form (Rubin (1996), Monken (1998))

$$\widetilde{\Psi}(k_{1x}, k_{2x}) \propto \int dk_{px}\, \widetilde{E}_p^*(k_{px})\, \mathrm{sinc}\left(\frac{d(k_{1x} + k_{2x} - k_{px})}{2}\right)$$

$$\times \mathrm{sinc}\left(\frac{L\left(\sqrt{k_1^2 - k_{1x}^2} + \sqrt{k_2^2 - k_{2x}^2} - \sqrt{k_p^2 - k_{px}^2}\right)}{2}\right), \tag{32}$$

where $\operatorname{sinc}(x) = \sin(x)/x$.

The function $\widetilde{E}_p^*(k_{px})$ and the first sinc-function under the integral on the right-hand side of Eq. (32) compete with each other in defining the region of k_{px} giving the main contribution to the integral. The role of these two functions depends on whether the crystal is wide or narrow compared to the pump laser beam.

A crystal is wide if its width d exceeds the width w_p of the pump beam, $d \gg w_p$. In this case the sharpest function under the integral in Eq. (32) is $\operatorname{sinc}[d(k_{1x} + k_{2x} - k_{px})/2]$. For integration over k_{px}, this function can be approximated by the delta function $\delta(k_{1x} + k_{2x} - k_{px})$ to give

$$\widetilde{\Psi}(k_{1x},\, k_{2x}) \propto \exp\left(-\frac{w_p^2(k_{1x} + k_{2x})^2}{8}\right)$$

$$\times \operatorname{sinc}\left(\frac{L\left(\sqrt{k_1^2 - k_{1x}^2} + \sqrt{k_2^2 - k_{2x}^2} - \sqrt{k_p^2 - (k_{1x} + k_{2x})^2}\right)}{2}\right), \tag{33}$$

where the pump-pulse envelope is taken Gaussian, $E_p(x) \propto \exp\left(-x^2/2w_p^2\right)$.

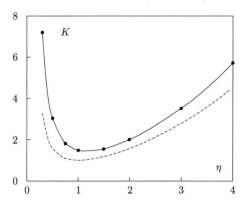

Figure 6. *The dependence of K on the PDC control parameter η, from the paper by Law and Eberly (2004), showing the similarity of the Gauss approximation (dashed line) to the more rigorous sinc function (solid line).*

Under the conditions of near-axis propagation of emitted photons ($|k_{1,2x}| \ll k_p$, $|k_{1,2}|$) and exact collinear phase-matching ($\widetilde{n} = n$, $k_1 = k_2 = k_p/2$) roots squared in the sinc−argument of Eq. (33) can be expanded in powers of k_{1x}, k_{2x} to reduce this equation to the form (Monken et al. (1998),Wallborn et al., (2003))

$$\widetilde{\Psi}_{w.c.}(k_{1x},\, k_{2x}) \propto \exp\left(-\frac{w_p^2(k_{1x} + k_{2x})^2}{8}\right) \operatorname{sinc}\left(\frac{L(k_{1x} - k_{2x})^2}{4k_p}\right), \tag{34}$$

where the subscript "w.c." means a wide crystal. By modelling the sinc−function by a Gaussian one with effectively the same width, we finally obtain

$$\widetilde{\Psi}_{w.c.}(k_{1x},\, k_{2x}) \propto \exp\left\{-\frac{w_p^2(k_{1x} + k_{8x})^2}{2}\right\} \exp\left\{-\frac{L(k_{1x} - k_{2x})^2}{8k_p}\right\}. \tag{35}$$

As an example of the qualitative similarities preserved in the sinc $\rightarrow$ Gauss approximation, we show in Figure 6 a comparison of two calculations of the Schmidt number K carried out for PDC under the conditions we have just described.

The double Fourier transformation of $\widetilde{\Psi}_{w.c.}(k_{1x}, k_{2x})$ of Eq. (35) gives the coordinate bipho-ton wave function

$$\Psi_{w.c.}(x_1, x_2) \propto \exp\left\{-\frac{(x_1+x_2)^2}{2w_p^2}\right\} \exp\left\{-\frac{k_p(x_1-x_2)^2}{2L}\right\}. \tag{36}$$

The wave functions $\widetilde{\Psi}_{w.c.}(k_{1x}, k_{2x})$ and $\Psi_{w.c.}(x_1, x_2)$ of Eqs. (35) and (36) belong to the general class of double-Gaussian wave functions of Eqs. (4) and (17) considered in Section 2. With the widths of the coordinate proto-functions defined as $a = w_p$ and $b = \sqrt{L/k_p}$, the other parameters of Eq. (4) are found to be

$$\alpha = \beta = \gamma = -\delta = 1, \tag{37}$$

which gives $D = -2$ and $K_{\min} = 1$.

Practically, it can be rather difficult to achieve the used above exact collinear phase matching condition, this can require a too precise orientation of a crystal. If this condition is not fulfilled, the refractive indices $\widetilde{n}$ and n are not equal exactly, though they can be rather close. Let us assume that $\widetilde{n} > n$ (which corresponds to a so called "negative crystal", where ordinary waves have a larger refractive index than the extraordinary one). In this case one can use a procedure (see, e.g., the paper by Burlakov et al, 1997) different from that used above for a simplification of the sinc–function argument in Eq. (33). First, one can find the "most probable" values of emitted photon transverse wave vectors, $k_{1,x}^{(0)}$ and $k_{2,x}^{(0)}$, at which both arguments of the Gaussian and sinc–function in Eq. (33) turn zero. These values can be easily found to be given by

$$k_{1x}^{(0)} = -k_{2x}^{(0)} = \pm\frac{\omega}{c}\sqrt{\widetilde{n}^2 - n^2} \equiv \pm\sin\theta\,\frac{\widetilde{n}}{2n}\,k_p \approx \theta\,\frac{k_p}{2}, \tag{38}$$

where θ is the scattering angle (which is small as long as $\widetilde{n} - n \ll n$)

$$\sin\theta = \sqrt{1 - \frac{n^2}{\widetilde{n}^2}} \approx \theta \approx \sqrt{2\left(1 - \frac{n}{\widetilde{n}}\right)} \ll 1. \tag{39}$$

And, second, one can linearize the argument of the sinc–function of Eq. (33) in deviations of k_{1x} and k_{2x} from $k_{1x}^{(0)}$ and $k_{2x}^{(0)}$, $q_1 = k_{1x} - k_{1x}^{(0)}$ and $q_2 = k_{2x} - k_{2x}^{(0)}$. Such a linearization reduces Eq. (33) to

$$\widetilde{\Psi}_{w.c.} \propto \exp\left[-\frac{w_p^2(q_1+q_2)^2}{8}\right] \operatorname{sinc}\left[\frac{L\theta}{2}(q_1-q_2)\right], \tag{40}$$

where $q_1 + q_2 = k_{1x} + k_{1x}$ and $q_1 - q_2 = k_{1x} - k_{1x} - 2k_{1x}^{(0)}$. Eqs. (40) and (34) differ from each other by giving different evaluations for the distribution width in the variable $k_1 - k_2$.

This width is on the order of $2\sqrt{k_p/L}$ at the exact-collinear phase matching and $2/L\theta$ for non-collinear phase matching. These two evaluations coincide at $\theta = \theta_0 = 1/\sqrt{Lk_p}$. At $\theta < \theta_0$ the collinear phase-matching conditions are fulfilled and one can use Eq. (34).

The opposite case, $\theta > \theta_0$, corresponds to the non-collinear phase matching and the appro-priate equation to be used is Eq. (40). A double Fourier transform of the function $\widetilde{\Psi}_{w.c.}(k_{1x}, k_{2x})$ of Eq. (40) gives the following expression for the biphoton wave function in the coordinate rep-resentation (in the case of a non-collinear phase matching)

$$\Psi_{w.c.}(x_1, x_2) \propto \cos\left(\frac{\theta k_p}{2}(x_1 - x_2)\right) \exp\left\{-\frac{(x_1 + x_2)^2}{2w_p^2}\right\}$$
$$\times \left[\text{sign}\left(1 - \frac{x_1 - x_2}{L\theta}\right) + \text{sign}\left(1 + \frac{x_1 - x_2}{L\theta}\right)\right], \tag{41}$$

where $\text{sign}(x) = x/|x|$. Though the sum of the sign–functions is not the same as the Gaussian function, qualitatively we can use for evaluating the Schmidt number K all the same formulas as described above but with $a = w_p$ and $b = L\theta$ and the parameters α, β, γ, δ given by Eq. (37).

Let us consider now the case of a narrow crystal or a wide pump beam, $d \ll w_p$. Such a situation is also possible and realizable experimentally, especially in investigation of interference effects investigated, e.g., by Burlakov et al. (1997). In this case the pump slow amplitude $E_p^*(x, z)$ is almost constant in the crystal. Its Fourier transform with respect to the x coordinate, $\widetilde{E}_p^*(k_{p\,x})$, is a much narrower function than the first sinc–function in Eq. (32). Approximation of $\widetilde{E}_p^*(k_{p\,x})$ by the delta-function $\delta(k_{p\,x})$ reduces Eq. (32) to the form used, e.g., by Burlakov et al. (1997).

$$\widetilde{\Psi}_{n.c.}(k_{1\,x}, k_{2\,x}) \propto \text{sinc}\left(\frac{d\Delta_x}{2}\right) \text{sinc}\left(\frac{L\Delta_z}{2}\right), \tag{42}$$

where n.c. means "narrow crystal" and the phase-matching detunings Δ_x and Δ_z are given by Eqs. (29). In contrast to Δ_x, the detuning Δ_z given by the second Eq. (29) is a nonlinear function of $k_{1\,x}$ and $k_{2\,x}$. Its linearization can be carried exactly in the same wy as described above for the wide-crystal case. Eqs. (38) and (39) remain valid, and the final result is given by

$$\widetilde{\Psi}_{n.c.} \propto \text{sinc}\left[\frac{d}{2}(q_1 + q_2)\right] \text{sinc}\left[\frac{L\theta}{2}(q_1 - q_2)\right]. \tag{43}$$

The corresponding position-dependent biphoton wave function received from $\widetilde{\Psi}_{n.c.}$ with the help of a double Fourier transformation has the form

$$\Psi_{w.c.}(x_1, x_2) \propto \cos\left(\frac{\theta k_p}{2}(x_1 - x_2)\right) \left[\text{sign}\left(1 - \frac{x_1 + x_2}{d}\right) + \text{sign}\left(1 + \frac{x_1 + x_2}{d}\right)\right]$$
$$\times \left[\text{sign}\left(1 - \frac{x_1 - x_2}{L\theta}\right) + \text{sign}\left(1 + \frac{x_1 - x_2}{L\theta}\right)\right]. \tag{44}$$

Again, though the sums of sign–functions in Eq. (44) are not the same as Gaussian functions, qualitatively we can use all the same evaluations of the Schmidt number K as received for double-Gaussian wave functions with $a = d$ and $b = L\theta$ and the parameters α, β, γ, δ given by Eq. (37).

Note that in he case of the ideal collinear phase-matching in a narrow crystal Eqs. (42) and (29) yield

$$\widetilde{\Psi}_{n.c.}^{(\theta=0)}(k_{1\,x}, k_{2\,x}) \propto \text{sinc}\left(\frac{d(k_{1\,x} + k_{2\,x})}{2}\right) \text{sinc}\left(\frac{L(k_{1\,x}^2 + k_{2\,x}^2)}{4k_p}\right). \tag{45}$$

This results differs from all that has been seen before because in this case a lineraization of the second sinc–function argument appears impossible.

6 Multiphoton pair production

Following the ideas of recent work by Fedorov et al. (2006), let us consider first the of electron-positron pair production by two coherent counter-propagating beams of a classical electromagnetic field with equal mean frequencies ω_0 in each beam (i.e., in the center of momentum frame).

Let the vector potentials of the beams be given by Gaussian superpositions of plane waves $\exp(i\vec{k}_{1,2}\vec{r})$ with the weight functions

$$
\begin{aligned}
A_1(k_1) &= \frac{A_{10}}{\pi^3 \Delta k_1 \Delta k_{1\perp}^2} \exp\left(-\frac{(k_1 - k_0)^2}{2\Delta k_1^2} - \frac{\vec{k}_{1\perp}^2}{\Delta k_{1\perp}^2}\right), \\
A_2(k_2) &= \frac{A_{20}}{\pi^3 \Delta k_2 \Delta k_{2\perp}^2} \exp\left(-\frac{(k_2 + k_0)^2}{2\Delta k_2^2} - \frac{\vec{k}_{2\perp}^2}{\Delta k_{2\perp}^2}\right),
\end{aligned}
\tag{46}
$$

where k_1 and k_2 are projections of $\vec{k}_1$ and $\vec{k}_2$ upon some given direction (the x-axis), $\vec{k}_{1\perp}$ and $\vec{k}_{2\perp}$ are components of $\vec{k}_1$ and $\vec{k}_2$ perpendicular to the x-axis, and $k_0 = \omega_0/c$. Let longitudinal and transverse widths of the functions $A_1(k_1)$ and $A_2(k_2)$ obey the conditions $\Delta k_{1,2\perp} \ll \Delta k_{1,2} \ll k_0$. These parameters are related to the coordinate width of electromagnetic pulses $\Delta x_{1,2} \sim 1/\Delta k_{1,2} \ll |\Delta \vec{r}_{1,2\perp}| \sim 1/\Delta k_{1,2\perp}$.

These are 'almost plane waves' having the form of wide and short packets. In such waves the x-components of the vector potential are small and can be ignored, and Fourier transformed transverse components are given by Eqs. (46).

Let us consider detecting electron-positron pairs with electrons and positrons propagating in the z-direction (i.e., with zero transverse components of the electron and positron momenta). In this case the distributions of the functions $A(k)$ over $\vec{k}_{1,2\perp}$ are important only for providing a finite pair-production probability, and are not important for investigation of entanglement. This produces a reasonable one-dimensional approximation.

The quantum-electrodynamical matrix element of a two-photon pair production is well known and can be found, e.g., in the book by Akhiezer and Berestetskii (1965). It consists of four δ-functions expressing energy and momentum conservation times a bispinor part $\mathcal{Q}$. The latter depends on photon and electron/positron momenta, but this dependence is slow compared to that resulting from the narrow Gaussian functions of Eqs. (46), so that $\mathcal{Q}$ can be considered as a constant not affecting the degree of entanglement. So, the matrix element for pair production, or the momentum-representation electron-positron wave function, are given by

$$
\begin{aligned}
\widetilde{\Psi}(p_-, p_+) &\propto 2 \int dk_1 \, dk_2 \exp\left(-\frac{(k_1 - k_0)^2}{2\Delta k_1^2} - \frac{(k_2 + k_0)^2}{2\Delta k_2^2}\right) \\
&\times \delta(p_- + p_+ - k_1 - k_2) \, \delta\left[E_- + E_+ - c\left(|k_1| + |k_2|\right)\right] = \quad (47) \\
\exp\left\{-\frac{[c(p_- + p_+) + (E_- + E_+ - 2\omega_0)]^2}{8c^2\Delta k_1^2}\right\} &\exp\left\{-\frac{[c(p_- + p_+) - (E_- + E_+ - 2\omega_0)]^2}{8c^2\Delta k_2^2}\right\},
\end{aligned}
$$

where $p_\mp$ and $E_\mp$ are the electron and positron momenta and energies, and $E_\mp = \sqrt{c^2 p_\mp^2 + m^2 c^4}$. In the limit $\Delta k_{1,2} \to 0$ the exponents in the last line of Eq. (47) turn into the δ-functions to give $E_\mp^{(0)} = \omega_0$ and $p_\mp^{(0)} = \pm\sqrt{k_0^2 - m^2 c^2}$. At small but finite values of $\Delta k_{1,2}$ the functions $E_\mp(p_\mp)$ can be linearized in $q_\mp \equiv p_\mp - p_\mp^{(0)}$: $E_\mp(p_\mp) \approx \omega_0 \pm v_0 q_\mp$, where $v_0 = c^2 \left|p_\mp^{(0)}\right|^2 / \omega_0$ is the mean electron (and positron) velocity. As a result, the wave function $\widetilde{\Psi}$ of Eq. (47) takes the form

$$
\widetilde{\Psi}(p_-, p_+) \propto \exp\left\{-\frac{[q_-(1 + \frac{v_0}{c}) + q_+(1 - \frac{v_0}{c})]^2}{8\Delta k_1^2} - \frac{[q_-(1 - \frac{v_0}{c}) + q_+(1 + \frac{v_0}{c})]^2}{8\Delta k_2^2}\right\}
\tag{48}
$$

The wave function in the coordinate representation is given by

$$\Psi(x_-, x_+) \propto \int dp_- dp_+ \, e^{i[p_- x_- + p_+ x_+ - (E_- + E_+)t]} \, \widetilde{\Psi}(p_-, p_+)$$

$$\propto \exp\left\{ -\frac{\Delta k_1^2 c^2}{8v_0^2} \left[x_- \left(1 + \frac{v_0}{c}\right) - x_+ \left(1 - \frac{v_0}{c}\right) - 2v_0 t \right]^2 \right\}$$

$$\times \exp\left\{ -\frac{\Delta k_2^2 c^2}{8v_0^2} \left[x_- \left(1 - \frac{v_0}{c}\right) - x_+ \left(1 + \frac{v_0}{c}\right) + 2v_0 t \right]^2 \right\}, \tag{49}$$

where x_- and x_+ are the electron and positron coordinates and the integrals are calculated with the energies $E_\mp(p_\mp)$ linearized again in $q_\mp = p_\mp - p_\mp^{(0)}$.

Clearly, Eqs. (48) and (49) are of the same form as the general equations of Section 2. In terms of the notations used in Eq. (4), $x_- \equiv x_1$ and $x_+ \equiv x_2$. The coordinate widths of the protofunctions and the control parameter can be conveniently defined now as

$$a = 2\frac{v_0}{c\Delta k_1} = 2\frac{v_0}{c}\Delta x_1, \quad b = 2\frac{v_0}{c\Delta k_2} = 2\frac{v_0}{c}\Delta x_2, \quad \eta = \frac{b}{a} = \frac{\Delta k_1}{\Delta k_2} = \frac{\Delta x_2}{\Delta x_1}. \tag{50}$$

Then the parameters α, β, γ, and δ of Eq. (4) take the form

$$\alpha = -\delta = \left(1 + \frac{v_0}{c}\right) \quad \text{and} \quad \gamma = -\beta = \left(1 - \frac{v_0}{c}\right). \tag{51}$$

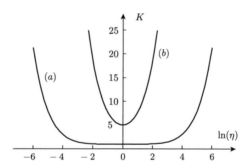

Figure 7. The Schmidt number K vs. ln(η) for (a) $v_0 = 0.9c$ and (b) $v_0 = 0.1c$, the picture from the paper by Fedorov et al., 2006.

This gives $D = \alpha\delta - \beta\gamma = -4v_0/c$ and the following expression for the Schmidt number K (15)

$$K(\eta) = \frac{\sqrt{\alpha^2\eta^2 + \gamma^2}\sqrt{\beta^2\eta^2 + \delta^2}}{\eta|D|}$$

$$= \frac{c}{4v_0\eta}\sqrt{\left(1 + \frac{v_0}{c}\right)^2 \eta^2 + \left(1 - \frac{v_0}{c}\right)^2}\sqrt{\left(1 - \frac{v_0}{c}\right)^2 \eta^2 + \left(1 + \frac{v_0}{c}\right)^2}. \tag{52}$$

The function $K(\eta)$ has a minimum achieved at $\eta_0 = \sqrt{|\gamma\delta/\alpha\beta|} = 1$ and the minimal value of the Schmidt number (Eq. (16)) is given by

$$K_{\min} = \frac{|\alpha\delta| + |\beta\gamma|}{|D|} = \frac{c}{2v_0}\left(1 + \frac{v_0^2}{c^2}\right). \tag{53}$$

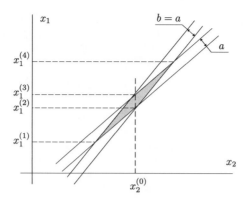

Figure 8. *Zones where the protofunctions (exponential functions in Eq. (49)) are not small, $a = b$ and $\eta = 1$, $|D| \ll 1$.*

The dependence of the Schmidt number on the control parameter is illustrated by two curves of Figure 7. The curve (a) is similar to the wider curve of Figure 2. In the ultra-relativistic case, $v_0 \approx c$, $K_{\min} \approx 1$, which means almost no entanglement. In contrast, in the nonrelativistic case, $v_0 \ll c$, $K_{\min}$ is large. This means that in the case of near-threshold pair creation, entanglement is always large, even when it is minimal, i.e., when $\eta = 1$. This feature distinguishes pair creation from all other above-considered examples of decaying bipartite systems. As mentioned in the introduction, such an unusual behavior of the function $K(\eta)$ originates from a decreasing value of the Jacobian D. This effect arises because as $D \to 0$ the $x - \xi$ transformation of variables (described by Eq. (6)) becomes almost degenerate, i.e., $\xi_2 \to \xi_1$.

The picture in Figure 8 shows in the $\{x_1, x_2\}$ map the zones where the protofunctions arising in this case are not small. The protofunctions are the exponential functions in Eqs. (4) and (49). The solid lines in Figure 8 are their boundaries determined by the condition that expressions in the exponents in Eqs. (4) and (49) are of the order of one. The picture describes the case when $a = b$ or $\eta = 1$.

The shaded area in Figure 8 is the region where protofunction zones overlap, i.e., the region where the total wave function $\Psi(x_1, x_2)$ is not small. The overlapping region is seen to have a rhombic shape. The vertical dashed line indicates some given value $x_2^{(0)}$ of the coordinate x_2, at which the x_1-dependent wave packet and its width are supposed to be measured in the coincidence scheme. Clearly, the result of such measurements will give $\Delta x_1^{(c)} \sim x_1^{(3)} - x_1^{(2)}$. Oppositely, in the case of single-particle measurements all points of the shaded area contribute to formation of the x_1-dependent wave packet and, hence, $\Delta x_1^{(s)} \sim x_1^{(4)} - x_1^{(1)} \gg \Delta x_1^{(c)}$, which means that entanglement is high.

Note that the zone structure of Figure 8 contrasts to that found earlier for all other processes, PDC Burlakov et al. (1997), photoionization-photodissociation (Fedorov et al., 2004) and spontaneous emission (Fedorov et al., 2005). In all these cases, entanglement could occur only if the aspect ratio of the overlapping area in the $\{x_1, x_2\}$ map is large. In the case under consideration the aspect ratio equals one, and nevertheless entanglement can occur because of the approaching degeneracy or a very small angle between the protofunction zone boundaries in the $\{x_1, x_2\}$ map.

If one of the two incident electromagnetic waves producing the pairs is much stronger and has a much lower frequency than the other, the process of pair production can occur via absorption of n photons of the stronger wave plus one photon of the weaker one. This situation corresponds to the famous SLAC experiment by the Stanford-Tennessee-Princeton-Rochester collaboration (Burke et al., 1997). All the ideas about electron-positron entanglement and all the features of the phenomenon described above hold good in the case of multiphoton absorption with only slight modifications to the formulas (Fedorov et al., 2006).

7 Conclusion

Our overview has been intended to unify the analysis of a wide variety of physical processes for which bipartite entanglement is an interesting issue. A number of these processes have close connections to the types of two-particle breakup that one can associate with short-pulse and intense-laser interactions of atoms and molecules. Intense fields are useful in obtaining observable experimental yields, of course. Short pulses are desirable for managing the complete break-up process before extraneous effects can interfere with the existing entanglement being studied.

A rather large number of breakup processes show a surprising similarity in their bipartite wave packets and the available degree of their entanglement. In many cases these can be rather well described by a double-Gaussian wave function. By concentrating attention on the unified description obtained in this Gaussian approximation, we have been able to point out some interesting features such as anomalously small products of conditional coordinate and momentum uncertainties, and relations between wave packet widths of different particles. For example, investigation of one particle's wave packet is seen to be sufficient for describing aspects of the other.

An open question is whether one can use the Gaussian approximation to further extend studies of quantum correlation in other processes such as electron double ionization, sequential vs. non-sequential (see earlier work of Liu, et al. (Liu et al., 1999)), or reverse-breakup processes such as the creation of entanglement in collisions, as recently discussed by Law (Law and Eberly, 2004). Another direction of generalization is related to consideration of more complicated systems of more than two particles, as well as to consideration of mixed states. These are questions at the frontier of contemporary quantum information studies. Furthermore, as mentioned earlier, interaction between particles can be an element responsible for decoherence, and analysis pursued in this direction (see the papers by Dodd and Halliwell (2004) and Dodd (2004)) can be both interesting and important. We hope to contribute to these discussions elsewhere.

Acknowledgement

The author acknowledges collaboration, useful discussions and correspondence with J.H. Eberly, K.W. Chan, C.K. Law, F.E. Kazakov, M.A. Efremov and P.A. Volkov. The work was supported partially by RFBR, Grant 05-02-16469, and by the U.S. Army International Technology Center - Atlantic / CRDF-GAP, Grant RUE1-1616-MO-05. I acknowledge the hospitality and support provided by the Organizing Committee of the 60th Scottish Universities Summer School in Physics at St. Andrews University, Scotland, UK, where this lecture was presented.

Bibliography

Akhiezer, A.I. and Berestetskii, V.B. 1965 *Quantum Electrodynamics* (New York: Interscience Publishers)

Arnaut, H. H. and Barbosa, G. A. (2000), *Phys. Rev. Lett* **85** 286

Bowen, W.P., Schnabel, R., Lam, P.K., and Ralph, T.C. (2004), *Phys. Rev.* A **69**, 012304

Brendel, J., Gisin, N., Tittel, W., and Zbinden, H. (1999), *Phys. Rev. Lett.* **82**, 2594

Burlakov, A.V., Chekhova, M.V., Klyshko, D.N., Kulik, S.P., Penin, A.N., Shih, Y.H., and Strekalov, D.V. (1997), *Phys. Rev.* A **56**, 3214

Burke, D.L. et al. (1997), *Phys. Rev. Lett* **79**, 1626

Chan, K.W., Law, C.K., and Eberly, J.H. (2002), *Phys. Rev. Lett.* **88**, 100402

Chan, K.W., Law, C.K., and Eberly, J.H. (2003), *Phys. Rev.* A **68**, 022110

Chan, K.W., Law, C.K., Fedorov, M.V., and Eberly, J.K. (2004), *J. Mod. Optics* **51**, 1779

D'Angelo, M., Kim, Y-H., Kulik, S.P., and Shih, Y.(2004), *Phys. Rev. Lett.* **92**, 233601

Dodd, P.J. (2004), *Phys. Rev.* A **69**, 052106

Dodd, P.J. and Halliwell, J.J. (2004), *Phys. Rev.* A **69**, 052105

Eberly, J.H. (2006), *Las. Phys.* **16**, # 6

Einstein, A., Podolsky, B., and Rosen, N. (1935), *Phys. Rev.* **47**, 777

Ekert, A. and Knight, P.L. (1995), *Am. J. Phys.* **63**, 415

Fedorov, M.V., Efremov, M.A., Kazakov, A.E., Chan, K.W., Law, C.K., and Eberly, J.H. (2004), *Phys. Rev.* A **69**, 052117

Fedorov, M.V., Efremov, M.A., Kazakov, A.E., Chan, K.W., Law, C.K., and Eberly, J.H. (2005), *Phys. Rev.* A **72**, 032110

Fedorov, M.V., Efremov, M.A., and Volkov, P.A. (2006), *Optics Comm.* to be published

Franke-Arnold, S., Barnett, S.M., Padgett, M.J., and Allen, L. (2002), *Phys. Rev.* A **65**, 033823

Freyberger, M. et al. (1999), *Adv. At. Mol. Opt. Phys.* **41**, 143

Grice, W.P., U'Ren, A.B., and Walmsley, I.A. (2001), *Phys. Rev.* A **64**, 063815

Grobe, R., Rzazewski, K., and Eberly, J.H. (1994), *J. Phys.* B **27**, L503

Howell, J.C., Bennink, R.S., Bentley, S.J., and Boyd, R.W. (2004), *Phys. Rev. Lett.* **92**, 210403

Huang, H. and Eberly, J.H. (1993) *J. Mod. Opt.* **40**, 915

Krekora, P., Su, Q., and Grobe, R. (2005) *J. Mod. Opt.* **52**, 489

Kwiat, P.G., Steinberg, A.M., and Chiao, R.Y. (1993), *Phys. Rev.* A **47** R2472

Law, C.K., Walmsley, I.A., and Eberly, J.H. (2000), *Phys. Rev. Lett.* **84**, 5304

Law, C.K. (2004), *Phys. Rev.* A **70**, 062311

Law, C.K. and Eberly, J.H. (2004), *Phys. Rev. Lett.* **92**, 127903

Liu, W.-C., Eberly, J.H., Haan, S.L., and Grobe, R. (1999), *Phys. Rev. Lett.* **83**, 520

Mair, A., Vaziri, A., Weihs, G., and Zeilinger, A. (2001), *Nature* **412**, 313

Monken, C.H., Souto Ribeiro, P.H., and Padua, S. (1998), *Phys. Rev.* A **57**, 3123

Nielsen, M.A. and Chuang, I.L. 2000 *Quantum Computation and Quantum Information* (Cambridge, England: Cambridge University Press)

Oemrawsingh, S.R., Ma, X., Voigt, D., Aiello, A., Eliel, E.R., 't Hooft, G.W., and Woerdman, J.P. (2005), *Phys. Rev. Lett.* **95**, 240501

Opatrny, T. and Kurizki, G. (2001), *Phys. Rev. Lett.* **86**, 3180

Opatrny, T., Beb, B., and Kurizki, G. (2003), *Phys. Rev. Lett.* **90**, 250404

Ou, Z.Y., Pereira, S.F., Kimble, H.J., and Peng, K.C. (1992), *Phys. Rev. Lett.* **68**, 3663

Plenio, M. and Virmani, S. (2005) arXiv: quant-ph/0504163

Rarity, J.G. and Tapster, P.R. (1990), *Phys. Rev. Lett.* **64**, 2495

Rau, A.V., Dunningham, J.A., and Burnett, K. (2003), *Science* **301**, 1081

Reid, M.D. and Drummond, P.D. (1988), *Phys. Rev. Lett.* **60**, 2731

Reid, M.D. (1989), *Phys. Rev.* A **40**, 913

Rubin, M.H. (1996), *Phys. Rev.* A **54**, 5349

Rzazewski, K. and Zakowicz, W. (1992), *J. Phys.* B **25**, L319

Schrödinger, E. (1935) *Naturwissenschaften* **23** 807, 823, 844; *Proc. Camb. Phil. Soc.* **31**, 555
Walborn, S.P., de Oliveira, A.N., and Monken, C.H. (2003), *Phys. Rev. Lett.* **90**, 143601
Wodkiewicz, K. and Englert, B.-G. (2002), *Phys. Rev.* A **65**, 054303
Wootters, W.M. (1998), *Phys. Rev. Lett.* **80**, 2245
Yonezawa, H., Aoki, T., Furusawa, A. (2004), *Nature* **431**, 430

13

Intense Field Quantum Electrodynamics

Nikolay Narozhny

Moscow Engineering Physics Institute (State University), Russian Federation

1 Introduction

Quantum electrodynamics (QED) is a theory of electron-photon interactions and is nowadays one of the most successful quantum field theories. Though the foundations of QED do not satisfy rigorous requirements of mathematics because to obtain physical results we have to operate with mathematically meaningless infinite quantities, the theory quantitatively (and with very high degree of accuracy) explains the value of electron's anomalous magnetic moment, the Lamb shift, the hyperfine structure of hydrogen energy spectrum, atomic radiative transition line widths, and many other phenomena.

We could say that in principle any electrodynamic effect can be appropriately described in the framework of modern QED, and thus from pragmatic point of view the construction of QED as a physical theory is completed. However, there exists one special class of effects which are rather poorly studied, at least experimentally. These are the effects which occur in the presence of very strong electromagnetic fields when the amplitudes of quantum processes manifest essentially nonlinear dependence on field intensity and QED generally becomes a theory with unstable vacuum. Certainly such effects are very interesting from theoretical point of view. However the latest progress in laser technology promises an exciting breakthrough in laser pulse intensity and thus investigation of Intense Field QED (IFQED) is becoming an urgent physical problem from an experimental point of view also.

1.1 What is "strong field" in QED?

The notion of "strong field" depends on a certain physical situation. For example, in atomic physics the field is compared to the field strength at the first Bohr orbit of a hydrogen atom $E_a \sim e/a_B^2$, where $a_B = \hbar^2/me^2$ is the Bohr radius. In such a field an electron gains across the distance a_B the energy of the order $eE_a a_B \sim I_{ion} \sim e^2/a_B$, which is the energy of ionization of a hydrogen atom.

This field strength is called the characteristic atomic field

$$E_a = \frac{m^2 e^5}{\hbar^4} \sim 5 \cdot 10^9 \mathrm{V/cm}\,, \tag{1}$$

and the fields with the strength $E \gtrsim E_a$ are considered to be strong in atomic physics. The external electromagnetic field with $E = E_a$ makes an atom an unstable system. The corresponding laser intensity, i.e. the intensity of a laser with the peak field equal to E_a is

$$I_a = \frac{c}{8\pi} E_a^2 \sim 3 \cdot 10^{16} \mathrm{W/cm}^2\,. \tag{2}$$

In order to understand how the notion of strong field is introduced with respect to an electron interacting with a laser field, we will consider the motion of a classical electron in the field of a plane monochromatic field. We will describe the field of a linearly polarized plane monochromatic field with 4-vector potential

$$A_\mu = a_\mu \cos\varphi\,, \tag{3}$$

where

$$\varphi = kx = k_0 ct - \mathbf{kr}\,, \quad k_0 = \omega/c;$$
$$k^2 = k_0^2/c^2 - \mathbf{k}^2 = 0\,, \quad ka = k_0 a_0 - \mathbf{ka} = 0\,.$$

Choose such reference frame that

$$k^\mu = \{k_0, \mathbf{k} = k\mathbf{e}_z\}\,, \quad k_0 = k = \omega/c\,,$$

and the gauge

$$a^\mu = \{0, a_x = a, 0, 0\}\,.$$

Then the electric and magnetic fields

$$\mathbf{E} = -\frac{1}{c}\frac{\partial \mathbf{A}}{\partial t} - \nabla A_0\,, \quad \mathbf{H} = [\nabla \mathbf{A}]$$

will look

$$\begin{aligned} E_x = F\sin\omega(t - z/c)\,, \quad E_y = E_z = 0;\\ H_y = F\sin\omega(t - z/c)\,, \quad H_x = H_z = 0\,, \end{aligned} \tag{4}$$

with $F = a\omega/c$.

Let us write down the equations of motion for an electron in a plane wave field in nonrelativistic (or dipole) approximation $v \ll c$. In this case we can neglect the Lorentz force in comparison with the electric force since

$$f_H \sim eFv/c \sim f_E v/c \ll f_E\,,$$

and omit dependence of the field on z since variation of the field phase due to the electron's displacement for the wave period $T = 2\pi/\omega$ is small

$$\omega\Delta z(T)/c \sim \omega v T/c \sim 2\pi v/c \ll 2\pi\,.$$

Then the equations of motion read

$$m\ddot{\mathbf{r}} = eF\mathbf{e}_x \sin\omega t\,.$$

The solution for these equations can be obtained easily

$$\dot{\mathbf{r}}(t) = -\frac{eF}{m\omega}\mathbf{e}_x \cos\omega t + \mathbf{v}_{in}\,,$$

$$\mathbf{r}(t) = -\frac{eF}{m\omega^2}\mathbf{e}_x \sin \omega t + \mathbf{v}_{in}t + \mathbf{r}_{in} \,,$$

where $\mathbf{r}_{in}$ and $\mathbf{v}_{in}$ are the initial position and initial velocity of the electron respectively.

So, the trajectory of a nonrelativistic electron in the field of a plane monochromatic field is a straight line with fast oscillations around it. The amplitudes of oscillations of the electron's coordinate and velocity can be represented as $r_0 = \eta\frac{\lambda}{2\pi}$, $v_0 = \eta c$, where

$$\eta = \frac{eF}{m\omega c} \tag{5}$$

is the field dimensionless intensity parameter. The expressions for the amplitudes are valid only for small η of course. But it is clear that at arbitrary η we will have

$$v_0/c = f(\eta)\,,$$

with function $f(\eta)$ satisfying the restriction $f(\eta) \approx \eta$, $\eta \ll 1$. Since we do not have any reasons to expect that function $f(\eta)$ has any singularities at $\eta \gtrsim 1$, we may conclude that at $\eta \gtrsim 1$ $v_0/c \sim 1$. Thus the fields of intensities $\eta \gtrsim 1$ are called the fields of relativistic intensities. The field strength corresponding to $\eta = 1$ is given by the expression

$$F_{rel} = \frac{m\omega c}{e}\,.$$

In a field of this strength an electron gains the energy equal to mc^2 at the distance equal to the wavelength of the field

$$eF_{rel}\lambda \sim mc^2\,.$$

The relation between the dimensionless parameter of intensity η and laser intensity I reads

$$\eta^2 = 3.7 \times 10^{-19}I\lambda^2$$

where I is in $\mathrm{W/cm}^2$ and λ in μm. Thus for a laser beam with $\lambda = 1\mu m$ $F_{rel} \sim 3 \cdot 10^{10}\mathrm{V/cm}$ and $I_{rel} \sim 3 \cdot 10^{18}\mathrm{W/cm}^2$.

In QED an external electromagnetic field is considered strong if it essentially changes properties of the electron-positron vacuum state. Vacuum in quantum field theory is not just an empty space but is a medium with specific properties. It can be seen as a sea of electron-positron pairs which keep being spontaneously created and annihilated. However the characteristic size for this process is smaller than the Compton length $l_C = \hbar/mc$ and therefore the particles of the pairs cannot be localized separately without destroying of the vacuum state. Therefore these particles are not observable and are called virtual.

The situation becomes different in the presence of an external electromagnetic field which can deform the electron-positron loops and orient them along a preferable direction as it happens with molecules in a dielectric. As a result, polarization of vacuum arises and one can introduce, in particular, the electric and magnetic susceptibility of the vacuum. This effect was first described by W. Heisenberg and H. Euler (1936) who have derived the vacuum polarization correction to the Lagrangian of the electromagnetic field, see also Schwinger (1951). If the electric field strength is such that it can perform the work of the order $2mc^2$ at the Compton length,

$$eF_S l_C = mc^2$$

the electron and the positron of a vacuum loop will be separated by a distance of the order of l_C and a real pair will be created. The field strength F_S

$$F_S = \frac{m^2 c^3}{e\hbar} = 1.32 \cdot 10^{16}\mathrm{V/cm}\,, \tag{6}$$

which was introduced by F. Sauter in connection with the Klein paradox (Sauter, 1931), is considered to be the characteristic field strength in QED. The corresponding laser intensity is

$$I_S = \frac{c}{4\pi} F_S^2 = 0.5 \cdot 10^{30} \text{W/cm}^2 \,. \tag{7}$$

1.2 How far we are from the QED "strong field"?

Intense field QED becomes a nonlinear theory when the field strength reaches the values of the order F_S. Nonlinear in the same sense as nonlinear optics: physical quantities reveal nonlinear dependence on the field strength. Two groups of nonlinear QED effects can be distinguished: vacuum polarization effects, and effects initiated by ultrarelativistic particles moving in a strong external field. The first group includes such effects as pair creation by a strong field in vacuum, nonlinear optical effects in vacuum (birefringence, dichroism, Cerenkov radiation), etc, while the second group, nonlinear Compton effect, pair production by a high energy photon, etc.

Vacuum polarization effects can be observed only if the field strength is of the order F_S in the laboratory frame. For observation of the effects initiated by relativistic particles it is enough if the field is of order F_S in the rest frame of the particle. The latter means that such effects are regulated by the parameter

$$\Upsilon = \frac{F_P}{F_S} \,, \tag{8}$$

which should be of order 1. Here F_P is the field strength in the rest frame of the particle. If one adopts head-on geometry for collision of an electron with a laser pulse, the relation between the laser field strengths in the rest and laboratory frames reads

$$F_P = F_L \sqrt{\frac{1 + v/c}{1 - v/c}} \,,$$

where $v = pc/\varepsilon$, p, ε are the electron velocity, momentum and energy. For ultrarelativistic particles $1 - v/c \ll 1$, or $\gamma = 1/\sqrt{1 - v^2/c^2} = \varepsilon/mc^2 \gg 1$, and we have $F_P \approx 2F_L\gamma$. Hence nonlinear behavior for effects from the second group can be seen at laboratory field strengths $F_L \ll F_S$.

The recent development of laser technology based on chirped pulse amplification (CPA) method have lead to increasing of laser intensity up to the value $I = 10^{22} \text{W/cm}^2$ (Tajima and Mourou, 2002). The corresponding peak laser field strength is of order $F \approx 10^{12} \text{V/cm}$. It is 4 orders of magnitude less the characteristic QED value F_S. Therefore vacuum polarization effects in such fields are very small and cannot be detected. However, to make parameter Υ of order 1 one need relativistic electrons with energy $\varepsilon/mc^2 \gtrsim 10^4$. Such electron energies are attainable at modern accelerators. For example, 50GeV electrons ($\gamma \approx 10^5$) are available at Stanford Linear Accelerator (SLAC). Therefore the nonlinear QED effects initiated by ultrarelativistic particles colliding with intense laser pulses can be observed with modern facilities. We will discuss in these lectures the first such experiments.

The situation with vacuum polarization effects also is not hopeless. Recently several papers (Shen and Yu, 2002; Bulanov et al, 2004) appeared where different methods to achieve intensities of the order $I = 10^{24-25} \text{W/cm}^2$ and even higher were suggested. Moreover, a path to reach an extremely high-intensity level $I = 10^{26-28} \text{W/cm}^2$ already in the coming decade has been suggested in Ref. (Tajima and Mourou, 2002). The field strength in such lasers is very close to F_S and thus investigation of vacuum polarization effects in intense laser fields is becoming an urgent physical problem from both theoretical and experimental points of view.

2 Quantum processes in a plane wave field

2.1 Furry picture

Evolution of a quantum system of interacting particles is usually described in quantum theory using either Heisenberg, or Shrödinger pictures. As it is known from the university course of quantum mechanics, state vector in Heisenberg picture does not depend on time

$$i\frac{\partial \Phi_H}{\partial t} = 0\,,$$

while time dependence of operators $\widehat{R}$ is given by the following equation of motion

$$\frac{d\widehat{R}_H}{dt} = i\big[\widehat{H}, \widehat{R}_H\big]\,.$$

Here $\widehat{H}$ is Hamiltonian for the system

$$\widehat{H} = \widehat{H}_0 + \widehat{V}_{int}\,.$$

$\widehat{H}_0$ is the Hamiltonian for free particles, and $\widehat{V}_{int}$ is responsible for interaction of the particles. For our purpose it is convenient to separate $\widehat{V}_{int}$ into two parts

$$\widehat{V}_{int} = \widehat{V}_{ext} + \widehat{V}_r\,,$$

such that $\widehat{V}_r$ describes interaction between particles and $\widehat{V}_{ext}$ describes interaction of particles with an external field.

Transition from the Heisenberg to Shrödinger picture is accomplished with the following unitary transformation

$$\Phi_S = U_S^{-1}\Phi_H\,, \quad \widehat{R}_S = U_S^{-1}\widehat{R}_H U_S\,, \quad U_S = e^{-i\widehat{H}t}\,.$$

It is easy to verify that in the Shrödinger picture operators are constant, while the state vector obeys the Shrödinger equation

$$\frac{d\widehat{R}_S}{dt} = 0\,, \quad i\frac{\partial \Phi_S}{\partial t} = \widehat{H}\Phi_S\,.$$

The probability amplitudes for quantum transitions are determined by matrix elements of S-matrix. They can be calculated in the framework of time-dependent perturbation theory. The easiest way to derive this perturbation expansion is to use the Dirac (or interaction) picture

$$\Phi_I = U_I^{-1}\Phi_H\,, \quad \widehat{R}_I = U_I^{-1}\widehat{R}_H U_I \quad U_I = e^{-i\widehat{H}_0 t}\,.$$

Time evolution of the state vector in interaction picture is determined by interaction Hamiltonian, while operators are ruled by equations for free particles.

$$i\frac{\partial \Phi_I}{\partial t} = \widehat{V}_{int}\Phi_I\,, \quad \frac{d\widehat{R}_I}{dt} = i\big[\widehat{H}_0, \widehat{R}_I\big]\,.$$

S-matrix is given in the interaction picture by the expression

$$S = T\exp\left(-i\int\limits_{-\infty}^{\infty} dt\widehat{V}(t)\right),$$

and matrix elements of S-matrix are calculated in the basis of eigenfunctions of the free Hamiltonian. Such an approach meets the requirements of experiment when generally an initial state at $t \to -\infty$ is prepared as a free particle state and then the final state at $t \to \infty$ is also analyzed in terms of free particles.

However, the described picture is not appropriate either if the initial and final states are bound states, or a quantum process is taking place in the presence of a strong external field which is not switching off at $t \to \pm\infty$. In these cases it is more convenient to use the Furry picture which is determined by the following transformation

$$\Phi_F = U_F^{-1}\Phi_H \,, \quad \widehat{R}_F = U_F^{-1}\widehat{R}_H U_F \quad U_F = e^{-i(\widehat{H}_0 + \widehat{V}_{ext})t} \,.$$

Then time evolution of the state vector is determined only by that part of the interaction Hamiltonian which is responsible for interaction between particles, and the operators obey equations of motion in an external field

$$i\frac{\partial \Phi_F}{\partial t} = \widehat{V}_r \Phi_F \,, \quad \frac{d\widehat{R}_F}{dt} = i\big[\widehat{H}_0 + \widehat{V}_{ext}, \widehat{R}_F\big] \,.$$

S-matrix in the Furry picture looks like

$$S = T \exp\left(-i \int_{-\infty}^{\infty} dt \widehat{V}_r(t)\right) ,$$

and matrix elements of the S-matrix are calculated in the basis of eigenfunctions of the Hamiltonian for a particle in an external field. One can construct perturbation expansion for the S-matrix in the Furry picture and develop the Feynman diagrammatic technique. It must be clear that the Feynman rules in the Furry picture differ from the analogous rules for the interaction picture by two points. All external electron lines in the Furry picture correspond to exact solutions of Dirac equation in an external field instead of plane waves for the interaction picture, and internal lines correspond to the exact Green function of an electron in the external field instead of the free electron propagator.

Certainly, to use perturbation theory in the Furry picture we must have exact solutions for Dirac equation in an external field. Unfortunately, exact solutions exist for only a few field configurations. The most important of these are the following: Coulomb field, constant electromagnetic field and the field of a plane electromagnetic wave. For laser physics the latter field is the most important, and we shall consider it in the next section.

2.2   Volkov solutions for Dirac equation for an electron in a laser field

We will write Dirac equation for an electron in an external electromagnetic field in the form[1]

$$\left[\gamma^\mu\left(i\frac{\partial}{\partial x^\mu} - eA_\mu(x)\right) - m\right]\psi = 0 . \tag{9}$$

Here $A_\mu(x)$ is 4-vector potential of the electromagnetic field, γ_μ are Dirac matrices obeying anticommutation relations

$$\gamma^\mu\gamma^\nu + \gamma^\nu\gamma^\mu = 2g^{\mu\nu} \,,$$

with metric tensor elements

$$g^{00} = 1, g^{11} = g^{22} = g^{33} = -1 \,.$$

---

[1] From now on we will use natural system of units $\hbar = c = 1$.

Most of the authors working with QED processes in intense laser fields use the model of a plane monochromatic wave to describe the electromagnetic field of a laser beam. Certainly, the evident advantage of this model is existence of exact solutions to Dirac equation for an electron in a plane wave electromagnetic field, the famous Volkov functions (Volkov, 1935). Volkov solutions, having simple quasiclassical form, allow one to obtain analytical, though not always simple, formulas for probabilities of quantum processes in a plane wave field. The plane wave approach can be successfully used for treating the processes in a focused laser beam also. It is clear that, if the formation length and time for a process are much less than the spatial and temporal inhomogeneities of the laser pulse, one may calculate the probability of the process at arbitrary point of the beam using the standard plane wave procedure and then obtain the total probability as the integral over the volume and duration of the pulse.

For a plane wave potential $A_\mu(x) = A_\mu(\varphi)$, $\varphi = \omega t - \mathbf{k}\mathbf{r}$ the Volkov solution read

$$\psi_p(x) = \left[1 + e\frac{(\gamma k)(\gamma A(\varphi))}{2pk}\right]e^{iS}u_p\,, \tag{10}$$

where

$$S = -px - \int\limits_0^\varphi d\varphi'\left[\frac{epA(\varphi')}{pk} - \frac{e^2A^2(\varphi')}{2pk}\right]$$

is a classical action of an electron in the plane wave field, and u_p is a free Dirac spinor satisfying equation $(\gamma p - m)u_p = 0$. Quantum numbers p^μ, $p^2 = m^2$, have the meaning of 4-momentum of the electron when the field is switched off.

Consider now a monochromatic plane wave field assuming circular polarization for definiteness. The 4-potential for such field reads

$$A(\varphi) = a_1\cos\varphi + a_2\sin\varphi\,, \tag{11}$$

$$ka_1 = ka_2 = a_1a_2 = 0\,, \quad a_1^2 = a_2^2\,.$$

After substitution of (11) into Eq. (10) we obtain

$$\psi_q(x) = \left[1 + e\frac{(\gamma k)(\gamma A(\varphi))}{2qk}\right]\exp\left(-iqx - i\frac{eqa_1}{qk}\sin\varphi + i\frac{eqa_2}{qk}\cos\varphi\right)u_p\,. \tag{12}$$

Quantum numbers q_μ which label the solutions (12) are defined as follows

$$q^\mu = p^\mu - \frac{e^2<A^2>}{pk}k^\mu = p^\mu + \frac{m^2\eta^2}{pk}k^\mu\,, \tag{13}$$

and

$$q^2 = m_*^2 = m^2(1 + \eta^2)\,. \tag{14}$$

It can be easily seen that transformation of solutions (12) under space-time translations

$$x^\mu \to x^\mu + \Delta^\mu\,, \quad k\Delta = 2\pi n\,, \quad n = 0, \pm 1, \pm 2, ...,$$

are given by the formula

$$\psi_q \to \psi_q e^{-iq\Delta}\,.$$

Therefore the 4-vector q_μ has the meaning of 4-quasimomentum, and m_* can be called an effective mass of an electron in a plane wave field.

It is seen from Eq. (12) that the quasimomentum q_μ completely determines a quantum state of an electron in the field of a monochromatic plane wave. Therefore, for better understanding of the nature of this quantity we will give the classical analysis of behavior of an electron in such a field.

We will write classical equations of motion of an electron in an electromagnetic field in covariant form (Landau and Lifshits, 1975).

$$\frac{d\pi^\mu}{d\tau} = \frac{e}{mc}F^\mu{}_\nu\pi^\nu\,, \quad \frac{dx^\mu}{d\tau} = \frac{\pi^\mu}{m}\,; \quad \pi^2 = m^2\,, \tag{15}$$

where $F^\mu{}_\nu$ is the tensor of the electromagnetic field, π_μ is kinetic 4-momentum of the electron and τ is its proper time. Since the electromagnetic tensor is transverse $k_\mu F^\mu{}_\nu = 0$, we easily get

$$k\pi \equiv kp = const\,,$$

and hence

$$\frac{d\varphi}{d\tau} = \frac{kp}{m} = const\,.$$

The latter means that the electron proper time up to a factor coincides with the plane wave phase φ and, since in our case $F^\mu{}_\nu$ is a function of only one variable φ, the equations of motion can be easily solved. In particular, for π_μ we obtain in the case of a circularly polarized wave (11)

$$\pi^\mu = p^\mu - eA^\mu + k^\mu\left(\frac{epA}{pk} - \frac{e^2A^2}{2pk}\right) = q^\mu - eA^\mu + \frac{eqA}{qk}k^\mu\,.$$

When averaged over phase, the potential vanishes $< A^\mu > = 0$, we get

$$< \pi^\mu > = q^\mu\,. \tag{16}$$

This means that the 4-quasimomentum of an electron has the meaning of the kinetic 4-momentum averaged over φ. Exactly as it was in nonrelativistic case (compare Section 1.1), the trajectory of an electron in the field of a plane monochromatic field is a straight line with fast oscillations around it.

We will proceed further in the reference frame in which the electron is at rest on average

$$\mathbf{q} = 0\,, \quad qk = m_*\omega\,, \quad qA = 0\,.$$

Let the wave propagate along the z axis, so that

$$k^\mu = (\omega, 0, 0, k_z = \omega)\,.$$

Then, in our reference frame

$$\pi_z = 0\,, \quad \boldsymbol{\pi}_\perp = -e\mathbf{A}\,.$$

The first equality means that the electron is moving in the plane $z = z_0 = const$, which we set equal to zero. Then $\varphi = \omega t$ and

$$\boldsymbol{\pi}_\perp = m\frac{d\mathbf{r}_\perp}{d\tau} = m_*\mathbf{v}_\perp = -e\mathbf{A}\,, \quad \mathbf{v}_\perp = \frac{d\mathbf{r}_\perp}{dt}\,.$$

From here we have

$$v_\perp = \frac{\eta}{\sqrt{1+\eta^2}}\,, \quad r = \frac{\eta}{\sqrt{1+\eta^2}}\frac{\lambda}{2\pi} \le \frac{\lambda}{2\pi}\,. \tag{17}$$

Note that when $\eta \ll 1$ we return to the formulas of Section 1.1.

The effective mass m_* is defined as the zero component of the quasimomentum q_μ in the reference frame where the electron is at rest at average. This means that in this reference frame

$$m_* = < \pi_0 >\,,$$

and the physical meaning of m_* is the kinetic energy of an electron in its proper reference frame $\mathbf{q} = \mathbf{0}$ averaged over fast oscillations at the wave frequency. It agrees with the equation

$$m_* = m\gamma_\perp = \frac{m}{\sqrt{1 - v_\perp^2}} = m\sqrt{1 + \eta^2}\,.$$

which directly follows from Eq. (17). In quantum theory fast electron oscillations, and hence the effective mass m_*, arise due to continual absorption and re-emission of laser photons by the electron.

It follows from Eq. (17) that the amplitude of electron oscillations is always less than the wavelength λ. Therefore the incident light interacting with electron cannot resolve the details of oscillations and the interaction process can be affected by the electron only in an average way. This explains why the solution (12) is labelled by quantum numbers q_μ.

3 Simplest processes (theory and experiment)

We will consider in this section the simplest processes which can be initiated by high-energy particles colliding with intense laser pulses, namely, nonlinear Compton scattering and pair photoproduction. The laser field will be considered as a plane monochromatic field and the matrix elements will be calculated in the Furry picture using the Volkov solutions (10),(12).

3.1 Nonlinear Compton scattering

The Feynman diagram for nonlinear Compton scattering is given in Figure 1.

Figure 1. *Feynman diagram for nonlinear Compton scattering*

The corresponding matrix element is

$$S_{i\to f} = -ie \int \bar{\psi}_{q'}(\gamma e'^*)\psi_q \frac{e^{ik'x}}{\sqrt{2k_0'}}dV\,dt\,,\tag{18}$$

where $\bar{\psi}_{q'} = \psi_{q'}^\dagger\gamma_0$ and k' is the 4-momentum of the emitted photon and e'_μ is its polarization vector. After substitution of the Volkov functions (12) we can expand the integrand in (18) in a Fourier series and perform an integration over the space coordinates and time. After this the matrix element acquires the form

$$S_{i\to f} = -ie(2\pi)^4 \sum_{s=1}^{\infty} M_{i\to f}^{(s)}\delta^{(4)}(sk + q - q' - k')\,,\tag{19}$$

of a sum of partial matrix elements each with its own conservation law

$$q + sk = q' + k'\,.\tag{20}$$

This allows one to interpret the quantities $M_{i \to f}^{(s)}$ as transition amplitudes for channels of the process of Compton scattering with absorption of s photons from the plane wave.

Using the conservation law (20) we can easily obtain the following formula for the frequency of the scattered photon in the average rest frame of the initial electron $\mathbf{q} = 0$, $q_0 = m_*$

$$\omega' = \frac{s\omega}{1 + (s\omega/m_*)(1 - \cos\theta)}.$$

It differs from the well known formula for linear Compton effect by the presence of the number of absorbed wave photons s and the effective mass m_* instead of the bare mass m.

We will also present the formula for the frequency of scattered photon as seen in laboratory frame

$$\omega' = \frac{4s\gamma^2\omega}{1 + 2\gamma^2(1 - \cos\theta) + (2s\gamma\omega/m + \eta^2/2)(1 + \cos\theta)}$$

since it allows us to derive the expression for the maximum energy of the backscattered photon

$$\omega'_{max} = \frac{4s\gamma^2\omega}{1 + \eta^2 + 4s\gamma\omega/m}, \tag{21}$$

and the minimum energy of the recoil electron ("Compton edge")

$$\varepsilon'_{min}(s,\eta) = \frac{\varepsilon_0}{1 + \frac{2s\varepsilon\omega}{m_*^2 c^4}(1 + \cos\theta)}. \tag{22}$$

The total probability of photon emission for the case of circular polarization of the laser field is (Narozhny et al., 1964)

$$W = \frac{e^2 m^2 n_e}{q_0} \sum_{s=1}^{\infty} \int_0^{u_s} \frac{du}{(1+u)^2} \left\{ -4J_s^2(z) + +\eta^2\left(2 + \frac{u^2}{1+u}\right)\left(J_{s-1}^2(z) + J_{s+1}^2(z) - 2J_s^2(z)\right) \right\},$$

where

$$u_s = \frac{2s\Upsilon_e}{\eta(1+\eta^2)}, \quad z = \frac{2\eta\sqrt{u(u_s - u)}}{u_1\sqrt{1+\eta^2}}. \tag{23}$$

Again it is a sum over partial probabilities corresponding to the absorption of a fixed number s of laser wave photons

$$W = W(\eta, \Upsilon_e) = \sum_{s=1}^{\infty} W_s(\eta, \Upsilon_e).$$

We see from Eq. (23) that the total probability W is determined by two parameters

$$\eta = \frac{ea}{m}, \quad \text{and} \quad \Upsilon_e = \frac{kq}{m^2}\eta,$$

compared with Eqs. (5) and (8). Let us consider the behavior of the probability at small η. Using the expansion formulas for Bessel functions we can present the partial probability W_1 in the form

$$W_1 = W_1^{(1)} + W_1^{(3)} + W_1^{(5)} + \cdots.$$

We shall not give here the explicit expressions for $W_1^{(2k+1)}$ for the sake of compactness. We will note only that $W_1^{(1)}$ coincides with the famous Klein-Nishina formula for the probability of (linear) Compton scattering and there exists the following relation

$$W_1^{(2n+1)}/W_1^{(2n-1)} \sim \eta^2.$$

Figure 2. *Feynman diagrams for the amplitude $M^{(1)}_{i \to f}$ of nonlinear Compton scattering*

These results become clearer if we take into account the partial amplitude $M^{(1)}_{i \to f}$ at small η which can be represented by the set of Feynman diagrams from Figure 2, where fermion lines, in contrast with Figure 1, correspond not to Volkov functions but to Dirac plane waves, dashed photon lines correspond to wave photons while the wiggly line to the backscattered photon. Certainly for every fixed k diagram which differ from those presented in Figure 2 by all possible permutations of photon lines should be taken into account. The amplitude $M^{(2)}_{i \to f}$ can be represented as a set of diagrams from Figure 3.

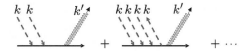

Figure 3. *Feynman diagrams for the amplitude $M^{(2)}_{i \to f}$ of nonlinear Compton scattering*

The parameter η does not contain the Plank constant $\hbar$ and is equal to ratio of the work done by the field at the wave length λ to electron rest energy mc^2. It can be also represented as the ratio of the work done by the field at the Compton length $l_c = \hbar/mc$ to the energy of a wave photon $\hbar\omega$. Therefore when η is small, the probability of a process with absorption of minimum quanta, say one, is dominant.

Corrections caused by absorption of additional quantum are small $\sim \eta^2$, and thus at $\eta \ll 1$ the probability corresponds to the probability calculated in the framework of the perturbation theory when the wave plays the role of a single photon. If $\eta \gtrsim 1$, the probabilities of absorption of different number of quanta become comparable and the processes become multiphoton. In this case the probability amplitude depends on the field strength nonlinearly. Therefore η is the classical parameter of nonlinearity.

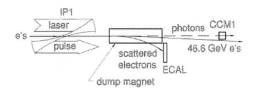

Figure 4. *Layout of the experiment on nonlinear Compton scattering (Fig. 2 from (Bula et al., 1996)).*

Parameter Υ_e contains $\hbar$, and is equal to the work done by the field at the Compton length in the electron rest frame related to mc^2. In other words, Υ_e is equal to E/F_S in the electron rest frame. Parameter Υ_e is responsible for nonlinear quantum effects. In particular, at Υ_e close to 1 the recoil of scattered electrons becomes significant. If $\Upsilon_e \ll 1$, the process of photon emission by an electron looks like classical radiation of an electron moving along a fixed trajectory independently of the value of parameter η.

Photon emission by an electron colliding with a high intensity laser pulse has been observed in an experiment carried out in the Final Focus test Beam at SLAC (Bula et al., 1996). In this experiment 46.6-GeV electrons collided with focused pulses from an intense Nd:glass laser with wavelength $\lambda = 1054$ and 527 nm, see layout of the experiment in Figure 4.

The peak focused laser intensity was $I \approx 10^{18} \mathrm{W/cm}^2$ at $\lambda = 1054$ nm. The corresponding value of parameter η was $\eta \approx 0.6$. For a 46.6 GeV electron colliding head-on with such a laser pulse, $\Upsilon_e \approx 0.13$. This meant that, in the rest frame of the electron beam, the energy of incident photon was 0.21 MeV (0.42 MeV at $\lambda = 527$ nm). So, the recoil of scattered electrons was significant and the process of photon emission could be considered as Compton scattering. Due to the high enough value of parameter η the process was multiphoton, so that the effect observed in the experiment (Bula et al., 1996) could be qualified as *nonlinear* Compton scattering.

There are at least two qualitative features which permit to distinguish nonlinear Compton scattering from the ordinary (linear) one experimentally.

First, the spectrum of scattered electrons in the presence of a strong laser field contains different harmonics each corresponding to absorption of definite number of laser photons.

Second, in a strong field the Compton edge (22) is shifted. When an electron of initial energy ε_0 is scattered by a laser pulse with intensity parameter η the spectrum of scattered electrons for all harmonics with $n > 1$ extends below the Compton edge $\varepsilon'_{min}(1, \eta)$ for the linear version of the effect. Besides, the $\varepsilon'_{min}(1, \eta)$ itself is shifted in comparison with the Compton edge for a very weak $\eta \ll 1$ field, $\varepsilon'_{min}(1, 0)$. This arises due to the presence of effective mass m_* in $\varepsilon'_{min}(1, \eta)$ instead of bare mass m in the expression for $\varepsilon'_{min}(1, 0)$.

It is worth noting that electrons with an energy below $\varepsilon'_{min}(1, \eta)$ also appear due to multiple Compton scattering when the electron independently scatters with absorption of n laser photons twice or more times as it traverses the laser focus. However this effect is physically distinct from nonlinear Compton scattering and was carefully taken into account while processing experimental data (Bula et al., 1996).

The normalized energy spectrum, $(1/N)(dN/dE')$, of scattered electrons was measured in the experiment of (Bula et al., 1996). Harmonics corresponding to absorption of up to four laser photons in a single scattering event were observed. The spectra agree within experimental uncertainty with the theoretical calculations (Narozhny et al., 1964) at two different laser wavelengths and over a wide range of laser pulse energies. This experiment was the first observation of a nonlinear IFQED effect. The field strength in the rest frame of relativistic electrons was only one order of magnitude lower than the "critical" for QED value E_S.

The experiment (Bula et al., 1996) was also an important step towards an experiment on pair production in collision of high energy photons with intense focused laser pulse performed by the same team of researchers (Burke et al., 1996).

3.2   Photoproduction of a $e^- e^+$-pairs

The Feynman diagram for pair production by a photon in a plane wave field is presented in Figure 5. The matrix element for this process

$$S_{i \to f} = -ie \int \bar{\psi}_{q'} (\gamma e') \psi_{-q} \frac{e^{-ik'x}}{\sqrt{2k'_0}} dV \, dt \,,$$

can again be presented as a sum of partial matrix elements

$$S_{i \to f} = -ie(2\pi)^4 \sum_{s=s_0}^{\infty} M_{i \to f}^{(s)} \delta^{(4)}(sk + k' - q - q'),$$

each with the conservation law

$$k' + sk = q + q'. \tag{24}$$

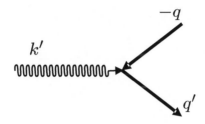

Figure 5. *Feynman diagram for $e^- e^+$-pair production by a photon.*

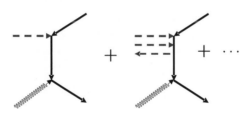

Figure 6. *Feynman diagrams for the amplitude $M_{i \to f}^{(1)}$ of pair photoproduction.*

In contrast with Compton scattering, there is a threshold for the process of pair photoproduction. It can be easily found from (24) that the process is possible only if the number of absorbed wave photons satisfies the relation

$$s \geq s_0 = 2\eta m_*^2 / m^2 \Upsilon_\gamma, \tag{25}$$

where

$$\Upsilon_\gamma = \eta \frac{kk'}{m^2}.$$

Note that for head-on collision of a photon with the laser wave

$$\Upsilon_\gamma = \frac{2\varepsilon_\gamma}{mc^2} \frac{F_L}{F_S}.$$

The probability of pair creation in a circularly polarized plane monochromatic electromagnetic wave was calculated (Narozhny et al., 1964) as

$$W = \frac{e^2 m^2 n_\gamma}{4\omega'} \sum_{s>s_0}^{\infty} \int_1^{u_s} \frac{du}{u\sqrt{u(u-1)}} \left\{ 2J_s^2(z) + +\eta^2(2u-1)\left(J_{s-1}^2(z) + J_{s+1}^2(z) - 2J_s^2(z)\right) \right\}, \quad (26)$$

where

$$u_s = s/s_0, \quad z = \frac{2\eta\sqrt{u(u_s - u)}}{u_1\sqrt{1+\eta^2}}.$$

The partial probabilities $W_s(\eta, \Upsilon_\gamma)$

$$W = W(\eta, \Upsilon_\gamma) = \sum_{s>s_0}^{\infty} W_s(\eta, \Upsilon_\gamma),$$

behave, at small η, as

$$W_s(\eta, \Upsilon_\gamma) \propto \eta^{2s}. \quad (27)$$

If Υ_γ is not too small, so that $\eta/\Upsilon_\gamma < 1$ and $s_0 < 1$, the amplitude $M_{i \to f}^{(1)}$ can be represented by the set of Feynman diagrams depicted in Figure 6.

The first diagram of Figure 6 describes the process of pair production by two colliding photons and $W_1^{(1)}$ is accordingly given by famous Breit-Wheeler formula. When Υ_γ is the smallest parameter in the problem, so that $\eta/\Upsilon_\gamma \gg 1$, $s_0 \gg 1$, the probability of pair production W is exponentially small. This is because Υ_γ is the quantum parameter of nonlinearity and hence the limit $\Upsilon_\gamma \to 0$ means transition to the classical limit which does not exist for the effect of pair production.

For observation of the effect of pair creation (Burke et al., 1996) basically the same experimental setup (see Figure 4) as in experiment of Bula et al. was used. The 46.6 GeV electron beam was colliding with terawatt pulses from a Nd:glass laser producing high energy photons due to nonlinear Compton scattering. The backscattered photons then propagated some distance in the same laser focus serving as probe photons for the multiphoton Breit-Wheeler process.

$$s\omega + e^- \longrightarrow e^- + \gamma \longrightarrow s\omega + \gamma \to e^- e^+.$$

The peak focused laser intensity for linearly polarized green (527 nm) pulses was $I \approx 1.3 \times 10^{18} \text{W/cm}^2$. The corresponding value of η was $\eta = 0.36$. The maximum backscattered photon energy from a 527 nm laser was 29.2 GeV. Then for parameter Υ_γ we have $\Upsilon_\gamma \approx 0.52\eta \approx 0.19$, and for the threshold s_0 we obtain $s_0 = 4.3$. Hence, five photons were required to produce a pair near threshold and the effect was quite observable.

In the experiment of Burke et al. data was collected at various laser intensities. Dependence of the positron rate per laser shot R_{e^+} on the laser field-strength parameter η was obtained.

Power law $R_{e^+} \propto \eta^{2s}$ with $s = 5.1 \pm 0.2$ fitted to the data (see Figure 7) in full agreement with the fact that the rate of multiphoton reactions (27) involving s laser photons is proportional to η^{2s} for $\eta^2 \ll 1$, and with the kinematic requirement derived above. The momentum spectrum of positrons was also measured. The data obtained is in reasonable agreement with the spectrum predicted by calculations. The experiment of Burke et al. was not only another verification of IFQED but the first laboratory observation of the effect of pair creation by real photons in vacuum.

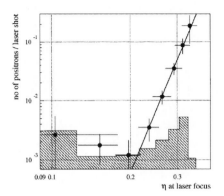

Figure 7. *Dependence of positron rate per laser shot on the field intensity parameter η (Fig. 4 from (Burke et al., 1997)).*

4 A short focused laser pulse

The results of experiments discussed in the preceding section confirm the validity of IFQED formalism and show that the observed rates for the multiphoton reactions are in agreement with the predicted values. This approves the approach of calculating effects of particles-laser interaction based on the plane wave model for the laser field. However, effects of interaction of charged particles with intense laser field exist which cannot be treated in the framework of the plane wave approach. Ponderomotive effect, which is indebted exclusively to the inhomogeneity of the electromagnetic field, serves as one of the examples. Another example is the effect of electron-positron pair creation by a strong electromagnetic field in vacuum. It is well known (Schwinger, 1951) that a plane electromagnetic wave does not create pairs. We will see that in both cases geometry of the focused laser pulse is extremely important.

It is well known that focusing of a plane monochromatic wave leads to the appearance of nonvanishing longitudinal components of the electric and magnetic fields in the focal region. Moreover, it affects also their transverse components. In particular, a linearly polarized (along the x axis) plane monochromatic wave propagating along the z axis is converted by an aplantic system into a converging spherical wave with nonvanishing y and x components of electric and magnetic fields, respectively.

As a result, the formulas describing the focused pulse usually look rather cumbersome and it is not easy to work with them. However, we will show in this section that it is possible to get a particular solution (Narozhny and Fofanov, 2000) of Maxwell equations which, being relatively simple, reproduces all peculiarities of a focused laser pulse.

We will begin with a superposition of monochromatic plane waves of frequency ω with wave vectors $\mathbf{k}$ ($\omega = |\mathbf{k}|$) directed so, that the angle between $\mathbf{k}$ and some preferable direction, say z axis, does not exceed the quantity Δ. The vector potential for such a field can be represented in the form

$$\mathbf{A}\left(\mathbf{r}, t\right) = \int d^3 k \mathbf{A}\left(\mathbf{k}\right) \exp\left[i\left(\mathbf{kr} - \omega t\right)\right], \qquad A_0 = 0, \tag{28}$$

with

$$\mathbf{A}\left(\mathbf{k}\right) = \frac{1}{\pi \Delta^2 \omega^2} \mathbf{a}\left(\vartheta, \alpha\right) \theta(\Delta^2 - \vartheta^2) \delta\left(\omega - \omega_0\right), \qquad \mathbf{ka}\left(\vartheta, \alpha\right) = 0.$$

Here α is the angle between the projection of vector $\mathbf{k}$ on a plane perpendicular to the z axis and the x axis, ϑ is the angle between vector $\mathbf{k}$ and z axis, and $\theta(x)$ is the Heaviside step function.

It is easy to see that the field (28) transforms, as it should do, into the field of a plane monochromatic wave when $\Delta \to 0$:

$$\mathbf{A}\,(\mathbf{r}, t) \to \mathbf{b} \exp\left[-i\omega_0(t - z)\right], \qquad \mathbf{b} = \frac{1}{2\pi} \int_{-\pi}^{\pi} d\alpha\, \mathbf{a}\,(0, \alpha)\ .$$

Being orthogonal to vector $\mathbf{k} = \omega \mathbf{n}$, vector $\mathbf{a}\,(\vartheta, \alpha)$ is determined by two scalar functions. We will define them as projections of vector $\mathbf{a}\,(\vartheta, \alpha)$ onto two orthogonal unit vectors

$$\mathbf{l} = \frac{[\mathbf{n}\mathbf{e}_z]}{\sin \vartheta}\,, \qquad [\mathbf{l}\mathbf{n}]\ ,$$

which constitute a basis in a plane perpendicular to vector $\mathbf{k}$:

$$\mathbf{a}\,(\vartheta, \alpha) = a_1\,(\vartheta, \alpha)\,\mathbf{l} + a_2\,(\vartheta, \alpha)\,[\mathbf{l}\mathbf{n}]\ .$$

It is easy to see that the Cartesian components of vector $\mathbf{a}\,(\vartheta, \alpha)$ can be expressed in terms of functions a_1 and a_2 as follows:

$$a_x = a_1\,(\vartheta, \alpha)\sin\alpha - a_2\,(\vartheta, \alpha)\cos\vartheta\cos\alpha,$$

$$a_y = -a_1\,(\vartheta, \alpha)\cos\alpha - a_2\,(\vartheta, \alpha)\cos\vartheta\sin\alpha, \tag{29}$$

$$a_z = a_2\,(\vartheta, \alpha)\sin\vartheta.$$

Generally speaking, the functions $a_i\,(\vartheta, \alpha)$ are arbitrary complex functions of ϑ and α. For the sake of simplicity we will assume that they do not depend on ϑ, and are linear functions of $\sin\alpha$ and $\cos\alpha$

$$a_i(\alpha) = c_i \cos\alpha + d_i \sin\alpha\,, \quad i = 1, 2\,, \tag{30}$$

where c_i and d_i are constants.

Now we can easily obtain the following expression for $\mathbf{A}\,(\mathbf{r}, t)$

$$\mathbf{A}\,(\mathbf{r}, t) \;=\; \frac{i\omega}{2\pi} \exp(-i\varphi) \int_{-\pi}^{\pi} d\alpha \left\{ \left[a_1(\alpha)\sin\alpha - a_2(\alpha)\cos\alpha \left(1 - i\Delta^2 \frac{\partial}{\partial\chi} \right) \right] \mathbf{e}_x \right.$$

$$\left. - \left[a_1(\alpha)\cos\alpha + a_2(\alpha)\sin\alpha \left(1 - i\Delta^2 \frac{\partial}{\partial\chi} \right) \right] \mathbf{e}_y - i\Delta a_2(\alpha)\frac{\partial}{\partial\nu}\mathbf{e}_z \right\} G(\nu, \chi; \Delta)\,,$$

$$\tag{31}$$

where

$$\varphi = \omega(t - z), \qquad \nu = \xi\cos(\alpha - \phi),$$

$$\xi = \omega r \Delta, \qquad \chi = \omega z \Delta^2, \tag{32}$$

$$r = \sqrt{x^2 + y^2}, \qquad \cos\phi = \frac{x}{r}, \qquad \sin\phi = \frac{y}{r},$$

and function $G(\nu, \chi; \Delta)$ is defined by the expression:

$$
G(\nu, \chi; \Delta) = 2 \int\limits_0^1 du \frac{\sin \Delta u}{\Delta} \exp \left\{ i\nu \frac{\sin \Delta u}{\Delta} - 2i\chi \frac{\sin^2 \left(\frac{\Delta u}{2} \right)}{\Delta^2} \right\}.
\tag{33}
$$

Beginning from Eq. (31) we will designate the field frequency by the letter ω without an index.

The integrals in Eq. (31) have been studied already in approximation of small aperture angles Δ in connection with the problem of the Fraunhofer diffraction by a circular opening, see, e.g., monograph (Born and Wolf, 1964) where they describe a focused light beam. Specifically, it is known that they have a maximum at the point $\xi = 0$ and $\chi = 0$, and as ξ and χ increase they oscillate with gradual decrease of the amplitude. The first zeros of the field amplitude arise for $\xi \sim 1$ and $\chi \sim 1$. Thus, the range of variation of the parameters ξ and χ

$$
|\nu| \lesssim 1, \qquad |\chi| \lesssim 1
$$

gives the region of space where the field differs most strongly from zero. To characterize the dimensions of that region, we will introduce the parameters

$$
R \equiv \frac{1}{\omega \Delta}, \qquad L \equiv \frac{1}{\omega \Delta^2},
\tag{34}
$$

having the meaning of the focusing radius and the diffraction (or Rayleigh) length respectively.

We will now discuss the question about polarization of a focused light beam described by Eq. (28). It is seen from Eq. (28) that vector potential $\mathbf{A}(\mathbf{r}, t)$ and hence both electric and magnetic vectors have nonzero longitudinal components. At the same time, the concept of polarization is commonly used with respect to transverse fields. Therefore, strictly speaking, one cannot ascribe some definite type of polarization to a tightly focused laser pulse. Nevertheless, for a weakly focused pulse ($\Delta \ll 1$), there always exists a region near the axis of the beam $r \ll R$ where the field properties are very close to those of a plane wave field. Indeed, if we put $\xi \equiv r/R = 0$, the longitudinal components of both fields, in virtue of (30), become equal to zero after integrating over α in Eq. (31), and the fields $\mathbf{E}, \mathbf{H}$ turn out to be transverse. It is reasonable to ascribe polarization of the beam field in this region, which we will call "the plane wave zone", to the beam as a whole. Thereafter we will refer to the field of the pulse as polarized just in this sense. For a very tightly focused beam the focal spot radius is of the order of the wavelength and the plane wave zone doesn't exist. Therefore, only polarization of the parental beam incident on the focusing optical system can be ascribed to the focused beam in this case.

It can be seen that an arbitrary field in our model may be represented as a superposition of e- and h-polarized waves, i.e. the waves respectively with vectors $\mathbf{E}$ and $\mathbf{H}$ perpendicular to the propagation direction of the beam. We will obtain the $e(h)$-polarized wave, if we put $a_2(\alpha)(a_1(\alpha)) = 0$ in (31). It will be convenient to characterize the relative contributions of e- and h-polarized waves to the resulting field by "asymmetry parameter" μ

$$
\mu = \frac{E_{x0}^h - E_{x0}^e}{E_{x0}^h + E_{x0}^e},
\tag{35}
$$

where $E_{x0}^{e,h}$ are the x components of the electric field strengths for e- and h-polarized waves at the focal point ($\mathbf{r} = 0$) for $\varphi = 0$. Note that, in contrast to the field strength amplitude, the quantities $E_{x0}^{e,h}$ can take both positive and negative values.

Formulas for the field strengths look simpler in terms of the functions

$$F_1(\xi, \chi; \Delta) = \frac{1}{2\pi} \int_{-\pi}^{\pi} d\alpha G(\xi \cos \alpha, \chi; \Delta),$$

(36)

$$F_2 = F_1 - \frac{2}{\xi^2} \int_0^{\xi} d\xi \xi F_1.$$

For example, for a circularly e-polarized wave we have

$$\mathbf{H}^e = \pm \omega b \exp(-i\varphi) \left\{ \left(1 - i\Delta^2 \frac{\partial}{\partial \chi}\right) [F_1 (\mathbf{e}_x \pm i\mathbf{e}_y) + F_2 \exp(\pm 2i\phi) (\mathbf{e}_x \mp i\mathbf{e}_y)] \right.$$

$$\left. + 2i\Delta \exp(\pm i\phi) \frac{\partial F_1}{\partial \xi} \mathbf{e}_z \right\},$$

(37)

$$\mathbf{E}^e = i\omega b \exp(-i\varphi) \left\{ F_1 (\mathbf{e}_x \pm i\mathbf{e}_y) - F_2 \exp(\pm 2i\phi) (\mathbf{e}_x \mp i\mathbf{e}_y) \right\}.$$

The expressions for the field strengths of a circularly h-polarized wave can be calculated according to the formulas

$$\mathbf{E}^h = \pm i\mathbf{H}^e, \qquad \mathbf{H}^h = \mp i\mathbf{E}^e.$$

(38)

Formulas (37),(38) for the electromagnetic field strengths of a focused light beam have been derived for the particular physical model formulated in the beginning of the present section. They describe 3-dimensional geometry of electromagnetic field in a focused light beam which could be realized experimentally, since they represent an exact solution of Maxwell equations. However, it is even more important that the formulas (37),(38) constitute an exact solution of Maxwell equations at any function F_1 which satisfies the equation

$$2i \frac{\partial F_1}{\partial \chi} + \Delta^2 \frac{\partial^2 F_1}{\partial \chi^2} + \frac{1}{\xi} \frac{\partial}{\partial \xi} \left(\xi \frac{\partial F_1}{\partial \xi}\right) = 0.$$

(39)

It was shown in (Narozhny and Fofanov, 2000) that there exists a nontrivial solution of Eq. (39) which at small $\Delta \ll 1$ acquires the form

$$F_1 = (1 + 2i\chi)^{-2} \left\{ 1 - \frac{\xi^2}{1 + 2i\chi} \right\} \exp\left\{ -\frac{\xi^2}{1 + 2i\chi} \right\},$$

(40)

$$F_2 = -\xi^2 (1 + 2i\chi)^{-3} \exp\left\{ -\frac{\xi^2}{1 + 2i\chi} \right\}.$$

Waves with envelopes of this type are ordinarily called Gaussian beams. However, the solutions (37),(38) with F_1 and F_2 from (40), being approximations to exact solutions of Maxwell equations, correctly reproduce polarization properties of focused light beams. We shall see in the succeeding sections that the latter is very important for explanation of some effects which could be observed (or are already observed) experimentally.

The field which we have considered thus far can be used as a model of a stationary laser beam. It can be checked straightforwardly that one can obtain equations for the field which models a short laser pulse after the following substitutions Eqs. (37),(38):

$$\exp(-i\varphi) \to if'(\varphi), \qquad \Delta \exp(-i\varphi) \to \Delta f(\varphi),$$

where the function $f(\varphi)$ contains, besides an ordinary phase factor, the temporal envelope $g(\varphi/\omega\tau)$ of the pulse

$$f(\varphi) = g\left(\frac{\varphi}{\omega\tau}\right)\exp(-i\varphi)\,.$$

The envelope g is assumed to be equal to 1 at the center of the pulse, $g(0) = 1$, and exponentially decreasing for $|\varphi| \gg \omega\tau$. Here τ is the duration of the laser pulse in the laboratory reference frame. We assume that

$$\tau \gtrsim R\,.$$

Such a field is no longer an exact solution of Maxwell equations, but it satisfies Maxwell equations with accuracy to terms of the order $1/(\omega R)^2$ and $(1/\omega\tau)(1/\omega R)$ inclusively.

It is worth noting that the limitation

$$\Delta = (\omega R)^{-1} \ll 1 \tag{41}$$

is not too severe. Indeed, Δ can be written as $\Delta = \lambda/2\pi R$ where λ is the laser wave length. So that, even at the minimum possible $R \sim \lambda$ (diffraction limit) $\Delta \sim 10^{-1}$. We will assume further that the condition (41) is always satisfied.

5 Ponderomotive effect

The ponderomotive effect arises in a rapidly oscillating field with amplitude slowly varying in space and time, and thus belongs to the group of effects that do no exist in the field of a plane monochromatic wave.

To analyze the ponderomotive effect we will consider the motion of a free classic relativistic electron in the electromagnetic field of a focused laser pulse using the model developed in the preceding section. But as the first step we will solve a simple one dimensional problem in classical mechanics.

5.1 Nonrelativistic analysis

Consider motion of a nonrelativistic particle under action of a fast oscillating force with slowly varying amplitude

$$m\ddot{x} = eF(x)\sin\omega t\,, \quad \Delta F(x)\big|_T = \frac{dF(x)}{dt}T \ll F(x)\,,$$

and $T = 2\pi/\omega$ is the period of fast oscillations. Assume that the trajectory is a smooth function of time with fast oscillations around it

$$x(t) = X(t) + \xi(t)\,,$$

$$<\xi>_T = \frac{1}{T}\int\limits_0^T \xi(t)dt = 0, \quad <X(t)>_T = X(t)\,.$$

If the amplitude of oscillations is not too large in comparison with displacement of the particle for the period, we can approximate the force as

$$F(x) \approx F(X) + \frac{dF(X)}{dX}\xi(t)\,,$$

and write the equation of motion in the form

$$m\ddot{X} + m\ddot{\xi} = eF(X)\sin\omega t + \frac{dF(X)}{dX}\xi\sin\omega t.$$

There are two groups of terms in this equation: smooth and oscillating. Evidently, they should mutually cancel in each group separately. For oscillating terms we have

$$m\ddot{\xi} = eF(X)\sin\omega t,$$

and hence

$$\xi(t) = -\frac{eF(X)}{m\omega^2}\sin\omega t.$$

If we now average the equation of motion over fast oscillations using the obtained expression for $\xi(t)$, we get

$$m\ddot{X} = -\frac{e^2}{m\omega^2}F(X)\frac{dF(X)}{dX} < \sin^2\omega t >_T = -\frac{e^2}{2m\omega^2}F(X)\frac{dF(X)}{dX}.$$

Finally this equation can be written in the form

$$m\ddot{X} = -\frac{dU_{eff}(X)}{dX},$$

where

$$U_{eff}(X) = \frac{e^2F^2(X)}{4m\omega^2} = < \frac{m\dot{\xi}^2}{2} >_T.$$

This result means that the averaged over oscillations motion of the particle looks like a motion in the field of effective potential U_{eff}. This potential is called ponderomotive and has the meaning of the average kinetic energy of fast oscillations.

5.2   Equations of average motion of a relativistic electron

To analyze the ponderomotive effect we will use covariant form (15) for classical equations of motion of a charged particle in an external electromagnetic field. Unlike the plane wave case, electromagnetic field tensor $F^\mu{}_\nu$ is supposed to be a function of the phase $\varphi = (kx)$ of the wave and the spatial coordinates x, y and z.

We will assume however that characteristic sizes of spatial and temporal inhomogeneities of the laser pulse amplitude are large compared to the wave length and the period of oscillations. In terms of the model, developed in the preceding section, this means that the radius R, the diffraction length L and the duration of the pulse τ satisfy the conditions

$$\omega R \gg 1, \quad L = \omega R^2 \gg R, \quad \tau \gg 1/\omega. \tag{42}$$

By analogy with the nonrelativistic problem considered in the previous section we shall seek the solutions of these equations in the form

$$\begin{cases} \pi = q(\varphi) + \pi', \qquad x = x^{(0)}(\varphi) + x', \\[2mm] \pi' = \sum_{n=1}^{\infty} \left(\pi^{(n)}(\varphi)\cos n\varphi + \widetilde{\pi}^{(n)}(\varphi)\sin n\varphi \right), \\[2mm] x' = \sum_{n=1}^{\infty} \left(x^{(n)}(\varphi)\cos n\varphi + \widetilde{x}^{(n)}(\varphi)\sin n\varphi \right), \end{cases} \tag{43}$$

where $q(\varphi), x^{(0)}(\varphi), \pi^{(n)}(\varphi), \widetilde{\pi}^{(n)}(\varphi), x^{(n)}(\varphi), \widetilde{x}^{(n)}(\varphi)$ are slowly varying functions of φ.

A search for a solution in this form corresponds to separation of the electron motion into a regular displacement along a smooth trajectory $x^{(0)}(\varphi)$ (average motion) and rapid oscillations around the trajectory with frequencies which are multiple of the frequency of the external field. The quantity $q(\varphi)$ we will call the average kinetic momentum.

It goes without saying that the representation of the solution for equations of motion (15) in the form (43) is justified only if the problem contains two substantially different time scales. Specifically, the time of flight of an electron over distances of the order of the dimensions of the field inhomogeneities, which in our case are determined by the focusing radius R and the diffraction length L, should be much greater than the period of the wave. For this, of course, the conditions (42) must be satisfied.

It is important that the solutions of the equations of motion for an electron in the field of a plane monochromatic wave can also be represented in the form (43). For that case the average kinetic momentum q and all coefficients in the oscillating parts of the functions $\pi(\varphi)$ and $x(\varphi)$ are constants, maximum one harmonic is present in the expansion for π', and maximum two in the expansion for x', compare, e.g., (Landau and Lifshits, 1975). For our case of a focused monochromatic wave, this means that the derivatives of the average momentum and all coefficients, equal to zero for a plane wave case, contain at least the first power of the small parameter Δ.

Substituting the representation (43) into Eqs. (15) and successively equating terms proportional to $\sin n\varphi$ and $\cos n\varphi$, $n = 0, 1, 2, \ldots$ to zero, we obtain the following equations for averaged motion of the particle[2]

$$\frac{d\mathbf{r}^{(0)}}{d\varphi} = \frac{\mathbf{q}}{kq}, \tag{44}$$

$$\frac{d\mathbf{q}_\perp}{d\varphi} = -\frac{m\omega}{kq}\frac{\partial U}{\partial \boldsymbol{\rho}}, \qquad \frac{dq_z}{d\varphi} = \frac{m\omega}{kq}\frac{\partial}{\partial\varphi}U\left(g(\frac{\varphi}{\omega\tau}), \boldsymbol{\rho}(\varphi)\right), \tag{45}$$

where U can be called the relativistic ponderomotive potential

$$U\left(g(\frac{\varphi}{\omega\tau}), \boldsymbol{\rho}(\varphi)\right) = \frac{e^2}{m\omega^2}\langle(ReE_x)^2 + (ReE_y)^2\rangle. \tag{46}$$

In Eqs. (44)-(46)

$$kq = \omega q_- = \omega(q_0 - q_z) = const,$$

$$q^2 = m_*^2 = m^2 + 2mU = m^2(1 + \eta^2), \quad \vec{\rho} = \vec{r}_\perp^{(0)}/R.$$

Note that the dimensionless intensity parameter η for a focused pulse is a slow function of coordinates and phase φ.

The concrete form of the ponderomotive potential strongly depends on polarization and on the type of the wave. For example in the case of a linearly polarized wave with arbitrary value of the asymmetry parameter (35) the electric field is given by the expression

$$\mathbf{E} = (1 - \mu)\mathbf{E}^e + (1 + \mu)\mathbf{E}^h,$$

and for the ponderomotive potential we have

$$U = \frac{m\eta_0^2}{2}g^2(\varphi/\omega\tau)\left\{|F_1|^2 + \mu^2|F_2|^2 + \mu\cos 2\psi\,(F_1 F_2^* + F_1^* F_2)\right\}, \tag{47}$$

where $\tan\psi = \varrho_x/\varrho_y$, and η_0 is the value of the dimensionless field intensity parameter at the focal point at the moment $\varphi = 0$.

---

[2]The details of derivation of Eqs. (44)-(46) see in (Narozhny and Fofanov, 2000).

Eq. (47) shows that the ponderomotive potential U depends on the azimuthal angle ψ, and hence, is, generally speaking, asymmetric. The potential U is symmetric only for the case $\mu = 0$, which justifies the appellation of the parameter. The shape of the ponderomotive potential in the plane $z = 0$ for $\varphi = 0$ is shown in Fig. 8 for the cases $\mu = 0$ and $\mu = 1.55$.

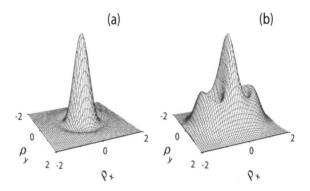

Figure 8. *Ponderomotive potential (47) in the plane $z = 0$ at $\varphi = 0$ for two values of the asymmetry parameter: (a) $\mu = 0$ and (b) $\mu = -1.55$.*

Fig. 8a represents the ponderomotive potential for the standard case of Gaussian beam commonly used in literature, while Fig. 8b illustrates dramatic difference between the cases $\mu = 0$ and $\mu \neq 0$. For $\mu \neq 0$, the ponderomotive potential, besides the central peak, possesses two extra maxima which are located at the polarization plane. They arise as a result of non-uniform intensity distribution in the plane $z = 0$ for $\mu \neq 0$. The location of the additional maxima, as well as their amplitudes, are determined by the value of the parameter μ. It is worth noting that the case $\mu = 0$ is the only case when for the pulse polarized along the x axis E_y and H_x components of the electric and magnetic fields remain to be equal to zero outside the plane wave zone.

5.3 Ponderomotive scattering of electrons by a focused laser pulse of relativistic intensity

The first direct observation of electron inelastic scattering from the ponderomotive potential of a laser pulse in vacuum was performed by Bucksbaum et al. in 1987 (Bucksbaum et al., 1987). Deflection of electrons out of the beam was observed, as well as the surfing effect when electrons gained energy when scattered from the temporal leading edge of the laser pulse and lost energy when scattered from the trailing edge. The laser intensity in experiment of Bucsbaum et al. was $I \approx 7 \times 10^{13} \, \mathrm{W/cm^2}$ which corresponded to ponderomotive potential with peak potential energy of 8 eV. The initial energy of electrons was below this value and the scattered electrons gained, or lost, about 0.2 eV.

Relativistic ponderomotive scattering was first observed in experiment of Malka et al. in 1997. The experimental setup is shown in Figure 9.

Electrons were created by the interaction of a creation laser beam (nanosecond laser pulse with $10^{12} - 10^{13}$ W/cm^2 intensity) with a thin plastic target, some distance away from the chamber center. Suprathermal electrons were expelled by Raman instability in a Boltzman-like distribution of a few tens of keV. The authors estimated that several 10^{12} electrons exited the plasma. Five hundred picoseconds after the creation beam, which was the time for keV electrons to arrive near the chamber center, the high intensity laser beam was focused at the chamber center.

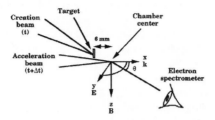

Figure 9. *Layout of the experiment on relativistic ponderomotive scattering (Fig. 1 from (Malka et al., 1997)).*

The parameters of the interaction beam were: $\lambda = 1.056\,\mu$m, pulse width $\tau = 300 - 500$ fs, $10\,\mu$m focal spot diameter, intensity up to 10^{19} W/cm^2.

The accelerated particles were detected by two electron mass spectrometers set at $\theta = 39°$ and $\theta = 46°$ of the laser propagation direction in the forward direction, in the horizontal plane, and at 15 cm from the chamber center. The laser polarization was horizontal, and the spectrometers were set in the (**Ek**) plane.

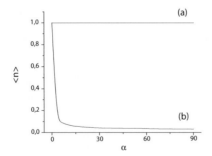

Figure 10. *The normalized number of scattered electrons with the energy $W \geq 0.9$ MeV as a function of the azimuthal angle α for (a) $\mu = 0$ and (b) $\mu = 1.55$.*

The maximum energy detected in the experiment was $W_{max} = 0.9 \pm 0.1$ MeV at $\theta = 39°$ for the laser intensity $I \approx 10^{19}$W/cm^2. Electrons with energy 0.63 ± 0.05 MeV at $\theta = 46°$ were also detected for the same laser intensity, and at $\theta = 39°$ for intensity $I \approx 5 \times 10^{19}$W/cm^2. For each case, more than 10^5 electrons in units of keV and sr were detected by the spectrometer.

Electrons with energies less than the maximum energies were also detected for these three cases. The energy distribution of the scattered electrons, as well as the relation between scattering energy and scattering angle, were in good agreement with theory. The most striking result of the experiment, however, was asymmetry in angular distribution of the accelerated electrons. All electrons were detected in the (**Ek**) plane, and not detected when polarization of the interaction laser beam was changed from horizontal to vertical.

We will now apply the theory developed in the preceding section to explanation of the results of experiment of Malka et al. The results of calculations are presented in Figure 10, where the normalized number of scattered electrons $\bar{n}$ with final energies $W \geq 0.9$ MeV is shown as a function of the azimuthal scattering angle α.

The values of $\bar{n}$ at any given α were obtained by averaging the number of scattered electrons over the range of α equal to the angular size of the detector used in the experiment. It is easily seen that the angular distribution of scattered electrons essentially depends on the parameter of asymmetry μ. At $\mu = 0$, which corresponds to the symmetrical Gaussian ponderomotive potential shown in Figure 8a, it is isotropic and purely radial.

At the same time, if $\mu = -1.55$ the number of scattered electrons detected in the $(\mathbf{E}, \mathbf{k})$ plane ($\alpha = 0$) is about 30 times higher than that in the $(\mathbf{H}, \mathbf{k})$ plane ($\alpha = \pi/2$). This result is in good quantitative agreement with observation of Malka et al.. The value of the asymmetry parameter, $\mu = -1.55$, has been chosen just for better fitting of our computational results to the results of the experiment.

The 30-fold anisotropy of accelerated electrons is clearly explained by asymmetry of the ponderomotive potential (47). The cross term in (47), besides asymmetric correction to the radial force, gives rise to a tangential force which is responsible for pushing electrons out of the plain perpendicular to the polarization plane. The latter corresponds to minimum of the ponderomotive potential as function of azimuthal angle ψ, while the perpendicular plane to its maximum. Therefore one could be surprised that the number of electrons scattered at the angle $\alpha = \pi/2$ is not equal to zero. Certainly, this is explained by complicated structure of the ponderomotive potential (47), namely by the fact that the cross term can change its sign at the periphery of the focus.

6 Pair creation by a focused laser pulse in vacuum

In this section we shall consider another physical effect for which the structure of the field near the laser focus is highly important, namely the effect of electron-positron pair creation by a focused laser pulse in vacuum. The effect of $e^- e^+$ pair creation was first considered for a constant electric field, see, e.g., (Schwinger, 1951).

The probability of pair creation in a constant electric field is known to acquire its optimum value when the electric field strength is of the order of QED characteristic value F_S. Clearly, such a field strength is unattainable for static fields experimentally in near future. Therefore attention of many researchers was focused on theoretical study of pair creation by time-varying electric fields. Since the latest achievements of laser technology promise very rapid growth of peak laser intensities the detailed study of the effect of pair creation in time-varying electromagnetic fields, in particular in the field of a focused laser pulse, is becoming an urgent physical problem from experimental point of view.

As it was shown in (Schwinger, 1951), a plane electromagnetic wave of arbitrary intensity and spectral composition does not create electron-positron pairs in vacuum because it has both field invariants $\mathcal{F} = (\mathbf{E}^2 - \mathbf{H}^2)/2$ and $\mathcal{G} = (\mathbf{E} \cdot \mathbf{H})$ equal to zero. This is not true of course for a focused quasimonochromatic laser pulse. We will use the model developed in Section 4 to describe the electromagnetic field of such a pulse choosing the variant of circular polarization for the sake of simplicity.

The method we will use for calculation of the number of pairs created by the laser pulse is based on the fact that the characteristic length of the process is determined by the Compton length $l_c = \hbar/mc$ which is much less than the wavelength λ of the laser field, $l_c \ll \lambda$. Therefore, at an arbitrary point of the pulse we can calculate the number of created particles per unit volume and unit time according to the Schwinger formula for a static homogeneous field and then obtain the total number of created particles as the integral over the volume V and duration

τ of the pulse

$$N = \frac{e^2 F_S^2}{4\pi^2 \hbar^2 c} \int_V dV \int_0^\tau dt\; \epsilon\eta \coth\frac{\pi\eta}{\epsilon} \exp\left(-\frac{\pi}{\epsilon}\right). \tag{48}$$

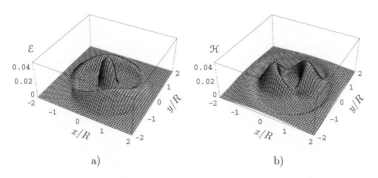

a)   b)

Figure 11. *The dependencies of $\mathcal{E}$ (a) and $\mathcal{H}$ (b) on spatial coordinates x and y for the time moment $t = 0$. $\mathcal{E}$ and $\mathcal{H}$ are measured in units of F_S, and the other parameters are chosen $E_0 = 0.1 F_S$, $z = 0$, $\Delta = 0.1$.*

Here $\epsilon = \mathcal{E}/F_S$, $\eta = \mathcal{H}/F_S$, and $\mathcal{E}$ and $\mathcal{H}$ are the field invariants which have the meaning of electric and magnetic fields in the reference frame where they are parallel. The invariants $\mathcal{E}$ and $\mathcal{H}$ can be expressed in terms of invariants $\mathcal{F}$ and $\mathcal{G}$, see, e.g., (Schwinger, 1951)

$$\mathcal{E} = \sqrt{(\mathcal{F}^2 + \mathcal{G}^2)^{1/2} + \mathcal{F}},$$

$$\tag{49}$$

$$\mathcal{H} = \sqrt{(\mathcal{F}^2 + \mathcal{G}^2)^{1/2} - \mathcal{F}}.$$

The field invariants $\mathcal{F}$, $\mathcal{G}$ can be easily calculated analytically using Eqs. (37), (38) and can be found in (Narozhny N.B. et al., 2004). Here we will give only the results of numerical calculations for the dependence of invariants $\mathcal{E}$ and $\mathcal{H}$ for a e-polarized wave on the spatial coordinates x and y for the time moment $t = 0$, see Figure 11.

The coordinate dependencies of the invariants $\mathcal{E}$ and $\mathcal{H}$ in the h-polarized wave are also given by Figure 11 if one makes the interchange $\mathcal{E} \leftrightarrows \mathcal{H}$.

It is clear that pairs are created mainly in the focal region where the field strength reaches its maximum value. Assuming that the spatial volume of the pulse is of the order of $\pi R^2 c\tau$ we can estimate the number of created pairs as

$$N \approx \frac{e^2 E_S^2}{4\pi^2\hbar^2 c}\pi R^2 c\tau^2 \bar{\epsilon}\bar{\eta}\coth\frac{\pi\bar{\eta}}{\bar{\epsilon}}\exp\left(-\frac{\pi}{\bar{\epsilon}}\right), \tag{50}$$

instead of Eq. (48).

Here $\bar{\epsilon}$ and $\bar{\eta}$ are the averaged over time values of dimensionless field invariants ϵ and η in the focus. One can see from Figure 11 that at the focus of the e-polarized pulse

$$\epsilon = (2\mathcal{F}^e/E_S^2)^{1/2}, \quad \eta = 0,$$

while for the h-polarized pulse we have

$$\epsilon = 0, \quad \eta = (2\mathcal{F}^e/E_S^2)^{1/2}.$$

So we conclude that pairs can be created only by a e-polarized but not h-polarized pulse. This statement is not exact however. We will see that h-polarized pulse also create pairs though their number is several orders of magnitude less than the case of an e-polarized pulse.

It follows from Eq. (50) that the number N_e of pairs created by a e-polarized focused pulse can be estimated as

$$N_e \approx \frac{1}{4\pi^3} \frac{\Omega_L}{\Omega_C} \bar{\epsilon}^2 \exp\left(-\frac{\pi}{\bar{\epsilon}}\right) , \qquad (51)$$

where $\Omega_L = \pi R^2 c\tau^2$ is the effective 4-volume of the laser pulse and $\Omega_C = l_C^4/c$ is Compton 4-volume where one pair is created with probability close to 1, if the field strength $E \sim F_S$. The ratio

$$\frac{1}{4\pi^3} \frac{\Omega_L}{\Omega_C} \sim 3 \times 10^{25}$$

at $R \sim \lambda \sim 1\mu m, \tau \sim 10 fms$. Hence, one pair per laser shot will be created at $\bar{\epsilon} \sim 5 \cdot 10^{-2}$. It is clear that $\bar{\epsilon} \lesssim E_0/F_S$, where E_0 is the peak field strength of the pulse, and we conclude that the effect of pair creation by a focused laser pulse in vacuum becomes observable at peak field strength essentially less then the characteristic value F_S.

The results of numerical calculations for the numbers of pairs $N_{e,h}$ created by e- and h-polarized waves for $\lambda = 1\mu m, \tau = 10 fms$ and different values of I and Δ are given in Table 1.

$I \times 10^{-27}$, W/cm^2	E_0/E_S	N_e, $\Delta = 0.1$	N_e, $\Delta = 0.05$	N_h, $\Delta = 0.1$
2	0.11	4.0(−11)	4.6(−42)	9.6(−23)
4	0.16	9.3(−2)	6.8(−24)	1.5(−10)
5	0.18	24	3.1(−19)	2.0(−7)
6	0.20	1.5(3)	8.8(−16)	4.4(−5)
10	0.25	3.0(7)	1.4(−7)	16
20	0.36	8.0(11)	32	8.5(6)
50	0.56	1.0(16)	1.3(9)	1.6(12)

Table 1. *The number of produced pairs for different values of laser pulse intensity and parameter Δ for e- and h-polarized waves.*

One can see from Table 1 that the observable number of pairs in agreement with our estimate is created at $I \sim 5 \cdot 10^{27}$W/cm$^2 \sim 10^{-2}I_S$, or at $E_0/F_S \sim 10^{-1}$. One can see also that numerical calculation yields nonzero result for the number of pairs created by a h-polarized pulse. This is explained by the fact that $\mathcal{E}$ is not equal to zero at the periphery of the focus of an h-polarized pulse, see Figure 11. However its values are suppressed by the exponents in functions F_1, F_2, see Eq. (40). Hence the number of pairs created by an h-polarized wave is several orders of magnitude less than by the e-polarized wave of the same intensity and the same value of parameter Δ. It is worth noting that dependence of the number of created pairs on parameter Δ is very sharp. Indeed it is seen from Table 1 that halving the value of Δ leads to many orders of magnitude decreasing of the number of created pairs.

Let us now compare the rest energy of created pairs with the total energy W carried by the laser pulse. By definition, the intensity I is

$$I = \frac{c}{4\pi} \bar{E}^2 \sim W/\pi R^2 \tau .$$

Assuming that $\bar{E}^2 \sim E_0^2/2$ we get[3] for accepted values of parameters λ and τ, and $\Delta = 0.1$

$$W \sim 5 \times 10^{21} \frac{I}{I_S} mc^2.$$

We see that the energy of the pulse with intensity $I \approx 0.2 I_S$ is of the order $W \sim 10^{21} mc^2$. The same order of magnitude we obtain from Eq. (51) for the energy of created pairs $W_p = 2mc^2 N_e \sim 10^{21} mc^2$. Thus, we conclude that the exploited method becomes inconsistent and one should take into account back reaction of the pair creation effect on the process of laser pulse focusing at such an intensity. In other words, one cannot consider the electromagnetic field of the pulse at near critical intensities as a given external field and should take into account depletion of the pulse due to pair production. This is the reason why we have limited the values of intensity in Table 1 by $I = 5 \times 10^{28} \text{W/cm}^2$. At larger intensities the exploited method of calculations ceases to be valid. This means that there exists a natural physical limit for attainable focused laser pulse intensities which is posed by the effect of pair creation. In the framework of our model this limit for the circularly e-polarized laser pulse with $\lambda = 1\mu m, \tau = 10 fms$, and $\Delta = 0.1$ is approximately $0.3 I_S$.

We have not taken into account the ponderomotive effect while considering the process of pair creation, though the value of the relevant ponderomotive potential could be very high compared to the energy of created particles. This is because the characteristic time for the effect of pair creation, $t_c = l_c/c$, is very small compared to the wave period T. As a result, the ponderomotive acceleration can be considered as the second stage of the process. It influences the energy and angular distributions which are not considered in the present lecture, but not the total number of created particles. It is worth noting that the inequality $t_c = l_c/c \ll T$ explains also why the created particles are characterized by the bare mass m instead of the effective mass m_*.

7 Conclusions

In the present lecture a brief review of IFQED was presented. We have discussed the general approach to treatment of quantum processes in the framework of IFQED. We have also considered several examples of effects initiated by relativistic particles interacting with strong laser fields, as well as the vacuum polarization effect. Certainly the review was not complete. Many interesting effects, such as light-by-light elastic scattering, splitting of one photon into two, self focusing effect in vacuum, etc., remained beyond the framework of this lecture.

The distinctive features of Intense Field QED are: (i) nonperturbative with respect to particle-field interaction nature of the theory, (ii) essentially nonlinear dependence of amplitudes of quantum processes on the field strength, and (iii) instability of vacuum in the presence of external fields of certain configurations. Study of IFQED constitute a new trend in high energy and laser physics: QED verification at strong fields, which is distinct from QED verification at small distances.

It is worth noting one more serious problem which appears at super intense fields. The probabilities of quantum processes in the field of a plane monochromatic wave transform at $\eta \to \infty$ into probabilities of the analogous processes in a constant crossed field. This can be easily understood if one notes that $\eta \sim eE_0/m\omega$, so that the limit $\eta \to \infty$ is realized either when $E_0 \to \infty$, or $\omega \to 0$. The processes initiated in a crossed field by a single electron are determined by a single parameter Υ. For an electron with $\gamma \approx 10^5$ colliding with a laser pulse of intensity 10^{28}W/cm^2 the value of parameter Υ is of the order 10^4. It was shown in (Narozhny, 1980) that expansion parameter of perturbation theory in IFQED at $\Upsilon \gg 1$ is $\alpha \Upsilon^{2/3}$. For the considered

---

[3]More accurate estimate for W can be found in (Narozhny et al., 2004)

values of parameters the interaction constant is of the order of unity. This means that IFQED ceases to be a theory with weak interaction and thus one cannot use perturbation theory with respect to radiation field as an instrument for calculation probabilities of quantum processes.

Hence we can conclude that, if T. Tajima and G. Mourou are right and the intensity level of 10^{26-28} W/cm^2 will be reached already in the coming decade, we certainly stand on the threshold of a new era of both experimental and theoretical study of nonlinear IFQED. If this happens, we will obtain a new powerful tool for study of fundamental physical problems at extremely high energy level, but we will be certainly compelled to revise the standard methods of QED.

Acknowledgments

I thank the organizers of the school for their hospitality and creating a stimulating environment. I am also grateful to Prof. Remo Ruffini for his hospitality and useful discussions at ICRANET in Pescara where a part of this manuscript was written. A.M. Fedotov is deeply acknowledged for many useful comments and suggestions on the manuscript.

References

Born, M. and Wolf, E. (1964), *Principles of Optics*, (Pergamon, New York).
Bula, C., et al. (1996), *Phys. Rev. Lett.*, **76**, 3116.
Bulanov, S.V., Esirkperov, T. and Tajima, T. (2004), *Phys. Rev. Lett.* **92**, 159901.
Burke, D.L., et al. (1997), *Phys. Rev. Lett.*, **79**, 1626.
Heisenberg, W. and Euler, H. (1936), *Z. Phys.* **98**, 714 .
Landau, L. D. and Lifshits, E.M. (1975), *The Classical Theory of Fields*, (Pergamon, Oxford).
Narozhny N.B., et al. (1965) *Sov. Phys. JETP* **20**, 622.
Narozhny N.B. and Fofanov M.S. (2000) *JETP* **90**, 753.
Narozhny N.B., et al. (2004) *Phys. Lett.* **A330**, 1.
Narozhny N.B. (2004) *Phys. Rev.* **D 21**, 1176.
Sauter, F. (1931) *Z. Phys.* **98**, 714.
Schwinger, J. (1951), *Phys. Rev.* **82**, 664.
Shen, B. and Yu, M. (2002), *Phys. Rev. Lett.* **89**, 275004.
Tajima, T. and Mourou, G. (2002), *Phys. Rev. STAB* **5**, 031301.
Volkov, D.M. (1994), *Z. Phys.*, **94**, 250.

14
Fundamentals of ICF Hohlraums

Mordecai ("Mordy") D. Rosen

Lawrence Livermore National Laboratory (LLNL)

1 Basic physics

Introduction

On the Nova Laser at LLNL, we have demonstrated many of the key elements required for assuring that the next laser, the National Ignition Facility (NIF) will drive an Inertial Confinement Fusion (ICF) target to ignition. The indirect drive (sometimes referred to as "radiation drive") approach converts laser light to x-rays inside a gold cylinder, which then act as an x-ray "oven" (called a hohlraum) to drive the fusion capsule in its center. On Nova a good understanding has been developed of the temperatures reached in hohlraums and of the ways to control the uniformity of the x-rays driving the spherical fusion capsules. Here we will review the physics of these laser heated hohlraums, recent attempts at optimizing their performance, and then return to the ICF problem in particular to discuss scaling of ICF gain with scale size, and compare indirect with direct drive gain.

In ICF, spherical capsules containing Deuterium and Tritium (DT) – the heavy isotopes of hydrogen – are imploded, creating conditions of high temperature and density similar to those in the cores of stars required for initiating the fusion reaction. When DT fuses, an alpha particle (the nucleus of a helium atom) and a neutron are created releasing large amounts of energy. If the surrounding fuel is sufficiently dense, the alpha particles are stopped and can heat it, allowing a self-sustaining fusion burn to propagate radially outward and a high gain fusion micro-explosion ensues.

To create those conditions the outer surface of the capsule is heated (either directly by a laser or indirectly by laser produced x-rays) to cause rapid ablation and outward expansion of the capsule material. A rocket-like reaction to the outward flowing heated material leads to an inward implosion of the remaining part of the capsule shell. The pressure generated on the outside of the capsule can reach nearly 100 megabar (100 million times atmospheric pressure [1b=10^6 cgs]), generating an acceleration of the shell of about 10 trillion times g which results in the shell reaching, over the course of a few nanoseconds, an implosion velocity of 300 km/sec. When the shell and its contained fuel stagnates upon itself at the culmination of the implosion, most of the fuel is in a compressed shell which is at 1000 times solid density. That shell surrounds a hot spot of fuel with sufficient temperature (roughly 10 keV or 100 million degrees) to ignite a fusion reaction.

The capsule must not only be driven hard, but also uniformly over its entire surface to cause uniform compression of the fuel to the center. With direct drive, this uniform heating of the capsule is caused by simultaneously illuminating the capsule from all sides with many laser beams and taking great care (via beam conditioning to avoid speckle etc.) to assure that 2 points close to one another on the capsule surface are driven with the same illumination. With indirect drive, the capsule is positioned in the center of a cylindrically symmetric container called a hohlraum. Laser beams enter the hohlraum through holes in the end caps to heat the walls of the cylinder, which then radiate soft x rays to fill the hohlraum with a bath of radiant energy. This energy causes the fuel capsule to implode. Typically, 70–80% of the laser energy can be converted to x-rays. The hohlraum concept leads to a natural, geometric uniformity of x-ray flux on the capsule surface, since two points close to one another on the capsule surface "look out" at the heated hohlraum walls and, for a wall to capsule radius ratio of order 4, see nearly identical large sections of the walls, thus making it insensitive to the non-uniformity of those sections and hence produce a nearly identical heat environment. We now proceed to study these hohlraums.

1.1   History of hohlraums

The year 2005 has been designated "World Year of Physics" to celebrate the centenary of Einstein's 1905 publication of 4 "miraculous" papers: "photo-electric" effect, Brownian motion, special relativity, and "$E = mc^2$". Interestingly, despite all the revolutionary implications of special relativity (and later of general relativity) Einstein's 1923 Nobel Prize cites the first paper – the photoelectric effect. Its role in the development of hohlraum physics is what concerns us here. So let us take a further step or two back in time:

Human beings have been using "hohlraums" for millennia, in the form of kilns – hot ovens that can harden clay pots. Their role in furthering human civilization through engendering storage, transportation, trade, and writing is well known. Jumping ahead to the late 19th century, scientific inquiry into the nature of these ovens matured. As we shall see, an important figure of merit for any system near thermal equilibrium is its "optical depth" ($\tau_{O.D.}$) – namely the ratio of a system length, L, to a typical mean free path ("m.f.p"). For a kiln, a m.f.p. of a typical photon inside the walls is about 0.1 μm. The heated depth of the wall is about 1 cm. Hence the walls of a kiln have an enormous $\tau_{O.D.}$ of about 10^5 and thus a kiln is expected to be very close to equilibrium.

The spectrum of the electro-magnetic (EM) radiation emerging from a small hole of a hot (1000 – 2000°) oven was measured and turned out to be independent of wall material. Kirchoff formulated his famous law that in equilibrium emissivity = absorptivity, so that the best absorbers make the best emitters. Thus a "black-body" whose name (in the optical range of photons) implies complete absorption (like an oven with only a small hole), will radiate more than any other system at the same temperature T. This universal blackbody (BB) spectrum of emerging intensity I_ν (power per unit area per unit solid angle) within a small interval of frequency $d\nu$ vs. ν was measured and found to have the following shape: for small ν, it rises with frequency as $\nu^2 T$, the so-called Rayleigh-Jeans (R-J) fit. It reaches a peak at $\nu = 2.8T$ (when appropriate units are used, to be discussed below). It then decays as $\nu^3 e^{(-\nu/T)}$, called the Wien fit. The quantity I_ν is related to the isotropic energy density U_ν by $I_\nu = cU_\nu/4\pi$ and thus both have the same spectrum.

Besides this experimental work, Wien's law, derived from purely thermodynamic arguments, declares that the spectrum must be in the form of $U_\nu = \nu^3 f(\nu/T)$. Note that Wien's fit clearly obeys this law, as does the R-J fit with $f(x) = 1/x$. An immediate consequence of Wien's law is that the total energy density which is an integral over ν, will be $U = aT^4$. This gives a total $I = (\sigma/\pi)T^4$, where $\sigma = ac/4$. If we ask what the one sided flux F out the small hole will be in any direction, say x, perpendicular to the plane of the hole we must integrate $2\pi I\mu$ over μ (where μ is the cosine of the angle between a ray and the x axis) from $\mu = 0$ to 1, to get the well known $F = \sigma T^4$.

Planck set out to derive this spectrum. Based on EM theory he found that within a given frequency interval $d\nu$, an oscillator of mass m, absorbs energy as $(\pi e^2/3m)U_\nu$ where U_ν is the sought after frequency behavior of the energy density of the radiation. He set it equal to the emission from the radiator, because the oscillators in the hohlraum wall are in equilibrium and emit as much as they absorb, which, again from EM theory is $(8\pi^2 e^2\nu^2/3mc^3)E_{osc}$, where E_{osc} is the energy of the oscillator. Thus

$$U_\nu = \frac{8\pi\nu^2}{c^3}E_{osc} \tag{1}$$

Planck then put forward some not-unreasonable arguments for the entropy of a many oscillator state, that resulted in an expression: $d^2S/dE^2 = -1/\nu E$. This combined with the classical thermodynamic relation $dS/dE = 1/T$, Eq. (1) and Wien's law, immediately resulted triumphantly in Wien's fit. But the triumph was short lived. Most of the data was in the large ν, Wien's fit regime (for a $1500°$ C oven, the spectral peak is at about a half of an eV, so the optical range (IR to UV of 1–3 eV where most instruments were available) is well past the peak. As technology progressed the far IR (low ν) regime began to give up its secrets, and the Wien fit did not fit so well anymore!

With the R-J law emerging from the new data, Planck noticed that $d^2S/dE^2 = -1/E^2$ led, as above, to the R-J fit. He was thus quickly led to the form $d^2S/dE^2 = -1/(\nu + E)E$ as a way to possibly cover the whole spectrum. So now, $dS/dE = 1/T$ leads to (and inserting 2 constants h & k that assure proper dimensionality within the exponential, as well as a good numerical fit to the R-J regime)

$$E_{osc} = \frac{h\nu}{e^{h\nu/kT} - 1}, \tag{2}$$

and via Eq. (1) to

$$U_{\nu BB} = \frac{(8\pi\nu^2/c^3)h\nu}{e^{h\nu/kT} - 1}. \tag{3}$$

This successful fit only begged the question – what was the theoretical justification for such a form of the entropy? Planck spent the next "two most difficult months of his life" answering this question. Part of the difficulty was his necessity to adopt an approach he previously rejected – Boltzmann's probabilistic view of entropy as $S = k\ln W$, where W measures the probability of being in a microstate with an energy corresponding to the macro-state with the entropy we seek, along with the notion that all microstates are equally probable. Thus we need to calculate the multiplicity of states to find S. Following in the path of an earlier calculation by Boltzmann, Planck considered a system with N oscillators. He assigned an energy to each, but only in artificially assumed integer units ("quanta") of size Δ. If the total energy of the system is $P\Delta$, then the average energy of an oscillator will be $E = P\Delta/N$.

Now for the combinatorics, which required the artificial integer assumption in order to proceed, a typical possibility of what the system could be, can be written as: "$|\ldots|..|\ldots..||\ldots|.|$ etc." Here we have assigned some number of energy units Δ (the dots) to each oscillator (the partitions $|$). If we consider this entire accounting enterprise as filling in $N + P$ blanks with P dots and N partitions, there are $(N + P)!$ ways to do so and we divide by $N!P!$ to avoid "double" counting choices that are the same but written in different order. This is what "W" represents in $S = k\ln W$. Using Stirling's approximation, defining $x = P/N = E/\Delta$ we find

$S/k = \{(1 + x)\ln(1 + x) - x\ln x\}$ and then via $dS/dE = 1/T$ we get $E = \Delta/(e^{\Delta/kT} - 1)$. Then Wien's law forces $\Delta = h\nu$ and the derivation is complete.

Planck stopped there, with finite, quantized Δ, because it gave the right answer. Had he followed Boltzmann he would have taken Δ "classically" to zero and "derived" $E = kT$ and thus via his Eq. (1) the R-J fit. Planck assumed that other workers would figure out the microphysics of oscillators and the meaning of these quanta. He spent over a decade trying to minimize the implications of this radical notion. Planck's paper was quite tersely written and difficult to follow, and remained rather neglected for 6 years. This sets the stage for 1905 and Einstein's "photo-electric" paper.

Einstein was troubled, as he would be throughout his career, with the lack of unity in physics – mechanics dealt with particles but EM theory dealt with waves. He interpreted Planck's work as still being in the classical camp. After all Planck's Eq. (1) is entirely from EM wave theory, so he mostly ignored him. Einstein did use Eq. (1) to immediately derive the R-J law (to him $E = kT$ was obvious) and to point out the "ultra-violet" catastrophe – that the R-J law, if assumed to hold for all ν, would make U diverge.

Einstein concentrated on the volume (V) dependence of the entropy S of the radiation field in a hohlraum. He defined $S_\nu d\nu = V f(U_\nu)d\nu$ and of course $E_{EM\nu}d\nu = VU_\nu d\nu$. He then just inverted the Wien fit to find $-(k/h\nu)\ln(U_\nu/\alpha\nu^3) = 1/T$. He then set that equal to dS/dE (sounds familiar?!) which is equal to df/dU_ν. This gives $f = -(kU_\nu/h\nu)[\ln(U_\nu/\alpha\nu^3) - 1]$, which gives $S_\nu d\nu = (kE\nu/h\nu)[\ln(E_\nu/V\alpha\nu^3 d\nu) - 1]$. He then considered a change in volume from V_0 to V, and the ensuing change in entropy

$$S_{BB} - S_0 = \frac{kE_\nu}{h\nu}\ln\left(\frac{V}{V_0}\right) \tag{4}$$

and compared it to an N particle classical ideal gas result

$$S_{ig} - S_0 = (kN)\ln\left(\frac{V}{V_0}\right). \tag{5}$$

Einstein then reminded us what Eq. (5) means in the $S = k\ln W$ context. The probability a *localized particle* will find itself in a volume V is (V/V_0), and that N of them will is $W = (V/V_0)^N$. Comparing Eqs. (4) to (5) immediately told Einstein that light of energy E_ν behaves as if it is made up of N localized particles (quantized photons), each with an energy $h\nu$.

It is only the next year, 1906, when Einstein turned his attention to calculating the T, ν behavior of S, that he understood enough of what Planck did to see the profound connection in his 1905 photon paper, namely: Planck's quantized oscillators emit Einstein's quantized photons – and thus Quantum Mechanics (QM) was born!

Since the notion of photons (vs. EM waves) proved far too radical for nearly everyone, further "hohlraum-inspired" developments also prove crucial. In 1907 Einstein reasoned that the vibrating molecules in a wall also radiate (far IR) and as oscillators should follow Eqs. (1) & (2). So he took dE/dT via Eq. (2) to find the specific heat and to predict it vanishing as T approaches zero. A wide community of physicists was quite interested in this, measured the prediction to be true, and QM became accepted. Planck still resisted for another decade, but eventually gave in, and in 1919 won the Nobel Prize for ... QM!

It was not until 1917 that Einstein developed a full theory of photon-matter interaction. The less well-known part of that paper deals with how atoms in motion tend towards an equilibrium with the photon field via momentum conserving interactions that prove the photon concept. This calculation is a forerunner (by 60 years!) of the laser-cooling concept. The better-known part of the paper deals with atoms at rest (where the excited states are Maxwell-Boltzmann

(MB) distributed in thermal equilibrium) tending to an equilibrium with the photon field via energy conserving interactions. This is a forerunner (by 45 years) of the laser concept. In it Einstein derives the Planckian (Eq. (3)) in 3 lines! In a two state atom (with energy levels 1 & 2) he balances (spontaneous & stimulated) emission with absorption

$$(A_{21} + B_{21}I_\nu)N_2 = B_{12}I_\nu N_1 \tag{6}$$

along with the MB distribution requirement

$$N_2/N_1 = e^{-E_{21}/kT}. \tag{7}$$

Solving this for I_ν, at the high T limit which forces $B_{21} = B_{12}$, we obtain

$$I_{BB\nu} = \frac{A/B}{e^{E_{21}/kT} - 1}. \tag{8}$$

Wien's law forces $E_{21} = h\nu$ and $A/B = \nu^3$ and then the constants are derived by having $I_{BB\nu}$ equal to the R-J fit in the classical low ν regime. The Planckian is thus derived. Another important consequence of this treatment, which is a perfect lead-in to our next section, is a useful form of Kirchoff's law: the spontaneous emission "source term" $J_\nu = A_{21}N_2$ is, by Eq. (6), related to the net absorption by

$$J_\nu = A_{21}N_2 = B(N_1 - N_2)I_{BB\nu} = \kappa' I_{BB\nu}. \tag{9}$$

Since J_ν is a property of the atom, if the atom is in LTE (collisionally thermalized levels) this "source term" result will hold for other I_ν choices. We now turn our attention to precisely that situation, where I_ν is not exactly the Planckian, and find how it propagates through matter.

1.2 Radiation transport

We now ask the question: "what if the radiation / matter were not exactly in equilibrium, but nearly so?" Consider the gain and loss of a beam of photons as it propagates along its trajectory "s".

$$\frac{dI_\nu}{ds} = J_\nu - \kappa' I_\nu = \kappa'(I_{BB\nu} - I_\nu). \tag{10}$$

Here, for simplicity, we neglect many things (such as scattering, higher order terms such as time derivatives, etc) that are treated at great length in textbooks. To make progress easier, indeed to reduce the problem to its core physics, we define an optical depth coordinate

$$\tau_\nu(L) = \int_0^L \kappa'_\nu ds. \tag{11}$$

This allows us to re-write Eq. (10) simply as

$$\frac{dI_\nu}{d\tau_\nu} = I_{BB\nu} - I_\nu. \tag{12}$$

Using a standard integrating factor $\exp(\tau_\nu)$, we integrate from $\tau_\nu(0) = 0$ to $\tau_\nu(L)$ to get

$$I_\nu(L)e^{\tau_\nu(L)} - I_{\nu 0} = \int_{\tau_\nu(0)}^{\tau_\nu(L)} I_{\kappa BB}e^{\tau_\nu}d\tau_\nu \tag{13}$$

or

$$I_\nu(L) = \int_{\tau_\nu(0)}^{\tau_\nu(L)} I_{\kappa BB}e^{\tau_\nu - \tau_\nu(L)}d\tau_\nu + I_{0\nu}e^{-\tau_\nu(L)}. \tag{14}$$

From this two lessons immediately emerge: 1) a source contributes from within approximately 1 optical depth and beyond this distance its influence/contribution rapidly diminishes exponentially; 2) an optically thick system (in our case $\tau_\nu(L)$ being large) naturally approaches a BB spectrum.

To see this, simplify Eq. (14) by having no incident flux, I_0, and let $I_{BB\nu}$ be constant in space to obtain

$$I_\nu(L) = I_{BB\nu}\left[1 - e^{-\tau\nu(L)}\right]. \tag{15}$$

Let us now return to Eq. (10) and solve it in the so-called "diffusion approximation". We restrict ourselves to a planar geometry where quantities vary with x. Thus a ray moving in an arbitrary direction "s" that makes an angle θ with the x axis, and define $\mu = \cos(\theta)$ so $ds = dx/\mu$, makes Eq. (10) reduce to

$$\mu\frac{dI_\nu}{dx} = \kappa'(I_{BB\nu} - I_\nu). \tag{16}$$

We assume that I_ν deviates slightly from the isotropic $I_{BB\nu}$: $I_\nu = I_{BB\nu} + \mu\psi$. Plug this into Eq. (16) and obtain immediately

$$I_\nu = I_{BB\nu} - \mu\frac{1}{\kappa'}\frac{dI_{BB\nu}}{dx}. \tag{17}$$

Note that the second term is smaller than the first by $1/\tau$. We can now calculate the net flux through an area perpendicular to the x-axis

$$F_\nu = \int_{-1}^{1}\mu I_\nu 2\pi d\mu = -\frac{4\pi}{3\kappa'_\nu}\frac{dI_{\kappa BB}}{dx} = -\frac{c}{3\kappa'_\nu}\frac{dU_{\kappa BB}}{dx}, \tag{18}$$

and integrate over ν, and define a Rosseland (ν averaged) mfp ($1/\kappa'$), which is weighted in the average by $dUBB\nu/dT$ to get

$$F = -\frac{c\lambda_R}{3}\frac{d(aT^4)}{dx} = -\frac{4\lambda_R}{3}\frac{d(\sigma T^4)}{dx} = -\frac{4}{3}\frac{d(\sigma T^4)}{d\tau}. \tag{19}$$

This is a principal result that we will (eventually) use in the equation that couples the radiation to the matter. It tells us that radiant heat diffuses in a way entirely expected from diffusion theory – namely that the flux is a diffusion coefficient times a gradient of an energy density, and that the diffusion coefficient is a free streaming velocity times a mfp divided by 3. The final form for F of Eq. (19) tells us an equivalent message in a useful form – namely that the diffusive flux is the free-streaming flux reduced by the number of mfps (the optical depth) of the system.

Before proceeding to some applications we prepare the way with one more exercise in formalism. We integrate Eq. (17) over ν, and define $\phi = \sigma T^4$ to obtain

$$I = \frac{\phi}{\pi} + \frac{3\mu}{4\pi}F = \frac{\phi}{\pi} - \frac{\mu\lambda}{\pi}\frac{d\phi}{dx}. \tag{20}$$

This form will be very useful in exploring the diffusion picture in a regime where in principle it does not belong – namely near a boundary. Thus we encounter what are known as the Milne boundary conditions.

Consider a source flux ϕ_D going from left to right, as it encounters matter that fills space from $x > 0$, and is heated to a depth x_F, and whose surface T at $x = 0$ is given by $(\phi_B/\sigma)^{1/4}$ (namely T_B). The source flux to the right at $x = 0$ is given by

$$\vec{F}_R = \int_0^1 2\pi\mu I d\mu = \frac{2\pi\phi_D}{\pi}\int_0^1 \mu d\mu = \phi_D$$

and the flux, to the left at $x = 0$, from the matter re-radiating is given by

$$\vec{F}_L = \int_0^{-1} 2\pi\mu I d\mu = \frac{2\pi\phi_B}{\pi}\int_0^{-1}\mu d\mu - 2\pi\frac{\lambda}{\pi}\frac{d\phi}{dx}\int_0^{-1}\mu^2 d\mu = \phi_B + \frac{2}{3}\lambda\frac{d\phi}{dx}$$

so the net flux at $x = 0$ is given by:

$$\vec{F}_{Net} = \vec{F}_R - \vec{F}_L = \phi_D - \phi_B - \frac{2}{3}\lambda\frac{d\phi}{dx}.$$

This net flux must be carried inward into the material via the diffusive flux that we have already calculated in Eq. (19) namely $F = -(4/3)\lambda d\phi/dx$. Equating these two expressions for flux gives

$$\phi_D - \phi_B = -\frac{2}{3}\lambda\frac{d\phi}{dx}, \Rightarrow \phi_D = \phi_B\left(1 + \frac{F}{2\phi}\right). \tag{21}$$

For a linear profile of $\phi(x) = \phi_B\{1 - (x/x_F)\}$, Eq. (21) tells us that $\phi_D = \phi_B\{1 + (2/3)\lambda/x_F\}$, which can be thought of in the following way: ϕ_D is a value higher than ϕ_B as if the linear profile if $\phi(x)$ was extended backward into the $x < 0$ regime by an amount $(2/3)\lambda$.

The same kinds of considerations can be applied to the surface of a star. Consider a half space $x < 0$ filled with matter and whose surface temperature at $x = 0$ is T_S. Then the flux to the right from the star is given by

$$\vec{F}_R = \int_0^1 2\pi\mu I d\mu = \frac{2\pi\phi_S}{\pi}\int_0^1\mu d\mu - 2\pi\frac{\lambda}{\pi}\frac{d\phi}{dx}\int_0^1\mu^2 d\mu = \phi_S - \frac{2}{3}\lambda\frac{d\phi}{dx}$$

and the flux to the left onto the star from the of vacuum of space is given by:

$$\vec{F}_L = \int_0^{-1} 2\pi\mu I d\mu = 0$$

thus the net flux from the star is:

$$\vec{F}_{Net} = \vec{F}_R - \vec{F}_L = \phi_S - \frac{2}{3}\lambda\frac{d\phi}{dx}.$$

But this net flux at $x = 0$ must be supplied by the diffusive flux from within the star, which as usual is $F = -(4/3)\lambda d\phi/dx$. Equating the two expressions for flux gives

$$\phi_S = -\frac{2}{3}\lambda\frac{d\phi}{dx} = \frac{F}{2}, \Rightarrow F_{Net} = 2\phi_S = 2\sigma T_S^4 = \sigma T_{Brite}^4. \tag{22}$$

This can be thought of again for a linear profile of $\phi(x) = \phi_s\{1 - [x/(2\lambda/3)]\}$, as if the surface that "truly" is emitting, is at a distance $x = -2/3\lambda$ within the star where $T_{Brite} = T(x = 0 - (2/3)\lambda) = 2^{1/4}T_S$. This result is consistent with our "lesson (1)" from above, namely that emission comes from within about 1 mfp.

With all of this formalism now established, we are finally in a position to apply the results from this section to a variety of ICF relevant situations. We begin by finding a highly relevant case where Eq. (19) plays a crucial role.

1.3 Solving the diffusion equation

The first application we seek is the most relevant to the ICF indirect drive problem. We consider a hohlraum illuminated by a laser of energy E_L. It enters the hohlraum (usually made of a high Z material such as Au) and is absorbed along the inner walls where it is aimed. The hot plasma that ensues is a copious source of x-rays. We parameterize this process by a conversion efficiency

η_{CE}. Thus we assume that $\eta_{CE}E_L$ of x-rays now floods the hohlraum and uniformly bathes the wall areas of all that it sees. Some of the x-rays are absorbed by the capsule, the goal of the exercise after all!, and some unfortunately leave the hohlraum through the laser entrance holes (LEH) necessary to get the laser into the hohlraum in the first place. Since the capsule ablator is normally of low Z material that does not re-radiate back much, it, like the LEH (which is the ultimate in energy sinks – a vacuum) absorb all the flux σT^4 that impinges on it. Thus we know immediately how to calculate those two energy loss channels: σT^4 times the area of the capsule and LEH respectively, integrated over time.

Our major challenge is to calculate the wall loss, since it does re-radiate energy to a significant degree and moreover, the majority of the area in the problem is the hohlraum wall. It is made of a high Z matter, which when heated by the flux of incident x-radiation will re-radiate much of it back. Typically, as we shall see, the re-radiation factor is <u>not</u> enough to disqualify the walls as the "chief energy loser" – the preponderance of wall area still make the wall the principal sink of energy in ICF hohlraums. Since the incident x-rays are absorbed by remaining bound electrons of the multiply ionized ions of Au, and then are re-radiated in a random direction as the electron finds a lower energy level to return to, we have a typical random walk / diffusive process. Collisions also keep the levels at near MB distributions (a situation called Local Thermodynamic Equilibrium (LTE)). This near total thermal equilibrium, near isotropic radiation field situation is precisely the conditions under which we derived the diffusive flux results in the previous section.

Thus our goal here is clear – to calculate the energy absorbed by a high Z wall subject to an external flux of x-rays. The time rate of change in internal energy (per unit volume) would be equal to the divergence of the diffusive flux. If the flux were divergence free then as much energy that entered a volume element of matter would then exit it and we'd expect no net change in the internal energy. For our system there are two fields of energy to consider: matter and radiation. The energy density of matter we write as ρe_{th}, where e is the specific energy, and we know what it is for radiation: aT^4. It turns out that the radiation energy density is negligible for the ICF problem. For the diffusive flux there are two contributions. We know what it is for radiation: $F = -(4/3)\lambda d\phi/dx$. For matter, the electron heat flux is similarly given by a Dd(energy density)$dx = (v_e\lambda_e/3)d(\rho e_{th})/dx$. It turns out that for above about 20 eV and density 1 g/cc the radiative flux dominates. Thus our energy equation should read:

$$\frac{\partial}{\partial t}(\rho e_{th}) = \frac{\partial}{\partial x}\left(\frac{c\lambda_R}{3}\frac{\partial}{\partial x}(aT^4)\right). \tag{23}$$

This is a complicated non-linear PDE. Before proceeding with any formal solutions, we can see how far a simple dimensional analysis can take us. We write down Eq. (23) dimensionally

$$\frac{\rho e}{t} \sim \frac{1}{x}\lambda\frac{caT^4}{x} \sim \frac{1}{x}\frac{1}{\kappa\rho}\frac{\sigma T^4}{x}. \tag{24}$$

Note a slight change in nomenclature – we follow from here on out the convention that opacity κ is defined as $1/\rho\lambda$ (whereas in Eq. (9) it was simply $1/\lambda$). It is still true that we always correct the absorption for the stimulated emission, so that there is net absorption. We can recast this in terms of the "Marshak front areal density" m_F

$$m_F^2 \equiv (\rho X_F)^2 \sim \frac{\sigma T^4 t}{\kappa e} \tag{25}$$

What we mean conceptually by this procedure is as follows: a non-linear heat wave of radiation (called a Marshak wave after R. E. Marshak (1958)) progresses diffusively through the material. Picture, at a given time, a flat-topped $T(x)$ profile that eventually takes a sharp nose-dive to zero at a front position x_F. We see that a $T(x) = T_0\{1 - [x/x_F(t)]\}^{1/4}$ is a good approximation to the actual solution. Since T is nearly flat and therefore at its surface value for most of x, the dimensional analysis works particularly well in giving us a feel for the time and T dependence of the front position x_F or m_F. To complete this analysis we need to know the T, ρ dependencies of κ and e. For Au we find that

$$\kappa = \kappa_0 \frac{\rho^{0.2}}{T^{1.5}} \quad \text{and} \quad e = e_0 \frac{T^{1.6}}{\rho^{0.14}}. \tag{26}$$

We note the remarkable near cancellation of dependencies of the product κe, which appears in the denominator of m_F in Eq. (25). Why is this? Well the basic reason is that the opacity κ depends on the number of <u>bound</u> electrons left in the partially ionized high Z ion, whereas the internal energy e depends on the opposite – the number of <u>free</u> electrons liberated from that high Z ion. (Note that e scales as $[(Z+1)/A]T$). We can be somewhat more precise. The T dependence of e can be understood as follows. We expect an ion to be ionized to the degree that its ionization potential I_p is of order the thermal temperature T (electron collisional ionization is the dominant process). But in a Bohr-like atom, I_p scales as Z^2. Thus expect Z to scale as $T^{1/2}$ and e as $T^{3/2}$ which it very nearly does! The opacity is quite complicated, but certainly it should decrease with T because its number of bound electrons is decreasing as T increases. For the density dependence we note that in a Saha equilibrium the higher the overall density, the lower the ionization state – it's as if the free states are more full and prevent bound electrons from occupying them – it's as if high density "pushes the electrons back into the atom". Thus the higher the ρ the lower the Z so e will scale as $(1/\rho)$ to a small power. Since lower Z means more bound electrons, the opacity κ will increase as ρ to a small power. In any event Eqs. (25) and (26) lead to

$$m_F \sim \frac{T^{1.95}t^{1/2}}{\sqrt{\kappa_0 e_0}\rho^{0.03}}. \tag{27}$$

With such a tiny ρ dependence you may think our task is done. But what we are really after is the energy (per unit area) in the wall which dimensionally is em_F and e has some small but non-negligible density dependence. So our task in this dimensional analysis exercise is not complete until we find an expression for the density in terms of T and t.

One way to do this is to assume we are deep into the supersonic regime – namely the heat wave progresses so fast through the material it hardly has time to move and thus the density is constant in space and time and at its initial value ρ_0. This will occur, at a given T and t, for very low densities (imagine a high Z Au foam). To see why, recast Eq. (27) in terms of $x_F = m_F/\rho$ and then differentiate with respect to time to get a velocity of the heat front. At a given T & t that speed will scale as $\rho^{-1.03}$. The sound speed c_s scales as $e^{1/2}$ and thus as $e_0^{1/2}T^{0.8}\rho^{-0.078}$, a very weak ρ dependence. Thus we can ensure that the heat front speed exceeds the sound speed for low ρ. Putting $\rho = constant = \rho_0$ into Eq. (27) and into e gives us a scaling for energy loss per unit area in the supersonic regime

$$E/A \sim (\rho e)X_F \sim em_F \sim \sqrt{\frac{e_0}{\kappa_0}} \frac{T^{3.55}t^{1/2}}{\rho_0^{0.17}}. \tag{28}$$

The exact opposite extreme is the highly subsonic regime, in which the heat front progresses very slowly through the solid density high Z material. What ends up happening is an isothermal rarefaction wave progresses through the heated material, significantly decompressing it and inducing hydrodynamic motion in this "blow-off" plasma.

A simplified (and dimensionally correct) way to proceed (Rosen (1979)) is to find an "average" density in the blow-off by reasoning that after some time t, an amount of mass (per unit area), $m_F(T, t, \rho)$, has been heated (reached by the heat front), and expands into the vacuum at the sound speed $c_S(T, t, \rho)$, so that $\rho = m_F/c_S t$. This accomplishes what we sought – a way to relate ρ (implicitly) to T & t. But since all ρ dependencies of m_F and c_S are power laws it is straightforward to solve for ρ explicitly in terms of T & t, and then solve for m_F and E/A. Before we do this, we note a particular piece of luck here. Since m is formally a running integral of ρ over x (it's basically a Lagrangian coordinate), then our simplified dimensional equation $\rho = m_F/c_S t$ is really an integral equation for ρ, with a solution that is exactly the isothermal rarefaction! So proceeding with the substitutions we eventually get

$$ m_F \frac{T^{1.91} t^{.52}}{(\kappa_0 e_0)^{0.48}} \Rightarrow \frac{E}{A} \; em_F \; \frac{e_0^{0.7}}{\kappa_0^{0.4}} T^{3.34} t^{0.6}. \tag{29} $$

This completes our dimensional analysis. The hard work of determining the exact coefficients is now briefly outlined. The full set of hydrodynamic equations, in Lagrangian format (m, not x) are

$$ (mass)\frac{\partial V}{\partial t} = \frac{\partial u}{\partial m}, \; (momentum)\frac{\partial u}{\partial t} = -\frac{\partial P}{\partial m}, \; (energy)\frac{\partial e}{\partial t} + P\frac{\partial V}{\partial t} = \frac{4}{3}\frac{\partial}{\partial m}\frac{1}{\kappa}\frac{\partial \sigma T^4}{\partial m}. \tag{30} $$

Here $V = 1/\rho$. We supplement these equations with equations of state, Eq. (26), along with $P = re/V$ where we find $r = 0.25$ to be reasonably accurate. For the record, we complete Eq. (26) by presenting the coefficients: $\kappa_0 = 7200.$ g/cm^2 and $e_0 = 3.4$ MJ/g. In Hammer & Rosen (H&R) (2003) we solve (30) by means of a perturbation technique with a small parameter $\epsilon = 1.6/(4 + 1.5) = 0.29$, the numbers being the power law T dependencies of e, aT^4, & κ respectively. We do so for a particular, self-similar assumption on the behavior of $T(x,t)$, namely $T = T_B t^k f(m/m_F(t))$. We find, to zero order, a T spatial profile $f = (1 - (m/m_F(t))^{1/4}$ where $m_F(t) = m_{F0} t^{(1+4k)/2}$. The ρ and u profiles are, to the same zero order, those of an isothermal rarefaction. The first order solutions differ from all these by quantities of order ϵ. In H&R we verify energy conservation, integral $E(x,t)dx = \int F(x = 0, t)dt$ through order ϵ^2, where E includes internal and kinetic energy, and F is the absorbed flux. In addition, in H&R we solve the simpler supersonic equation for arbitrary $T_B(t)$.

We quote here the results for two useful choices of k: 0 & 0.18. The scaling of m & E/A are precisely those of Eq. (29), as they must be! The coefficients are $m_{F0} = (9.9, 7.4)10^{-4}$ g/cm^2 respectively, and $E/A = (0.58, 0.39)$ hJ/mm^2 respectively. The absorbed flux is given by $F = F_0 T^{3.34} t^{-0.41}$ with coefficients $F_0 = (0.34, 0.46)$ hJ/ns/mm^2. Note that E/A is simply the time integral of F. Also be aware that for the $k = 0.18$ case you must remember to put the time dependence of $T = T_0 t^{0.18}$ into all of the equations. Thus for example the E/A (for $k = 0.18$) $= 0.39 T_0^{3.34} t^{1.2}$ hJ/mm^2.

It seems unfair to present all of these results without giving the reader a sense of how they come about. So we now proceed to show an illustrative example of an exact solution for the supersonic (constant ρ) case. Thus we start with

$$ \frac{\partial}{\partial t}T^\beta = C\frac{\partial^2}{\partial x^2}T^{4+\alpha}, \quad \text{with} \quad C = \frac{16}{4+\alpha}\frac{g\sigma}{3f\rho^{2-\mu+\lambda}}, \tag{31} $$

where here we generalize the T, ρ power law behavior of κ and e and write them as $\kappa = (1/g)T^{-\alpha}\rho^\lambda$ and $e = fT^\beta\rho^{-\mu}$. Thus for Au (see Eq. (26)) $\alpha = 1.5$, $\beta = 1.6$, $\lambda = 0.2$, $\mu = 0.14$, and the coefficients are $(1/g) = 7200.$ g/cm^2 and $f = 3.4$ MJ/g. In the early 1950's Louis Henyey found 2 exact solutions of Eq. (31). This work remains unpublished so in H&R we devoted an appendix to present it. The simpler of the 2 solutions is what we now present. Henyey took a power law time dependence t^p for the boundary temperature with $p = 1/(4 + \alpha - \beta)$, which for Au is approximately 1/4. He then finds:

$$ T(x,t) = T_B t^{1/(4+\alpha-\beta)}\left(1 - \frac{x}{x_{F0}t}\right)^{1/(4+\alpha-\beta)} \text{ with } \quad x_{F0}^2 = C\frac{4+\alpha}{4+\alpha-\beta}T_B^{4+\alpha-\beta}. \tag{32} $$

The reader should verify this by plugging Eq. (32) into (31) and experience the fun of terms miraculously canceling and verifying that indeed Eq. (32) solves Eq. (31) exactly. Note that this solution looks very much like the advertised zero order solution of H&R, namely a spatial profile that looks like $\{1 - (x/x_F(t))\}^{1/4}$ and with the $T = T_B t^{k=1/4}$, we get the front advancement behavior $x_F(t) = x_{F0}t^{(1+4k)/2}(= x_{F0}t^1$ here). Since $p = 1/4$ is a small number, it is easy to accept that other (non-exact solution) cases that might be of interest such as $k = 0$, will simply be of order ϵ different from this solution. Using this exact solution we can derive some interesting quantities, especially the very useful one of wall reflectivity, albedo.

1.4   Solving for the Albedo

With an exact solution in hand let us proceed to calculate the fundamental quantity of diffusion processes, the optical depth τ. To simplify the notation set $y = x/x_{F0}t$. Then, in this supersonic (constant ρ) regime, $\lambda = (\kappa_0\rho_0)^{-1}T^\alpha$, thus

$$
\tau = \int \frac{dx}{\lambda} = \kappa_0\rho_0 \int \frac{dx}{T^\alpha} = \left(\frac{\kappa_0\rho_0 x_F t}{T_B^\alpha t^{\alpha p}}\right) \int_0^y \frac{dy}{(1-y)^{\alpha/(4+\alpha-\beta)}}
$$
$$
= \left(\frac{\kappa_0\rho_0 x_F t}{T_B^\alpha t^{\alpha p}}\right)\left(\frac{4+\alpha-\beta}{4-\beta}\right)\left[1 - (1-y)^{\frac{4-\beta}{4+\alpha-\beta}}\right] \equiv \tau_F\left[1 - (1-y)^{\frac{4-\beta}{4+\alpha-\beta}}\right].
$$
(33)

Note that τ_F represents the total number of mfps within the Marshak wave. Note that it increases with time as $t/t^{1.5/4} = t^{0.6}$. We can invert Eq. (33) to get:

$$
1 - y = \left(1 - \frac{\tau}{\tau_F}\right)^{\frac{4+\alpha-\beta}{4-\beta}} \quad \text{and thus} \quad T(\tau) = T_B(1-y)^{\frac{1}{4+\alpha-\beta}} = T_B\left(1 - \frac{\tau}{\tau_F}\right)^{\frac{1}{4-\beta}}.
$$
(34)

In Figure 1 we show the $T(y)$ and $T(\tau)$ profiles for a situation ϵ different than our exact solution, namely a "NIF-like" 250 eV, 7.7ns "flat top" ($T_B = T_0t^0$) drive. It shows shapes very much like those predicted by Eq. (34). For the record, the H&R solutions for that case are $T(y) = T_B(1 - y^{1.2})^{0.25}(1 - 0.05y)$ and $T(\tau) = T_B[1 - (\tau/\tau_F)]^{0.42}[1 - 0.1(\tau/\tau_F)]$, which indeed only differ by ϵ from Eq. (34). Also for the record, putting in the constants we find for that case that $\tau_F = 8.4T_0^{0.65}t^{0.4}$ (with T in heV and t in ns) $= 34$ by the end of the 7.7 ns pulse.

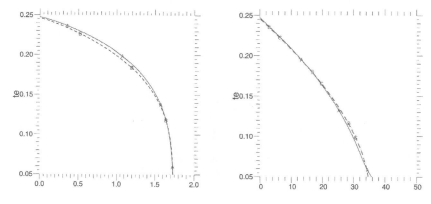

Figure 1. *a) $T(m)$ profile & b) $T(\tau)$ profile for a 250eV 7.7 ns drive. (Dotted lines are analytic theory and solid lines are numerical solutions.)*

With the temperature profile now written in its fundamental form, in terms of τ, we can now straight away solve the radiation transport equation to find how much this "source profile" $I = \sigma T(\tau)^4/\pi$ will radiate back into the vacuum. Since we know most of the radiation will come from near the first mfp, we can simplify the calculation by expanding $T(\tau)^4$ near the surface (τ/τ_F small), and we now proceed to integrate the emission $I(\mu)$ all along a ray that makes an angle θ with the x axis with $\cos\theta = \mu$:

$$
\begin{aligned}
I_\mu &= \left(\frac{\sigma T_B^4 t^{4p}}{\pi}\right) \int_0^{\tau_F/\mu} \left(1 - \frac{4}{4-\beta}\frac{\tau}{\tau_F}\right) e^{-\tau/\mu} d\tau/\mu \\
&\approx \left(\frac{\sigma T_B^4 t^{4p}}{\pi}\right)\left[1 - \left(\frac{4}{4-\beta}\right)\frac{\mu}{\tau_F}\right] + O(e^{-\tau_F}).
\end{aligned}
\tag{35}
$$

Note that because all of our quantities vary with x, we have defined τ here as $\tau(x)$, not as the simpler $\tau(s)$ as before, hence the τ/μ terms in Eq. (35) vs. Eq. (14).

The total reemission can then be found by integrating over all angles of emission:

$$
I_{Tot} = \int_0^1 2\pi\mu I_\mu d\mu = \sigma T_B^4 t^{4p}\left[1 - \frac{4}{4-\beta}\frac{2/3}{\tau_F}\right].
\tag{36}
$$

Another immediate outgrowth of Eq. (34) is our ability to immediately calculate the absorbed flux (through the $x = \tau = 0$ front surface) from the last form of F of Eq. (19)

$$
F = -\frac{4}{3}\frac{d(\sigma T^4)}{d\tau} = \frac{4}{3}\sigma T_B^4 t^{4p}\frac{4}{4-\beta}\frac{1}{\tau_F}.
\tag{37}
$$

When considering Eqs. (36) and (37) we are struck with a paradox. We would expect the sum of the re-radiated flux (Eq. (36)) plus the absorbed flux (Eq. (37)) to add up to the incident flux $\sigma T_B^4 t^{4p}$. But they do not! The $1/\tau_F$ terms have coefficients $-2/3$ and $+4/3$. What went wrong?

The answer lies in the Milne boundary conditions. We have defined all of our Marshak wave solutions in terms of T_B which is the <u>matter</u> boundary temperature. This makes no statement whatsoever as to what is the required <u>external</u> drive temperature T_D needed to achieve that boundary temperature. That is what the Milne boundary conditions tell us. In fact Eq. (21) tells us that $T_D^4 = T_B^4[1 + (F/2T^4)]$. If we employ that relationship into our energy accounting all turns out well, as we now demonstrate.

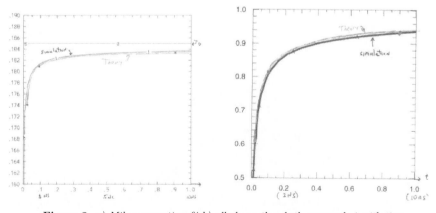

Figure 2. *a) Milne correction & b) albedo as they both approach 1 with time.*

Using Eq. (37) we can re-write Eq. (36) as re-emitted flux $= \phi_B[1 - (F/2\phi)]$, and Eq. (37) as absorbed flux $F = \phi_B(F/\phi)$, and from the Milne boundary condition, incident flux $= \phi_B[1 + (F/2\phi)]$. Now it is self-evident that energy is conserved and that incident flux $=$ absorbed $+$ re-emitted flux. Now with energy conservation safely within our grasp we can consider albedo α as the ratio of re-emitted flux to incident flux that conserves energy: $\alpha = 1 - (F/\phi)$. Since F/ϕ is proportional to $1/\tau_F$, and τ_F grows in time, we expect both the albedo and the Milne correction to approach 1 as time increases. This is shown rather clearly in Fig. 2. The conditions of Fig. 2 are a Ta_2O_5 40 mg/cc foam driven by a flat T_D of 185 eV.

Before proceeding, let's make sure we did not get lost in all of the math. Let us ask what we might have guessed the formula for emitted flux to be. Based on our "lesson number 1" we know that emission comes from about the first mfp. To fine-tune it a bit from our Milne boundary condition calculations it seemed like 2/3 of a mfp was the more precise answer. Thus we would expect re-emission to be T^4 but T is evaluated at 2/3 of a mfp into the Marshak profile. Thus using Eq. (34)

$$T^4(\tau = 2/3) = T_B^4(t)\left[1 - \frac{2/3}{\tau_F}\right]^{\frac{4}{4-\beta}} \approx T_B^4(t)\left[1 - \frac{4}{4-\beta}\frac{2/3}{\tau_F}\right], \tag{38}$$

which is precisely the re-emission formula we calculated in Eq. (36). One more "reality check" is in order. Eq. (35) tells us to expect the re-emission to be angle dependent so experimentalists must be aware of this as they choose angles of detection. The physics of this may well be called "Marshak limb brightening". Basically, if you look normal to a wall, the emission coming from 1 mfp comes from deep in the wall, and thus at a low T within the Marshak wave $T(x)$ profile. If you look at a shallow angle to the wall, then along that ray you will reach 1 mfp in, yet you will hardly be different than $x = 0$, so you will be getting emission from a high T very near the surface of the $T(x)$ Marshak wave profile. Fig. 3 illustrates this principle for the same conditions as Fig. 2.

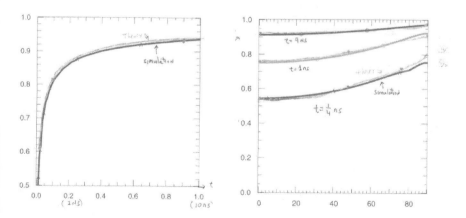

Figure 3. *a) Angle averaged albedo vs. time & b) Albedo vs. angle for 3 different times.* $\text{Cos}(48^0) = 2/3$ *represents the angle averaged result*

As a final application of this albedo section we compare predictions based on our analytic work with full simulations and data. Jones et al (2004) measured the albedo of a large hohlraum wall that was fed by a smaller hotter hohlraum. The flux out of the exit hole of the small hot hohlraum was measured and that is the incident flux impinging onto the large hohlraum's wall.

The flux re-emitted by that larger hohlraum was also measured and the ratio was interpreted as an albedo. There are 4 data points. The first two are Au at 70 eV and at 100 eV (both at 1.5 ns) with experiment and simulation giving albedoes of 0.53 and 0.63 respectively. Using $\alpha = 1 - (F/\phi)$ with F given by H&R for the hohlraum relevant $k = 0.18$ case (more about that in the next section), namely $F/\phi = 0.46/T^{0.7}t^{0.53}$ (T in heV, t in ns) we would predict albedoes of 0.53, 0.63 respectively as well! For a $Au_{0.2}Dy_{0.2}U_{0.6}$ "cocktail" (more about these in the next section) at 100 eV and 1.5 ns we predict a slight decrease in opacity and an albedo of 0.62, which again is what the experiment measures and the simulations predict. For $U_{0.86}Nb_{0.14}$ the experiment shows an albedo of 0.65 but the simulations predict 0.53. We still do not understand that result. Our model matches the simulation (that's the best it can do in any event!) with a prediction of 0.54. Thus our simple formulae do a very good job of matching complex simulations (which in turn, often match complex experimental data).

1.5   Solving for the hohlraum temperature

We are now very well positioned for our principal application – for a given hohlraum geometry and incident laser pulse, predict the hohlraum temperature. We will then compare that with data so let us first describe the measurement techniques (Kaufman (1994)). We typically obtain T by two different methods.

We measure the x-ray flux through a hole in the can (sometimes on the side, sometimes a view through the LEH) via a dozen or so roughly 100 eV wide ("Dante") channels that span an energy from 50 eV to 5 keV with most of the energy in the 100-1000 eV range where most Planckian spectra lie for temperatures of order 100-300 eV. A Planckian is fit to the resultant spectrum, and it usually matches in both brightness (intensity) and color (shape). Examples of Dante spectra are given in Figure 4.

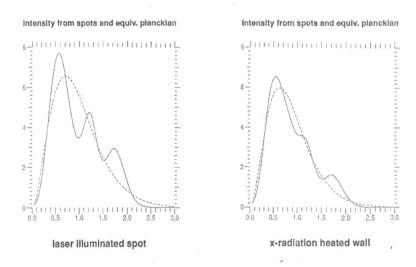

Figure 4. *Observed spectra from a) a laser illuminated spot b) an un-illuminated spot on the hohlraum wall.*

Since these channels have decent time resolutions we can also get $T(t)$ information. The second

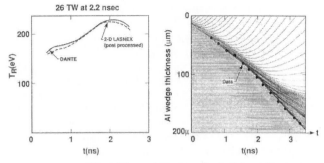

26 TW at 2.2 nsec

Pulse is 2.2 nsec, 1:3 foot to main pulse power ratio

Figure 5. *a) Time history of a shaped pulse from Dante and from Lasnex simulation. b) Applying the Lasnex simulation to an Al wedge witness plate predicts the experimentally observed temporal history of the shock break out.*

method involves an Al "witness plate" on the side of the hohlraum. Radiative flux absorbed sub-sonically in the Al launches a shock. When the shock breaks out the back of the Al it lights up in the visible range, and a streaked optical signal detector records the time history of the shock breakout. For a stepped witness plate we can derive a shock speed, hence a pressure, and a drive T. For a wedge shaped witness plate we can even derive a more sophisticated pulse shaped $T(t)$. In general the 2 methods match, as do the Lasnex (Zimmerman(1975)) simulations. See Fig. 5.

To calculate all of this analytically we adopt a simple "source = sink" model. The source is the laser energy E_L, and as described above in the beginning of Section 1.3, we assume it is converted to x-rays with some conversion efficiency, so that now $\eta_{CE}E_L$ worth of x-rays bathe the hohlraum walls. If the absorption fraction is less than one, we should also take this into account. We will set the source equal to the energy sinks, which for a very simple hohlraum (no capsule) is the wall loss (which we have calculated as E/A of the previous two sections times the area of the walls, and the LEH loss, which is the time integral of σT^4 times the area of the laser entrance holes.

Before presenting these results, let us first introduce convenient "radiation hohlraum units" ("rhu") in which T is measured in hectovolts (hundreds of eV), area in mm^2, time in ns, mass in gm and energy (a bit clumsily) in hectojoules. With these units, $\sigma = 1$, (well, 1.03 to be exact) and normalized irradiance is 10^{13} W/cm^2 ($=$ hJ/mm^2 ns $= 10^2$ J /10^{-2} cm^2 10^{-9} s) and similarly, normalized power is 10^{11} W ($=$ hJ/ns $= 10^2$ J /10^{-9} s). Thus a 100 eV ($=$ 1 heV) hohlraum radiates out a hole with a flux of $\sigma T^4 = 1$ (in rhu) $= 10^{13}$ W/cm^2.

As an example we take the following "scale 1" hohlraum illuminated on the Nova laser at LLNL in the 1990's. It was a gold cylinder of length $L = 2.5$ mm, and radius $R = 0.8$ mm, and on each end a disk sealed the cylinder. Each disk had a "50% LEH" namely a laser entrance hole of radius 0.4 mm. One immediately calculates the wall area $A_W = 15.6$ mm^2 and $A_{LEH}=1$ mm^2. The source energy, a "flat top" laser power of 100 – 300 hJ/ns for a duration of 1 ns. ($=$ 10-30 TW). Our simulations predict a conversion efficiency to x-rays of $\eta_{CE} = 0.7\ t_{ns}^{0.2}$. The efficiency increases with time in part because the albedo behind the conversion layer builds up with time. This time behavior helps explain an important experimental observation, that T rises as $t^{0.18}$, hence our interest as quoted above with the $k = 0.18$ case. We can understand that result in the following way. Equating the x-ray source, $\eta_{CE}E_L = \eta_{CE}P_L t$ which scales as $t^{1.2}$ to the principal x-ray sink, the wall, E_W which scales as $T^{3.3}t^{0.6}$ (Eq. (29)) we see that these two terms will balance if $T = T_0 t^{0.18}$! Conversely, if we wish to have a truly flat $T = T_0 t^0$, we need a $P_L(t)$ that "droops" in time.

Let us now proceed directly to the calculation. For the 30 TW experiment, the source of x-rays (at 1 ns) will be $0.7 P_L t = (0.7)(300)(1) = 210$ hJ. The wall loss E_W will be (using the $k = 0.18$ results from H&R quoted in the discussion that followed Eq. (30) above) $0.39 T_0^{3.3} t^{1.2} A_W = 0.39 T_0^{3.3}(1)(15.6) = 6.24 T_0^{3.3}$ at 1 ns. We must also calculate the LEH loss. The flux out the LEH will be $T^4 A_{LEH}$ so we integrate $t^{4(0.18)}$ in time and get $E_{LEH} = 0.58 T^4$ at 1 ns. Solving $210 = 6.24 T_0^{3.3} + 0.58 T^4$ results in a $T = 2.75$ with 176 hJ of wall loss and 34 hJ of LEH loss (justifying our claim that most of the loss is in the walls). The resulting prediction of 275 eV matches data and simulations quite well. Repeating this calculation for say 10 TW (70 replaces 210) yields a $T = 1.99$ again in agreement with data and simulation. The results for the entire database are shown in Figure 6, and shows that our simple model of source=sink with sinks calculated by H&R organize the database very nicely.

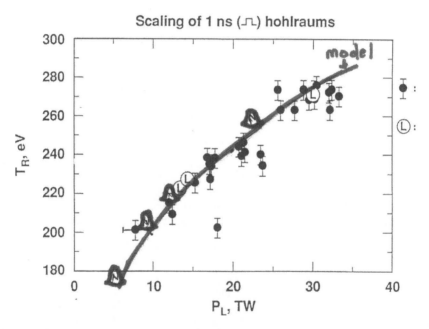

Figure 6. *Our simple source=sink model (line) organizes the 1 ns database very well. Solid points are from Nova. L means Lasnex 2-D simulation. N means NIF early light experiments (2 ns pulses, with T here extracted from that data at t = 1 ns).*

As a final application of what we have learned so far, consider scaling to bigger systems (say from Nova to NIF to a reactor scale driver). We seek to improve the coupling efficiency defined as energy absorbed by the capsule to incident laser energy. Since most of the energy is absorbed and turned into x-rays, and most of these are lost into the walls, this principally reduces to a question of the capsule loss divided by the wall loss. Now the capsule ablator is low Z, therefore it just sucks up radiation energy and does not re-emit much, and it absorbs mostly like an LEH, therefore E_{cap} scales roughly as $A_{cap} T^4 t$. The wall loss E_W we now know well, scales roughly as $0.5 A_W T^3 t^{1/2}$. Thus the ratio, the coupling efficiency, scales as $(A_{cap}/A_W)2.T t^{1/2}$. To improve the coupling which of these four factors can we change? The wall ratio is mostly fixed by geometric smoothing symmetry arguments. The value for the drive T is mostly fixed by hydrodynamic instability and laser plasma interaction physics.

However, as we increase the driver energy scale, the size scale will increase as will the time scale. So that is at least one natural way to improve the coupling. An alternate route to improving the ratio would be to change the number 2. Namely, are there wall loss mitigation tricks that we can test on present lasers that can be used on future laser driven hohlraums to improve the coupling efficiency? Answering that question is the subject of the next section.

2 Hohlraum optimization

Introduction

There is always, in practice, pressure on the scientists to keep a laser below its damage threshold. This means using less laser energy. So we ask the question: Can the NIF reach ignition using only 1 MJ of energy rather than 2 MJ? To do so we need to find ways to make hohlraum walls less lossy than the standard solid Au. In this section we discuss the state of research trying to answer this question.

2.1 Cocktails

When we consider Eq. (29) we see that in order to lower the E/A of a wall loss, we need to lower e and to raise k. Since e scales as Z/A, the higher the Z_N (or A) the lower will be the ionization state Z and hence e (at a given T). Thus, either replacing Au with U or at least mixing in higher Z_N elements into the wall will lower e. Moreover, if we do mix in higher Z_N elements, at a given T, they will have different atomic levels and structures and their opacities, if Z_N is chosen properly, will be high at frequencies when Au is low. Overall this "cocktail" of materials accomplishes both things. A $U_{0.6}Dy_{0.2}Au_{0.2}$ cocktail is an example that at NIF-like temperatures of about 300 eV can, by Eq. (29), save nearly 20% in wall loss. We have chosen to test the concept via a baseline Au hohlraum at the Omega Laser at URLLE. A "scale 0.75" cylinder ($L = 2.06$ mm, $R = 0.6$ mm, with 66% LEH ($R = 0.4$ mm) so that $A_W = A_{endcaps} + A_{cylwall} = 1.2 + 7.8 = 9.0$ mm^2 and $A_{LEH} = 1$ mm^2. The incident flattop power was 20 TW for 1 ns. As in the discussion of Section 1.5 we use the $k = 0.18$ results of H&R. We infer (from experiments) an approximately 8% reflectivity, and therefore with a 68% conversion efficiency we get as a source at 1 ns 101 hJ. We set this equal to the wall loss $3.5T_0^{3.3}$ and LEH loss $0.6T_0^4$, solve for T and get $T = 2.55$ with 76 hJ wall loss and 25 LEH loss. The 255 eV is very close to the observed temperature.

Early experiments with cocktails (Orzechowski (1996)) compared the burn-through times t_{bt} of Au foils (Rosen(1996)) placed across a hole in the side of a 260 eV hohlraum, to those of AuGd cocktail foils. A delay in burn-through signal for the cocktail is clearly seen. From Eq. (28) we expect (again for a $k = 0.18$ case) that t_{bt} should scale as $mT_0^{-2}(e\kappa)^{1/2}$, so the higher κ of the AuGd cocktail causes the delay. Since then we have tried to measure the rise in T for a full cocktail (vs. Au) hohlraum at the same laser drive.

We fit our latest opacity/ EOS theory of Au as $\kappa = 6544\rho^{0.18}/T^{1.43}$ (cm^2/g) and $e = 3.33T^{1.54}/\rho^{0.15}$ (MJ/g), and of $U_{0.6}Dy_{0.2}Au_{0.2}$ is $\kappa = 5670\rho^{0.10}/T^{0.90}$ (cm^2/g) and $e = 0.95e_{Au}$. The cocktail has a "flatter", less sensitive T, ρ behavior because it averages over several elements. We also note that the opacity of cocktails does not exceed that of Au until past 130 eV. Using that input, H&R predicts for $k = (0, 0.18)$, for Au an $E/A = (0.598, 0.398)T^{3.3}t^{0.6}$ (hJ/mm^2) respectively and for the $U_{0.6}Dy_{0.2}Au_{0.2}$ $E/A = (0.604, 0.407)T^{3.1}t^{0.57}$ (hJ/mm2), respectively. Thus, at 270 eV and 1 ns, the wall loss ratio (cocktail/Au) is (0.84,0.85) respectively while a full multi-group simulation gives (0.85,0.87), very close to H&R theory but differing mostly because the opacity is hard to fit with a single power law. All of these were for T_B scaling as t^k. If we simply let T_D scale as t^k we have the Milne conditions kicking in and the ratios are (0.84,0.88) respectively. For U mixed with 6% Nb by weight (= 14 atom %) add 1% to all these ratios.

Another outgrowth of these scaling laws is to notice that the wall loss ratio scales as $T^{-0.22}t^{-0.05}$ for $k = 0.18$ and even for the "flat top" $k = 0$ case the wall loss ratio has a $t^{-0.02}$ ratio. Thus to the degree that the Omega experiments are not either at the full NIF temperature of 300 eV, nor at the NIF pulse length of 3-4 ns, then the results from such experiments will be pessimistic in showing a wall loss ratio advantage of a cocktail hohlraum over Au than would a NIF ignition hohlraum. The ratio for NIF is about 0.83. All of these time behaviors stem from the fact that early in time the lower T parts of the Marshak wave profile are relatively more important, and for low T the cocktail is actually worse than Au.

So let us redo the Omega hohlraum calculation for T with cocktail walls (actually a shot with $U_{0.86}Nb_{0.14}$) and thus our E/A wall loss is $0.416T^{3.1}$ at 1 ns vs. Au $0.39T^{3.3}$. The solution now to $101 = 3.7T^{3.1} + 0.6T^4$ is $T = 2.62$, so we expect a 7 eV hotter hohlraum than the 255 eV Au hohlraum. Many shots were done with Au end plates and just a cylinder body of cocktail. Redoing that we must solve $101 = 0.49T^{3.3} + 3.2T^{3.1} + 0.6T^4$ we get 2.61 thus we expect a 6 eV improvement for those type of cocktail hohlraums. However, until very recently there was barely any (less than 3 eV) difference between Au and cocktail hohlraums – so what went wrong?

We formulated the following hypothesis – perhaps oxygen contaminated the cocktail walls in the process of making them. Since cocktail foils don't necessarily get leached from their substrates, cocktails hohlraums certainly do and the leaching process may be the key to the contamination. While Au does not bind to O, U & Dy certainly do – they are in fact often used as O getters! The trouble with O in the cocktail is that they are fully ionized so contribute about twice the Z per unit weight than the high Z elements, and thus raise e by raising the specific heat.

For atomic numbers between 6 and 71 and for T between 1 & 3 heV we find the following fits for the specific energy e. The ideal gas law would give $e_{ig} = 15[(Z + 1)/A]T_{heV}$ in MJ/g. Here the ionization state Z is fit by $Z = (Z_N/71)^{0.6}16T_{heV}^{0.6}/\rho_{g/cc}^{0.14}$. There are non-ideal gas law contributions, principally from ionization energy, and we fit those by a multiplier "mult" $= 2.5(Z_N/71)^{0.1}$. Thus $e = (e_{ig})$ ("mult") which we can write as δ_N/A_N. The reader can check that this gives a number reasonably close to the one we quoted for Au above. For a mixture of j elements we take the ratio $e = \Sigma\delta_{Nj}/\Sigma A_{Nj}$. (not, $\Sigma(\delta_{Nj}/A_{Nj})$). So for example for a AuNdDy cocktail to be discussed shortly, we find $e_c/e_{Au} = 1.06$ with no oxygen, 1.08 with 4% O, and 1.22 with 40% O.

A vivid example of the O problem came from a re-analysis of another burn-through experiment, that of Olsen et. al. (2003). A AuNdDy cocktail foil burn-through time ($t_{b.t.c}$) was compared to that of an Au foil's in a 160 eV hohlraum with a T that rose as $t^{0.1}$. This would lead us to predict that (via the equivalent of Eq. (29)) the ratio of $(\kappa e)_c/(\kappa e)_{Au}$ would equal $[(\rho\Delta x)_{Au}/(\rho\Delta x)_c]^{2.08}$ times $[t_{b.t.c}/t_{b.t.Au}]^{1.5}$. Plugging in the data that equals $[1.9/1.5]^{2.08}$ times $[1.3$ ns / 1.45 ns$]^{1.5} = 1.38$. Now theoretically the $(\kappa e)_c/(\kappa e)_{Au}$ ratio should be 1.22, which disagrees with the data. But if there were an O for each Dy and for each Nd, then theoretically the ratio is 1.36, which is quite close (and well within error bars) to the data. Thus due to this re-analysis, we "post-dicted" that the sample was fully oxygenated. We then had the target fabrication records examined and indeed that was precisely the case! Of course, now that we are aware of this issue, future targets can be carefully made without O.

Assuming the older hohlraum shots at Omega were also fully oxygenated (we'll never really know – the state of those hohlraums at shot time was not characterized) we can redo our source=sink model once again but with a higher wall loss due to the high e as a result of the oxygen. Now the E/A coefficient is 0.44 and, with Au end plates, we solve $101 =$

$0.49T^{3.3} + 3.4T^{3.1} + 0.6T^4$ to get $T = 2.575$, thus we, in retrospect, should have only expected a 2.5 eV difference from the 255 eV pure Au hohlraums, in rather close agreement with what was observed. The good news is that very recent shots in which great care has been taken to avoid oxygenation, have shown a ΔT much closer to our original expectations for cocktails. Part of the way to prevent the oxygen from getting into the sample is to coat it with a thin layer of Au. (roughly 0.2 nm). Even that layer lowers ΔT somewhat, since we are replacing cocktail with Au. However, for the NIF, the same thickness of Au can still do the Oxygen prevention job, while being an utterly trivial fraction of the Marshak depth and hence not cause any worrisome detriment to the cocktail wall loss advantage. Thus, with a few more confirmatory shots, we are well within grasp of proving the cocktail principle as an energy saver for NIF.

2.2 Foam-walled hohlraums

We now ask the following question: Can we save on driver energy by making hohlraum walls out of low density high Z foams, which have less hydrodynamic motion (namely less radiation heated and ablated material that streams back into the hohlraum interior as a low density isothermal blow-off) and hence, reduces net absorbed energy by the walls? We answer this question using our HR analytic theory, as well as by numerical simulations. To the degree that the "pure" HR theory diverges from the simulations we derive non-ideal non-self-similar corrections to the theory that bring it into agreement with the simulations (Rosen (2005)). We show that low-density high Z foams can indeed produce a savings of $\sim 20\%$ in the required driver energy. Remarkably, this reduction is universally independent of drive T and its pulse-duration t. We derive an analytic expression for the optimal density (for any given T and t) that will achieve this reduction factor and which agrees very well with numerical simulations. Such an approach might allow more routine operation of the National Ignition Facility (NIF) with laser energy further away from the optics damage threshold, and still provide the nominal energy (as originally designed with solid wall hohlraums) to the fuel capsule. Reduced hydrodynamic motion of the wall material may also reduce symmetry swings, as found for heavy ion beam targets.

For the sake of brevity and clarity we will restrict our study here to a drive that is constant in time for a duration t. Solving the supersonic Marshak wave, (Eq. (31) for Au, our small parameter, $\epsilon = 0.291$ and the constant C is given by $4.08 \times 10^{-7}/\rho^{2.06}$ cm^2/ns. For those parameters we find $x_F^2 = [(2 + \epsilon)/(1 - \epsilon)]CT^{4+\alpha-\beta}t$, which, for our case gives $x_F(cm) = 0.0012T^{1.95}t^{.5}/\rho^{1.03}$. Our solutions for the energy per unit area, E/A, (in MJ / cm^2) absorbed by the gold wall is:

$$E/A = 0.0029T^{3.55}t^{.5}/\rho^{0.17} \quad \text{(for the pure supersonic regime).} \tag{39}$$

For the sub-sonic solution, we repeat what we presented above: For $k = 0$, we found $m_F(t) = m_0 T_S(t)^{1.91}t^{0.52}$ with $m_0 = 9.90 \times 10^{-4}$ g/cm^2 and (in MJ / cm^2)

$$E/A = 0.0058T^{3.35}t^{.59} \quad \text{(for the pure subsonic regime).} \tag{40}$$

Comparing Eqs. (39) and (40) we see (for a typical drive of $T = 2.5$ heV and duration $t = 2$ ns) that for densities in the neighborhood of 0.3 gm/cc there is clearly less wall loss for the supersonic case. Lowering densities further decreases opacity and increases specific heat, both in the undesirable direction of more loss to internal energy (as per Eq. (28)). Raising densities would be desirable as it would lower wall losses even further, but unfortunately it would take us into the subsonic regime. The sound speed, C_S, at 250 eV in gold is about 60μm/ns, which (using the expression for x_F that precedes Eq. (39)) exceeds the supersonic heat front velocity, dx_F/dt, at 2 ns when ρ_0 is about 0.5 gm/cc.

In Fig. 7 we plot E/A vs. initial ρ_0 of the wall from Eq. (39) (for $T = 2.5$ and $t = 2$.) and we plot the subsonic ("infinite density") result of Eq. (40) as well. We also plot the numerical simulation results. Note that Eq. (39) closely matches the full physics numerical simulations, deep in the supersonic regime (at very low ρ_0) when little hydrodynamic motion is expected. When hydrodynamic motion is artificially turned off in the numerical simulations (not shown here), Eq. (39) closely matches the artificial simulations for all densities.

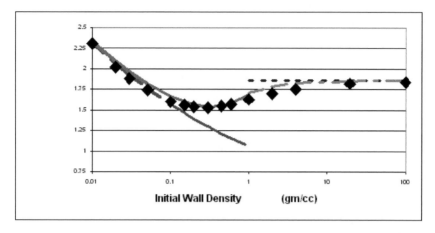

Figure 7. *Wall loss (0.1 MJ/cm^2 vs. initial wall density (g/cc). Diamond points are simulation results. Dashed line: Eq. (39). Dotted line: Eq (40). Solid line: Eq. (41). Dot-Dashed line: Eq. (42). Drive Conditions: $T = 250$ eV; Duration $t = 2$ ns.*

To account for the divergence of the full physics simulations from our self similar solutions we reason as follows: In the supersonic regime but at higher ρ_0, rarefactions do in fact begin to eat into that portion of the heated wall nearest the drive boundary and hydro motion ensues. Consider an isothermal rarefaction wave propagating leftward (at speed C_S) into an $x < 0$ half space of temperature T and original constant density ρ_0, which results in low density material blowing out, rightward. If we define $z(x,t) = 1 + (x/C_S t)$, then the density profile is given by $\rho(x,t) = \rho_0 \exp(-z)$ and the velocity profile is given by $U(x,t) = zC_S$. Within the rarefaction the kinetic energy per unit area at any given time t can easily be found by integrating $0.5\rho U^2$ over x from $-C_S t$ to infinity and is equal to $\rho_0 C_S^3 t$. This calculates to $0.0024\rho_0^{0.79}T^{2\cdot}t$ (MJ/cm^2) and matches the full physics simulation of the kinetic energy. Also the lower density profile within the rarefaction leads to a higher specific heat. This too can be easily found by doing a similar integral for ρe. We find that the lower density profile contributes an additional $\mu/(1-\mu)$ fraction of internal energy to that portion ($= C_S t/x_F$) of the heated matter overtaken by the rarefaction front (the portion not overtaken is still at its original density). For our value of μ this becomes $0.0011\rho_0^{0.79}T^{2.4}t$ (MJ/cm^2). These two effects together now lead to a corrected E/A (in MJ / cm^2) or the supersonic regime

$$E/A = 0.0029T^{3.55}t^{.5}/\rho_0^{0.17} + 0.0035\rho_0^{0.79}T^{2.4}t \quad \text{(full supersonic regime)}. \quad (41)$$

The solid curve in Fig. 7 is Eq. (41) calculated out to the high ρ_0 edge of the supersonic regime and largely reproduces the E/A full physics numerical simulation curve throughout the entire supersonic regime. While these additional energy sinks reduce the full "bonus" of being supersonic that Eq. (39) naively promises, we still note a nearly 20% reduction from the solid wall result.

Note too that in Fig. 7, Eq. (40) closely matches the full physics numerical simulation at the very high end of the initial-wall-density x-axis, deep in the subsonic regime. However, in the lower density part of the subsonic regime the simulations differ from the infinite density result. We believe that is due to the period of time early in the simulation when indeed the heat wave is supersonic and therefore less lossy. As the initial density, ρ_0, decreases, an increasingly longer early-time duration of supersonicity exists. We can correct for this by first finding $t_{catch} = 0.17T^{2.3}\rho_0^{-1.9}$, the time when the rarefaction front, $C_S\,t$, catches up to the heat front (the $x_F(t)$ that precedes Eq. (41)). We then subtract the subsonic $E/A(t = t_{catch})$ of Eq. (40)) from $E/A(t)$ of Eq. (40) and add in its stead the supersonic $E/A(t = t_{catch})$ of Eq. (41)). For our gold parameters, the procedure outlined above leads to a simple expression for the correction

$$E/A(MJ/cm^2) = 0.0058T^{3.35}t^{.59} - 0.002T^{4.7}/\rho_0^{1.1} \quad \text{(full subsonic regime)}. \quad (42)$$

This result largely reproduces the E/A simulation curve throughout the entire subsonic regime, as seen in the dot-dashed curve of Fig. 7.

As the minimum of E/A vs. ρ_0 occurs at densities low enough to be within the supersonic regime, we can easily take the E/A derivative with respect to ρ_0 in Eq. (41) and find the optimal density, ρ^*:

$$\rho^* = 0.17T^{1.2}t^{-0.5}g'^{0.52}f'^{-1.04}. \quad (43)$$

In the above we explicitly included the scaling of the Au opacity and specific heat coefficients (the prime denotes a scaling to their nominal values which were given above) so that this formula can be used more generally. Plugging this back into Eq. (41) gives us the minimum E/A^*

$$E/A^* = 0.0048T^{3.35}\tau^{.59}f'^{0.68}g'^{0.41}. \quad (44)$$

Comparing this to the E/A of the very high density (solid and above) regime of Eq. (40) we see that they scale exactly the same way. Thus their ratio implies a universal (independent of T and t) saving of 17% when the optimal ρ^* is chosen as the initial wall density. Also, self-consistently, ρ^* "universally" falls within the supersonic regime.

Fig. 8 shows results from the $T = 1.25$ heV simulation set, where we plot the resultant E/A curves (normalized to their values at solid density) vs. initial density, for 3 pulse lengths varying from 1 to 64 ns. We clearly see the "universal" nature – the energy savings (vs. a solid wall) at the optimal density for each t is the same value (of about 16%), very close to our predictions.

Full 2-D Lasnex simulations also show that combining both schemes works best, namely foam cocktail hohlraums. This idea has been tried in detail for a heavy ion reactor scale hohlraum by Callahan (2000). It was optimized via tedious full 2-D simulations and a foam density of 0.1 gm/cc is arrived at. It is Au-Gd and as such $g' = 0.62$, $f' = 1.04$, with $T = 2.5$ and $t = 8$, leading to our prediction of an optimal density of 0.13 gm/cc quite close to the optimized design value.

We have recently tested this foam wall as an energy saver concept using a cylinder of Ta_2O_5 made of either 4 gm/cc or 0.1 gm/cc, each with a gold ring hit by the laser that served as the x-ray source to drive the rest of the cylinder walls. They were performed by P. Young of LLNL at URLLE. A drooping pulse produced about a 100 eV flat-topped drive. A Dante viewed the walls, and the 0.1 gm/cc foams were about 15% brighter in accord with 2-D simulations and in accord with the "source = sink" approach discussed here, when albedo effects are taken into account. More experiments are planned in a more fully enclosed hohlraum geometry. More work will be needed to extend this idea to shaped pulses, for which perhaps graded density foams may have to be invoked.

2.3 Hohlraums with axial shine shields

Another "trick" to save energy is to insert axial shields (small Au disks) to block the capsule's view of the cold LEH, published by Amendt et. al. (1996). The laser beams enter the cylinder

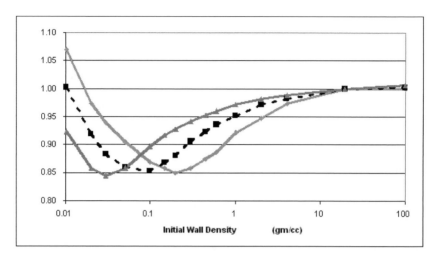

Figure 8. *Simulated wall loss (normalized by wall loss at solid initial density) vs. initial wall density (gm/cc). Drive Conditions: T = 125eV; Duration t: triangles: 64 ns. squares: 8 ns. diamonds: 1 ns.*

through entrance holes in the end cap as usual, but propagate through the outer "cold exterior" ("ce", not to be confused here with "conversion efficiency") section, into the central "hot interior" ("hi") section of the hohlraum, and impinge on the walls there. The outer parts are cooler, and the aperture through which the radiation flows from the middle ("hi") to the ends ("ce") is the annulus between the on-axis shine shield disk and the cylinder wall.

The LASNEX "observables" to be explained here are the 228 eV for a hohlraum with no shine shields vs. a 241 eV drive on capsule for one with shine shields. Why would a hohlraum that introduces about 500J of more wall loss via the shine shield disks, actually produce a hotter hohlraum rather than a cooler one? The answer (Rosen (1996)) to the paradox is essentially that we have created an inside out hohlraum, in which the central section is a "hot interior (hi) hohlraum", and drives the capsule. Indeed, the outer sections of these hohlraums are "cold exterior (ce) hohlraums", and are predicted by LASNEX to be only about 215 eV. We will derive all of these numbers presently, following the calculations of Rosen (1994b).

The cylinder is 0.8 mm in radius with a laser entrance hole (LEH) of 0.6 mm radius. The half length (from mid plane to end cap) of the can is 1.15 mm. The shine shield is 0.3 mm in radius and placed on axis 0.55 mm from the midplane, (thus, 0.6 mm from the LEH). The capsule has a radius of 0.27 mm. Thus the half-area of the hot interior hohlraum walls (from mid-plane to shine shield and including the shine shield disk area) A_{hi} is 3.04 mm^2, of the cold exterior A_{ce}, 4.2 mm^2 , and of the half-capsule A_C, 0.46 mm^2. The LEH area A_{LEH} is 1.13 mm^2, and the annular area between shine shield and wall, A_{hi-ce}, through which the radiation flows from the hot interior to the cold exterior, is 1.73 mm^2. A slight complication here is that some (about 33%) of the laser energy is deposited along the cylinder wall between the axial position of the shine shield and the end of the can; namely there is some of the laser source in the cold exterior region. Nonetheless we can generalize our treatment of source=sink to account for this entire situation, and we will define E_L to be made up of two parts E_{Lhi} and E_{Lce} for those amounts absorbed in the hi and ce regions respectively. In our case, for half of the incident 20KJ (= 200 hJ) going into the half of the hohlraum being calculated, and at a conversion efficiency of 70%, half of 140 hJ of x-ray energy is available. If 33% is created in the ce, then 47 hJ is available in the hi and 23 hJ in the ce. Defining $y = T_{ce}/T_{hi}$ and appealing as we have done throughout this paper to energy balance, and using our Au wall loss results for $k = 0.18$ as usual, we find the following, The hot interior equation reads

$$\eta E_{Lhi} = E_{Whi} + E_{hi-ce} + E_C \quad \text{or}$$
$$\eta E_{Lhi} = 0.4T_{hi}^{3.3}A_{hi} + 0.6A_{hi-ce}T_{hi}^4(1 - y^4) + 0.6A_CT_{hi}^4 \tag{45}$$

while the cold exterior equation reads

$$\eta E_{Lce} + E_{hi-ce} = E_{Wce} + E_{LEH} \quad \text{or}$$
$$\eta E_{Lce} + 0.6A_{hi-ce}T_{hi}^4(1 - y^4) = 0.4T_{hi}^{3.3}y^{3.3}A_S + 0.6A_{LEH}T_{hi}^4y^4. \tag{46}$$

Adding Eq. (46) to Eq. (45) we get

$$\eta E_{Lce} = \eta(E_{Lhi} + E_{Lce})$$
$$= 0.4T_{hi}^{3.3}A_{hi}[1 + (A_{ce}/A_{hi})y^{3.3}] + 0.6A_CT_{hi}^4[1 + (A_{LEH}/A_C)y^4]. \tag{47}$$

Inserting the values for the areas (discussed above) Eq. (47) reduces to

$$70 = 1.27T_{hi}^{3.3}(1 + 1.38y^{3.3}) + 0.28T_{hi}^4(1 + 2.5y^4) \tag{48}$$

while Eq. (46) reduces to:

$$23 + 1.04T_{hi}^4(1 - y^4) = 1.67T_{hi}^{3.3}y^{3.3} + 0.68T_{hi}^4y^4. \tag{49}$$

Equations (48) and (49) can be solved iteratively with a solution quickly converging to $T_{hi} = 2.43$ and $y = 0.89$, namely $T_{ce} = yT_{hi} = 2.16$. These values for the hot interior and cold exterior temperatures are in excellent agreement with LASNEX (2.41 & 2.15 respectively). Moreover, had we considered a simple geometry (no shine shields) we would be solving:

$$\eta E_L = E_W + E_{LEH} + E_C \quad \text{or}$$
$$\eta E_L = .4T^{3.3}A_W + 0.6T^4A_L + 0.6A_CT^4. \tag{50}$$

For the simple, no shine shield, geometry we find $A_W = 6.7$ mm^2 and A_L and A_C are as above. Thus Eq. (50) reduces to:

$$70 = 2.7T^{3.3} + 0.95T^4 \tag{51}$$

whose solution is $T = 2.31$ heV, again in excellent agreement with the LASNEX result for no shine shield of 228 eV. The enhancement of temperature, taking the Lasnex numbers for example of 2.41 vs. 2.28, represent about a 20% energy savings. The reader can confirm this by putting these two values into Eq. (51) and getting 81 and 67 hJ respectively.

This concept has been tested successfully and published. The drive increase in the hot interior hohlraum has been measured via the decreased implosion time of a capsule. It has also been noticed in these experiments that the axial shine shields provides yet another "knob" to control the symmetry of the illumination onto the capsule.

Thus we have presented at least three ways to reduce wall losses, each by nearly 20%, and each, in principle, could be used in conjunction with the other schemes, leading to an overall energy saving of 0.8^3 or about 0.5. This will allow NIF to operate quite far from its damage threshold, and still provide the required drive to the capsule in the center of the hohlraum. We invite the reader to dream up any other energy savings schemes!

3   Applications of hohlraums to ICF

Introduction

Space limitations here do not allow a complete description of the ICF problem. We refer the reader to Rosen (1999) for a rather complete tutorial on the subject. We in fact intend this section to be an auxiliary to that review, by completing / supplementing it with the next two sections. For completeness though, we'll just summarize the main points of the tutorial.

An ICF 1 GW reactor, with a 10% efficient driver (supplying 6 MJ pulses @ 5 Hz to the target chamber) must have a target with gain $G > 100$. The inertial confinement time of assembled fuel of radius R is given by $R/4C_S$. The burn-up fraction f_B of the fuel is given by $\rho R/(\rho R + 70)$ in MKS units. We expect a target to operate optimally near $f_B = 1/3$ or a ρR of 30. With that fixed, and using "nature's bounty" the Q of DT being $3.4 \ 10^{14}$ J/kg, we can see that to have the expected and containable output of 600 MJ we must compress DT 1000 fold so that its mass will be about $5 \ 10^{-6}$ kg. This target has a momentum of 80 kg m/s or the impact an average person would have walking into a wall at average walking speed, which is obviously quite containable.

A spherical implosion is the least stressing way to compress matter to this extent. A hohlraum allows for good symmetry for such an implosion as described at the very beginning of this chapter, where we described the method of implosion – basically a rocket. Thus there are two coupling efficiencies that get us from driver energy E_D to the thermal energy of the assembled fuel: η_C the coupling efficiency of the driver to thermal energy on the surface of the capsule. The hot gas is the exhaust of the rocket, which delivers a moving payload with an efficiency η_H. The kinetic energy is converted to the thermal energy of the assembled self-stagnating fuel.

Heating the entire fuel assembly to 10 keV to start the fusion process in earnest requires (1.5) (10 keV) (4) / 5 AMU $= 10^{12}$ J/kg, a factor 1/340 less than the fusion pay-off Q_{DT} quoted above. The ("4") of the above equation is due to the need to heat the deuteron, the triton, and the electron that come with each. However, we only burn 1/3 of this, and for a hohlraum, typical coupling efficiencies are $\eta_C = 0.2$ and $\eta_H = 0.2$, thus the gain is only $G = (340)(0.2)(0.2)(1/3) = 5$, far too low for a reactor. The secret to high gain is to only heat a small central hot spot to 10 keV, and then let the alpha particle produced by the DT reaction stop within the surrounding fuel and do the heating. Typically the hot spot will have 10% of the surrounding dense fuel shell density, and its radius will be at about $R/2$. Thus it will have a negligible 1% of the mass of the fuel.

The requirements of a hot spot, are to be 10 keV, and to have a ρR of 3 (kg/m²). This is the range of an alpha in a 10 keV plasma. Moreover, the f_B formula implies that the hot spot will burn 5% of its fuel. This is precisely the amount needed to supply enough alphas to the first thin shell (a layer inside the dense fuel which also has a ρR of 3 kg/m²) surrounding the hot spot to get it up to 10 keV, and thus launch the propagating thermonuclear burn-wave.

Energy is also required to compress the cold dense fuel to its high-density state in the shell that surrounds the hot spot. It is Fermi Degenerate matter, so its pressure will be $P_{FD} = 2.2 10^6 \alpha \rho^{5/3}$ (Pa) and the energy required is $E_{FD}(J) = 3.3 \times 10^6 \alpha \rho^{2/3} M_{DT}$ (kg). Here α measures how far off the isentrope we are. We are now in a position to calculate the gain.

3.1   Gain calculation: conventional and fast ignitor

We assume an incident pulse of $E_D = 6$ MJ. With the coupling and rocket efficiencies each of 0.2 means $E_F = 0.24$ MJ is in the assembled fuel. Let us assume the hot spot radius is 50 μm, namely $R_{HS} = 5 \times 10^{-5}$ m. This turns out to be near optimal. Since the hot spot requires $\rho R = 3$, then ρ_{HS} must equal 6×10^4 kg/m³. Then the mass of the hot spot is easily calculated to be 3×10^{-8} kg, and its thermal energy (at the required 10 keV) is $E_{HS} = 3.6 \times 10^4$ J. Its pressure $P_{HS} = 2n_ikT$ (2 because of electrons) is 4.6×10^{16} Pa.

This self-stagnated assembly is isobaric, the hot spot pressure stops the cold fuel shell from imploding further, and pressures equilibrate $P_{HS} = P_C = P_{FD}$, which can then tell us what the cold density is: $\rho_C = 1.6 \times 10^6$ kg/m^3, indeed a 1000 fold denser than solid DT. We can plug that into the formula for the energy available for the cold dense shell: $E_{FD} = E_C = E_F - E_{HS} = 2.4 \times 10^5 - 3.6 \times 10^4 = 2.04 \times 10^5$ J, to obtain the mass of the shell, $M_C = 4.8 \times 10^{-6}$ kg , in line with our assumption for the mass of the reactor target. We can then set $(4/3)\pi\rho_C(R^3 - R_{HS}^3) = M_C$ to find $R = 9.5 \times 10^{-5}$ m, thus $\Delta R = R - R_{HS} = 4.5 \times 10^{-5}$ m thus $\rho\Delta R = 70$ kg/m^2 thus $f_B = 70/(70 + 70) = 1/2$. Putting that altogether yields $G = f_B M_C Q_{DT}/E_D = (1/2)(4.8 \times 10^{-6}\text{kg})(3.4 \times 10^{14}\text{J/kg})/(6 \times 10^6\text{J}) = 130$. Thus we have accomplished our "mission" of showing that ICF can in principle supply a gain 100 target to sustain a 1 GW power reactor.

Before we leave this section let us compare this calculation with a fast ignitor scheme. In that approach we create a hot spot via an additional driver that creates the hot spot on the exterior of the assembled fuel. We do it quickly, so there is no pressure equilibrium between hot spot and cold dense fuel. There is simply density equality (isochoric) instead. As we shall see, this will allow the assembly to be larger but less dense, hence in general much less stressing to the implosion symmetry and stability requirements. So proceeding as before we note the requirements on the fast ignitor driver – it must supply $E_{HS} = 3.6 \times 10^4$ J in about a disassembly time of 10 ps, hence it must be of order 3 to 4 PW. Since now the densities are equal, $\rho_C = 6 \times 10^4$ kg/m^3, so again solving for the mass of the cold assembly, $M_C = 4 \times 10^{-5}$, an order of magnitude larger than the conventional scheme!

With M_C and ρ_C known, we calculate $R = 5.5 \times 10^{-4}$, and $\Delta R = 5 \times 10^{-4}$ so $\rho\Delta R = 30$ and thus $f_B = 0.3$. The gain is then $G = f_B M_C Q_{DT}/E_D = (0.3)(4. \times 10^{-5}\text{kg})(3.4 \times 10^{14}\text{J/kg})/(6 \times 10^6\text{J}) = 670$, quite an improvement!

Another way to look at the advantage of a fast ignitor is to see that it gives large gain even for a smaller driver. Large drivers mean large initial capital investment expense – so the smaller the better. We invite the reader to re-do this calculation for $E_D = 1$ MJ rather than 6 MJ. The optimal hot spot scales as $E_D^{0.5}$, so take $R_{HS} = 20\mu$ this time. Find $\rho_{HS} = 1.5 \times 10^5$ kg/m$^3 = \rho_C$, $M_{HS} = 5 \times 10^{-9}$ kg, $E_{HS} = 6$ kJ, $Ec = .04$ MJ - 6 kJ $= 3.4 \times 10^4$ j, thus $M_c = 3.7 \times 10^{-6}$ kg, $R = 1.61 \times 10^{-4}$, $\rho\Delta R = 24$, $f_B = .256$, and $G = 312$. Obviously the great advantage of the fast ignitor is mitigated by the challenging physics of creating this externally driven hot spot.

Let us return then to the conventional approach, and deal with one final topic: pulse shaping.

3.2 Pulse shaping

If we analyze our conventional hot spot gain example from the previous section, we reach some important conclusions. We can ask: what was the required kinetic energy of the dense shell as it implodes in order to supply the thermal energy of the fuel when it stagnated and fully assembled? Normally we allow an "ignition margin" of a factor of 2, so we set $(1/2)M_cV_{imp}^2$ equal to $2E_F$ which is almost $= 2E_C$. A cold density of order 10^6, as we calculated in our example, gives $V_{imp} = 3.6 \times 10^5$ m/s.

Other "external" requirements flow from this: Since the final fuel radius was about 100μ, and a typical convergence ratio to get to these high densities is of order 30, we get an initial radius of the capsule of 3 mm. Then implosion time is $R/v_{imp} = 10$ ns, and thus a power requirement of 600 TW, and an irradiance on target of order 600 TW / 4π (3 mm)$^2 = 5 \times 10^{14}$ W/cm^2, or about 270 eV.

Another question to ask is what pressure is required to push the original shell to the required kinetic energy. We set $KE = P\Delta V$, the "PdV" work done in implosion. Typically we start pushing at radius R_0, and stop when the shell is imploded to $r = R_0/2$, since after that there is not much volume left to exploit for PdV work. The driver turns off and the shell coasts inward until stagnation. For a shell that starts at a thickness ΔR_0, our equation leads to $P = 3 \times 10^{13}(\Delta R_0/R_0)$. For stability reasons we avoid thin shells and keep $(\Delta R_0/R_0)$ at about $1/5$. This tells us that the pressure doing the pushing is 6×10^{12} Pa $= 60$ Mbar.

The problem is we cannot apply this pressure to the original capsule! Consider the FD isentrope of solid DT. $P_{FD} = 2 \times 10^6 \rho^{5/3} = 1.4 \times 10^{10}$ Pa for $\rho_0 = 200$ kg/m^3. Thus 60 MB is way off the isentrope and would make it energetically very difficult to proceed with the implosion. Hence the need for pulse shaping.

If we compare the hugoniot relations for shocks vs. isentropic compression we learn a valuable lesson. For a jump of pressure $Y = P_1/P_0$ there will be a shocked density jump $X = \rho_1/\rho_0$. The Hugoniot relations tell us that $X = (4Y + 1)/(Y + 4)$. Let us compare that to $X_{isen} = Y^{1/\gamma}$ for $\gamma = 5/3$. For $Y = 1$ they are equal. For $Y = 2$, $X = 1.5$ and $X_{isen} = 1.51$. For $Y = 4$, $X = 2.13$ and $X_{isen} = 2.3$. For $Y = 8$, $X = 2.75$ and $X_{isen} = 3.5$. For infinite Y, $X = 4$ and X_{isen} is infinite. So as long as the pressure jumps are less than 4, the drift off the isentrope via the sequential shock method will be less than 10%. Thus our pulse shaping strategy should be clear. The first shock will necessarily be a strong one, hence $X = 4$, hence the post-shock density will be 800 kg/m^3. Thus the P_{FD} for this density will be 1.4×10^{11} Pa $= 1.4$ Mb, which be should precisely the magnitude of our first shock, to match that and stay on the FD isentrope. After this we launch 3 more shocks, each 3.5 times bigger than the previous one. Our final pressure will be $(3.5)^3$ (1.4 Mb) $= 60$ Mb as required, and now the shell has compressed to the proper high density to remain on the FD isentrope as we push on it at 60 Mb and accelerate it to the required implosion velocity of 3.6×10^5 m/s on its way to a successful thermonuclear implosion. We must of course carefully time these shocks so that they coalesce at close to the frozen DT shell so that most of the fuel remains cold Fermi Degenerate fuel.

With this explanation we also can understand the need for a frozen DT shell and not a gaseous target. If it were gas, its original density would be low, so even at $X = 4$, the isentrope will have a very low P, and hence require much larger dynamic range on the low side for the driver to have. Moreover, as we must keep say 97% of the fuel on the cold FD isentrope, it means the shocks should coalesce at about $0.3R_0$. So now when we are ready for the final push to accelerate the cold (and now properly dense) fuel to the proper velocity we do not have very much volume left to do PdV work. Thus the final push must be much larger than 60 Mb, so we again stress the dynamic range of the driver this time on the high side! All told this is an untenable situation, and we must live with the complexities of frozen DT solid shelled targets.

4   Conclusions

This concludes our ICF overview. How coupling and gain scale with driver size is covered in Rosen (1999). Comparison with direct drive is also covered there. Direct drive has a much better coupling efficiency, but is more challenging on the hydro stability issue, which is beyond the scope of this review. This too is covered in Rosen (1999). Direct drive is making great strides in solving its stability issues, and in principle could have higher gains than indirect drive. HIF indirect drive targets have no LEH, so the coupling efficiency is better than laser driven hohlraums, which enables a gain 100 at about 3 MJ driver. They show great promise as well since the drivers are high efficiency and in principle high reliability.

In summary we have shown the principles of ICF hohlraums. We showed how they naturally lead to an increase in the efficiency to be reached of coupling initial laser energy to absorption by the fuel capsule as we increase in scale from Nova to NIF to reactors. This helps motivate the predictions of the complex LASNEX simulations that NIF will achieve moderate gains of 10-20, and reactors will achieve higher gains allowing them to be competitive energy sources for the next century.

Acknowledgement

We gratefully acknowledge the warm hospitality of St. Andrews University and the NATO sponsored Scottish University Summer School 60 under the directorship of Prof. Dino Jaroszynski and ably assisted by Prof. Alan Cairns and Ms. Giselle Smith. Useful conversations with our LLNL theory and design colleagues J. Hammer, O. Jones, L. Suter, J. Edwards, A. Szoke, J. Castor, S. MacLaren, R. Thiessen, and P. Amendt, as well as with our AWE colleagues B. Thomas, P. Thompson, M. Stevenson and S. McAlpin. Our experimental collaborators are also gratefully acknowledged: J. Schein, P. Young, E. Dewald, N. Landen, S. Glenzer, K. Campbell, R. Turner, R. Kauffman, C. Darrow, H. Kornblum, J. Porter, T. Orszechowski, and W. Hsing. The computational and data base assistance was ably provided by G. Zimmerman, J. Harte, D. Bailey, M. Marinak, J. Albritton, B. Goldstein, and B. Wilson, for which we are grateful. Target fabrication work by R. Wallace, J. Gunther et al of LLNL's ICF program and J. Kass, A. Hamza et al of LLNL's C&MS department were invaluable, as was the target fabrication efforts of colleagues at GA: H. Wilkens, A. Nikroo, J. Kaae, T. Back, J. Kilkenny, and M. Campbell. The collaborations with our Sandia colleagues R. Olson, G. Rochau, and R. Leeper are also appreciated. The encouragement and support of B. Hammel, J. Lindl, E. Moses, G. Miller, C. Verdon, B. Goodwin, and M. Anastasio are also gratefully acknowledged. The privilege of having J. Nuckolls as a mentor is hereby acknowledged as being valuable beyond measure. This work as performed under the auspices of the US. Department of Energy, by the Lawrence Livermore National Laboratory under contract W-7405-ENG-48.

References

Amendt P et al., 1996, Novel Symmetry Tuning in Nova Hohlraums Using Axial Gold Disks, *Physics of Plasmas* **3** 4166

Callahan D and Tabak M, 2000, Progress in target physics and design for heavy ion fusion, *Phys. Plasmas* **7** 2083

Hammer J H and Rosen M D, 2003, Exact solutions to the diffusive radiation transport equations, *Physics of Plasmas* **10** 1829

Jones O S et al., 2004 Measurement of the Absolute Hohlraum Wall Albedo under Ignition Foot Drive Conditions, *Physical Review Letters* **93** 065002

Kauffman R L et al., 1994, High Temperatures in Inertial Confinement Fusion Radiation Cavities Heated with 0.35 μ Light, *Physical Review Letters* **73** 2320

Marshak R E, 1958, Effects of Radiation on Shock Wave Behavior, *Phys. Fluids* **1** 24

Olsen R et al., 2003, Time and spatially resolved measurements of x-ray burnthrough and re-emission in Au and Au:Dy:Nd foils, *Rev. Sci. Instr.* **74** 2186

Orzechowski T J et al., 1996, The Rosseland Mean Opacity of a Mixture of Gold and Gadolinium at High temperatures, *Phys. Rev. Lett.* **77** 3545

Rosen M D, 1979, Scaling law for radiation temperature, Laser Program Annual Report (1979), Lawrence Livermore National Laboratory, Livermore, CA, UCRL-50055-79, pp. 2-37 to 2-46 (unpublished). This work was extensively quoted and presented in J. Lindl, Phys. Plasmas, **2**, 3933, (1995)

Rosen M D, Lindl J D, and Kilkenny J D, 1994a, Recent results on Nova, *Journal of Fusion Energy* **13** 155

Rosen M D, 1994b, Scaling laws for non-conventional hohlraums, *Bulletin of the American Physical Society* **39** 1684

Rosen M D, 1996, The science applications of the high-energy density plasmas created on the Nova laser, *Phys. Plasmas* **3** 1803

Rosen M D, 1999, The physics issues that determine ICF target gain and driver requirements: A tutorial, *Phys. Plasmas* **6** 1690

Rosen M D and Hammer J H, 2005, Foam walled Hohlraums, submitted to *Phys. Rev E.*

Zimmerman G B and Kruer W L, 1975, The LASNEX simulation code, *Comments in Plasma Physics and Controlled Fusion* **2** 85

15

Dense Plasma Physics – the Micro Physics of Laser-Produced Plasmas

S. J. Rose

Imperial College, Prince Consort Road, London, UK
e-mail: s.rose@imperial.ac.uk

1 Introduction

The interaction of high-power laser radiation with matter produces plasma of varying characteristics. For solid targets the plasma produced in the blow-off is at a high temperature and relatively low density (in comparison with solid) whilst the shock generated by the ablation pressure moves into the target leaving material behind the shock front at relatively low temperature (for irradiances of $< 10^{12}$ Wcm^{-2} this can be below the melting point of the solid) and higher than solid density. On the microscopic scale the variety of plasmas formed by high-power lasers presents great difficulty to the theorist who endeavours to provide a general model of plasma properties. For low temperature plasmas (at temperatures lower than the dissociation energy of the molecules involved) the microscopic description involves aspects of solid-state physics and plasma chemistry. At somewhat higher temperatures, which are considered in this lecture, the plasma involves only ions (potentially partially-ionised) and electrons, but even this simplification still results in great difficulty in the construction of a model of the microscopic physics involved.

2 Non-equilibrium microphysics

High-power laser radiation is highly monochromatic and can interact with the free electrons giving up energy by inverse bremsstrahlung. At high intensity the electromagnetic field can interact with bound electrons, ionising them by multiphoton or tunnel ionisation. The free electrons are produced by a variety of processes: electron collisional ionisation, photoionisation by soft X-rays, multiphoton and tunnel ionisation by the laser field. In each of these cases the free electron distribution so produced is far from equilibrium; indeed the distribution is the product of the ionising energy spectrum (free electron or photons) and the cross-section for ionisation. The time-scale for equilibration is, however so short that in most models it is an equilibrium distribution that is assumed. However there are cases where the distribution is not in equilibrium. For example, with ultra-short-pulse heating by sub-picosecond high-power lasers the timescale for equilibration is comparable to the timescale for ionisation which suggests that the electron velocity distribution may show, for example a depleted high-energy tail above the ionisation energy. The equilibration time for ions is also very short and an equilibrium distribution is again assumed in most models. The characteristic ionic temperature can be very different from the electronic temperature as the equilibration between the distributions is generally slow in comparison with the equilibration within either the electronic or ionic distributions.

The bound electrons exhibit a huge number of possible discrete states of excitation and ionisation and these equilibrate principally by electron and photon collisional processes. For high-energy transitions (typically in the X-ray region) the spontaneous radiative decay rate (A value) is so large that it may dominate the collisional de-excitation rate. This results in the excitation and de-excitation not being in detailed balance resulting in a departure from equilibrium. The high A value results in ions which have lower excitation and ionisation than is the case for equilibrium. Of course, at sufficiently high density, the collisional downward rate dominates and equilibrium is established at the same temperature as the free electron distribution because the bound electrons are linked to the free electron system by the collisional processes. It is also possible that in the transient ionisation that occurs in ultra-short-pulse heating by sub-picosecond high-power lasers, the bound states will not be in equilibrium because of the short timescales involved. It should be pointed out that short-pulse laser interaction may also drive the system from equilibrium by virtue of the multiphoton and tunnel ionisation leading to non-Maxwellian free-electron distributions with the possibility of consequent non-thermal excitation and ionisation rates.

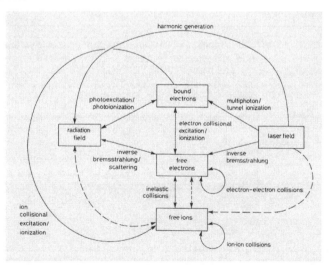

Figure 1. *The diagram shows interactions between the different sub-systems in a laser-produced plasma (taken from Rose 1994). The full lines represent the named interaction mechanisms whereas the dashed lines represent mechanisms not usually considered, that the author has suggested can also be operative. These mechanisms involve the transient encounters between partially-ionised ions that can only be understood using electronic wavefunctions that spread over more than one nucleus.*

Photons of higher energy than the driving laser pervade laser-produced plasmas: line and continuum emission typically create soft X-ray emission by the plasma. Low-Z materials are optically thin over most of the spectrum, with only a few intense lines being optically thick and the emission is far from Planckian. Higher-Z materials produce X-ray emission which has fewer spectral features and is closer to a Planckian spectrum. The heating of material on ultra-short timescales by sub-picosecond high-power lasers also gives rise to a non-Planckian radiation field because the timescale for the development of the radiation field ($\sim \lambda(v/c)$, where λ is the mean free path for photon absorption and c is the speed of light) is comparable to the heating time.

A plasma consisting solely of ions and electrons can exhibit non-equilibrium behaviour in each of its subsystems: free electrons and ions, bound electrons and the radiation field. It is interesting to consider how one would set up a model which properly describes the non-linear interaction between the sub-systems. Such a model would need to account for the interactions between the different sub-systems, as shown in figure 1.

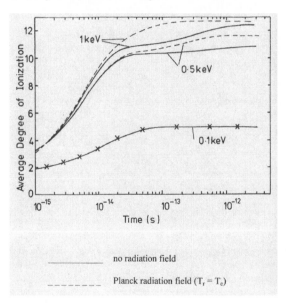

Figure 2. *The calculated time evolution of the average degree of ionisation of an iron plasma at 1 gcm^{-3} in which the electron temperature is set at $t = 0$ to 100 eV, 500 eV and 1000 eV (taken from Edwards and Rose, 1993).*

For simplicity we consider here a zero-dimensional plasma and so no effect of transport by electrons, radiation or hydrodynamic motion is included, only the microphysics of the interactions between the different sub-systems. Nonetheless, even to account for this complex set of interactions is very demanding and in the calculations that are regularly performed, so as to simplify the modelling, equilibrium is usually assumed in one or more of the sub-systems. For example, in long-pulse irradiation to produce shock-heated and compressed material the density of the material directly behind the shock front is sufficiently high (above solid) and the temperature is low enough (typically of order 10 eV) that the equilibration time for free electrons, ions and bound electrons is much shorter than the timescale for plasma evolution. It is therefore usual to assume that the free electrons, ions and bound electrons are each in an equilibrium distribution, characterised by a single temperature.

The radiation field is usually assumed to play no part in the behaviour of the plasma. For the material in the coronal blow-off plasma produced in a long-pulse interaction, the densities are much smaller and the temperatures are much higher and similar arguments of equilibration timescales and optical depths suggest that, the free electrons and ions can be taken to be in equilibrium (though not with one another), yet the bound electrons and the radiation field are far from being described by an equilibrium distribution. Understanding the physics in each of these regimes is of great importance to, for example, the study of Inertial Confinement Fusion (Atzeni and Meyer-ter-Vehn, 2004).

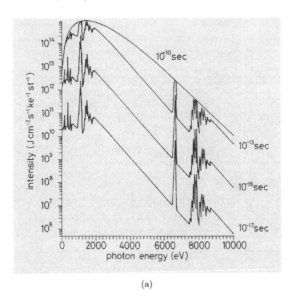

(a)

Figure 3. *The time evolution of the radiation field associated with an iron plasma with the free and bound electron distributions in equilibrium at 500 eV and 1 gcm^{-3}. The model calculates the time evolution of the radiation field for a plasma that is optically thick over the whole spectrum. As a consequence the steady-state solution (shown here to be achieved by 10^{-10} sec) is a Planckian. At t = 0 the radiation field is absent and the calculation shows the assembly of the radiation field over time.*

Whilst the assumptions made in the long pulse experiments are probably justified, the situation is more dubious for short-pulse interactions. A proper description would involve following the time-evolution of the distributions of each of the sub-systems coupled together along the lines previously described. However this is beyond current computational capabilities and simplifications are usually made. For example, the time-dependant ionisation of material close to solid density and heated on ultra-short timescales has been investigated by, for example Edwards and Rose (1993) and the results for aluminium are presented in figure 2. However in this model the electron distribution was taken to be Maxwellian and the radiation field to be either zero or Planckian. As can be seen from figure 2, the finite ionisation time cannot be ignored for a plasma heated on the sub-picosecond timescale. The time dependence of the radiation field was investigated in a study by the author (figures 3(a), 3(b) and 3(c)) which also considered material close to solid density. However in this case the bound electron distribution was taken to be in equilibrium. Figure 3 shows that the radiation field is still non-Planckian a picosecond after the heating pulse is applied. Whilst such studies are useful as model calculations, we await a full calculation to see the complex non-linear physics involved.

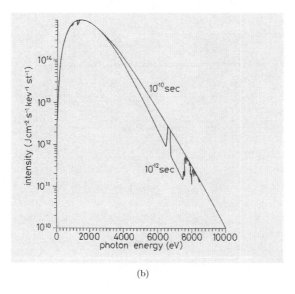

(b)

Figure 3. *As 3 a)*

3 Equilibrium microphysics

In addition to difficulties encountered in describing the microphysics of laser-produced plasmas because of non-equilibrium effects, even in equilibrium, laser-plasmas present considerable difficulties to the theorist. The Debye-Hückel model provides the first-order correction to an ideal gas description of the free electrons and ions. The model describes the average arrangement and screening of ions and electrons by one another in a plasma. It gives the average potential around any ion as

$$V(r) = \frac{Z^*e}{4\pi\epsilon_0 r} \exp(-r/\lambda) \tag{1}$$

where r is the distance from the ion, Z^* is the average degree of ionisation and λ is the Debye length which is given by

$$\frac{1}{\lambda^2} = \frac{1}{\lambda_e^2} + \frac{1}{\lambda_i^2} \tag{2}$$

λ_e and λ_i are the electron and ion Debye lengths respectively and are given by

$$\lambda_e = \sqrt{\frac{\epsilon_0 kT}{n_e e^2}} \tag{3}$$

$$\lambda_e = \sqrt{\frac{\epsilon_0 kT}{Z^* n_i e^2}} \tag{4}$$

n_e and n_i are the average electron and ion number densities respectively and T is the temperature (in the usual formulation of the model the ionic temperature T_i is the same as the temperature of the free electron distribution T_e). It should be noted that for most laser-plasmas Z^* is much greater than 1 and so $\lambda \sim \lambda_i$. The model, and the physics underlying it, is used extensively in plasma physics in areas such as the equation of state where Debye-Hückel theory provides a simple correction to the ideal gas model (Cooper, 1966). However the model assumes that

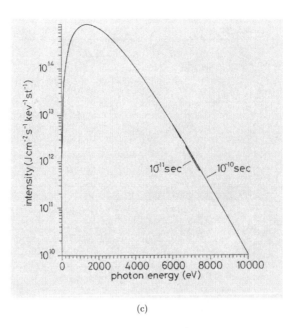

(c)

Figure 3. *As 3 a)*

the interaction energy between the particles in the plasma is substantially smaller than kT. Although the model is valid for many astrophysical and low-density high-temperature (such as magnetically-confined) plasmas, this cannot be assumed for laser-plasmas. As to whether or not the assumptions underlying the Debye-Hückel theory hold can be judged by considering a dimensionless number called the strong-coupling parameter, Γ, defined as

$$\Gamma = \frac{Z^* e^2}{4\pi\epsilon_0 r_{ii} kT} \tag{5}$$

Where r_{ii} is the average distance between ions in the plasma. It can be seen that for $\Gamma \ll 1$ the interaction energy between the ions is much less than their kinetic energy (and that the same holds for the electrons) and so the Debye-Hückel theory holds. However for $\Gamma \gg 1$ the opposite is true and the Debye-Hückel theory breaks down. Boyd and Sanderson (2003) show that the effect of two-particle correlation can be ignored if there are many particles in the Debye sphere (a sphere of radius equal to the Debye length) which can be easily shown to be equivalent to the situation in which $\Gamma \ll 1$.

Another area in which equilibrium laser plasmas can depart substantially from the usual low-density picture is in the electron distribution. In equilibrium at low-density the distribution is Maxwellian, but under high-density and low-temperature conditions the effect of quantum mechanics can be seen in the distribution. For high-temperature low-density plasmas the DeBroglie wavelength of the free electrons is much less than the inter-electron separation. However for low-temperature and high density the opposite can be true and in this case, because the electrons are fermions, this results is a partially degenerate Fermi-Dirac distribution. The probability of occupancy of a free electron state of energy ϵ is then

$$p = \frac{1}{\exp\left(\frac{\epsilon}{kT} - \eta\right) + 1} \tag{6}$$

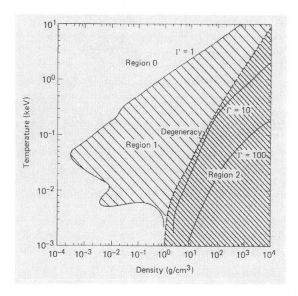

Figure 4. *Contours of different* Γ *values and the contour* $\eta = 0$ *(marked "Degeneracy") for an aluminium plasma in thermal equilibrium (taken from More, 1981). Plasma in Region 0 is neither strongly-coupled nor degenerate and is therefore able to be described by conventional low-density models. Region 1 is non-degenerate, but strongly-coupled, whereas Region 2 is both degenerate and strongly-coupled.*

η is a normalisation factor determined by the average free electron number density

$$n_e = \frac{4\pi}{h^3}\left(2m_e kT\right)^{3/2} I_{1/2}(\eta) \tag{7}$$

where $I_{1/2}(\eta)$ is the Fermi-Dirac function

$$I_{1/2}(x) = \int_0^\infty \frac{y^{1/2}dy}{e^{y-x}+1}. \tag{8}$$

Equation (3) arises from an integration over electron energies and for $\eta << 0$

$$p = \exp(\eta)\exp(-\epsilon/kT) \tag{9}$$

which is just the non-degenerate Maxwellian distribution, whereas for $\eta >> 0$

$$\begin{aligned} p &= 1 \quad \epsilon/kT < \eta \\ p &= 0 \quad \epsilon/kT > \eta \end{aligned} \tag{10}$$

which is the fully degenerate case.

Figure 4 shows the contours of constant $\Gamma(1, 10, 100)$ and the contour of $\eta = 0$ for the case of an aluminium plasma under a range of conditions of density and temperature, many of which can readily be achieved using high-power lasers. For example, for aluminium that has been shock-compressed by high-power laser irradiation, the compressed density will be several times that of solid density and the temperature achieved will be a few eVs. It can be seen that for these conditions neither the Debye-Hückel model nor the assumption of a Maxwellian distribution holds. Corrections due to these effects have been derived by a number of authors and rather than repeat them here the reader is referred to an earlier SUSSP lecture by the author in which they are collected and described in detail (Rose, 1988; but see also Lee and More, 1984 and Ichimaru, 1992).

One other area in which laser-produced plasmas can depart from the low-density picture is in the description of the bound electronic structure. Low-density plasmas support bound electronic excitation to (depending on the density) high principal quantum numbers. As the density increases the separation between the ions reduces and high principal quantum number states can extend over a larger distance than the mean inter-ion separation. Because of this, these high-lying states cannot be supported by the plasma. The effect has been long recognised even in low density plasmas (indeed it has to be invoked to cut off the summation in the partition function involving bound electronic states of, for example, the H atom, and thereby ensure that the sum does not diverge). However the effect becomes much more pronounced at the high densities typically encountered in laser-produced plasmas and even low-lying excited states can be removed. As before the reader is referred to Rose (1988) for a simple description and some relevant calculations. For a much more complete description see Ebeling, Kremp and Kraeft (1976).

4 Conclusion

It has been the aim of this lecture to acquaint the newcomer to the field with some of the difficulties in describing the microscopic physics of the dense plasmas produced by high-power laser irradiation. The author's aim has been to show the reader that a complete model of the plasma physics encountered with laser-produced plasmas is very complex and that simplifications are regularly made. Indeed these can be very successful if they are applied in the appropriate circumstances, but the newcomer must always check as to whether the model that is being applied is valid.

5 Acknowledgements

The author would like to thank AWE, Aldermaston for the award of a William Penney Fellowship.

References

S Atzeni and J Meyer-ter-Vehn, *The Physics of Inertial Fusion*, Oxford University Press (2004).

T J M Boyd and J J Sanderson, *The Physics of Plasmas*, Cambridge University Press (2003).

J Cooper, Rep. Prog. Physics., **24**, 35 (1966).

J Edwards and S J Rose, J.Phys. B: Atom. Molec. and Opt. Phys., 26, L523 (1993).

S Ichimaru, *Statistical Plasma Physics*, volume I, Addison-Wesley (1992).

Y Lee and R M More, Phys. Fluids., 27, 1273 (1984).

W Ebeling, D Kremp and W D Kraeft, *Theory of Bound States and Ionisation Equilibrium in Plasmas and Solids*, Akademie-Verlag, Berlin (1976).

R M More, Lawrence Livermaore National Laboratory report UCRL-84991 (1981).

S J Rose, 1988, "Dense Plasma Physics", Laser-Plasma Interactions IV, SUSSP Publications, University of Edinburgh (1989).

S J Rose, "Atomic and radiation physics of hot dense plasmas", Laser-Plasma Interactions IV, SUSSP Publications, University of Edinburgh (1995).

16
Progress in Fast Ignition

Peter A. Norreys

Central Laser Facility, CCLRC, Rutherford Appleton Laboratory, United Kingdom

1 Introduction

These lecture notes will begin with a very brief review of the concept of inertial confinement fusion as they are covered at length in other lectures, particularly those by Professor MD Rosen (Rosen 2006). They will then provide a brief history of the ideas that lie behind the cone-guided/focused fast ignition scheme before discussing topics that are important to present day investigations regarding fast electron energy transport in solid and compressed plasmas.

1.1 Background

The energy crisis is real and looming. The burning of fossil fuels and the consequent rise in greenhouse gas emissions is becoming a major problem for the environment. The search for alternative sources of electricity that are both clean and safe is becoming increasingly urgent for society and remains one of the major challenges for science and engineering today. Renewable alternatives such as solar, wind and the tidal power schemes, such as the Severn River barrage, appear attractive in terms of safety, but are not without environmental and ecological costs and are unlikely to fully satisfy the world's energy requirements in the coming decades.

An alternative that has been pursued for some time is that of controlled thermonuclear fusion. This promises a source of electricity that is environmentally benign and ecologically friendly. However, developing nuclear fusion into a useful source of electricity has taken a lot longer to achieve than originally envisaged. The principle obstacle is that nuclear fusion requires temperatures of tens of millions of degrees centigrade. The hydrogen isotopes must have sufficient kinetic energy to overcome the repulsive Coloumb force between them so that the strong nuclear force can fuse them into new elements. Clearly, no vessel can withstand direct physical contact with matter at such temperatures - the heated fuel must be confined in some manner to prevent this.

Two possibilities have been pursued to overcome this obstacle. The magnetic confinement fusion approach is based on the principle that the fusion fuel is contained by strong magnetic fields suitably arranged to confine the fuel for many seconds. Alternatively, in the laser - fusion

Figure 1. *The National Ignition Facility building, LLNL. Credit Lawrence Livermore National Laboratory*

scenario, the fuel is compressed to ultra-high density and fusion occurs on a time-scale so fast that the dense fuel does not have time to disassemble.

The physics community's confidence that the obstacles to controlled thermonuclear fusion can be overcome has led to the design of machines that will test physics issues related to ignition in both routes to controlled fusion. The final decision on the construction of the ITER machine - the international thermonuclear test reactor based upon magnetic confinement - has been taken and Cadarache, France has been chosen for the site. The physics base of the alternative route - laser fusion - has developed to the extent that the construction of multi-Mejajoule lasers, such as the "National Ignition Facility" in the United States and the "Laser MegaJoule" in France, are already underway. It is expected that these laser-fusion machines will be completed before the end of the decade.

1.2   Inertial confinement fusion

The invention of the laser in the early 1960's promised the capability of depositing huge energy densities onto matter. By the early 1970's, laser pulses could be focused onto target with intensities above 10^{15} Wcm^{-2}, which is equivalent to electric fields in the laser focus of a billion Volts per cm. Clearly, any atom placed within a field of this magnitude will be ionised within one or two oscillations of the laser electric field. This means that the laser pulse interacts not with solid density material, but with a plasma. It was quickly realised that, because of the enormous energy density, laser-plasmas could be heated to millions of degrees centigrade at 1/10th - 1/100th of solid density. These densities and temperatures generate pressures of millions of atmospheres, albeit for the duration of the laser pulse, which is typically between 1 - 20 nanoseconds.

These pressures generated by laser-plasmas can be used to compress a shell containing deuterium-tritium fuel to ultra-high densities by spherical convergence. The concept is conceptually simple - think of how a rocket escapes the earth gravitational field - exhaust gases are expelled from the rocket at high velocity - conservation of momentum then demands that the rocket rises against gravity. In the same way, the plasma generated on the outer surface of the spherical shell containing deuterium-tritium fuel expands rapidly into the surrounding vacuum, imparting momentum to the remaining shell such that it implodes with high velocity. The implosion stops only when the internal plasma pressure (i.e. the density and temperature product of the compressed fuel) prevents further compression (Nuckolls *et al.*, 1972).

When the laser pulse generates the plasma on the surface of the shell, a shock wave is driven through the shell before the material starts to accelerate. But here is a contradiction - a strong shock will always heat material behind the shock front, whereas to compress matter to high densities a low temperature is required. To start the fusion process, a number of carefully timed shocks of increasing amplitude are allowed to collapse simultaneously at the centre of the fuel to generate a hot spark, while minimising the shock heating of the fuel. Provided a sufficient number of alpha-particles (generated during the fusion of deuterium-tritium in the spark) are stopped in the surrounding cold fuel, localised heating will occur and a fusion burn wave can propagate through the rest of the cold dense fuel, leading to controlled energy gain.

The early years of inertial confinement fusion were somewhat optimistic in their assessment of the amount of laser energy required to achieve ignition and burn propagation. At the time, there was little experience of the effect of the Rayleigh-Taylor instability on the implosion performance. This hydrodynamic instability grows during both the acceleration and deceleration phases of the implosion and has the effect of mixing the hot spark material with the surrounding dense fuel and shell materials, thereby quenching the hot spark. Balancing the contradictory requirements of low-entropy compression and spark formation, while minimising the growth of the hydrodynamic instabilities, is expensive energetically and demands the use of the multi-Mega-Joule class lasers now under construction (Lindl, 1995).

1.3 The fast ignition concept

Now imagine that it is now possible to separate the two processes of compression and heating. What are the consequences? This idea was first proposed in 1994 by Max Tabak and colleagues from the Lawrence Livermore National Laboratory and stirred immediate and intense interest around the world (Tabak *et al.*, 1994). Firstly, more fuel can be compressed for the same laser drive energy, as the requirement to drive a series of shock through the fuel can be reduced in number - and with it shock preheating. Since the fusion energy gain depends only on the amount of fuel present, once the spark has been formed, the energy gain is substantially higher - ×300 or more compared with ×15 or so for indirect drive. Quantitatively, the fusion energy gain in fast ignition is

$$G = 3 \times 10^4 \eta \left(\frac{\eta E}{\alpha^3} \right)^{0.4} \tag{1}$$

where η is the coupling efficiency, E is the drive energy and α is is the factor by which the cold fuel is nondegenerate.

Secondly, if more fuel can be compressed during the implosion, then the "in-flight aspect ratio" (the ratio the thickness of the remaining shell and fuel during the acceleration phase of the implosion to the initial shell radius) can be much larger than the conventional scheme. This substantially reduces the implosion symmetry needed to assemble the fuel to high density - because the growth rate of hydrodynamic instabilities (that deleterously mix the fuel and shell materials) is known to significantly increase the larger the in-flight aspect ratio becomes.

These are attractive features: reduced laser-drive energy to sub-Megajoule class laser systems; substantially increased fusion energy gain that can offset the inefficiencies in driving the fusion process; and much lower drive symmetry requirements to assemble the fuel to high densities. But how is it possible to separate the compression and heating stages? It is only with the very recent advent of picosecond duration, ultra-intense, PW-class lasers that this possibility of Fast Ignition can be realised. Rapid progress has been made in the past few years - new results have been obtained that add much greater confidence to this exciting concept.

In the original fast ignition scenario, three separate sets of laser pulses were required. Nanosecond pulses from a first laser compress the deuterium-tritium fuel in the normal way. Shortly before peak compression, pulses from a second laser (that last several hundred picoseconds and have an intensity in the range 10^{17} - 10^{19} Wcm^{-2}) are used to form a channel in the plasma atmosphere. During this "hole-boring" phase the "critical surface" is pushed towards the dense core of plasma. When the implosion stops, an ignitor pulse from a third PW (petawatt = 10^{15} Watt) laser that lasts about 10 ps is driven into the channel formed by the second laser pulse. The huge energy density of the focused light accelerates copious numbers of electrons at the critical surface to energies similar to the ponderomotive potential energy of the pulse - 1 MeV or more. The electrons then propagate into the dense plasma where they are stopped. This process leads to strong local heating and the ignition of a fusion burn wave.

1.4   The cone-guided and cone-focused concepts

One particular aspect of the original scheme appears problematic - the hole-boring phase. Serious questions have been raised about whether the channel will remain empty and stable enough for the PW laser pulse (and/or the multi-MA current that is generated) to propagate into the dense core without being deflected by plasma instabilities.

A very attractive fast-track path has been explored that by-passes the hole-boring phase altogether - the advanced fast ignition concept. At the Second "International Workshop on the Fast Ignition of Fusion Targets", held in Garching, Germany in 1997, Professor Steven Rose (Oxford University) proposed the concept of driving a partial shell down the inside of a hollow gold cone. This allows the PW laser pulse to propagate completely free of plasma before interacting with the solid density material the tip of the cone. The fast electron beam would then be able to propagate into the compressed plasma and deposit its energy. At the same time, Max Tabak and Steve Hatchett were working out the details of a similiar approach at LLNL - the insertion of a hollow gold cone into a hollow shell. This allows the compressed plasma to stagnate at the apex of the cone and the electron beam (generated by the PW laser pulse interacting at the tip of the cone) does not have far to propagate.

In 1999 and 2000, an Anglo-Japanese team tested these unusual ideas in an concerted experimental campaign for the first time. They coupled a 100 TW short-duration laser pulse to these novel target geometries that allowed simultaneous fuel compression and fast heating, first at RAL (Norreys *et al.*, 2000) and then with great success at Osaka University (Kodama *et al.*, 2001). A hollow Au cone was inserted into a hollow deuterated plastic shell. The cone kept a channel completely clear of plasma exhaust as the shell imploded and the short pulse laser interacted with the wall of the cone at its tip. The dense plasma was formed very close to the tip of the cone (a few tens of microns away) so that the MeV electrons did not have far to propagate. The team observed 2×10^5 deuterium-deuterium thermonuclear neutrons at 2.45 MeV. This corresponded to a conversion efficiency of short pulse laser energy to the thermal energy of the compressed plasma of 27 ($\pm$7)%.

The next step was to conduct experiments at a laser power equivalent to the level required in a full-scale ignition system, albeit with a pulse duration 1/20 of that needed. The first near-equivalent power experiments using a PW laser system as the heating source were performed at Osaka University, Japan in a June 2002 by the same international team. The team (comprising physicists from Osaka University, the Rutherford Appleton Laboratory, Imperial College London, the Queen's University of Belfast and the University of Essex) demonstrated significant enhancement of fusion products by three orders of magnitude (from 10^4 to 10^7 neutrons without and with the heating pulse on). The results confirmed that the high heating efficiency observed previously is maintained as the short-pulse laser power is substantially increased to the PW level (Kodama *et al.*, 2002). Measurements using perfectly synchronised lasers for the fast heating and the plasma compression indicate the possibility of using longer pulses ($\geq$10ps) to achieve

Figure 2. *A cone-shell target used for integrated fast ignition experiments.*

ignition. These new results add much greater confidence to this exciting concept.

2 Fast electron energy transport

These are very encouraging results, but there remain many unanswered questions related to this concept, particularly those related to fast electron energy transport. The number of experiments in compressed matter is limited, due to the complexity of the experimental arrangements and limited shot rate. To gain greater knowledge of the field, modelling of present day laser-solid interactions is necessary. However, it must be emphasised that the density and temperature regime that these experiments can access is of limited applicability to fast ignition (due to the resistivity regime that can be accessed between 10 eV and a few hundred eV), but is essential to gain an understanding of the physical processes that may become important there.

2.1 Background

The first measurements of energy transport of electron beams generated by multi-TW laser pulses showed collimated electron flow patterns in both deuterated plastic targets (Tatarakis *et al.*, 1998) and large area glass slabs (Borghesi *et al.*, 1999). However, it soon became clear that these observations were the exceptions rather than the rule. Gremillet *et al.*, for example, used glass slab targets and observed both filamented jets moving into the target with a velocity close to c followed by a slower moving hemispherical ionisation front (at c/2). It was concluded that the energy within the jet was quite small, otherwise the transverse probe used to observe the effect would be inaccessible to the ionised region around it, and that the bulk of the energy was within the ionised "cloud" region (Gremillet *et al.*, 1999). Since then, various degrees of electron beam divergence have been reported in the literature. Optical images of the rear surface of aluminium targets have shown a divergence pattern of 34° from optical transition radiation and 25° from Planckian thermal radiation using the LULI and GEKKO XII 100TW laser facilities, respectively (Santos *et al.*, 2002), (Kodama *et al.*, 2001).

In experiments at the Vulcan and LULI 100 TW facilities, a spherically bent Bragg crystal monochromatic 2-D x-ray imaging technique was used to record the origin of K_α photons created in a 20 μm thick buried Ti or Cu fluor layer in a planar Al or CH target. The photons were emitted following K-shell ionization by fast electrons. The X-rays were sufficiently energetic to escape through the surrounding target material, permitting the observation of the transverse dimension of the laser-generated electron beam with a resolution at the target of about 12 μm and depths inside the target up to several hundred microns. The results indicated that the K_α source size was always larger than the focal spot dimensions and that the electron beam diverged with an angular spread of 40° (Stephens et al., 2004).

Recent experiments have measured the background electron temperature of the bulk material by resonance line emission from buried Al signature layers inside solid plastic targets (Koch et al., 2002). It is interesting to note here that the energy transport pattern from the NOVA PW laser facility was annular, as measured by X-ray pinhole camera imaging, and that the background electron temperature measured there may have been affected by resistivity mismatching between the plastic and metal layers. Taking this effect into account in hybrid code computer simulations, the measurements on the Vulcan PW (Evans et al., 2005) indicated that there was an energy barrier on the front surface of the target that reduced the fast electron energy transport into the bulk of the target, resulting in smaller background electron heating. More recently, a double ring pattern has been observed with low density foam targets irradiated by PW laser pulses from the Vulcan laser facility (Jung et al., 2005).

It is clear from this survey that the experiments are indicating a complex mix of competing processes that influence the beam divergence pattern. These processes are becoming more amenable to experimental, analytical and theoretical investigation, as discussed below.

2.2 Absorption processes

In long scale-length plasmas, the laser pulse propagates up to the turning point given by $n = n_c cos^2\theta$, where n_c is the critical density. Plasma waves generated at the turning resonate at the critical density surface, resulting in an amplification of their amplitude at that point. The waves then break (in a process analagous to ocean waves on the seashore) resulting in the acceleration of large numbers of energetic electrons into the target. The absorption is maximised at an angle θ from the target normal that is dependent on the laser frequency ω and density scale-length L given by $sin\theta \approx 0.8(c/\omega L)^{1/3}$, where c is the speed of light.

However, resonance absorption becomes inefficient when the density scalelength is L is less than the excursion distance of the oscillating electrons in the laser field. The excursion length is obtained from $v_{osc} = dx_{osc}/dt = (eE_0/m\omega)$. In this case, the Brunel-type resonance absorption process occurs (Brunel, 1987). For oblique incidence, p-polarised irradiation, the electric field is normal to the target surface at the critical density surface. Electrons that are within the skin depth of the target experience the electric field of the laser pulse, are accelerated into the vacuum during one half of the cycle and back into the target in the following half cycle.

The $j \times B$ absorption process can be understood by consideration of the motion of a single electron in an intense infinitely plane polarised laser field. The electron exhibits a figure of eight motion due to the $v \times B$ term in the Lorentz force $F = -e(E + v \times B)$. At relativistic intensities, electrons are accelerated in the direction of k twice every laser cycle, due to the phase lag between the velocity of the electron v and the magnetic field B. The kinetic energy the electron acquires is roughly proportional to the ponderomotive potential energy that is given by (Wilks et al., 1992)

$$U_p \approx kT_{hot} = \left(\gamma^2 - 1\right) mc^2 = 511 \left[\left(1 + I_{10^{18}Wcm^{-2}}/1.37\right)^{1/2} - 1 \right] keV \qquad (2)$$

A simple substitution is needed for more electrons in the plasma, since $j = nev$. This absorption process is particularly important for solid density plasmas that have a large density scalelength generated by the pedestal of the laser pulse.

Still larger scalelengths generate electrons accelerated to energies much larger than the ponderomotive potential energy either from wave breaking via the self-modulated laser wakefield acceleration process (Modena *et al.*, 1995) or, alternatively, from oscillations of electrons that are reinjected into in the laser field via the very large self-generated magnetic fields associated with the interaction - the so-called B-loop acceleration mechanism (Pukhov *et al.*, 1998). These larger electron temperatures were measured in NOVA PW laser-plasma interactions, in addition to a distribution that had the ponderomotive potential energy (Cowan *et al.*, 1999).

Experiments have shown that resonance absorption is an important absorption process for intensities up to 10^{19} Wcm^{-2} on target (Beg *et al.*, 1997) when the intensity contrast ratio between the main pulse and the pedestal is $1:10^{-7}$. At higher intensities, it appears that there is a transition to Brunel-type resonance absorption where the electron temperature is given by the ponderomotive potential energy (Clark *et al.*, 2000). The main difference is the angular direction of the generated fast electron beam: resonance absorption-type processes accelerate electrons in the direction of the target normal, whereas the $\boldsymbol{j} \times \boldsymbol{B}$ process is along the laser axis. It is interesting to note that there is evidence for $\boldsymbol{j} \times \boldsymbol{B}$ absorption and classical/Brunel-type resonance absorption to co-exist, since the process is time dependent and profile steepening occurs during the laser pulse itself due to the intense ponderomotive pressure (Santala *et al.*, 2000), (Brandl *et al.*, 2003).

2.3 Alfvén-Lawson current limit

It is necessary to understand how much current can propagate in an intense beam before the self-generated magnetic field prevents propagation. The relativistic gyro-radius of an electron in a uniform magnetic field is given by

$$r_g = \frac{mc^2}{eB} \left(\gamma^2 - 1\right)^{1/2} (cgs) = 1.7 \times 10^3 \left(\gamma^2 - 1\right)^{1/2} B^{-1} cm \tag{3}$$

where $\gamma = (1 - \beta^2)^{-1/2}$. In vacuum, the self-generated magnetic field will reverse the trajectory of the electrons and prevent propagation of the electron beam above the Alfvén-Lawson current limit given by

$$I_A = \left(\frac{mc^3}{e}\right) \beta_z \gamma(cgs) = 1.7 \times 10^4 \beta_z \gamma(A) \tag{4}$$

where $\beta_z = v_z / c$.

This current is limited to tens of kA, but multi-MA currents are generated in PW laser-plasma interactions. Fortunately, the Alfvén-Lawson current limit can be overcome by propagating the beam in a plasma with a density that is sufficiently large to allow the return current to compensate the fast electron current.

2.4 Weibel instability

The Weibel instability is one of the plasma instabilities thought to be responsible for electron current filamentation, anomalous resistivity and the generation of chaotic magnetic field structures. The instability arises when the hot forward electron flow is balanced by returning electrons such that there is no net current. If a sinusoidal magnetic field arises from noise, then the magnetic term in the Lorentz force, $-e\boldsymbol{v} \times \boldsymbol{B}$, bends the oppositely directed electrons to spatially different points. The resulting current flows are phased to reinforce the B-field and the perturbation

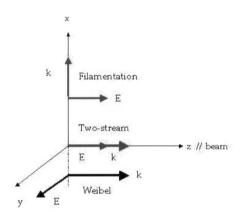

Figure 3. *The Weibel, two-stream and filamentation modes.*

grows. In this instability, the unstable waves have a wave vector parallel to the beam direction and normal to the electric field, i.e., k∥beam, k⊥E.

The threshold of the instability is strongly dependent on the ratio of the background electron density to the fast electron beam density and scales as

$$\frac{n_b}{n_e} > \gamma_0 \left\{ \frac{p_\perp}{p_\parallel} \right\} \tag{5}$$

n_b and n_e are the fast electron and background electron densities respectively, $p_\perp$ and $p_\parallel$ are the momenta of the fast electrons perpendicular and parallel to the beam. The expression indicates that the threshold for the instability to grow increases with higher intensity and $p_\perp$ of the beam. Indeed, the instability growth has been calculated from relativistic kinetic theory using a waterbag distribution in the momenta perpendicular to the main direction of the beam. Using this model, the maximum growth rate ω_{im}^2 is (Silva *et al.*, 2002):

$$\omega_{im}^2 = \frac{\beta_{z0}^2}{(1 - \beta_{z0}^2)^2} \left[(1 + \beta_{z0}^2)k_c^2 + 2\Gamma - 2\sqrt{(k_c^2 + \Gamma)(\beta_{z0}^2 k_c^2 + \Gamma)} \right] \tag{6}$$

and the fastest growing mode k_{max} for a cold background plasma is given by:

$$k_{max}^2 = \frac{1}{(1 - \beta_{z0}^2)^2} \left[(1 + \beta_{z0}^2)(\sqrt{(k_c^2 + \Gamma)(\beta_{z0}^2 k_c^2 + \Gamma)} - \Gamma) - 2k_c^2 \beta_{z0}^2 \right] \tag{7}$$

where z is the direction perpendicular to the beam, $\gamma_z \beta_{z0} = p_{z0}/m_e c$ is the normal Lorentz term, k_c is the wave number of the cold background plasma and Γ is an expression relating the background and beam electron velocities, taking into account the ratio n_b/n_e.

When the return current temperature is high the growth rate favours higher-order modes of the instability. This results in small-scale filaments whose transverse dimensions are the collisionless skin depth or c / ω_p, where c is the speed of light and ω_p is the electron plasma frequency. However, when the return current temperature is lower this can favour lower-order mode growth, and results in a smaller number of larger filaments. This could important for fast ignition as the lower order modes allow less current to propagate. However, it is very important to emphasise that the growth rate for the purely transverse mode of the instability has been shown not to be important for realistic fast ignition conditions, as the tranverse energy spread reduces the growth rate quite substantially or even eliminates it. Two dimensional particle-in-cell (PIC) simulations appear to support this conclusion, as the Weibel induced filamentation only grows close to the target surface. On the other hand in cone-guided fast ignition, there is a density discontinuity at the end of the guide cone in the gap before the compressed fuel where this instability may be important.

It is interesting to note that good agreement with this theory has been obtained in an experiment with two-sided irradiation of thin foil targets (Wei *et al.*, 2004) where the scalelength on the back of the foil (and therefore the background electron density) was varied in a controlled manner and the filamented electron beam was observed in radiochromic film.

2.5   Two-stream and filamentation instabilities

The two-stream instability arises when electrons (with a velocity v_e) pass through a potential step $\phi(x)$ that is finite in space. The electrons are accelerated by the force $-\partial\phi(x)/\partial x$ and become faster in the step than electrons outside, where $\phi=0$. This means that the electron number n_e in the step is reduced because, in a steady state, the electron flux $n_e v_e$ has to be constant at every point. The reduction in the electron density inside the step results in a further increase in the potential and the perturbation is amplified.

In the case of an intense laser-plasma interaction, the second "stream" is the return current generated to compensate the fast electrons. In contrast to the Weibel instability described in the previous section, the longitudinal unstable waves generated by the two-stream instability have a wave vector parallel to the beam direction and the electric field, i.e. $k\|beam$ and $k\|E$.

The filamentation instability, on the other hand, generates unstable waves that have a wave vector normal to both the beam and the electric field, i.e. $k\perp beam$ and $k\perp E$. This instability is the analogous to whole beam self-focusing, but in this case density perturbations (that have a spatial scale smaller than that of the beam and that arise from noise) results in the local enhancement of the pondermotive pressure within the beam and eventually its break-up into a number of small-scale filaments.

In a series of seminal papers, Bret, Firpo and Deutsch have shown that the fast growing instability modes are those that are intermediate between the two-stream and the filamentation modes (Bret *et al.*, 2004, 2005). The maximum growth rate is

$$\delta_m = \frac{\sqrt{3}}{2^{4/3}} \left(\frac{\alpha^{1/3}}{\gamma_b} \right) \omega_p \qquad (8)$$

where $\alpha = n_b/n_p$ and, in a plasma with a warm background temperature with a small α, this occurs for a wavevector given by

$$k_M \approx \left(\pm \frac{\omega_p}{c} \sqrt{V_b/V_{tp}} \pm \frac{\omega_p}{V_b} \right) \qquad (9)$$

V_b and V_{tp} are the beam and thermal plasma velocities, respectively.

It is very interesting to know the spatial scale of the filaments L_f that are generated as a result of this instability. This is:

$$L_f \approx \pi \frac{c}{\omega_p} \sqrt{V_b/V_{tp}} \tag{10}$$

This calculation of the spatial scale compares well with filaments observed in experiment (Tatarakis *et al.*, 2003). In that experiment, the 5×10^{19} Wcm^{-2}, 1 ps duration laser pulse irradated an aluminium target and the generated fast electron beam propagated into a low density gas located behind the target. The filamentation was observed using tranverse optical shadowgraphy. The spatial scale of the filaments was approximately 20 μm when the background electron density was $\approx 10^{20}$ cm^{-3}. This compares well with equation (10) when a background electron temperature of 100 eV is used.

2.6 Filament coalesence and stochastic electron motion

Two further processes have been proposed to provide an explanation for anomalous resisitivity: filament coalesence (Honda *et al.*, 2000) and stochastic electron motion (Sentoku *et al.*, 2003). Both mechanisms have been explored using multi-dimensional particle-in-cell simulations.

The first describes the evolution of the Weibel instability. Initially, the fast electron beam filaments into a large number short-wavelength modes. In this situation, each beamlet (that supports its own fast electron current) is surrounded by the return current in an annular shaped pattern. The hot electron current flow in each filament is given by the Afvén-Lawson criteron. Later, the filaments begin to merge into larger filaments as a result of magnetic attraction. However, the merger process itself is extremely violent, principally because the larger filaments cannot support the total current flow contained in the smaller filaments before merger. So after each merger event, the current that can propagate into the target is significantly reduced. This results in transverse heating of the electrons and the dissipation of the fast electron beam energy into the ion population via the formation of a collisionless shock wave.

The second arises as a result of a similar beam-filamentation process. In this case, the merger of the beamlets results the generation of electromagnetic turbulence as a result of stochastic scattering of cold electron current from the magnetic perturbations. The electrostatic fields that are generated enhance the return current but also decelerate the fast electrons. The mean stopping power of the electrons in the beam is given by

$$\rho R \approx 0.89 \times 10^5 T_{hot} \left(\frac{n_e}{n_{cr}} \right) \left(gcm^{-2} \right) \tag{11}$$

where T_{hot} is given in MeV, n_e and n_{cr} are the background and critical densities, respectively. This equation indicates that the mean penetration of the fast electron beam can be reduced by up to $\times$ 1000.

2.7 A two-step process for ion heating in fast ignition

An additional step in the explanation for the high PW laser to thermal energy conversion efficiency observed at Osaka University in the integrated fast ignition experiments has recently been proposed (Mendonça *et al.*, 2005). It is based on the excitation of the two-stream instability (discussed above) that leads to the generation of plasma turbulence. The resulting plasma waves have relativistic phase velocities and can easily become modulationally unstable by decaying into heavily damped ion acoustic oscillations. This results in the occurrence of anomalous ion heating process which efficiently dissipates the energy transferred to the ion oscillations without significant background electron heating.

It can be shown that the maximum growth rate of the ion instability is given by

$$\Gamma_{max} = \left(\frac{m_e}{16m_i}\right)^{1/3} \left(\frac{n_f}{n_e}\right)^{2/3} \omega_{pe} \tag{12}$$

where n_f and n_e are the fast and background electron densities, respectively. This process will grow only when the the energy in the electron perturbations is transferred to the ions faster than it is transferred to electron heating. This requires the growth rate to exceed the electron collision frequency and allows an estimate for the intensity above which the instability will grow.

$$I > 10^{20} \left(\frac{0.5}{f_{abs}}\right)^2 \left(\frac{\lambda}{\mu m}\right)^2 \left(\frac{n_e}{4 \times 10^{22} cm^{-3}}\right)^{25/11} \left(\frac{Z_s ln\Lambda}{4}\right)^{6/11} W cm^{-2} \tag{13}$$

where λ is the laser wavelength, 4×10^{22} cm^{-3} is the number density of CH_2 in polyethylene and $Z_s ln\Lambda$ comes from the electron scattering term.

Confirmation of this instability threshold has come from neutron time of flight signals that were obtained using a high resolution neutron spectrometer. Plastic sandwich targets containing a deuterated layer were irradiated with PW laser pulses at the Rutherford Appleton Laboratory (Norreys et al., 2005). The neutron spectra that were deduced from the signals had two elements: a high-energy component generated by beam-fusion reactions and a thermal component around 2.45MeV. The ion temperatures calculated from the neutron signal width demonstrated a dependence on the front layer thickness and were significantly higher than electron temperatures measured under similar conditions (Evans et al., 2005). The ion heating process was not observed with laser intensities on target below 10^{20} Wcm^{-2}, in agreement with Equation (13). The measurements suggested that a very thin layer of plasma on the front surface of the target, containing ≈ 1 % of the incident laser energy, was heated to temperatures of approximately 40 keV. More measurements are needed to characterise this interesting energy transfer phenomenon more closely. In this context, it is interesting to note that similar ion temperatures have been measured via neutron spectroscopy with the interaction of an ultra-intense laser pulse with an underdense deuterium gas (Fritzler et al., 2002).

The measurements do suggest an alternative approach to hot spark formation in fast ignition where one relies on this instability to heat the core rather than collisions. Malkin and Fisch have suggested a similar approach to hot spark formation based on Langmuir turbulence (Malkin and Fisch, 2002). This might be achieved by using a higher intensity femtosecond rather than a picosecond laser pulse, provide that this is compatible with the growth rate of the two-step instability. The stopping of fast electrons due to plasma turbulence is a subject of continued and active study. Indeed, the coronal ignition model of Hain and Mulser may become viable for fast ignition (Hain and Mulser, 2001).

2.8 Collimation, beam hollowing and annular transport

The experiments described in section 2.1 indicating a variety of beam divergence patterns require some discussion. To a first approximation (i.e. neglecting the filamentation instabilities discussed above in the first instance), when the fast electron beam enters a solid density target, an electric field **E** is set-up that opposes the current of fast electrons **j** fast due to charge separation. A return current is drawn such that $j_{cold} = -\mathbf{E}/\eta$, where η is the plasma resistivity. To a first approximation $\mathbf{j}_{fast} = -\mathbf{j}_{cold}$. A magnetic field **B** is then generated, according to $\partial\mathbf{B}/\partial t = -\nabla\times(\eta\mathbf{j}_{fast})$, azimuthally around the beam that then acts to collimate it. The return current

Ohmically heats the background plasma, thereby lowering the resistivity. The ability of the magnetic field to collimate the fast electrons is determined by the ratio of the beam radius R to the fast electron Lamor radius r_g. Collimation is predicted to occur if $R/r_g > \theta_{1/2}$, where $\theta_{1/2}$ is the cone half angle, that is to say, if the magnetic field is strong enough to bend the fast electron trajectory through an angle $\theta_{1/2}$ (Bell and Kingham, 2003). In the limit of substantial resistive heating, collimation is expected to occur when $\Gamma > 1$, where:

$$\Gamma = 0.13 n_{23}^{3/5} Z^{2/5} ln\Lambda^{2/5} P_{TW}^{-1/5} T_{511}^{3/10} (2 + T_{511})^{-1/2} R_{\mu m}^{2/5} t_{ps}^{2/5} \theta_{1/2}^{-2} \qquad (14)$$

n_e is the electron density in units of 10^{23} cm^{-3}, lnΛ is the Coloumb logarithm, t_{ps} is the pulse duration in ps, T_{511} is the electron temperature in units of 511 keV.

Equation (14) indicates that raising the laser power will lead to a reduction in collimation, and this goes some way to explain the variety of electron beam divergence measurements reported in the literature. However, the large number of variables means that it is somewhat difficult to make precise comparisons between the different experiments. It is also important to stress that, in reality, only a fraction of the fast electron population will ever be collimated by the self-generated magnetic field. Typically, 30%–40% of the laser energy is contained within the focal spot - the rest is spread out into a much larger area. With high intensities, the wings of the focal spot are still intense enough to generate energetic electrons that will influence the observed fast electron pattern using K_α, thermal and optical imaging techniques.

Beam hollowing can also occur under some conditions, particularly when the fall in resistivity with target heating is taken into account. A simple model of this has been given by Davies (2003), based on Ohm's law E $= \eta\mathbf{j}$ for the electric field, where η is resistivity, and Faraday's law for the growth rate of the magnetic field. Ohmic heating is given by $\eta\mathbf{j}^2$, so if the resistivity falls faster than linearly with temperature then the electric field will change from increasing to decreasing with current density once the temperature has increased significantly. This leads to a change in sign in the magnetic field generation, causing the magnetic field to fall and eventually to change sign, so it acts to hollow the beam instead of focus it. This effect would be expected to occur since the Spitzer resistivity applies at sufficiently high temperatures and, most importantly, it would be expected to occur above a given intensity since heating would be expected to increase with intensity. Experimental evidence for this process occuring in plastic targets has recently been presented (Norreys *et al.*, 2006). It is unlikely that this mechanism will be a major factor in realistic fast ignition conditions.

There are a number of other possible mechanisms that can give rise to an annular transport patterns. Magnetic fields generated in the coronal plasma can convectively transport the fast electrons along the surface via an $\mathbf{E} \times \mathbf{B}$ drift (Forslund and Brackbill, 1982), (Wallace, 1985). These fast electrons eventually enter the target some distance away from the focal spot region.

Fountain fields can also give rise to an apparant annular transport pattern. Some of the fast electrons, having entered and travelled through thin foil targets, escape into the vacuum from the rear surface. The target potential eventually prevents further electron losses and the following fast electrons are returned into the target by its potential. In the process, an azimuthal magnetic field is generated (Pukhov, 2002).

Filamentation at a wavelength equal to the diameter of the beam would cause it to hollow. Hollowing of an electron beam in numerical modelling has been observed (Taguchi *et al.*, 2001). In those simulations, a beam with a radius comparable to the skin depth was considered, so it could only support a limited number of modes, and with a square profile, which strongly seeds the hollowing mode. Nevertheless, those simulations showed that the annular beam profile filaments and the tearing mode and two-stream instabilities develop. The filamented transport pattern remains somewhat stable thereafter (Taguchi *et al.*, 2004)

Electric field inhibition of the beam may also give rise to this effect. The inhibition is greatest on-axis because the higher laser intensity at that point would be expected to lead to a higher current density (Bell *et al.* 1997). One may expect this to occur for femtosecond-duration laser pulses where this effect will be most pronounced.

3   Computational modelling

The challenge of modelling these fascinating, but complex, plasma physics problems is to self-consistently solve the electromagnetic fields through Maxwell's equations and the trajectories, currents and charge densities of the particles using relativistic equations of motion.

One approach is to use the fully explicit particle-in-cell (PIC) technique where the full set of Maxwell's equations are solved together with the individual trajectories of each plasma particle. The main difficulty with PIC codes, particularly the solutions to two and three dimensional problems, is that they are also computationally demanding. They therefore need to be run on large computing clusters using parallelised algorithms on even the simplest problems.

Hybrid codes, where the fast electrons are modelled kinetically while treating the background electron and/or ion populations as fluids, is a prefered method of tackling these difficult problems at the moment. Describing the background electrons as a fluid is computationally much faster than modelling them kinetically since the thermal velocity of the background electrons does not have to be resolved and the time-step is set by the relatively large characteristic fast electron times, rather than the much shorter times characterising the kinetic behaviour of the thermal electrons. This allows some aspects of the problem to become solvable within the present day computational resources.

Relativistic Vlasov-Fokker-Planck codes treat all electrons, both fast and cold, as part of the same population. The Fokker-Planck description overcomes the incorrect assumptions made in hybrid codes that the fast and cold electrons are distinct populations, and that non-local and non-Maxwellian effects are unimportant for cold electrons and electrons with intermediate energies. In the most advanced versions of these codes (Bell and Kingham, 2003) the ions are treated as a stationary fluid with constant Z, i.e. ionization processes are neglected. The code is two dimensional and cylindrical in space and three dimensional in momentum. Momentum space is represented by a grid in magnitude of momentum, with the angular coordinates represented by an expansion in spherical harmonics.

4   Conclusions

Great progress has been made in characterising the fast ignition process since the concept was invented in 1994. There is evidence for a high PW laser - thermal energy conversion efficiency in integrated fast ignition experiments. Both analytical and numerical modelling of these experiments has provided much greater confidence to the veracity of these experimental results (Campbell *et al.*, 2005, Mendonça *et al.*, 2005).

Experimental investigations into laser interaction with solid targets has also taken great strides. There is experimental evidence for transport inhibition and beam hollowing occuring in petawatt laser-plasma interaction experiments at solid density. A rich variety of transport patterns have been reported in the literature. The first quantitative agreements with numerical modelling is now also allowing, for the first time, future predictions of energy transport to be made with more confidence.

Of course, there is still much to do. Equation (14) tells us that the fast electron energy transport depends upon a host of parameters, each one of which needs to be varied in a controlled manner.

We can look forward to even greater progress towards the end of the decade in integrated experiments. Multi-kJ PW facilities are being constructed in the United States (OMEGA EP), Japan (FIREX), France (LIL PW) and also in the UK (Orion). These facilities may bring us close to the equivalent conditions for energy breakeven using deuterium-deuterium reactions (which have a 2-order of magnitude smaller cross section that deuterium-tritium reactions, but no radioactive handling issues).

Plans for ignition-scale lasers systems are being considered in Europe (the HiPER proposal) and Japan (FIREX II). This is a truly exciting prospect, and we can begin to see the outline of the roadmap for the commercial realisation of fusion energy via fast ignition.

Acknowledgments

I would like to acknowledge my colleagues from Osaka University, Imperial College London, the Queens University Belfast, the Universities of Oxford, York and Essex. Special thanks to the Physics Group and all other staff of the Central Laser Facility, Rutherford Appleton Laboratory for their encouragement and assistance.

References

Bell A.R., Davies J.R., Guerin S., and Ruhl H., (1997) *Plasma Phys. Control. Fusion* **39**, 653
Bell A.R., Kingham R.J. (2003)*Phys. Rev. Lett.* **91**, 035003
Beg F.N., Bell A.R., Dangor A.E., Danson C.N., Fews A.P., Glinsky M.E., Hammel B.A, Lee P., Norreys P.A. and Tatarakis M (1997) *Phys. Plasmas* **4** 447.
Brandl F., Pretzler G., Habs D. and Fill E (2003) *Europhys. Lett.* **61**, 632.
Brunel F. (1987) *Phys. Rev. Lett.* **59**, 52.
Borghesi M., Mackinnon A.J., Bell A.R., Malka G., Vickers C., Willi O., Davies J.R., Pukhov A. and Meyer-ter-Vehn J. (1999) *Phys. Rev. Lett.* **83**, 4309.
Bret A., Firpo M-C and Deutsch C. (2004) *Phys. Rev. E* **70**, 046401.
Bret A., Firpo M-C and Deutsch C. (2005) *Phys. Rev. Lett.* **94**, 115002.
Campbell R., Kodama R., Melhorne T.A., Tanaka K.A., Welch D.R.(2005) *Phys. Rev. Lett.* **94**, 055001.
Clark E.L., Krushelnick K., Zepf M., Beg F.N., Machacek A., Norreys P.A., Santala M.I.K., Tatarakis M., Watts I. and Dangor A.E. (2000) *Phys. Rev. Lett.* **85**, 1654.
Cowan T.E., Perry M.D., Key M.H., Ditmire T.R., Hatchett S.P., Henry E.A., Moody J.D., Moran M.J., Pennington D.M., Phillips T.W., Sangster T.C., Sefcik J.A., Singh M.S., Snavely R.A., Stoyer M.A., Wilks S.C., Young P.E., Takahashi Y., Dong B., Fountain W., Parnell T., Johnson J., Hunt A.W., Kuhl T. (1999) *Laser and Particle Beams* **17**, 773.
Davies J.R. (2003) *Phys. Rev. E* **68**, 056404.
Evans R.G., Clark E.L., Eagleton R.T., Dunne A.M., Edwards R.D., Garbett W.J., Goldsack T.J., James S., Smith C.C., Thomas B.R., Clarke R., Neely D. and Rose S.J. (2005), *Appl. Phys. Lett.* **86**, 191505.
Fritzler S., Najmudin Z., Malka V., Krushelnick K., Marle C., Walton B., Wei M.S., Clarke R.J., Dangor A.E. (2002) *Phys. Rev. Lett.* **89**, 165004.
Gremillet L., Amiranoff F., Baton S.D., Gauthier J.-C, Koenig M., Martinolli E., Pisani F., Bonnaud G., Lebourg C., Rousseaux C., Toupin C., Antonicci A., Batani D., Bernardinello A., Hall T., Scott D., Norreys P., Bandulet H. and Pepin H. (1999) *Phys. Rev. Lett.* **83**, 5015.
Hain S. and Mulser P. (2001) *Phys. Rev. Lett.* **86**, 1015.
Honda M., Meyer-ter-Vehn J. and Pukhov A., *Phys. Rev. Lett.* **85**, 2128.
Jung R., Osterholz J., Lowenbruck K., Kiselev S., Pretzler G., Pukhov A., Willi O., Kar S., Borghesi M., Nazarov W., Karsch S., Clarke R. and Neely D. (2005) *Phys. Rev. Lett.* **94**, 195001.
Koch J.A., Key M.H., Freeman R.R., Hatchett S.P., Lee R.W., Pennington D., Stephens R.B. and Tabak M. (2002) *Phys. Rev. E* **65**, 016410

Kodama R., Norreys P.A., Mima K., Dangor A.E., Evans R.G., Fujita H., Kitagawa Y., Krushelnick K., Miyakoshi T., Miyanaga N., Norimatsu T., Rose S.J., Shozaki T., Shigemori K., Sunahara A., Tampo M., Tanaka K.A., Toyama Y., Yamanaka Y. and Zepf M. (2001) *Nature* **412**(6849), 798.

Kodama R., Shiraga H., Toyama Y., Fujioka S., Azechi H., Jitsuno T., Kitagawa Y., Krushelnick K.M., Lancaster K.L., Mima K., Nagai K., Nakai M., Nishimura H., Norimatsu T., Norreys P.A., Sakabe S., Tanaka K.A., Youssef A., Zepf M. and Yamanaka T. (2002) *Nature* **418**(6091), 933.

Lindl J.D. (1995) *Phys. Plasmas* **2**, 3933.

Malkin V.M. and Fisch N.J. (2002) *Phys. Rev. Lett.* **89**, 125004.

Mendonça J.T., Norreys P., Bingham R. and Davies, J.R. (2005),*Phys. Rev. Lett.* **94**, 115002.

Modena A., Najmudin Z., Dangor A.E., Clayton C.E., Marsh K.A., Joshi C., Malka V. Darrow C.B., Danson C., Neely D. and Walsh F.N. (1995) *Nature* **377**(6550), 606.

Norreys P.A., Allott R., Clarke R.J., Collier J., Neely D., Rose S.J., Zepf M., Santala M., Bell A.R., Krushelnick K.M., Dangor A.E., Woolsey N.C., Evans R.G., Habara H., Norimatsu T. and Kodama R. (2000) *Phys. Plasmas* **7**, 3721.

Norreys P.A., Lancaster K.L., Habara H., Davies J.R., Mendonï£¡a J.T. Clarke R.J., Dromey B., Gopal A., Karsch S., Kodama R., Krushelnick K., Moustaizis S.D., Stoeckl C., Tatarakis M., Tampo M., Vakakis N., Wei M.S. and Zepf M. (2005),*Plasma. Phys. Controlled Fusion* **47**, L49.

Norreys P.A., Green J.S., Davies J.R., Tatarakis M., Clark E.L., Beg F.N., Dangor A.E., Lancaster K.L., Wei M.S., Zepf M., and Krushelnick K. (2006) *Plasma Phys. Control. Fusion (in press).*

Nuckolls J. Thiessen A, Wood L and Zimmerman G. (1972) *Nature* **239**(5368), 139.

Pukhov A. and Meyer-ter-Vehn J. (1998) *Phys. Plasmas* **5**, 1880.

Pukhov A. (2002) *Phys. Rev. Lett.* **86**, 3562

Rosen, M.D. (2006) this volume.

Santos J.J., Amiranoff F., Baton S.D., Gremillet L., Koenig M., Martinolli E., Le Gloahec M.R., Rousseaux C., Batani D., Bernardinello A., Greison G. and Hall T. (2002) *Phys. Rev. Lett.* **89**, 025001.

Santala M., Clark E., Watts I., Beg F.N.,Tatarakis M., Zepf M. Krushelnick K., Dangor A.E., McCanny T., Spencer I., Singhal R.P., Ledingham K.W.D., Wilks S.C., Machacek A., Wark J.S., Allott R., Clarke R.J. and Norreys P.A. (2000) *Phys. Rev. Lett.* **84**, 1459.

Sentoku Y., Mima, K., Kaw P. and Nishikawa K. (2003) *Phys. Rev. Lett.* **90**, 155001.

Silva, L.O., Fonseca R.A., Tonge J.W., Mori W.B. and Dawson J.M. (2002) *Phys. Plasmas* **9**, 2482.

Stephens R.B., Snavely R.A., Aglitskiy Y., Amiranoff F., Andersen C., Batani D., Baton S.D., Cowan T., Freeman R.R., Hall T., Hatchett S.P., Hill J.M., Key M.H., King J.A., Koch J.A., Koenig M., MacKinnon A.J., Lancaster K.L., Martinolli E., Norreys P., Perelli-Cippo E., Le Gloahec M.R., Rousseaux C., Santos J.J. and Scianitti F. (2004) *Phys. Rev. E.* **69**, 066414.

Tabak M., Hammer J, Glinsky M.E., Kruer W.L., Wilks S.C. Woodworth J., Campbell E.M. Perry M.D. and Mason R.J. (1994) *Phys. Plasmas* **1**, 1626.

Taguchi T., Antonsen Jr. T.M., Liu C.S. and Mima K. (2001)*Phys. Rev. Lett.* **86**, 5055.

Taguchi T., Antonsen Jr. T.M. and Mima K. (2004)*Comp. Phys. Commun.* **164**, 269.

Tatarakis M., Lee P., Davies J.R., Beg F.N., Bell A.R., Norreys P.A., Haines M.G. and Dangor A.E., (1998) *Phys. Rev. Lett.* **81**, 999.

Wei, M.S., Beg F.N., Clark E.L., Dangor A.E., Evans R.G., Gopal A., Ledingham K.W.D., McKenna P., Norreys P.A., Tatarakis M., Zepf M., and Krushelnick, K. (2004) *Phys. Rev. E* **70**, 056412.

Wilks S.C., Kruer W.L., Tabak M. and Langdon A.B. (1992) *Phys. Rev. Lett.* **69**, 1383.

17
Fusion Targets

Damian C. Swift

Los Alamos National Laboratory, U.S.A.

1 Introduction

In inertial confinement fusion (ICF) (Lindl 1998), thermonuclear reactions are induced by compressing the fuel dynamically, through an implosion. This leads to the high compressions and temperatures necessary for nuclear fusion reactions to occur, transiently before the compressed material re-expands. "Laboratory" fusion experiments using lasers to implode the fuel are energy-limited, which means that relatively high implosion convergence ratios – i.e., the ratio of initial to imploded radius – are needed to induce reactions. Implosions are unstable with respect to azimuthal variations in the drive or the material response, so great care is needed to assure adequate symmetry.

The most prominent aspect of the U.S. fusion program is the programme to demonstrate the ignition of a thermonuclear burn wave, with the first experiments planned to start in 2010 at the National Ignition Facility (NIF). This series of lectures presents some of the considerations in the development of the fuel capsule for thermonuclear ignition targets at NIF. One family of targets designs at NIF uses indirect-drive – i.e., a hohlraum – to implode a spherical capsule containing deuterium-tritium (D-T) fusion fuel. In the absence of an additional ignition source, such as the ion pulse in a fast igniter design, the implosion is designed to assemble the fuel in a high-density state ($\sim$20 g/cm^3) through which a thermonuclear burn wave can propagate, and to create a central hotspot ($\sim$10 keV) to initiate the reactions. The fuel is initially maintained at cryogenic temperatures and a few tens of atmospheres, to give a layer of solid D-T on the inside of the capsule and a region of D-T gas in the centre. For efficient compression, the bulk of the fuel should remain as cold as possible, i.e., close to an isentrope: thus it should not be subjected to strong shock waves as these greatly increase the temperature.

The NIF laser comprises 192 beams, delivering pulses up to 20 ns long, with a high degree of control over the pulse history, and a total energy of 1.8 MJ at 351 nm (three times the fundamental frequency, for more efficient coupling to matter). The laser power history is designed to induce a hohlraum temperature history which implodes the fuel capsule through a series of shocks, as an approximation to isentropic compression. The laser beams generally generate some higher-energy X-rays, in the kilovolt regime: these can subject the fuel capsule to X-ray heating as well as hydrodynamic loading.

It is difficult and expensive to obtain enough energy in laboratory experiments to implode the target vigorously enough to ensure ignition. Thus the implosion must convert hohlraum energy as efficiently as possible to kinetic energy in the capsule. This constraint leads to a relatively high convergence ratio for the implosion of around 40. The implosion is susceptible to the ablative Rayleigh-Taylor instability during the acceleration phase, and to the hydrodynamic Rayleigh-Taylor instability as the capsule and fuel decelerate on stagnation. It is also potentially unstable to the Richtmeyer-Meshkov instability as shocks propagate across the capsule/fuel interface. This sensitivity to instabilities means that the capsule and fuel must be spherical to a high degree of accuracy, and the hohlraum drive must also be highly uniform. Otherwise, capsule material could mix in with the fuel, reducing its ability to burn; cold fuel could mix in with the central hotspot, cooling it; and the hydrodynamic compression and heating achieved in the fuel and the hotspot respectively could be insufficient for a thermonuclear burn wave to ignite.

2   Ablator material and microstructures

Copper-doped beryllium is the ablator most likely to allow ignition, as it couples most efficiently to the hohlraum drive and its high thermal conductivity produces a uniform layer of D-T fuel inside. For the most efficient coupling to the hohlraum radiation field, the composition of the ablator should vary through its thickness: pure at the inside and outside, with the copper concentrated toward the inside surface. Low copper levels at the outside allow the hohlraum radiation to penetrate more deeply; the region of elevated copper concentration prevents preheat from kilovolt-level X-rays; pure beryllium at the inside reduces the risk of high-Z contamination of the fuel. Beryllium is toxic, making it relatively difficult to fabricate and characterize. Beryllium is a crystalline metal, and its crystals have anisotropic properties: the amount of expansion and contraction under heating or compression is significantly different along each crystal axis. Under the loading that occurs as the capsule is imploded, or under X-ray heating, this anisotropic response may cause spatial variations in stress and velocity. It is important to control these spatial variations so that they do not seed instabilities large enough to prevent ignition. Microstructures are thus a consideration in choosing methods of fabricating the capsule.

Compared with most other engineering problems, components in laser experiments are generally thin. Historically, many target components have been produced by deposition onto a mandrel of the appropriate shape, eliminating the need for machining. Deposition is also the only practical technique for varying the composition through the capsule wall. However, it is difficult to prevent the formation of voids in the deposited material. The deposition technique being used for NIF ablators is sputtering (Figure 1). Sputtering tends to produce columnar grains oriented radially through the shell. The microstructure can be varied by controlling parameters in the sputtering process. In particular, the composition can be varied by changing the relative source rates of copper compared with beryllium. One non-obvious control is to use a beam of inert ions such as argon to break up the nucleation of oriented grains (Figure 2).

The principal alternative fabrication technique is to cast bulk material and then machine the capsule from the cast bulk. Pressing from initially-powdered material typically leaves more pores than casting, though cast material may have large grains (Figure 3). A refinement is to force the pressed material repeatedly through a tube with a bend, a process known as equal-channel angular extrusion (ECAE – Figure 4). The resulting plastic deformation reduces the grain size of material. After a heat-treatment stage to relieve internal stresses, the material has a very homogeneous grain structure and essentially no voids (Cooley 2005) (Figure 5). However, this technique is not suitable for fabricating capsuled with a graded copper dopant: the composition must be uniform. It is also complicated to machine hollow spherical hemishells from the bulk material and bond the halves.

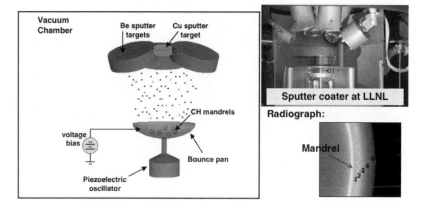

Figure 1. *Schematic of capsule manufacture by sputtering.*

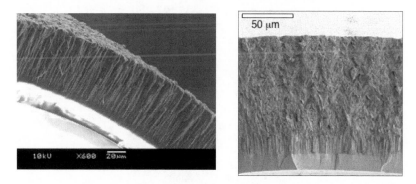

Figure 2. *Capsule microstructures obtained from sputtering: columnar growth (left), and twisted growth with ion current (right).*

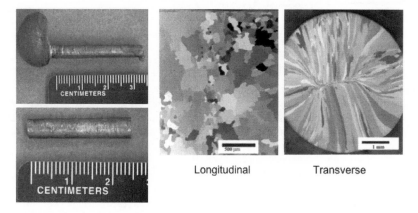

Figure 3. *Bulk material and grain sizes produced by casting beryllium.*

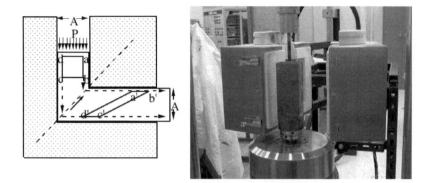

Figure 4. *Schematic of the equal-channel angular extrusion process, and press equipment.*

Figure 5. *Microstructure produced by ECAE followed by annealing. Example for 4 ECAE passes at 650-850° C, followed by annealing for 1 hour at 775° C: equiaxed 20 μm grains.*

In either case, the capsule must be filled with the D-T fuel. For the NIF experiments, this is likely to be done through a laser-drilled hole in the capsule wall. After filling, the hole must also be re-filled, either with a separate plug welded into place or by the use of a laser or particle beam to push material from the rest of the capsule into the hole. Femtosecond lasers have been found capable of drilling holes less than 5 μm in diameter through beryllium shells at least 125 μm thick (Nobile 2005). Any residual hole, or the plug itself, may perturb the implosion or fuel purity; experimental and simulation studies are being performed to study the effects.

If the grain size can be made small enough, spatial variations in the response of the beryllium should not seed instabilities severe enough to prevent ignition: this is one palliative to the anisotropic response of beryllium. The anisotropic behaviour is not understood well enough under NIF ignition conditions for the maximum grain size to be specified on the basis of simulations.

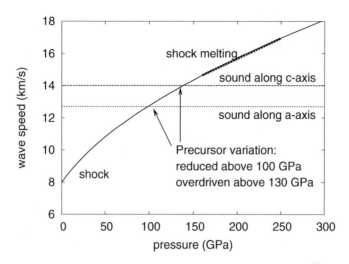

Figure 6. *Palliatives against microstructural effects. If cold capsule material is driven at a low pressure, differences in the elastic wave speed in different orientations of beryllium crystal cause spatial variations which may seed instabilities. If the shock pressure exceeds 100 GPa, the orientation-dependent differences start to fall. If driven strongly enough, the capsule material will melt behind the first shock – though the pressure for shock-melting is not known accurately. X-ray heating of the capsule may lead to melting at a lower pressure.*

There are limits to the extent to which the microstructure can be controlled during sputtering, or the grain size refined by ECAE. Another palliative is to melt the ablator material by X-ray or shock heating. However, if the temperature does not rise fast enough then the anisotropic thermal response may still seed instabilities. Even if heating is prompt or the shock is strong enough to cause immediate melting, some small velocity perturbations may possibly be caused. (Figure 6.)

The ablator characterization task in the U.S. National Ignition Campaign addresses the risk of instability seeding from the microstructure through research to improve our understanding of the dynamic behaviour of beryllium, the validation of simulations of the heterogeneous response of polycrystalline beryllium, and experiments where instabilities are seeded and their growth measured for comparison with simulations.

3   Continuum mechanics and shock physics

The operation of an ICF target is complicated: capsule performance is an integrated, time-dependent response to the laser drive. To give any reasonable prediction of the behaviour of an ICF target, its response must be simulated with spatial and temporal resolution. In practice this means discretising the components: dividing them into cells, in each of which the state (velocity, mass density, ion temperature, electron temperature, radiation temperature, ionisation, pressure, etc) is assumed to vary in a simple way, often as a constant in space. For the behaviour of the capsule, particularly any microstructural processes, much of the physics of radiation transport can be neglected – treated as a source term rather than explicitly – and the resulting continuum mechanical simulations are performed so as to include shear stresses as well as pressures.

For a given drive history and target configuration, the response of the capsule is an initial value problem: given a distribution of material in some initial state of mass density, velocity, and specific internal energy $\{\rho, \vec{u}, e\}(\vec{r}, t_0)$ over some region $\{\vec{r}\} \in \mathcal{R}$, what is the time-evolution $\{\rho, \vec{u}, e\}(\vec{r}, t > t_0)$? Heat conduction can usually be neglected in most shock-wave scenarios in condensed matter. In Lagrangian form, i.e. in a co-ordinate system deforming with the material, the continuum equations are

$$\frac{\partial \rho(\vec{r}, t)}{\partial t} = -\rho(\vec{r}, t)\mathrm{div}\,\vec{u}(\vec{r}, t) \tag{1}$$

$$\frac{\partial \vec{u}(\vec{r}, t)}{\partial t} = -\frac{1}{\rho(\vec{r}, t)}\mathrm{grad}\,p(\vec{r}, t) \tag{2}$$

$$\frac{\partial e(\vec{r}, t)}{\partial t} = -p(\vec{r}, t)\mathrm{div}\,\vec{u}(\vec{r}, t), \tag{3}$$

$$\tag{4}$$

subject to boundary conditions in pressure $p(\vec{r}, t)$ and/or velocity $\vec{u}(\vec{r}, t)$ for $\{\vec{r}\} = d\mathcal{R}$. This set of equations is incomplete – there are more unknowns than equations – and is closed using the inherent properties of the material, in particular its equation of state (EOS), here in the form $p(\rho, e)$, to allow time-integration to proceed.

The EOS is a material-dependent property, relating a thermodynamic potential to its natural parameters, e.g. $de = Tds - pdv$ for the specific internal energy in terms of specific entropy and specific volume. The EOS is here the relation $e(s, v)$ for a material. Derivatives and other functions of the EOS give other quantities: pressure p, temperature T, sound speed c^2, specific heat capacity c_v, and so on.

The most familiar example is the perfect gas EOS:

$$p = nk_BT \quad p = (\gamma - 1)\rho e; \tag{5}$$

with a suitable calculation of the number density of particles n this is often used in the plasma regime. However, condensed matter is *not* accurately represented as a perfect gas. The equilibrium phase diagram is usually presented in temperature-pressure space: the phase plane comprises contiguous regions of structure. The structures include solid and liquid/vapour, ultimately becoming plasma at high temperatures. Many solids, particularly metals and ceramics, are crystalline. Some solids adopt a single crystal structure – i.e., symmetry of the lattice of atoms – over the whole of the solid region, but many adopt a variety of structures, with solid-solid phase boundaries between them. The detailed structure depends on the composition of a material. Beryllium may occur in the hexagonal or body-centered cubic structures.

In principle, the EOS may be affected by a range of non-equilibrium processes, including electron-ion equilibration and metastability of condensed-matter structures. Over a wide range of states and time scales, electron-ion equilibration can be ignored. Condensed-matter phase changes cannot in general be ignored. Given a knowledge of time scale and the material, it may be possible to consider a single EOS which includes phase changes that occur relatively rapidly, or it may be possible to use a separate EOS for each structure together with a rate law for transitions between structures, i.e. for phase changes. Thus it is possible to define unique EOS surfaces which apply for geophysics, hypervelocity impact, terrestrial explosions, ICF, and so on. The range of applicability is, in rough terms $-\mathrm{GPa}$ (tensile stress for material failure) $< p < \mathrm{PPa}$ (or Gbar); $\sim 10\,\mathrm{K} < T < \mathrm{MeV}$; $\mathrm{ps} < t < \mathrm{Gyr}$. With few known exceptions, the specific internal energy of a material can be described accurately by splitting into terms for cold compression, thermal motion of the ions, and thermal excitation of the electrons:

$$e(\rho, T) = e_{\mathrm{cold}}(\rho) + e_{\mathrm{ion\text{-}thermal}}(\rho, T) + e_{\mathrm{electron\text{-}thermal}}(\rho, T). \tag{6}$$

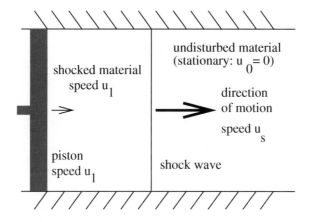

Figure 7. *Schematic of a shock wave driven by a piston.*

The EOS can be derived from $e(\rho, T)$ using thermodynamics.

High-pressure states can be explored by quasistatic experiments, using a mechanical press such as a diamond anvil cell to exert a steady pressure on the sample. The sample may be simultaneously heated or cooled. However, there is a limit to the pressure that can be exerted before the components of the press deform, and this problem becomes worse on heating or cooling. Maximum quasistatic pressures are in the region of a couple of hundred gigapascals, though experiments become very difficult above a few tens of gigapascals. Temperatures range from cryogenic to a couple of thousand kelvin, but are limited to around 1000 K above about 100 GPa. Alternatively, high pressures can be induced dynamically, and studied rapidly before they dissipate or the sample disassembles. Dynamic high pressures are generally better-suited to characterize materials for dynamic loading applications such as ICF, as the material response can be explored or verified on the appropriate time scale for the application. The archetypal dynamic high-pressure experiment employs a steady shock wave.

The canonical shock wave experiment consists of a rigid piston pushing into a sample enclosed by rigid lateral walls. The piston, initially at rest, rapidly accelerates to a speed u_p. Material pushed ahead of the piston also moves at a speed u_p. However, it takes a finite time for the piston to affect material any given distance ahead; the effect of the piston is transmitted by a shock wave moving through the undisturbed material at a speed $u_s > u_p$. At any instant of time, material ahead of the shock wave is in its original, undisturbed state, and material behind the shock wave moves at u_p and is in a compressed, heated state (Figure 7). If a shock wave experiment is performed on a time scale in which physical processes in the material either occur much more quickly or much more slowly, then shocked material is in a constant state and the shock travels at a constant speed.

This steady shock wave scenario is a simple problem in continuum mechanics. Considering the conservation of mass, momentum and energy as material passes through the shock, the Rankine-Hugoniot relations are obtained to relate the unshocked and shocked states with the shock speed:

$$u_s^2 = v_0^2 \frac{p - p_0}{v_0 - v}, \tag{7}$$

$$u_p = \sqrt{[(p - p_0)(v_0 - v)]}, \tag{8}$$

$$e = e_0 + \frac{1}{2}(p + p_0)(v_0 - v) \tag{9}$$

– three equations in five unknowns. If a shock wave experiment is performed, two quantities must therefore be measured to determine the material state. For a given initial state, the locus of shocked states which can be generated by the passage of steady shock waves of different strengths is called the shock Hugoniot. The Hugoniot for material initially at STP is called the principal Hugoniot; this term is often used to refer to material which is initially free of pores, i.e., at the theoretical maximum density. Shock compression adds entropy to the shocked material, so the Hugoniot is hotter than the isentrope.

For experiments on condensed matter, real materials are not strong enough to make a piston or rigid wall in the idealised shock wave picture. Traditional shock wave experiments employ the impact of a projectile ("flyer") to make an "inertial piston". The diameter of the experiment is much greater than its thickness in the impact direction, so that edge effects do not perturb the axial region during the measurement of the shocked state, giving the effect of a rigid lateral boundary. Components are thus thin right-circular cylinders. On impact, shock waves are induced in the flyer and target; if these are made of the same material then the material speed is half the flyer speed u_f, thus $u_p = u_f/2$. An example of a measurement to determine a state on the Hugoniot would be to measure the flyer speed u_f and the shock speed u_s from the transit time of the shock through the target.

As conventionally defined, the EOS describes the scalar response of a material to loading and heating: the pressure as a function of compression and temperature. The response is more accurately described in terms of the stress tensor s as a function of the strain tensor e and temperature T, plus additional parameters such as defect densities describing the microstructural state. Generally, the strain tensor is not isotropic: the deformation experienced by material subjected to dynamic loading is a combination of compression and shearing. For instance, a one-dimensional planar shock compresses material in its direction of propagation but not in the perpendicular directions; this deformation can be expressed as an isotropic compression plus a shear. In a solid subjected to gradually increasing strain, the response is at first elastic, until the shear stresses become large enough to drive plastic deformation processes such as slip. At the level of the atomic lattice, plastic strains act to decrease the elastic shear strain and thus reduce the stored elastic energy. The tensor EOS – often referred to as a constitutive or strength model – can be represented in a number of ways of varying sophistication, consistency and rigour. Here we adopt a relatively rigorous and self-consistent representation. The local state is represented by the specific internal energy or temperature T, the elastic strain tensor e, and an equivalent plastic strain ϵ_p used to estimate work-hardening. The instantaneous stress tensor is calculated from the elastic strain tensor using a set of elastic constants C (a $3 \times 3 \times 3 \times 3$ matrix):

$$s = Ce. \tag{10}$$

Each crystal structure has a different set of unique elastic constants. In practice, C is decomposed as an explicit scalar EOS and a modified set of deviatoric constants C'; this ensures that the tensor EOS has the correct scalar behaviour at high compressions, where it is usually dominated by the bulk compression.

Dynamic deformation is described by the gradient of the velocity field, $\partial \vec{u}/\partial \vec{r}$. This 3×3 matrix can be decomposed into symmetric and antisymmetric matrices. The antisymmetric matrix is rigid-body rotation. The symmetric part is the strain rate tensor; its trace is the rate of compression or expansion and the remainder is the deviatoric strain rate. Often in continuum mechanics, tensor quantities are represented using Voigt notation: the unique elements of a symmetric 3×3 tensor are represented as a 6-vector, in the order $(11, 22, 33, 23, 31, 12)$. The elastic constants are then represented by a 6×6 matrix instead of $3 \times 3 \times 3 \times 3$.

We have treated plastic flow in a number of ways as described below, including a model representing the mobility of the lattice defects mediating plastic flow. A simple but powerful alternative is to consider the flow stress: the stress necessary to sustain a given plastic flow rate. The flow stress is similar to the yield stress, which is the stress at which deformation becomes plastic instead of elastic. The flow stress is relatively straightforward to infer experimentally. The velocity history at the free surface of a sample impacted by a flyer often exhibits a clear inflexion when plastic flow occurs (Bushman 1993).

4 Properties of the ablator material

Since the capsule microstructure could seed implosion instabilities, and the palliatives include fabricating the capsule with grains too small to couple to unstable modes or driving a shock wave strongly enough to melt the beryllium capsule, it is important to understand the processes of microstructural plasticity and shock-induced melting.

The high-pressure properties of beryllium and its response to dynamic loading have been investigated previously on longer time scales: quasistatic compression and shock waves induced by the impact of flyers millimetres thick. The phase diagram is interesting as it exhibits a solid-solid phase transition from the ambient hexagonal (hex) structure to body-centered cubic (bcc), before melting (Young 1991). The EOS has been deduced empirically from the principal Hugoniot, and the flow stress from free-surface velocity histories. However, phase changes and plastic flow are time-dependent processes: the pressure at which a phase change occurs, and the shear stress at which plastic flow occurs, depend on the rate of deformation. The time scales relevant in the NIF ignition capsule are much shorter than the time scales explored by previous impact experiments – nanoseconds compared to microseconds – so the deformation rates are much greater. We have performed a range of experimental and theoretical studies to investigate the response of beryllium to dynamic loading on nanosecond time scales.

4.1 Equation of state and phase diagram of beryllium

EOS for beryllium have been developed previously, using shock Hugoniot data (Steinberg 1991, Holian 1984). Empirical EOS such as these typically do not have accurate temperatures. They are likely to be particularly poor for beryllium because the vibrational frequencies of its atoms are unusually high for a metal. The vibrational modes are not all active at room temperature, which reduces the specific heat capacity. An accurate thermal EOS is desirable because the NIF fuel capsule is subjected to a combination of loading and X-ray heating which may be changed easily with relatively subtle changes in the hohlraum design and drive beam histories: it is not adequate simply to determine, for instance, melting on the principal Hugoniot.

Quantum mechanics was used to calculate an independent EOS, including rigorous temperatures, using a procedure applied previously to a range of substances including aluminium, silicon and the alloy NiAl (Swift 2001 and 2005). The cold compression energy e_c was calculated from the ground state of the electrons with respect to stationary atoms, the lattice-thermal energy e_l from the phonon modes of the crystal lattice, and the electron-thermal energy by calculating the excitation of electrons from the ground state to higher energy bands, using the zero-temperature band structure. Lattice cells were modelled using a pseudopotential to represent the potential from each nucleus and the inner two electrons, which were assumed to be unperturbed by compression or heating in the regime of interest. The pseudopotentials used were non-local (varying with the angular momentum of each state) and generated using the Troullier-Martins method (Troullier 1991) to reproduce the scattering properties of the outer electrons in all-electron calculations. Compressions were applied by changing the lattice parameters. The wavefunctions $\psi_i(\vec{r})$ of the outer electron states were expanded in a plane wave basis set,

$$\psi_i = \sum_j \alpha_{ij} \exp(\vec{k}_j . \vec{r}), \tag{11}$$

for which the amplitudes α_{ij} were optimized to find the ground state. The number of basis states was finite, corresponding to an energy cutoff in the expansion. The calculations were performed with periodic boundary conditions, representing an infinite crystal. The ground-state wavefunctions, and hence the electron band structure, were calculated by minimising the energy of the Schrödinger Hamiltonian

$$H\psi_i = E_i\psi_i \quad : \quad H \equiv -\frac{\hbar^2}{2m_e}\nabla^2 + V_r(\vec{r}; \{\psi_i\}) \tag{12}$$

so the outer electrons were represented in a non-relativistic manner. The local density approximation (LDA) (Hohenberg 1964, Kohn 1965, Perdew 1992, White 1994) was used to model exchange and correlation.

The electron ground state and band structure was calculated as the lattice cell was subjected to compression and shear distortions. The electron charge density was used to calculate the stress on the lattice cell – and hence the pressure and shear stress – and the force on the atoms as one was displaced from equilibrium. The hex structure is defined by two lattice parameters, a and c. This structure thus has an internal degree of freedom: the c/a ratio. In general, if an isotropic stress is applied to the lattice cell, the c/a ratio changes. When calculating the EOS, it is not sufficient to apply an isotropic strain, in which the c/a ratio stays constant. We chose a series of values of a, and for each varied c until the ground state stress was isotropic. In this way, the c/a ratio was predicted as a function of compression. At STP, the c/a ratio for beryllium is quite far from the ideal value for hard spheres: 1.567 compared with 1.63, caused by angular forces in the electron orbitals (Ashcroft 1976). The theoretical calculation reproduced the observed c/a ratio well (Figure 8), and is a powerful demonstration of the accuracy of this method since the electronic structure calculations include no normalizations to the structure of beryllium.

The EOS was calculated as a tabular relation $e(\rho, T)$, from which the specific entropy $s(\rho, T)$ was found by integrating the second law of thermodynamics along each isochore (line of constant volume)

$$ds = \frac{de}{T}\bigg|_v, \tag{13}$$

giving the Helmholtz free energy $f \equiv e - Ts$ and hence the pressure $p = -\partial f/\partial v|_T$. This procedure was repeated for each relevant crystal structure to give an independent EOS for each, and the equilibrium structure or mixture of structures was found using Maxwell constructions, i.e. minimising f at each (ρ, T) state. The LDA typically gives an error of around 1% in lattice parameter, so a pressure offset was used to bring the theoretical EOS into agreement with a single experimental value: the mass density at STP. EOS calculated in this way can be extremely accurate (Swift 2005).

The beryllium EOS was used to calculate the principal Hugoniot, and compared with experimental data (Marsh 1980) and with Hugoniots calculated using empirical EOS (Steinberg 1991, Holian 1984). The electronic structure calculations were performed over a relatively coarse

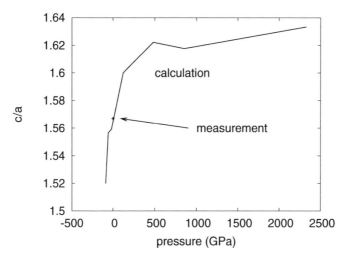

Figure 8. *c/a ratio predicted quantum mechanically for beryllium, and observed experimentally at STP.*

tabulation in mass density, and the predicted Hugoniot exhibited a slight bump at pressures of a few tens of gigapascals. In pressure-density space, or any other space except temperature, the EOS were all similar and all matched the experimental data reasonably well. At very high compressions, the quantum-mechanical (QM) EOS matched the SESAME EOS – which, at these pressures, was based on a different form of electronic structure calculation – more closely than the Steinberg EOS, based on the experimental shock data, which did not extend above a few hundred gigapascals. In pressure-temperature space, the EOS predicted different Hugoniots. The QM EOS followed the SESAME EOS at low pressures, was more similar to the Steinberg EOS at intermediate pressures, then lay between the two empirical EOS at terapascal pressures. It is extremely difficult to measure temperatures during shock wave experiments, and the discrepancy between the different EOS is so far unresolved. Assumptions of fairly simple thermodynamic behaviour were made when constructing temperatures in the empirical EOS, so the QM EOS is valuable as a much more rigorous way of predicting the temperature, within the limitations of its overall accuracy. (Figure 9.)

Another way of exploring differences between the EOS is by considering other loading paths, such as isentropic compression. It is advisable for ICF design simulations to use an EOS which is valid along the isentrope, as it is relatively easy to alter the loading history between shock and isentropic compression by relatively small changes in the laser power history. At low pressures, the Hugoniot lies very close to the isentrope starting at the same initial state. In the regime of interest for the ICF capsule – up to a few hundred gigapascals – the Hugoniot and isentrope diverge significantly. The QM EOS predicted a principal isentrope which was considerably softer than the isentropes from the other EOS (Figure 10). As discussed below, this difference is testable using isentropic compression experiments.

The elastic constants of beryllium were predicted as a function of compression by performing several series of electronic structure calculations for compression under isotropic stress – i.e., as used to predict the EOS – and calculations with a uniaxial strain or a shear strain applied to the isotropic state. Given each electronic ground state, the stress tensor was calculated from the charge density via the Hellman-Feynmann approximation, and the elastic constants were found

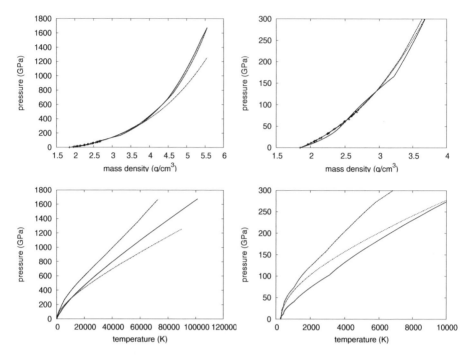

Figure 9. *Comparison between shock Hugoniots predicted using different equations of state (quantum mechanical: solid; SESAME: dashed; Steinberg: dotted) and previous experimental data (Marsh 1980: points).*

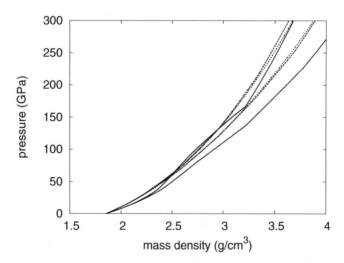

Figure 10. *Comparison between shock Hugoniots and isentropes predicted using different equations of state (quantum mechanical: solid; SESAME: dashed; Steinberg: dotted). In each case, the Hugoniot lies above the isentrope.*

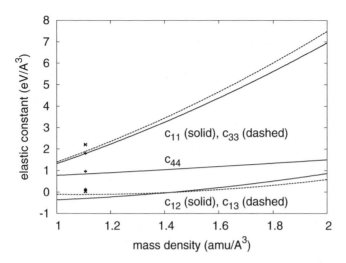

Figure 11. *Elastic constants for beryllium: predictions as a function of compression, and measurements at STP.*

from the rate of change of stress with strain. For the hex structure, the unique elastic constants are (in Voigt notation) c_{11}, c_{33}, c_{12}, c_{13} and c_{44}. These were found by performing ground-state calculations with uniaxial strains along the $[10\bar{1}0]$ and $[0001]$ crystal directions and shear strains in the $(01\bar{1}0)$ plane.[1] The elastic constants of beryllium have been measured at STP, and for a few other temperatures at atmospheric pressure. The predicted elastic constants reproduced the observed values fairly well, giving confidence in the likely accuracy of the values predicted under compression (Figure 11).

The equation of state and dynamics of melting are being investigated experimentally with shock and quasi-isentropic loading experiments at the Z pulsed power facility at Sandia National Laboratories. At Z, electrical energy stored in capacitor banks is discharged through a coaxial conductor. Magnetic pressure and the Lorentz force act on the conductors, forcing the outer conductor, made from aluminium, away from the inner conductor. The current density is high, so the surfaces of the conductors ablate, which also contributes to the outward force. The force on the outer conductor can be used to launch a projectile attached to the conductor and induce a shock wave by impact with a stationary target. The force can also drive a compression wave into a sample attached directly to the conductor, the compression wave loading the sample along a path close to an isentrope.

In the isentropic compression experiment, beryllium samples of different thickness were attached to the outer conductor, and the velocity history was measured using laser Doppler velocimetry, at the interface between each sample and a lithium fluoride window. The window provided an approximate impedance match to the sample. By looking at the compression wave after it had propagated through different thicknesses of beryllium, the rate at which it steepened could be measured and hence the longitudinal sound speed could be determined as a function of pressure. Corrections must be made to account for the impedance mismatch with the window, so it is easier to present the results as comparisons between the measured velocity histories and simulations using different EOS. We used beryllium samples 250 and 500 μm thick, and

---

[1]For convenience in indexing faces, directions and plane normals in hexagonal crystals are conventionally represented by four rather than three indices: $x, y, -(x + y), z$.

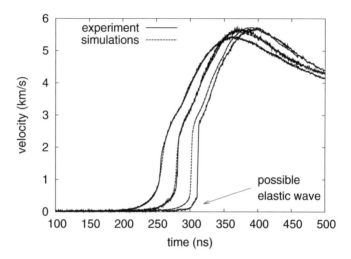

Figure 12. *Velocity histories from isentropic compression experiment, compared with simulations. The three groups of curves are for zero, 250, and 500 μm of beryllium. The curves from the different equations of state were not distinguishable.*

an additional measurement of the velocity history of the aluminium conductor, also through a lithium fluoride window. We found that, though the alternative EOS varied significantly along the principal isentrope, this variation had little effect on the velocity history. All the EOS appear similarly accurate for isentropic compression, so these isentropic compression measurements did not provide an indirect way to suggest which shock temperatures were more accurate. (Figure 12.)

In our shock wave experiments, copper projectiles 200-400 μm thick and 9 mm in diameter impacting a target assembly consisting of a copper baseplate 200 μm thick, a beryllium sample 500 μm thick, and an aluminized lithium fluoride window. The velocity history at the interface between the sample and the window was monitored using laser Doppler velocimetry. This assembly covered half of the area of the flyer; a second, uncoated window was mounted flush with the impact surface of the baseplate to allow the acceleration of the flyer and the time of impact with the target assembly to be monitored, again with laser Doppler velocimetry (Figure 13). The projectile speed was governed by its thickness, the range used giving speeds of 7 to 9 km/s. The EOS of copper and lithium fluoride are known accurately, so their contribution can be taken into account to infer the properties of beryllium.

Typical experimental data consisted of one velocimetry record for the acceleration history of the projectile and its deceleration on impact with the window, and another showing, for the sample/window interface, the rapid acceleration to a constant speed as the shock passed, followed some time later by a deceleration as the release wave from the back of the projectile relieved the shock pressure.

As discussed above, a Hugoniot state can be inferred from two simultaneous measurements. Here we had three – the projectile speed, the shock speed (from its transit time) and the interface speed – so the Hugoniot state could be calculated in different ways. The most accurate states were inferred from the projectile and interface speeds. There was a systematic error in states inferred from the shock transit time, which was most likely caused by the impact faces of the baseplate and the impact window not being adequately co-planar. Otherwise, the EOS were consistent

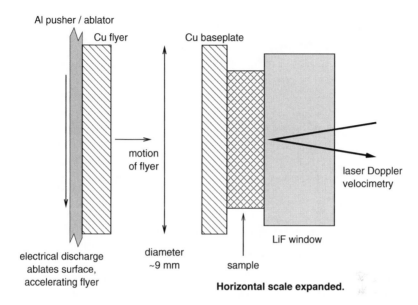

Figure 13. *Schematic of projectile impact experiments at Z.*

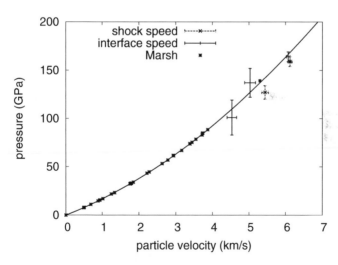

Figure 14. *Shock Hugoniot for beryllium: comparison between equation of state (solid line), previously-published data (Marsh 1980) and Z projectile impact data.*

with the measured states, and were hence verified in the range of 130-200 GPa. (Figure 14.)

We used the detailed velocity histories to look for evidence of melting in these experiments. If the shock pressure is not quite high enough to melt the sample, it may melt on release, and this may show up as a region of constant speed or more gradual deceleration during the release portion of the record. In practice, the experimental records were more complicated: the shock

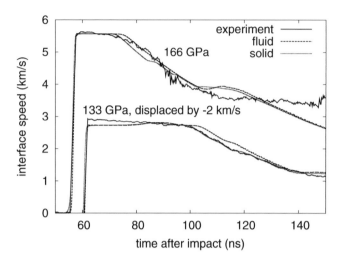

Figure 15. *Velocity history from projectile impact experiments at 133 and 166 GPa, compared with simulations omitting elastic stresses. The histories for 133 GPa were shifted downwards for clarity.*

pressure was not perfectly constant because the projectiles were still accelerating slightly on impact, and a region of constant velocity on release was caused by the presence of some residual aluminium from the conductor still attached to the back of the projectile on impact. Another indication of melting was the duration of the shock, before the onset of deceleration marking the arrival of the release wave. Release waves travel more rapidly in solid than molten material, all else being equal, because elastic stress contributes to the speed. Simulations assuming no elastic stress were a good match to the shock duration in the experiment with the highest shock pressure (166 GPa), whereas the duration was longer than observed at the next lower shock pressure (133 GPa), suggesting that shock melting occurred between these pressures (Figure 15).

One way of considering the seeding of velocity perturbations from the microstructure is from the different speeds of elastic waves along different crystal axes, caused by the anisotropic elasticity of beryllium. At sufficiently low pressures, the speed of elastic waves along any crystal direction is faster then the plastic shock wave: an elastic precursor wave runs ahead of the shock wave. If a wave in one crystal runs faster than than in its neighbour, it will induce lateral velocity components. The higher the driving pressure, the faster the plastic shock. At a sufficiently high pressure, the shock will run faster than the elastic wave in any crystal direction, so there is no elastic precursor – this is known as *overdriving* the elastic wave. The anisotropic elastic response is limited by plastic flow, itself a time-dependent and anisotropic process. The absence of an elastic precursor in an overdriving shock does not guarantee an absence of lateral velocity components, as the anisotropy in flow stress can also cause shear stresses even behind an overdriven shock wave. The elastic wave speeds are high in beryllium compared with most other materials, and relatively high shock pressures are needed to overdrive them. In fact, overdriving for all crystal directions occurs only for shock pressures approaching the pressures predicted for shock melting.

At atmospheric pressure, the hex to bcc phase transition occurs just below the melt temperature. As the pressure increases, the transition temperature decreases. These observations were made at relatively long time scales; the hex-bcc transition has not been reported in shock experiments. This phase transition could be important in fusion capsules, as daughter phases generally nucleate in conditions of low elastic strain: if there were time for the phase change to occur in the ICF capsule it could well reduce the elastic stresses and thus the lateral velocity components.

Melting is often thought of as instantaneous – taking as little as a few atomic vibrations, i.e. picoseconds – and thus that it should always occur once material crosses the equilibrium phase boundary. However, more careful analysis of impact-induced shock experiments indicates that melt does not follow the equilibrium phase boundary, and therefore that the time scale of an experiment or application may make a large difference, though melting may occur rapidly once it starts (Luo 2003). It is possible that solid ablator material could superheat by as much as 30% before melting (Luo 2005). Time-dependence is potentially an important effect in the hex-bcc phase transition as well.

4.2 Plasticity of beryllium

Plastic flow occurs through the motion of defects in the crystal lattice: dislocations and stacking faults. For a given applied stress, the defects in a crystal move on average with some characteristic speed. Thus plastic flow is a time-dependent process, and the stress needed to induce large-scale flow depends on the time scale of interest. Plastic flow models for beryllium have been developed using projectile impact experiments with samples several millimetres thick. For the NIF ignition capsule, with a wall thickness of $150\,\mu$m, the flow stress is likely to be greater. Defect mobility depends on the crystal orientation with respect to the applied stress, and can also be very sensitive to impurities in the crystal and to the crystal size. Thus plasticity will generally vary depending on the composition and the fabrication method.

We are using the TRIDENT laser facility at Los Alamos National Laboratory to perform shock-loading experiments investigating plasticity and phase changes. Measurements of the elastic wave are being used to develop and calibrate a microstructurally-based model of plasticity.

At TRIDENT, an intense laser pulse was used to induce a shock wave in each sample by ablation into vacuum. The laser pulse was 1 to 2.5 ns long. The intensity history was designed to induce a constant shock pressure, and was measured on each shot so that the actual intensity history could be predicted by radiation hydrodynamics simulations. The laser energy was delivered at a wavelength of 527 nm, and phase plates were used to distribute the energy uniformly over a spot of known diameter. Most experiments so far have been performed with a spot 5 mm in diameter, and pressures up to a few tens of gigapascals. The response of the sample has been measured with a range of diagnostics, principally line-imaging laser Doppler velocimetry, which is straightforward to compare with simulations (Swift 2004). We have performed some experiments using a laser-heated plasma X-ray source to perform X-ray diffraction during the passage of the shock wave. X-rays of wavelengths suitable for diffracting from beryllium – several kilo-electron volts – are absorbed over several hundred micrometres of beryllium, so the diffraction signal was integrated over the whole thickness of the sample, complicating the interpretation. We have also performed a series of experiments using the polarisation-dependent reflectivity (ellipsometry) as a possible indicator of phase transitions. (Figure 16.)

In the X-ray diffraction experiments, shock compression and other deformations change the crystal lattice spacing, which changes the diffraction angle for the X-rays and hence the position of diffraction lines on the detector. In the most interesting diffraction experiments so far, the samples were single crystals cut parallel to the (0001) planes. These crystals contained large defect densities compared with single crystals of other materials we have investigated. Diffraction lines from unshocked material would ideally be vanishingly narrow compared with the field-of-view of the detectors; for the beryllium crystals, they were typically $2°$ wide (full width, half maximum). The shift in the diffraction angle was typically a few degrees, so the dynamic record

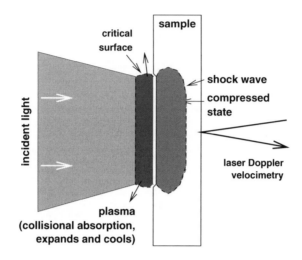

Figure 16. *Schematic of laser-loading experiments at TRIDENT.*

was deconvolved using the measured line shape for unshocked material in order to find the distribution of lattice parameters. The different lattice parameters were interpreted using simulations and velocimetry data as a guide. We found a line corresponding to isotropic compression at the peak shock pressure and a line corresponding to uniaxial compression in the elastic precursor. We also found an additional, unexpected line corresponding to the same compression as found in the elastic precursor, but expressed isotropically rather than uniaxially – i.e., a smaller change in lattice parameter. In some shots, we found additional lines which may correspond to higher compressions at early times or possibly to material in the bcc phase, but these lines cannot be identified unambiguously without additional shots. (Figure 17.)

Surface velocimetry showed a strong elastic precursor which was sensitive to the sample texture: the shape and amplitude were different in rolled foils compared with crystals. Interestingly, in the crystals cut parallel to (0001) planes, the surface velocity decreased sharply following the arrival of the elastic precursor, before the acceleration from the plastic shock. This behaviour indicates a decrease in flow stress, suggesting time-dependence of plastic flow during the period of the experiment. We have performed a series of simulations using different models of plastic flow to explore the type of model necessary to reproduce this behaviour adequately for use in simulations of the response of a polycrystalline microstructure. Obviously, it is desirable to build on previous work as far as possible: experiments on, and calibrations for, longer time scales. Using parameters developed for multiple-millimetre scales and microsecond time scales, the rate-independent Steinberg-Guinan model (Steinberg 1991) predicted an elastic precursor with the correct arrival time but far too low an amplitude, and did not reproduce the deceleration after the initial elastic peak. Adjusting the appropriate low-pressure flow stress parameters in this model, the amplitude could be reproduced, but not the shape. With precursors of the right magnitude, the acceleration history marking the arrival of the plastic wave exhibited large steps: these are caused by elastic waves reverberating between the free surface and the approaching plastic wave. It is interesting to note that the peak free surface velocity in the plastic wave varies with the flow stress, and some of the short-period variations observed experimentally around the peak velocity may be real wave reverberations. The rate-dependent Steinberg-Lund model (Steinberg 1991) predicted a response which was wholly elastic on these time scales. Rather than adjusting the large set of parameters in this model, we investigated a simpler model representing defect

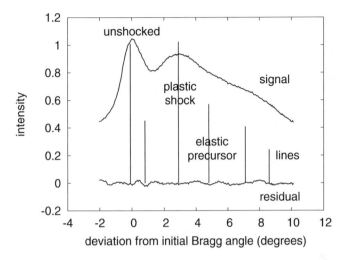

Figure 17. *Example results from X-ray diffraction experiment. The unshocked line is not at precisely zero because of the finite accuracy with which the sample was rotated in the target chamber.*

mobility under an applied stress, but omitting terms describing work-hardening and pressure-hardening. As a defect moves through a crystal, it is subject to a potential which is periodic in displacement, where the period is the Burgers vector $\vec{b}$. The motion of defects was treated as a thermally-activated jumping process, bidirectional for an unstrained crystal, but biased by the strain field in an elastically-distorted crystal. The plastic strain rate is then

$$\dot{\epsilon}_p = \alpha(\rho) \left[e^{-T^*(1-\gamma)/T} - e^{-T^*(1+\gamma)/T} \right] \tag{14}$$

where

$$\gamma(\epsilon_e) \equiv \sin^2 \left(\frac{\pi e_e}{|\vec{b}(\rho)|} \right), \tag{15}$$

is the strain-induced bias, e_e is the elastic strain and T^* is the Peierls barrier energy to motion, divided by Boltzmann's constant. The Burgers vector and Peierls barrier are well-known material properties. For beryllium, we found that the initial elastic amplitude and the subsequent deceleration were reproduced by taking a prefactor α of 2000/ns, which was physically plausible based on the vibration frequency for beryllium atoms at STP and the density of defects in the single crystal as estimated from the width of the diffraction line. The rising part of the plastic wave did not reproduce the experiment accurately, probably because of the omission of work-hardening. Again, the simulation exhibited short-period oscillations around the peak velocity, and the experiment showed evidence of such oscillations. We are now extending this model as a complete crystal plasticity treatment, with the mobility calculated for each slip system and interactions between defects on different systems, which is a primary contribution to work-hardening. (Figure 18.)

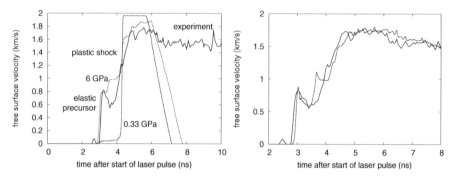

Figure 18. *Velocity history at the free surface of a (0001) beryllium crystal, showing the elastic precursor and the plastic wave. (a) Experiment (solid line) compared with simulations using an elastic-plastic model and different flow stresses. These simulations did not include a model of material damage in tension, leading to spall, thus the late-time velocity was too low. (b) Comparison with the defect mobility model (dashed line). These simulations included a simple spall model.*

5 Instability seeding from the microstructure

To predict the seeding of instabilities from a given ablator microstructure, and hence to specify constraints on the loading and heating conditions and the microstructure, we are developing a capability to predict the spectrum of velocity perturbations using resolved-microstructure simulations and the resulting instabilities. We are adding a radiation hydrodynamics capability to a multidimensional continuum mechanics code, allowing realistic loading conditions to be imposed. A variety of validation experiments have been performed. By developing diagnostics which measure the velocity and displacement history of a surface with spatial resolution, the detailed wave structure has been investigated in bicrystals of a prototype anisotropic material, NiAl, at the TRIDENT laser facility. Experiments on the seeding and growth of instabilities with a "NIF-like" hohlraum drive are being conducted at the OMEGA laser facility, including studies of the X-ray heating and instability growth from sinusoidally-perturbed surfaces.

5.1 Experiments resolving microstructural response

As discussed above, we can develop models for the response of individual crystals in the capsule by a combination of theory (for the EOS and elasticity) and experiment (to confirm the EOS and measure plastic flow). It would be too much of a leap of faith to use these models in simulations of instability seeding without first testing their accuracy in predicting the initial variations in velocity which may or may not ultimately seed instabilities. We have performed shock-loading experiments on a range of materials including beryllium, collecting data on the spatial response for different microstructures to measure the actual behaviour for comparison with simulations.

Velocity histories were collected using an imaging laser Doppler velocimeter of the "velocity interferometry for surfaces of any reflectivity" (VISAR) type (Barker 1972). A laser beam was directed at the surface of the sample, and the reflected light passed through a misaligned Mach-Zehnder interferometer to generate a pattern of interference fringes which shift as the velocity changes. The fringe pattern was recorded on a streak camera, providing the velocity history as a function of position along a line. Differential displacement histories were obtained by using the sample as the mirror in one arm of a misaligned Michelson interferometer (Greenfield 2004). A

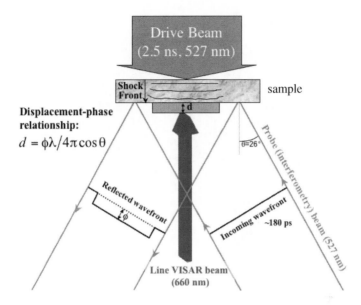

Figure 19. *Schematic of simultaneous velocity and displacement measurements with spatial resolution.*

series of laser pulses was passed through the interferometer, and the resulting fringe pattern was recorded using a framing camera. The fringes in the interferometer distort to reflect variations in the surface relief. (Figure 19.)

As a simplified microstructure, we studied the response of bicrystals: pairs of differently-oriented single crystals in intimate contact, with the grain boundary running at different angles with respect to the surface of the sample (Figure 20). The intermetallic compound NiAl was chosen as an anisotropic prototype material for various reasons including its low toxicity compared with beryllium. Using ablative laser loading, we were able to record clear spatial variations from shock propagation through the different crystal orientations, and a region affected in a non-linear way by the grain boundary (Figure 21).

We have also performed experiments on beryllium with different microstructures: single crystals compared with rolled foils and sputtered samples. The rolling process changes the orientation of grains in the starting material so their [0001] axes point preferentially out of the foil: similar to the single crystal though with a wider distribution of orientations and with a greater amount of initial plastic strain. Even so, the shape of the elastic wave is significantly different from the wave in a single crystal (Figure 22). We have seen spatial variations in the response from a foil – no variations were seen in single crystals – where in some regions the velocity history appeared more like that seen in a (0001) crystal, exhibiting a deceleration before the arrival of the plastic wave (Figure 23).

5.2 Simulations of microstructural effects

We have performed two types of simulation so far, and are working to bring them together. In resolved-microstructure simulations, an example microstructure is generated and represented explicitly in a continuum mechanics simulation, including the anisotropic properties of the material.

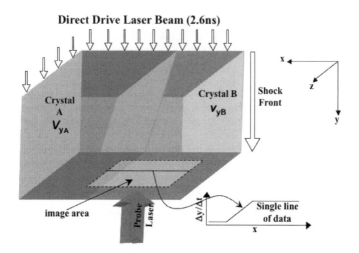

Figure 20. *Schematic of bicrystal, used as a simple prototype microstructure.*

A loading history is applied in the simulation, and the response of the microstructure is simulated including all spatial variations, to predict the spatial variation in velocity and displacement. In perturbed-velocity simulations, an initial variation in velocity is imposed on the material but its properties are treated as isotropic. It is harder to carry resolved-microstructure simulations far in time, so we have used them so far to predict the spectrum of velocity perturbations from a given microstructure and loading history (Figure 24). We have used perturbed-velocity simulations to calculate the seeding and growth of instabilities from the initial velocity perturbation. Conveniently, although there are differences in the early evolution of an instability from a velocity perturbation compared with a surface perturbation of the same wavelength, at later times the instabilities converge (Hoffman 2003). This makes it possible to define an "equivalent surface perturbation" for a given velocity perturbation, and hence to take advantage of the extensive studies performed of combinations of surface perturbations to estimate the effect of velocity perturbations without having to perform the full corresponding set of simulations. (Figure 25.)

So far, our resolved-microstructure simulations have used an applied pressure or applied velocity boundary condition to simulate the effect of the radiation drive. We would like to carry these simulations further in time, to see at least the early growth of instabilities and check the validity of applying a spectrum of velocity perturbations to simulations where the material is modelled as isotropic. It is thus highly desirable to incorporate some form of radiation hydrodynamics in the continuum mechanics simulations. Therefore, we are developing radiation hydrodynamics schemes to link intimately with continuum mechanics. This is a difficult problem in several respects. The best-established radiation transport methods operate on a logically-orthogonal mesh, whereas the continuum-mechanical mesh should follow the grain boundaries and is thus naturally described by meshes with arbitrary connectivity and $D+1$-hedral cells (where D is the number of space dimensions, i.e., tetrahedral meshes in 3D). This problem in algorithm and code development is being addressed in two parallel ways: we are implementing a dual-mesh method in which a logically-orthogonal mesh is used for radiation transport (Moses 2004), mapping to and from the arbitrarily-connected continuum mechanics mesh; and by developing non-orthogonal local-support operators for radiation transport, which would allow the continuum-mechanics mesh to be used directly (Morel 1998). The dual-mesh technique is a powerful starting-point, as it will be trivial to use a different mesh resolution for radiation transport compared with the continuum

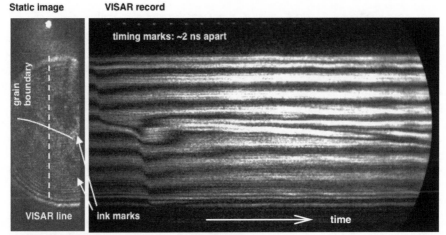

Static image   **VISAR record**

Laser drive: 527 nm, 11 J over 2.4 ns, 5 mm diameter.
VISAR: ~50 ns by ~0.6 mm; 800 m/s/fringe.

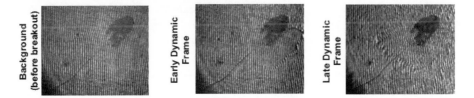

Figure 21. *Example results from NiAl bicrystal experiments. Top: line-imaged velocity history, showing difference in shock arrival in each crystal (acceleration appears as fringe motion in the spatial direction). Bottom: frames from area-imaged displacement history (a flat surface would have straight fringes). Waves can be seen propagating from the grain boundary, and the included grain is tipping out of the plane of the surrounding grain.*

mechanics part of the calculation. The radiation meshing could be finer normal to the sample near its surface, providing a more accurate calculation of ablation, and it could be much coarser – in the extreme limit, only a single cell wide – parallel to the surface, avoiding the need to resolve low spatial modes in the radiation field with the full continuum mesh, which is unnecessarily fine for the radiation field.

5.3   Experiments measuring instability growth

The final test of our understanding of instability seeding is to observe the growth of perturbations in a NIF-like environment. We are using the OMEGA laser facility at the University of Rochester, NY, U.S.A., to induce a loading history representative of NIF ignition experiments and to measure the response of capsule material. Without the full capabilities of the NIF – in particular its energy – it is not practical to attempt a completely-representative drive. OMEGA has 60 beams delivering 30 kJ in total, so we used a "halfraum" geometry with a thin, planar sample of ablator material opposite the laser entrance hole (Figure 26). The most NIF-like halfraum temperature history was achieved by firing beams in two groups at different times, with two pulse shapes:

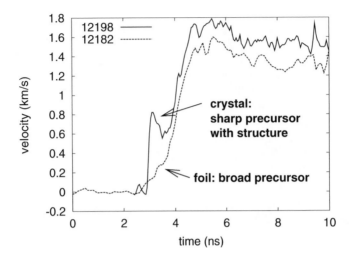

Figure 22. *Comparison between free surface velocity history observed in a beryllium* (0001) *single crystal compared with a rolled foil, from TRIDENT experiments. Both samples were 30 μm thick and were shocked to around 10 GPa by laser ablation.*

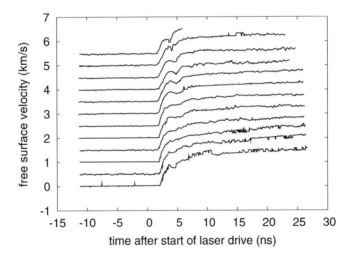

Figure 23. *Free surface velocity history at different locations along a line 500 μm across, for a rolled foil of beryllium 13 μm thick shocked to 7 GPa by laser ablation. Traces have been offset vertically for clarity.*

a bow-tie shape in 3 early beams to simulate the NIF foot and the early part of the main drive, followed by up to 12 beams with a triangular pulse shape to simulate the main NIF drive (Figure 27). It was important to simulate the temperature as accurately as possible in the foot, as this part of the drive interacts most with the microstructure, imprinting the perturbations which may then grow as instabilities. The main drive was chosen to make the instabilities grow

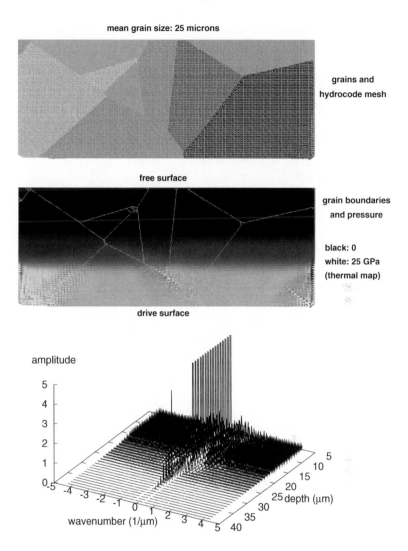

mean grain size: 25 microns

grains and
hydrocode mesh

free surface

grain boundaries
and pressure

black: 0
white: 25 GPa
(thermal map)

drive surface

Figure 24. *Example frame from microstructural simulations, and through-sample velocity spectrum. The sample was 100 μm wide by 50 μm thick, driven from below with a constant pressure of 20 GPa. Grains of different orientation are shaded differently. The pressure distribution shows spatial variations caused by the propagation of the elastic wave with different speeds and amplitudes through different grains.*

as much as possible given the beam energy and orientations available. Compared with the NIF drive, the foot was shorter and the main drive cooler in the OMEGA experiments. Laser power histories were designed for two different temperature histories, intended to give a foot pressure of 100 and 200 GPa respectively: below and above some estimates of the pressure for shock-melting in beryllium (Figure 28). The corresponding halfraum temperatures were about 65 and 85 eV in

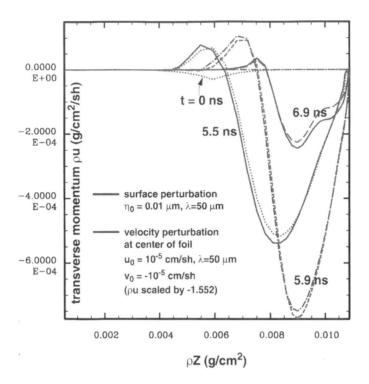

Figure 25. *Growth of velocity and surface perturbations. The different types of perturbation grow slightly differently at early times, but converge at later times. This example shows that for a wavelength of 50 μm, a velocity perturbation of 10 m/s is equivalent to a surface perturbation of 6.44 nm. (1 sh = 10 ns: a convenient unit for these simulations.)*

the foot; the peak temperature was predicted to be around 180 eV (Cobble 2006).

The later part of the halfraum temperature history was inferred by the spectrum of photons emitted through the laser entrance hole – the DANTE instrument. DANTE measurements were estimated to be inaccurate by tens of percent for temperatures below 100 eV, so the foot drive was tested by laser Doppler velocimetry (line-imaging VISAR) of the free surface of the sample. Thermal emission from the sample was measured with a single channel: the "streaked optical pyrometer" (SOP) instrument. The amount of energy lost by stimulated emission from the halfraum was measured from the resulting backscatter – a correction at the 10% level for the hotter portion of the history, and much less in the foot. The principal diagnostic for instability growth was axial radiography; side-on radiographs were also made to observe the ejection of the sample from the end of the halfraum. Up to 10 additional beams were used to drive the one or two backlighter foils.

Early experiments used a vacuum halfraum. VISAR measurements indicated that the foot drive was slightly stronger than predicted in advance, but in reasonable agreement. Temperature histories measured using DANTE agreed well with the halfraum simulations. The vacuum halfraum platform was thus tested quite carefully. We found subsequently that gold from the halfraum walls obscured the axial radiographs at later times, at least with the aluminium backlighters desirable for contrast with 0.9% copper in the beryllium. Subsequent experiments have

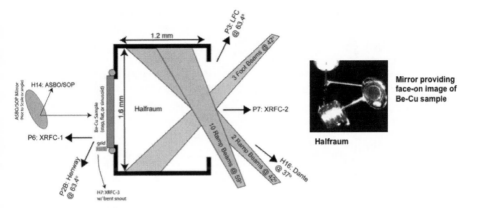

Figure 26. *Schematic of halfraum target for instability growth experiments, and photograph of target showing sample reflected in mirror. The mirror allows simultaneous on-axis radiography and VISAR measurements.*

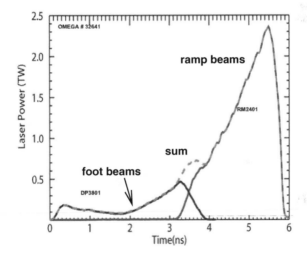

Figure 27. *Laser pulse shapes for different groups of drive beams in instability growth experiments.*

explored the use of different gases in the halfraum to hold back the gold.

Experiments have been performed using copper-doped beryllium samples with machined-in perturbations to observe the growth of instabilities from a well-controlled seed. The perturbations were sinusoidal trenches with a wavelength of $100\,\mu$m and an amplitude of 5 or $0.5\,\mu$m, compared with a sample thickness of around $30\,\mu$m. The perturbations were found to grow enough to be visible radiographically during the experiment (Figure 29), though various details – such as the degree of growth and the late-time speed of the sample – remain to be reconciled with the simulations. Samples so far have mainly been for development purposes and cut from powder-pressed material; we will use NIF-quality ECAE and sputtered material in future experiments. On the radiographs, instabilities originating from the microstructure may be visible as perturbations

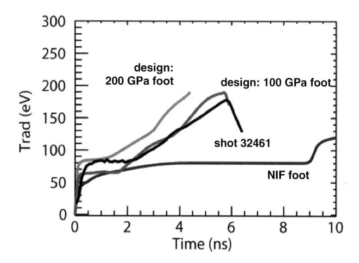

Figure 28. *Halfraum temperature histories for instability growth experiments: histories predicted to produce a 100 and 200 GPa foot, the history inferred from halfraum emission in an example experiment, and comparison with the NIF foot.*

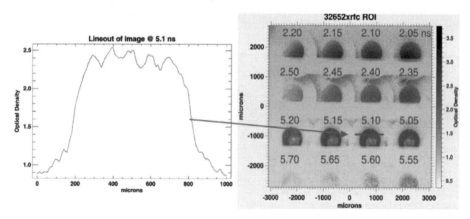

Figure 29. *On-axis X-ray framing camera images from an instability growth experiment with machined-in perturbations. The example profile shows the growing perturbations, with evidence of bubble-and-spike asymmetry between minima and maxima.*

superimposed on the sinusoids. We expect to see a difference in the growth of microstructural perturbations with the low-pressure foot compared with a foot strong enough to cause prompt melting.

One interesting and unexpected aspect of the experiments was that the VISAR signal dropped earlier than expected, and the SOP indicated quite high temperatures correspondingly early in time. This behaviour was found to correlate with the last part of the drive in the early beams, when the power was increased to start the main part of the drive. These observations have not been explained completely, in part because of uncertainties in the absolute temperature calibration of the SOP. However, by looking at the time at which the reflectivity dropped and

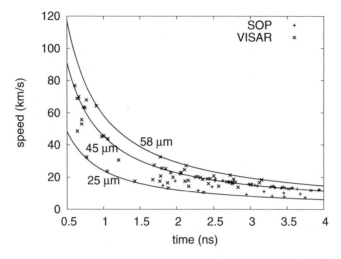

Figure 30. *Apparent X-ray heating of the beryllium-copper sample in instability seeding experiments. The time at which the surface reflectivity or emission reached a given level was measured as a function of sample thickness, and plotted as an apparent speed. Reflectivity and emission measurements fell on consistent curves. The reciprocal-time behaviour observed is characteristic of heating by attenuating radiation, and the magnitude is consistent with kilovolt-level X-rays. The curves agreed well with the measurements for each sample thickness.*

the emission rose to a given level, as a function of sample thickness, we inferred that the likely cause was X-ray heating of the sample by M-band emission (kilovolt energies) from the gold (Tierney 2006, Figure 30). M-band emission is highly non-linear in laser irradiance, requiring a threshold level to activate, and this depends on the speckle structure of the laser beam. The OMEGA experiments seemed to have a significant amount of X-ray heating, which must be taken into account to determine when and where melt occurs. This is an additional complication when applying results from these experiments to the NIF.

6 Conclusions

With current technology, ICF implosions are highly prone to instabilities, and the microstructure of the ablator and fuel capsule is one possible source, particularly for a strongly anisotropic material like beryllium. The coupling of the microstructure can however be understood using theoretical and experimental tools, and it is likely that microstructure-induced instabilities can be controlled by achievable constraints in manufacture and on the laser pulse driving the implosion.

Quantum mechanics can be used to predict thermodynamically-complete equations of state, phase diagrams, and compression-dependent elasticity. The predicted properties are not generally perfect, but they should certainly be adequate to understand the average and microstructurally-resolved response of beryllium. Flyer impact and quasi-isentropic compression experiments have been performed for beryllium at the Z pulsed-power facility. The experimental data complement and validate the various equations of state available for beryllium in terms of the pressure-volume behaviour, but do not discriminate between the large variations in temperature that these equations of state exhibit. Evidence of high-pressure melting was observed in the flyer impact experiments, suggesting that the shock pressure necessary to melt beryllium initially at room temperature lies between 130 and 170 GPa. Plastic flow properties for beryllium on relevant (nanosecond) timescales were deduced from ablatively-driven shock experiments at the TRIDENT laser facility. Plasticity in this regime was found to be quite complicated, but consistent with microstructural physics, i.e. the motion of defects in the crystal lattice.

Continuum mechanics simulations were performed in which representative microstructures were represented explicitly, to predict the velocity perturbations induced. These simulations used simplified models of the hohlraum loading conditions; radiation hydrodynamics will be added to improve the fidelity of the simulations. Spectra of velocity perturbations were used in radiation hydrodynamics simulations – with isotropic models for beryllium – to predict the growth of instabilities. It was possible to associate an equivalent surface roughness with a given velocity perturbation, allowing predictions of microstructurally-seeded instabilities to take advantage of the large body of work already performed on roughness-induced instabilities.

Spatially-resolving velocity and displacement diagnostics were developed to look at spatial variations in the response to loading which may seed instabilities. Bicrystals of a model anisotropic material were studied as a prototype grain boundary; the intermetallic compound nickel aluminide was used. Spatial variations in shock arrival and amplitude, and in the displacement history following shock breakout, were clearly observed. Non-simple interactions were also observed close to the grain boundary, where the elastic waves interact with the boundary.

Hohlraum-driven experiments were used to study the growth of instabilities in a regime relevant to NIF. The hohlraum drive was designed to produce a pressure foot, followed by a strong ramp. Radiation hydrodynamics simulations of the hohlraum drive were validated against measurements of the velocity history at the free surface of the beryllium sample for the foot, and against measurements of the X-ray spectrum emitted by the hohlraum for the main part of the drive. In experiments with an initial machined-in surface perturbation, the perturbation grew to be visible on dynamic radiographs. Both thermal emission from the sample and loss of reflectivity for laser Doppler velocimetry pointed to significant X-ray preheating as the main drive started up. The X-ray environment is important in order to relate the drive to NIF experiments.

Acknowledgements

The Los Alamos ICF project involves significant efforts by a large number of people, who made essential contributions to these lectures. John Lindl (Lawrence Livermore National Laboratory) provided the overview of NIF ignition target design and non-microstructural tolerances. Arthur Nobile Jr (MST-7) provided information, diagrams and micrographs of material preparation. Scott Evans (P-24) prepared some of the beryllium samples. Marcus Knudson (Sandia National Laboratories) and the Z support team were instrumental in performing the flyer and isentropic compression experiments. James Cobble, Sheng-Nian Luo and Dennis Paisley (P-24), and Thomas Tierney IV (P-22), performed experiments at OMEGA and TRIDENT. Roger Kopp, Robert Goldman, David Tubbs, Barbara Devolder, Ian Tregillis and Nelson Hoffman (X-1) performed radiation hydrodynamics simulations and other design and analysis work in support of OMEGA and TRIDENT experiments. Allan Hauer and George Kyrala (P-24), and Justin Wark (University of Oxford) developed the original X-ray diffraction capability at TRIDENT. Scott Greenfield and Aaron Koskelo (C-ADI) developed and fielded the displacement-measuring imaging Michelson interferometer. Randall Johnson, Samuel Letzring, Robert Gibson, Ray Gonzales, Fred Archuleta, Nathan Okamoto, Sha-Marie Reid, Danielle Pacheco, Tom Hurry, Tom Ortiz and Tom Shimada (P-24) ran the TRIDENT laser and provided target area support. Tom Sedillo (P-24) provided support for X-ray streak camera measurements. Jason Cooley (MST-6) prepared ECAE material and characterised samples. Robert Margevicius (MST-6) prepared pressed material. Pedro Peralta and Eric Loomis (Arizona State University), and Kenneth McClellan and Darrin Byler (MST-8) prepared and characterised the NiAl samples. Additional target fabrication was provided by Robert Day and Ronald Perea (MST-7). David Bradley and David Braun (Lawrence Livermore National Laboratory) provided valuable insights from their related studies of halfraum-driven samples.

The project was funded U.S. Department of Energy under NNSA Campaign 10 (Inertial Confinement Fusion), with a contribution from the Laboratory-Directed Research and Development program at Los Alamos National Laboratory for the NiAl project. Allan Hauer, Cris Barnes, Steven Batha and Nelson Hoffman supported the beryllium characterisation projects. David Watkins supported the NiAl project; the principal investigator was Aaron Koskelo.

References

Ashcroft, N. W. and Mermin, N. D. (1976) *Solid State Physics* (Holt-Saunders).

Barker, L. M. and Hollenbach, R. E. (1972), *J. Appl. Phys.* **43**, 4669.

Bushman, A. V. et al (1993) *Intense Dynamic Loading of Condensed Matter* (Taylor and Francis).

Cobble, J. A., et al (2006), *Phys. of Plasmas*, to appear.

Cooley, J. [Los Alamos National Laboratory] (2005) *unpublished*.

Greenfield, S. R. et al (2004) in *Proceedings of the American Physical Society Topical Conference on Shock Compression of Condensed Matter, Portland, Oregon, U.S.A., 20-25 June 2003*, American Institute of Physics.

Hoffman, N. M. and Swift, D.C. (2004) in *Proceedings of the American Physical Society Topical Conference on Shock Compression of Condensed Matter, Portland, Oregon, U.S.A., 20-25 June 2003*, American Institute of Physics.

Hohenberg, P. and Kohn, W. (1964), *Phys. Rev.* **B136**, 3.

Holian, K. S. (1984) *T-4 Handbook of Material Property Data Bases, Vol 1c: Equations of State*, Los Alamos National Laboratory report LA-10160-MS.

Kohn, W. and Sham, L. J. (1965), *Phys. Rev.* **A140**, 4.

Lindl, J. D. (1998) *Inertial Confinement Fusion* (Springer-Verlag).

Luo, S.-N. et al (2003), *Phys. Rev.* **B68**, 134206.

Luo, S.-N. and Swift, D. C. (2005) *Equilibrium and dynamic melting: implications for melting of beryllium capsule in inertial confinement fusion*, Los Alamos National Laboratory report LA-UR-05-2988.

Marsh, S. P. (1980) *LASL Shock Hugoniot Data* (University of California).

Morel, J. E. et al (1988), *J. Comp. Phys.* **144**, 1751.

Moses, G. [University of Wisconsin – Madison] (2004), *unpublished*.

Nobile, A. [Los Alamos National Laboratory] (2005) *unpublished*.

Perdew, J. P. et al (1992), *Phys. Rev.* **B46**, 6671.

Steinberg, D. J. (1991), *Equation of State and Strength Properties of Selected Materials*, Lawrence Livermore National Laboratory report UCRL-MA-106439.

Swift, D. C. et al (2001), *Phys. Rev.* **B64**, 214107.

Swift, D. C. et al (2004), *Phys. Rev.* **E64**, 036406.

Swift, D. C. et al (2004), *Phys. Rev.* **E69**, 056401.

Swift, D. C. et al (2005), *Phys. Rev.* **B**, submitted.

Tierney, T. E. et al (2006), *Phys. Rev. Lett.*, to be submitted.

Troullier, N. and Martins, J. L. (1991), *Phys. Rev.* **B43**, 1993.

White, J. A. and Bird, D. M. (1994), *Phys. Rev.* **B50**, R4954.

Young, D. A. (1991) *Phase Diagrams of the Elements* (University of California).

18
Direct-Drive Inertial Fusion: Basic Concepts and Ignition Target Designing

Valeri N. Goncharov

Laboratory for Laser Energetics, Department of Mechanical Engineering, University of Rochester, Rochester, New York, USA

1 Introduction

Inertial confinement fusion (ICF) is an approach to fusion that relies on the inertia of the fuel mass to provide confinement. The confined fuel must reach high temperature and density to produce enough reactions $D + T \rightarrow \alpha(3.5 \text{ MeV}) + n(14.1 \text{ MeV})$, so the total energy released in the reactions is much larger than the driver energy required to compress the fuel. In an ICF implosion, the capsule, a spherical cryogenic deuterium-tritium (DT) shell filled with DT vapor, is irradiated either directly by laser beams (direct-drive approach) or by x-rays emitted by a high-Z enclosure (hohlraum) surrounding the target (indirect drive) (Lindl, 1998). Only a small portion of the fuel is heated to ignition conditions in a typical ignition target. This part of the fuel forms a hot spot that initiates a burn wave which ignites the remaining fuel.

In the direct-drive approach, the following stages of an implosion can be identifies: At the beginning of the laser pulse, the outer portion of the pellet heats up and expands outward, creating a plasma atmosphere around the pellet. Then, a critical electron density $n_{cr} = \pi m c^2 / e^2 \lambda_L^2$ is established outside the cold portion of the shell. Here, m is the electron mass, c is the speed of light, e is the electron charge, and λ_L is the laser wavelength.

The laser energy is absorbed in a narrow region near the critical surface via the inverse bremsstrahlung, and the absorbed energy is transported, mainly by electrons, toward the colder portion of the shell. The cold material, heated by the thermal conduction, expands outward. Such an expansion creates an ablation pressure that, similar to the rocket effect, compresses the pellet.

At the beginning of implosion, the ablation pressure launches a shock wave that propagates ahead of the thermal ablation front and increases the fuel entropy. Then, as the first shock breaks out at the rear surface of the shell, the transmitted shock is formed. It converges through the vapor to the capsule center. After reflection from the center, the shock moves outward and interacts with the incoming shell. At this point the velocity of the inner portion of the shell starts to decrease, reversing its sign at stagnation.

This is a crucial point of the implosion since no more "pdV" compression work can be done to the hot spot after the stagnation, and the only remaining source of heating is the energy deposition of α-particles produced by fusion reactions inside the hot spot (α heating). At the deceleration phase of the implosion, the kinetic energy of the shell is transferred into the internal energy of the hot spot. To ignite the fuel, the energy gain due to the pdV work of the imploding shell and α heating must be larger than the energy losses due to thermal conduction and radiation. This requirements sets a minimum value for the implosion velocity v_{imp} of the shell.

This paper will review the basic concept of ICF ignition (Sec. 2). Section 3 presents the simplest direct-drive ignition target design. The stability issues will be discussed in Sec. 4.

2 Basic concepts

To burn a substantial fraction of the fuel mass, the fuel density at stagnation must be very large. This can be easily shown if we assume that the main fuel at the peak compression is assembled as a uniform-density sphere with radius R_f and density ρ_m (Lindl, 1998; Rosen, 1999). The reaction rate is given by $dn/dt = n_D n_T \langle \sigma v \rangle$, where $\langle \sigma v \rangle$ is the average reactivity, and n_D and n_T are the density of deuterium and tritium, respectively. Assuming a 50/50 DT mixture, $n_D = n_T = n_0/2 - n$, where n_0 is the initial density. To calculate the total number of reactions, we integrate $N = \int_0^{t_d} V(t)(dn/dt)dt$ over the burn duration time t_d, where $V(t)$ is the volume of burning fuel.

The burn time is determined by the fuel disassembly rate. Since there is no external force to keep the fuel together after the stagnation, the outer region of the fuel expands, launching a rarefaction wave toward the center. The rarefaction wave propagates at the local sound speed c_s, and it takes approximately $t_d = R_f/c_s$ for the whole sphere to decompress and cool down, quenching the fusion reactions. During the decompression, only the high-density portion of the fuel inside the radius $R(t) = R_f - c_s t$ is burning. Since the total number of reactions is proportional to the time integral of the burning-fuel volume, we can define an effective confinement time t_c as $\int_0^{t_d} V(t)dt = V_0 t_c$, where $V_0 = 4\pi R_f^3/3$. The integration gives $t_c = t_d/4$. If the number of fused atoms is small, $n \ll n_0$, then the total number of reactions becomes $N = n_0^2 \langle \sigma v \rangle V_0 t_c/4$.

The ratio $f = N/N_0$ is commonly referred to as a burn fraction, where $N_0 = n_0 V_0/2$ is the initial number of DT pairs. Substituting N into f gives $f = n_0 R_f \langle \sigma v \rangle/8c_s = \rho_m R_f/(8c_s m_{DT}/\langle \sigma v \rangle)$, where m_{DT} is the DT ion mass. The combination $8c_s m_{DT}/\langle \sigma v \rangle$ has a minimum value of 6 g/cm^2 at the ignition conditions; thus, to have an efficient burn, the fuel must reach $\rho_m R_f > 1$g/cm^2 at peak compression (more accurate calculations show that $f = 0.3$ at $\rho_m R_f = 3$g/cm^2). Using such an estimate, we can determine the maximum density of the assembled fuel and convergence ratio of the shell at ignition. Assuming that a fraction f_H of the laser energy E_L goes into the shell kinetic energy $M v_{\mathrm{imp}}^2/2$, the shell mass M can be expressed as $M = 2f_H E_L/v_{\mathrm{imp}}^2$. The fraction f_H is a product of the hydrodynamic efficiency (defined as a ratio of the shell kinetic energy to the absorbed laser energy, typically $\sim 10\%$ for a direct-drive implosion) and the laser absorption fraction ~ 0.6. This yields $f_H \sim 0.06$.

The fuel mass at stagnation can be rewritten as $M = 4\pi(\rho_m R_f)^3/3\rho^2$. Equating the two expression for the mass, we obtain the fuel density at the peak compression

$$\rho_m = v_{\mathrm{imp}} \sqrt{\frac{2\pi(\rho_m R_f)^3}{3f_H E_L}}. \tag{1}$$

The value of $\rho_m R_f \simeq 2$g/cm^2 is fixed by the fuel burn-up fraction. The implosion velocity cannot be much less than 3×10^7cm/s in order to have a temperature increase inside the hot spot during the shell deceleration (energy gain exceeds the energy losses at such a velocity). Using these values gives $\rho_m(\text{g/cm}^3) \simeq 160/\sqrt{E_L(\text{MJ})}$, where the laser energy is measured in Mega Joules. For an $E_L = 1$MJ facility, $\rho_m = 160\text{g/cm}^3 \simeq 630\rho_{DT}$, where $\rho_{DT} = 0.25\text{g/cm}^3$ is the density of cryogenic uncompress DT mixture at $T = 18$K.

To find the shell convergence ratio $C_r = R_0/R_f$ required to achieve such high density, we write the mass conservation equation $4\pi\rho_{DT}R_0^3(1 - f_A)/A_0 = 4\pi\rho_m R_f^3/3$, where $A_0 = R_0/\Delta_0$ is the initial aspect ratio, R_0 and Δ_0 are the inner radius and thickness of the undriven shell, and f_A is the fraction of the shell mass ablated during the implosion. For a typical direct-drive ignition design $f_A \simeq 0.8$. This leads to

$$C_r \simeq \left(\frac{5}{3} A_0 \frac{\rho_m}{\rho_{DT}}\right)^{1/3}. \tag{2}$$

Taking $A_0 = 4$ and $\rho_m/\rho_{DT} = 630$ gives $C_r = 16$.

So far, we considered only conditions for the high-ρR fuel assembly. To initiate the burn wave inside the main fuel, as mentioned earlier, the hot spot must reach ignition conditions first. Since the reaction rate is proportional to p^2 ($dn/dt \sim n^2\langle\sigma v\rangle$ and $\langle\sigma v\rangle \sim T^2$ for $T > 6$keV give $dn/dt \sim p^2$), high pressure must be achieved inside the hot spot.

We can estimate the pressure evolution inside the hot spot during the deceleration phase by considering an adiabatic compression of a gas by a spherical piston. The adiabatic condition relates the pressure and density as $p \sim \rho^{5/3}$. Then, the mass conservation yields $\rho R^3 = \rho_d R_d^3$ or $\rho \sim R^{-3}$, where ρ_d and R_d are the mass density and radius at the beginning of deceleration. This gives

$$p = p_d \left(\frac{R_d}{R}\right)^5. \tag{3}$$

Strictly speaking, the hot-spot compression cannot be considered as adiabatic during the deceleration phase because of thermal conduction effects. A detailed calculations including thermal conduction losses (Betti et al., 2002), however, show that an R^{-5} law is valid in a more general case (not including α deposition). This can be easily explained if we consider pressure as an internal-energy density.

The thermal conduction deposits part of the hot-spot energy into heating the inner layer of the surrounding cold shell. The heated layer, then, ablates and the ablated mass returns the lost energy back into the hot spot. With the help of Eq. (3), we can calculate the shell trajectory during the deceleration using Newton's law

$$M\frac{d^2R}{dt^2} = 4\pi p R^2. \tag{4}$$

Integrating the latter equation gives

$$v^2 = v_{\text{imp}}^2 + \frac{4\pi p_d R_d^3}{M}\left(1 - C_{rd}^2\right), \quad \text{where } C_{rd} = \frac{R_d}{R}. \tag{5}$$

At stagnation, $v = 0$ and Eq. (5) yields the total convergence ratio

$$C_{rd} \sim \sqrt{1 + \frac{E_k}{E_i}}, \tag{6}$$

where

$$E_k = \frac{Mv_{\text{imp}}^2}{2}, \quad E_i = \frac{4\pi}{3}\frac{p_d}{\gamma - 1}R_d^3 = 2\pi p_d R_d^3$$

is the shell kinetic energy and the total internal energy of the gas at the beginning of the deceleration phase, respectively , γ is the ratio of specific heats [we used $\gamma = 5/3$ in Eq. (6)].

The relation (6) is satisfied only approximately, since it does not take into account the pressure increase due to α deposition. Such an effect is important, however, only near the stagnation. Using the limit $E_k \gg E_i$, the maximum pressure takes the form

$$p_m = \frac{1}{p_d^{3/2}} \left(\frac{E_k}{3V_d} \right)^{5/2} , \qquad (7)$$

where $V_d = 4\pi R_d^3/3$.

Equation (7) shows that, to achieve the hot-spot ignition (high pressure), we must minimize, for a given kinetic energy, the gas pressure and radius R_d at the beginning of shell deceleration (when the reflected shock starts interacting with the shell). The next section will explain how such a minimization is achieved in a direct-drive ignition target.

3   Direct-drive ignition target design

High density and convergence-ratio requirements put a limitation on the maximum entropy increase in the shell during an implosion. Let us assume that the pressure increases from its initial value p_0 inside the shell to the ablation pressure p_a at the maximum laser intensity. The density ρ is maximum when the shell entropy increase is zero ($\Delta s = 0$) during the compression (adiabatic implosion), $\rho/\rho_0 = (p_a/p_0)^{1/\gamma} e^{-\Delta s/c_p}$, where c_p is the specific heat at constant pressure, and ρ_0 is the initial density. It is not feasible, however, to drive a perfectly adiabatic implosion. Shock waves, radiation preheat, hot electron preheat, etc. increase the entropy during the shell compression.

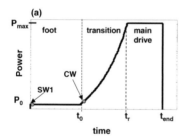 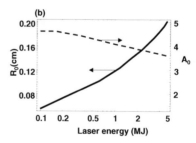

Figure 1. *(a)Laser drive pulse for direct-drive target. (b) Dimensions of direct-drive ignition targets versus laser energy ($v_{\mathrm{imp}} = 4.5 \times 10^7\,cm/s$).*

We can minimize hydrodynamic part in the entropy change, nevertheless, by accurately timing all hydrodynamic waves launched into the target during the laser drive. The shell entropy is commonly characterized by the adiabat $\alpha = p/p_F$ defined as the ratio of the shell pressure p to the Fermi-degenerate pressure (mainly due to electrons) p_F calculated at the shell density. The Fermi pressure of an electron gas has the form $p_F = \mu\rho^{5/3}$, where $\mu = (3\pi^2)^{2/3}\hbar^2 Z^{5/3}/5m_i^{5/3}m$, m and m_i are electron and ion masses, respectively, $\hbar$ is the Planck constant, and Z is the ion charge. For fully ionized DT, $\mu = 2.15[\mathrm{Mbar}/(\mathrm{g/cm^3})^{5/3}]$. Since the shell entropy during an implosion is a crucial parameter, the target design in ICF is characterized by the adiabat value. For example, an "$\alpha = 3$ design" means that the pressure inside the shell during the implosion in such a design is three times larger than the Fermi-degenerate pressure of the fully-ionized DT.

Next, we consider the simplest direct-drive target design which consists of a spherical DT-ice shell filled with DT gas. The main fuel in the shell is kept at cryogenic temperatures ($\sim$ 18K) in order to maximize the fuel mass and to minimize the shell entropy. The initial shell thickness is Δ_0, and the inner shell radius is $R_0 \gg \Delta_0$. The target is driven by the laser pulse that consists of three distinct regions: low-intensity foot, transition region, and the main drive [see Fig. 1(a)].

The main parameters of the laser pulse are the foot length t_0, the end of the rise time t_r, the end of the pulse t_{end}, the foot power P_0, and the maximum power P_{max}. During the foot, a shock wave (SW1), launched at the beginning of the pulse, propagates through the shell. The adiabat of the post-shock material depends on the shock strength, which, in turn, is a function of the laser intensity. Thus, the foot power P_0 is chosen to set the post-shock material on a desired adiabat α.

Analytical models (Manheimer and Colombant, 1982; Lindl, 1998) predict that the drive pressure and the laser intensity I are related as $p(\text{Mbar}) = 40(I_{15}/\lambda_L)^{2/3}$, where I_{15} is measured in 10^{15} W/cm^2, and λ_L is the laser wavelength measured in microns. Substituting $p = \alpha\mu\rho_1^{5/3}$ into the latter equation shows that the laser intensity scales with α in a DT shell driven by $\lambda_L = 0.351\mu$m laser as $I_{12} = 4\alpha^{3/2}$, where the intensity is measured in TW/cm^2, and the post-shock density is $\rho_1 \simeq 4\rho_{DT} = 1$g/cm^3 in the strong-shock limit. The 3/2 power law does not take into account an intensity dependence in the laser absorption. When the latter is included (with the help of numerical simulations), the required laser intensity becomes $I_{12} = 7\alpha^{5/4}$. Using this result, we obtain a relation between the foot power and the shell adiabat $P_0(\text{TW}) = 90R_0^2 \left(1 + 1/A_0\right)^2 \alpha^{5/4}$.

Next, as the laser intensity increases during the transition time, a compression wave (CW) is launched into the shell. To prevent an adiabat increase, such a wave should not turn into a strong shock inside the shell. This determines the intensity slope in the transition region. To maintain an adiabatic compression during the rise time, one can use Kidder's solution (Kidder, 1974) for the drive pressure. This gives the power history during the transition time

$$P = \frac{\hat{P}}{[1 - (t/T_c)^2]^\omega},\tag{8}$$

where T_c and $\hat{P}$ are normalization coefficients determined from matching P with P_0 at $t = t_0$ and $P = P_{\text{max}}$ at $t = t_r$.

Numerical simulations show that moderate variations in ω ($2 < \omega < 7$) do not significantly affect the shell adiabat. Since there are limitations on the maximum power imposed by both instabilities due to the laser-plasma interaction and damage threshold issues, the laser power cannot follow Eq.(8) at all times. Thus, we assume that the laser power becomes flat after time t_r when P reaches the peak power P_{max} [see Fig. 1(a)].

Next, we consider what determines the values of t_0 and t_r. The CW, as any sound wave in hydrodynamics, travels with the local sound speed. It eventually catches up with the SW1. After coalescence, both the shock strength and the adiabat of the post-shock material increase. Minimizing the shell adiabat, we must prevent the CW and SW1 coalescence inside the shell. This sets the minimum value of t_0. On the other hand, if t_0 is too large, the SW1 and CW will be well separated in time. As the SW1 breaks out at the shell's rear surface, the surface starts to expand, launching a rarefaction wave (RW) toward the ablation front.

The RW establishes some velocity, pressure, and density gradients in its tail. Since the CW and RW travel in opposite directions, they meet inside the shell. After such time the CW propagates through the hydrodynamic gradients established by the RW. It is well known in hydrodynamics that a sound wave traveling along a decreasing density turns into a shock sooner than a sound wave traveling through a uniform or increasing density. As the CW turns into a shock, the latter will excessively increase the shell entropy, reducing the shell convergence ratio.

The foot duration t_0 can be related to the shell adiabat α and the initial shell thickness Δ_0. Indeed, using Hugoniot jump conditions across the shock, one can easily obtain the shock velocity, neglecting spherical convergence effects, $U_s(p_1 \gg p_0) = \sqrt{(\gamma + 1)p_1/2\rho_0}$, where p_0 and p_1 are the initial and post-shock pressure, respectively. Then, the SW1 transit time through the shell becomes $t_{sh} = \Delta_0/U_s$. The CW travels through the shock-compressed shell with thickness $\Delta_c = \Delta_0\rho_0/\rho_1$, where ρ_0 and ρ_1 are initial and post-shock density, respectively. The CW propagation time is $t_{cw} = \Delta_c/c_1$, where $c_1 = \sqrt{\gamma p_1/\rho_1}$. Thus, the foot length takes the form $t_0 = t_{sh} - t_{cw}$. Using expressions for the shock velocity and the adiabat yields the following equations for t_0 and t_{sh}:

$$t_0(\text{ns}) = 0.016\frac{\Delta_0(\mu m)}{\sqrt{\alpha}}, \quad t_{sh}(\text{ns}) = 0.03\frac{\Delta_0(\mu m)}{\sqrt{\alpha}}, \tag{9}$$

where $\gamma = 5/3$.

Similar to t_0, the rise time t_r is also determined by avoiding an additional strong shock formation. The transition region cannot be too short, the CW otherwise turns into a shock. The time t_r, as well, can not be too large. This is due to formation of a second shock wave (SW2) inside the shell. It is easy to show that an adjustment wave (AW) is formed as the leading edge of the RW breaks out at the ablation front. Indeed, each fluid element inside the RW is accelerated according to $dv/dt = -\partial_x p/\rho$. At the head of the RW, ρ is equal to the shell density compressed by the SW1 and CW. When the leading edge reaches the ablation front, the density suddenly drops, creating a large gradient in acceleration of fluid elements. This forms a local excess in the pressure that starts to propagate in the form of a compression (adjustment) wave along a decreasing density profile of the RW tail. As mentioned earlier, the AW traveling along a decreasing density turns into a shock inside the shell (Betti et al., 1998). Thus, the SW2 is formed even for a constant-intensity laser pulse.

The formation of the AW can also be shown by comparing the following density profiles: In the first case, the profile is created by a rarefaction wave traveling across a uniform-density foil. For the second profile, we take a solution of the motion equation for an accelerated foil. The solution for a rarefaction wave profile can be found using a self-similar analysis (Landau and Lifshitz, 1982). Calculated at the breakout time of the leading edge, the density becomes

$$\rho = \rho_{a1}\left(\frac{\bar{x} + d_1}{d_1}\right)^3, \tag{10}$$

where $\bar{x}$ is the position in the frame of reference moving with the ablation front ($\bar{x} = 0$ at the ablation front), $\gamma = 5/3$, and d_1 is defind as $\rho(-d_1) = 0$. On the other hand, during the shell acceleration, the density profile can be found by solving Newton's equation

$$\rho g = \frac{\partial p}{\partial x}. \tag{11}$$

Assuming that the entropy is uniform across the entire shell, the pressure is related to density as $p = \mu\alpha\rho^{5/3}$, and the solution of Eq. (11) takes the form

$$p = p_a\left(1 + \frac{\bar{x}}{d_2}\right)^{5/2}, \quad \text{where} \quad d_2 = \frac{5\mu^{3/5}\alpha^{3/5}p_a^{2/5}}{2g}. \tag{12}$$

Then, the adiabatic relation $\rho = (p/\mu\alpha)^{3/5}$ gives

$$\rho = \rho_{a2}\left(\frac{\bar{x} + d_2}{d_2}\right)^{3/2}. \tag{13}$$

The comparison of Eqs. (10) and (13), keeping the same mass for the two profiles, gives a relation between the shell thicknesses

$$d_2 = \frac{5}{8} \left(\frac{p_{a1}}{p_{a2}} \right)^{3/5} d_1. \tag{14}$$

Since $p_{a1} \leq p_{a2}$ (the shell is accelerated during the main pulse where intensity reaches the maximum value), we conclude that the shell during the acceleration should be more compact than produced by a rarefaction wave, $d_2 < d_1$. This is possible only if a compression wave is launched into the shell at the beginning of the acceleration.

Even though the SW2 cannot be avoided, its strength and an effect on target performance can be minimized by appropriately choosing t_r. An intensity rise prior to the leading edge of the RW breakout at the ablation front helps to steepen the density profile, reducing the AW strength. If t_r is too large and the SW2 formation occurs during the drive-pressure rise, the SW2 will be too strong, raising the pressure in the vapor. Thus, t_r must be between the SW1 breakout at the rear surface and the leading edge of RW breakout at the ablation front (in other words, the laser must reach the peak power while the leading edge is inside the shell). This concludes the pulse shape specification for a simple direct-drive ignition design.

Next, we determine the optimum target radius R_0 and shell thickness Δ_0 for a given laser energy. The following effects must be considered: After the SW1 breaks out at the back of the shell, a transmitted shock is formed in the vapor. This shock reaches the center, reflects, and, eventually, interacts with the incoming shell. During the reflected-shock propagation through the shell, the shell decelerates (deceleration phase of an implosion). If R_0 is too large, the laser is turned off well before the beginning of the deceleration phase. The shell, then, coasts inward between the end of the drive pulse and beginning of deceleration. Both the front and back surfaces of the shell expand during the coasting phase, reducing the shell density. Thus, to maximize the density, duration of the coasting phase must be minimized. This sets an upper limit on R_0. In the opposite case, when R_0 is too small, the reflected shock interacts with the shell while the laser pulse is still on. The pressure inside the shell in this case increases above the drive-pressure value, and the shell acceleration, therefore, goes to zero, preventing an effective transfer of the laser absorbed energy into the kinetic energy of the shell. Thus, the end of the pulse in an optimized design should occur after the main shock reflection from the target center, but before the beginning of the deceleration phase.

Taking into account the optimization arguments listed earlier, we plot in Fig. 1(b) the shell radius and initial aspect ratio $A_0 = R_0/\Delta_0$ versus the incident laser energy, keeping the maximum laser intensity at $I_{max} = 10^{15}$ W/cm^2 and the implosion velocity at $v_{imp} = 4.5 \times 10^7$ cm/s. Calculations show that the target dimensions do not have a strong dependence on the shell adiabat. The shell radius is fitted well with a 1/3-power law $R_0 = 0.06[E_L(\mathrm{MJ})/0.1]^{1/3}$. The shell thickness, on the other hand, has a stronger energy dependence, $\Delta_0 = 0.012[E_L(\mathrm{MJ})/0.1]^{1/2.6}$. A deviation from the $E_L^{1/3}$ scaling in Δ_0 (which is expected from a simple $E_L \sim D^3$ argument, where D is a scale length of the problem) is due to a scale-length dependence in the laser absorption (the smaller the target, the steeper the density scale length, and the smaller, therefore, the absorption fraction). This results in an increased initial aspect ratio for the lower laser energies, as shown in Fig. 1(b).

Using the obtained target dimensions and laser pulse shapes, the gain curves and the maximum shell ρR can be calculated with a one-dimensional hydrocode. These are shown in Fig. 2. Figure 2(a) indicates that the designs with the shell adiabat up to $\alpha = 6$ are expected to ignite on the NIF Laser System with the incident laser energy $E_L = 1.5$MJ.

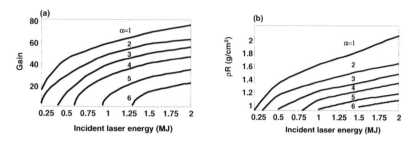

Figure 2. *Gain curves (a) and maximum ρR (b) for ignition direct-drive targets.*

4 Stability

Hydrodynamic instabilities put severe constraints on target designs because they limit the maximum convergence ratio and temperature of the hot spot (Lindl, 1998; Atzeni and Meyer-Ter-Vehn, 2004). The dominant hydrodynamic instability in an ICF implosion is the Rayleigh-Taylor (RT) instability. The RT instability inevitably occurs in systems where the heavier fluid is accelerated by the lighter fluid (Chandrasekhar, 1961). Such conditions arise during the shell compression in ICF implosions where the heavier shell material is accelerated by the lighter blowoff plasma. The RT instability growth amplifies the shell distortions seeded by initial surface roughness and laser nonuniformities (laser "imprint"). If allowed to grow to substantial amplitudes, the shell nonuniformities reduce the shell ρR and the neutron yield.

Fortunately for ICF implosions, the thermal conduction that drives the ablation process creates several stabilizing effects that reduce both the nonuniformity seeding and the RT growth rates (Sanz, 1994; Goncharov et al., 1996). Indeed, seeding due to the laser nonuniformity is determined by how quickly the plasma atmosphere is created around the imploding shell. The laser irradiation is absorbed at some distance from the cold shell. The larger this distance (the conduction zone), the larger the smoothing effect of the thermal conductivity within the conduction zone and the smaller the laser imprint. The RT modes are also stabilized by the thermal conductivity that drives the mass ablation of the shell material. The ablation process is characterized by the ablation velocity V_a, which is defined as the ratio of the mass ablation rate to the shell density $V_a = \dot{m}/\rho_{sh}$. The larger value of the ablation velocity, the larger the ablative stabilization.

Taking thermal smoothing and ablative stabilization into account, one can make a general statement that the higher the initial intensity of the drive laser pulse (larger P_0), the smaller the nonuniformities and the more stable the implosion. Indeed, the higher intensity tends to create the conduction zone in shorter time, reducing the laser imprint. In addition, the initial shock launched by the higher-intensity pulse is stronger, resulting in higher shell adiabat. This reduces the shell density, increasing the ablation velocity. Furthermore, a lower density leads to an increase in the shell thickness and a reduction in the perturbation feedthrough from the ablation front to the shell's rear surface (which becomes unstable during the deceleration phase of the implosion).

There is a price to pay, however, for the greater stability. As the stronger shock propagates through the shell, it places the shell material on higher adiabat. This leads to a reduction in target gain and shell ρR, as shown in Fig. 2. A common practice in designing direct-drive targets is to find the delicate balance between reduction in the target performance due to an increase in the adiabat and the increase in shell stability.

In optimizing the target design, one can take into consideration that the RT modes are surface modes peaked at the ablation surface of the shell. Therefore, to reduce the instability growth, it is sufficient to raise the adiabat only at the outer region of the shell, which ablates during the implosion. If the inner portion of the shell is kept on a lower adiabat, the shell and vapor compressibility will not be reduced during the final stage of implosion, and the neutron yields will be unaffected by this selective adiabat increase (adiabat shaping).

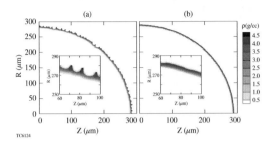

Figure 3. *Isodensity contours of the standard (a) and picket (b) target designs.*

The shell adiabat is shaped by launching a shock whose strength decreases as it propagates through the shell (Goncharov et al., 2003). This places an adiabat gradient directed toward the ablation front. Time variation in the shock strength is imposed by replacing the foot with an intensity picket in front of the main-drive pulse. The picket launches a strong shock that propagates through the shell. As the laser intensity drops at the end of the picket, the shocked material starts to expand and a rarefaction wave is launched toward the shock. After the rarefaction and the shock coalesce, the shock strength decays, reducing the adiabat of the shock-compressed material. Implementing the adiabat shaping to the direct-drive ignition target designs shows a significant improvement in shell stability, without compromising the target gain. This is illustrated in Fig. 3 where two target designs with a foot (a) and an intensity picket (b) in front of the main drive are shown at the end of the acceleration phase. The isodensity contours, shown in Fig. 3, are obtained using two-dimensional hydrocode $ORCHID$ (McCrory and Verdon, 1991). An improvement in the shell stability increases our confidence in success of the direct-drive ignition campaign on NIF.

5   Acknowledgements

This work was supported by the U.S. Department of Energy Office (DOE) of Inertial Confinement Fusion under Cooperative Agreement No. DE-FC03-92SF19460, the University of Rochester, and the New York State Energy Research and Development Authority. The support of DOE does not constitute an endorsement by DOE of the views expressed in this article.

References

Atzeni S and Meyer-Ter-Vehn J, 2004, *The Physics of Inertial Fusion* (Clarendon press, Oxford).

Betti R, Lobachev V, and McCrory R L, 1998, *Phys. Rev. Lett.* **81** 5560.

Betti R, Anderson K, Goncharov V N, et al., 2002, *Phys. Plasmas* **9** 2277.

Chandrasekhar S, 1961, *Hydrodynamic and Hydromagnetic Stability* (Clarendon, Oxford).

Goncharov V N, Betti R, McCrory R L at al., 1996, *Phys. Plasmas* **3** 1402.

Goncharov V N, Knauer J P, McKenty P W at al., 2003, *Phys. Plasmas* **10** 1906.

Kidder R E, 1974, *Nucl. Fusion* **14** 53.

Landau L D and Lifshitz E M, 1982, *Fluid Mechanics* (Pergamon, New York).

Lindl J D, 1998, *Inertial Confinement Fusion* (Springer, New York).

Manheimer W M and Colombant D G, 1982, *Phys. Fluids* **25** 1644.

McCrory R L and Verdon C P, 1991, in *Computer Applications in Plasma Science and Engineering*, editor Drobot A T (Springer-Verlag, New York).

Rosen M D, 1999, *Phys. Plasmas* **6** 1690.

Sanz J, 1994, *Phys. Rev. Lett.* **73** 2700.

19

Ablative Richtmyer-Meshkov Instability: Theory and Experimental Results

Valeri N. Goncharov

Laboratory for Laser Energetics, Department of Mechanical Engineering, University of Rochester, Rochester, New York, USA

1 Introduction

In inertial confinement fusion (ICF) a spherical shell filled with DT is compressed to very high densities and temperatures to achieve thermonuclear burn. An ignition target design in ICF relies on accurate control of the main fuel entropy and the asymmetry growth due to the hydrodynamic instabilities developed during the implosion (Atzeni and Meyer-Ter-Vehn, 2004). Both shell entropy and instability seeding are set during the early stage of implosion. The laser drive pulse for a direct-drive ignition target design (McKenty et al., 2001) consists of a low-intensity "foot" with $I \sim 2 - 7 \times 10^{13}$ W/cm^2 followed by the main drive pulse with peak intensity $\sim 10^{15}$ W/cm^2. The shock wave launched at the beginning of the foot pulse controls the shell entropy. The shell asymmetry evolution during the shock transit determines the initial conditions for the Rayleigh-Taylor (RT) instability (Chandrasekhar, 1961) developed at the ablation surface during the shell acceleration. Thus, an accurate modeling of plasma conditions at the early stage is crucial for designing a robust ignition target.

At the beginning of the drive pulse, the laser energy is absorbed mainly in the outer layer of the shell. Then, this layer heats up lunching a heat wave into the colder portion of the target. The heated material then expands outward, creating a mass flow (mass ablation) and the ablation pressure ("rocket effect") that drives a shock wave into the shell. The speed of the heat (ablation) front relative to the unablated material of the shell is referred to as an ablation velocity V_a. As the hot ablated plasma expands, a critical electron density $n_{cr} = \pi m c^2 / e^2 \lambda_L^2$ is established outside the cold portion of the shell. Here, m is the electron mass, c is the speed of light, e is the electron charge, and λ_L is the laser wavelength. The laser is absorbed in a region outside the critical surface where $n < n_{cr}$. Thus, a zone (conduction zone) is formed between the ablation front and where the laser energy is deposited.

The conduction zone size D_c is one of the main parameters which affect the early-time perturbation evolution in ICF implosions. Short wavelength modes with the wave numbers $kD_c > 1$ are stable and experience oscillatory behavior (Velikovich et al., 1998; Goncharov, 1999; Aglitskiy et al., 2001), while the long wavelengths with $kD_c < 1$ grow due to the Richtmyer–Meshkov (RM) (Ritchmyer, 1960) and the Landau–Darrieus (LD) instabilities (Landau and Lifshitz, 1982; Piritz and Portugues, 2003).

It is straightforward to show similarities between the instability drive in the classical RM configuration and an ICF foil. In the case of RM instability, a planar shock interacts with a corrugated interface between two fluids. The transmitted shock created after such an interaction becomes distorted, producing a pressure modulation in the shocked region. Such a modulation accelerates the interface ripple amplitude before vanishing on a time scale of the sound wave transit across the perturbation wavelength. For a rippled surface in an ICF target, the shock launched at the beginning of the pulse is also distorted. Resulting modulations in the pressure accelerate the ablation-front amplitude, similar to the classical RM case. The presence of thermal conduction in an ICF plasma, however, creates a restoring force (Sanz, 1994; Goncharov et al., 1996). Such a force is a consequence of a slightly enhanced heat flux near the tip of the ablation-front ripple (which is the closest to the energy-deposition region) due to larger temperature gradient in comparison with the gradient at the valley. An enhanced heat flux leads to an increase in the ablation and blowoff velocity and the dynamic pressure. Thus, the rocket effect becomes larger at the peaks and lower at the valleys; this leads to a restoring force.

The RM instability drive lasts only for a short period of time δt_p while a sound wave traverses a distance of the order of the perturbation wavelength. The dynamic overpressure force, on the other hand, is present at all times, stabilizing the perturbation growth after δt_p. To be precise, the restoring force totally stabilizes only the ablation-surface modes localized inside the conduction zone ($kD_c > 1$). Longer wavelengths are expected to grow exponentially with the growth rate $\gamma_L \simeq \sqrt{\rho_a/\rho_{bl}}kV_a$ (Landau and Lifshitz, 1982), where ρ_{bl} is the density of coronal plasma outside the critical surface.

In this lecture we present a model describing the perturbation evolution during the shock transit through the shell. The results of the model are compared against the experiments carried out on the OMEGA Laser System (Boehly et al., 1997).

2 One-dimensional flow

We begin with a description of the hydrodynamic profiles in the absence of asymmetries. For simplicity, we consider a planar foil of the initial thickness d_0. During the shock transit time across the foil, the following four regions can be identified: (I) $-d_0 < x < x_s$, the undriven foil with uniform density ρ_0 and pressure p_0; (II) $x_s < x < x_a$, the shock-compressed region with uniform density ρ_1 and pressure p_1; (III) $x_a < x < x_a + L_0$, the ablation region; and (IV) $x > x_a + L_0$, the blowoff plasma. Here, the x axis is pointing from the foil toward the blowoff plasma (shock travels in negative x direction), and x_s is the position of the shock front.

To relate the hydrodynamic quantities on both sides of the shock front, we use the ideal gas equation of state and the standard Hugoniot relations (Landau and Lifshitz, 1982). In the frame of reference moving with the shock front, the fluid velocities in regions I and II are $U_0^s = U_s = c_0\sqrt{\Pi(\gamma+1)/2\gamma+1}$, $U_1^s = \rho_0 U_0^s/\rho_1$, respectively. Here, $\Pi = p_1/p_0 - 1$ is the shock strength, U_s is the shock speed, γ is the ratio of specific heats, and $c_0 = \sqrt{\gamma p_0/\rho_0}$ is the sound speed in the undriven foil. The superscript "s" denotes quantities calculated in the shock-front frame of reference. The density jump across the shock takes the form

$$\frac{\rho_1}{\rho_0} = \frac{\Pi(\gamma+1)+2\gamma}{\Pi(\gamma-1)+2\gamma}. \tag{1}$$

Observe that the density ratio for a strong shock ($\Pi \gg 1$) asymptotes to a finite value $\rho_1/\rho_0 = (\gamma+1)/(\gamma-1)$.

3 Perturbation evolution: theory

In this section, we study stability of the planar foil with a corrugated interface at the laser side. The initial amplitude of the ripple is η_0 and the perturbation wave number is k. To clarify the physical mechanisms driving the perturbation growth, we start the analysis with a simplified problem, replacing the ablation region with a "classical" free surface, neglecting all the effects due to thermal conduction. Then, introducing the finite thermal conductivity into the analysis, we study the stabilizing and destabilizing mechanisms due to the ablative process.

3.1 Classical case without ablation

The analysis of the perturbation evolution is significantly simplified in the limit of $\gamma \to 1$ and $\Pi \gg 1$. We start our study with this case as the physical mechanism driving such an evolution becomes more transparent.

$\gamma \simeq 1$, strong-shock limit

In the limit $\gamma \simeq 1$, the density ratio across the shock front is very large [$\rho_1/\rho_0 \sim \Pi$ for $1/(\gamma-1) > \Pi \gg 1$ and $\rho_1/\rho_0 \sim 1/(\gamma-1)$ for $\Pi > 1/(\gamma-1)$, see Eq. (1)], and the post-shock velocity is much smaller than the shock velocity $U_1^s = \rho_0/\rho_1 U_s \ll U_s$. Then, the Mach number in region II can be considered as a small parameter of the problem $M_1 = U_1^s/c_1 \simeq \sqrt{U_1^s/U_s} \ll 1$, where $c_1 = \sqrt{\gamma p_1/\rho_1}$ is the sound speed in region II. If the shock front is rippled, a lateral velocity $\tilde{v}_y$ is created in the shock-compressed region [see Fig. 1(a)] due to the conservation of the velocity component tangential to the front.

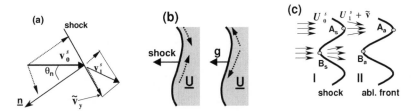

Figure 1. *(a) Fluid flow for an oblique shock in the shock-front frame of reference. (b) Direction of the lateral component of the fluid flow after the shock (left) is opposit to the direction in the RT case where the fluid is subject to a gravitational field g (right). (c) Sketch of the two-dimensional flow in the case of the rippled shock propagation. Shock frame of reference is used.*

It is easy to see from Figs. 1(a) and 1(b) that the mass in region II flows from ripple spikes into the bubbles. It is interesting to note that the direction of the established flow is opposite to the direction in the case of the RT instability where the mass flows from the bubbles into the spikes [Fig. 1(b)]. Because of this fact, the shock ripple is stable (D'yakov, 1954). Indeed, the flow in the post-shock region is incompressible [since the flow Mach number is small, $\partial_t \tilde{\rho}/\rho_1 \sim M_1^2 \nabla \tilde{\mathbf{v}}$] and, because of the mass flow into the bubbles, the shock velocity near the bubble is increased and decreased near the spikes. This reduces the shock amplitude η_s.

The ripple overshoots, however, the position of zero amplitude and starts to grow in the opposite direction. After this point the bubble becomes the spike and vice versa. The evolution, therefore, continues as a series of oscillations. Quantitatively, the equation describing the ripple can be found by considering the time derivative of the perturbed shock velocity. Since the shock front coincides with the material interface in the limit of $U_s \gg U_1^s$, the perturbed shock velocity is calculated as a velocity perturbation $\tilde{v}_x$ in region II taken at the shock front,

$$d_t \eta_s \equiv -\tilde{U}_s \simeq \tilde{v}_x(x_s). \tag{2}$$

Taking time derivative of Eq. (2) (in the shock-front frame of reference), with the help of the momentum conservation equation, yields

$$d_t^2 \eta_s = \partial_t \tilde{v}_x = -\partial_x \tilde{p}/\rho_1 - U_1^s \partial_x \tilde{v}_x. \tag{3}$$

The right-hand side of Eq. (3) contains the following two terms: the force term due to the pressure gradient behind the shock and the convective term. Effect of the convective term will be considered first. The spatial derivative of $\tilde{v}_x$ can be obtained from the incompressibility condition in region II, $\partial_x \tilde{v}_x \simeq -\partial_y \tilde{v}_y = -ik\tilde{v}_y$. Here, we used the Fourier form of the perturbed quantities $\tilde{Q}(x, y, t) = \tilde{Q}(x, t)e^{iky}$. The lateral component $\tilde{v}_y$, as mentioned earlier, is a result of the tangential velocity continuity.

Neglecting the unperturbed flow in the post-shock region II, the continuity condition gives $\tilde{v}_y \simeq U_s \sin \theta_n$, where θ_n is the angle of the vector $\mathbf{n}$ normal to the front [see Fig. 1(a)]. The shock-front ripple amplitude $\eta_s(y, t)$ is assumed to be small, $k\eta_s \ll 1$; thus, $\theta_n \simeq \partial_y \eta_s \ll 1$ and

$$\tilde{v}_y \simeq ik\eta_s U_s. \tag{4}$$

Keeping only the convective term, Eq. (3) reduces to $d_t^2 \eta_s + k^2 c_1^2 \eta_s = 0$, where we used the relation $U_s U_1^s = c_1^2$ valid for $\gamma \simeq 1$. This yields the ripple oscillations $\eta_s = A_s \sin \omega_s t + B_s \sin \omega_s t$ with the frequency $\omega_s = kc_1$, in agreement with the qualitative picture described earlier. The amplitude of such oscillations, however, is not constant in time. The amplitude decays as a consequence of the pressure force in Eq. (3). The pressure modulation at the shock front $\tilde{p}_s$ can be related to the perturbed shock velocity. Since $U_s \simeq \sqrt{p_1/\rho_0}$ for $\gamma \simeq 1$, we obtain

$$d_t \eta_s \equiv -\tilde{U}_s = -\frac{1}{2} \frac{\tilde{p}_s}{\rho_1 U_1^s}. \tag{5}$$

The normalized position $\hat{x} = kx$ changes inside region II from $\hat{x} = 0$ at the free surface to $\hat{x} = -M_1 \tau$ at the shock front (in the frame of reference moving with the post-shock fluid), where $\tau = \omega_s t$. Since $M_1 \ll 1$, we can expand $\tilde{p}(\hat{x}, \tau)$ in Taylor series near $\hat{x} = 0$, $\tilde{p}(\hat{x}, \tau) = \bar{p}_0(\tau) + \hat{x}\bar{p}_1(\tau) + O(\hat{x}^2)$. Using the boundary condition at the free surface $\tilde{p}(0, \tau) = 0$, the first term in the expansion goes to zero, $\bar{p}_0(\tau) = 0$. Thus, keeping only the first non-vanishing term, we write $\tilde{p} \simeq \hat{x}\bar{p}_1(\tau)$. The undetermined function $\bar{p}_1$ is defined by the boundary condition (5),

$$\bar{p}_1(\tau) = 2k\rho_1 c_1^2 \frac{d_\tau \eta_s}{\tau}. \tag{6}$$

Combining Eqs. (3), (4), and (6) the following equation for the shock ripple is derived:

$$d_\tau^2 \eta_s + \frac{2}{\tau} d_\tau \eta_s + \eta_s = 0. \tag{7}$$

Solution of the latter equation takes the form

$$\eta_s = \eta_0 \frac{\sin kc_1 t}{kc_1 t}, \tag{8}$$

where η_0 is the initial amplitude. Equation (8) shows that the ripple front amplitude decays in time as t^{-1} for $\gamma \simeq 1$ in the strong-shock limit.

The free-surface modulation of $\eta_a^{\rm cl}$ evolves according to

$$d_t^2 \eta_a^{\rm cl} = \partial_t \tilde{v}_x(0, t) = -\partial_x \tilde{p}/\rho_1. \tag{9}$$

With the help of Eqs. (6) and (7), the latter equation can be reduced to

$$d_\tau^2 (\eta_a^{\rm cl} - \eta_s) = \eta_s. \tag{10}$$

The solution of Eq. (10), taking into account the initial conditions $\eta_a^{\rm cl}(0) = \eta_0$ and $d\eta_a^{\rm cl}/dt(0) = 0$, takes the form

$$\eta_a^{\rm cl}(\tau) = \eta_s(\tau) + \int_0^\tau d\tau' \int_0^{\tau'} d\tau'' \eta_s(\tau''). \tag{11}$$

To find the asymptotic behavior of $\eta_a^{\rm cl}$ at $\tau \gg 1$, we use the limiting value of the double integral $(\tau \to \infty) \int_0^\tau d\tau' \int_0^{\tau'} d\tau'' \sin \tau''/\tau'' \to \pi\tau/2 - 1$ to obtain

$$\eta_a^{\rm cl}(kc_1 t \gg 1) = \eta_0 \left(\frac{\pi}{2} kc_1 t - 1 \right). \tag{12}$$

Thus, as in the classical RM case, the interface perturbation grows as a linear function of time.

To summarize, the shock ripple is stable and oscillates in time with a decaying amplitude. The oscillations are due to the mass flow from the perturbation spikes to the bubble behind the shock front. The amplitude decay is caused by the pressure force, which is proportional to the ripple velocity. The decaying rate depends on the ratio of specific heats γ and the shock strength. The oscillation amplitude decreases as t^{-1} for $\gamma \simeq 1$. When $\gamma > 1$, as shown in the next subsection, the decaying rate asymptotes to $t^{-1/2}$ for $M_s^4 > kc_1 t \gg 1$ and to $t^{-3/2}$ in the limit $kc_1 t \gg M_s^4$, where $M_s = U_s/c_0$ is the shock Mach number.

$\gamma > 1$, arbitrary shock strength

We start, as in the previous subsection, with deriving the boundary conditions at the rippled shock front. The lateral velocity is determined by the continuity of the velocity component tangential to the shock front,

$$\tilde{v}_y = (U_0^s - U_1^s)\partial_y \eta_s = ik\eta_s c_1 \frac{\Pi}{\gamma(\Pi + 1)} \frac{c_1}{U_1^s}. \tag{13}$$

Equation (13) shows that the flow behind the convex part of the shock front [point B_s in Fig. 1(c)] is divergent, and the flow behind the concave part [point A_s in Fig. 1(c)] is convergent. The convergent flow speeds up the shock and increases the pressure at point A_s. This results in the shock amplitude reduction and amplification of the free-surface amplitude. Indeed, because the shock speed depends on the shock strength, the velocity of the shock ripple is related to the pressure modulation as

$$d_t \eta_s \equiv -\tilde{U}_s = -\frac{dU_s}{d\Pi}\tilde{\Pi} = -\frac{\gamma + 1}{4} \frac{\tilde{p}_s}{\rho_0 U_s}. \tag{14}$$

Here, $\tilde{p}_s = \tilde{p}_1(x_s)$. In addition, the perturbed component $\tilde{v}_x$ and $\tilde{\rho}$ at the shock front can be calculated with the help of momentum conservation equations. The result is

$$\tilde{v}_x(x_s) = -\frac{M_s^2 + 1}{2M_s^2}\frac{\tilde{p}_s}{\rho_0 U_s}, \quad \tilde{\rho}_1(x_s) = \frac{\tilde{p}_s}{(M_s U_1^s)^2}. \tag{15}$$

At the free surface, the pressure is equal to the drive pressure p_1; thus, the pressure perturbation must vanish $\tilde{p}_1(x_a) = 0$. The modulation in p is finite at the shock front and zero at the free surface; the gradient in the perturbed pressure, therefore, creates a force that accelerates the free surface at point A_a and decelerates it at B_a [see Fig. 1(c)].

Next, the derived boundary conditions are applied to the solution of the wave equation for the pressure perturbation in region II

$$\partial_t^2 \tilde{p}_1 - c_1^2 \partial_x^2 \tilde{p}_1 + k^2 c_1^2 \tilde{p}_1 = 0. \tag{16}$$

Equation (16) is solved by introducing dimensionless variables (Zaidel, 1965) $r = \sqrt{\tau^2 - \hat{x}^2}$ and $\theta = \text{arctanh}(\hat{x}/\tau)$, where $\tau = kc_1 t$ and $\hat{x} = kx$. The solution of Eq. (16) bounded at $r \to 0$ takes the form $\tilde{p}_1/\rho_1 c_1^2 = k\eta_0 \sum_\mu (M_\mu \cosh \mu\theta + N_\mu \sinh \mu\theta) J_\mu(r)$, where $J_\mu(r)$ is the Bessel function. Constants M_μ and N_μ can be determined from the boundary conditions at the shock front and the free surface. After lengthy manipulations, the shock amplitude takes the form (Ishizaki and Nishihara, 1997)

$$\eta_s \simeq \eta_0 \left[J_0(r_s) + \frac{2(M_s^2 + 1)}{3M_s^2 + 1} J_2(r_s) \right], \tag{17}$$

where $r_s = \tau\sqrt{\Pi(\gamma + 1)/2\gamma(\Pi + 1)}$. Equation (17) shows that the shock amplitude decays as $t^{-1/2}$. As discussed in (2005), however, Eq. (17) is valid for $r_s < M_s^4$. At larger times, when $r_s \gg M_s^4$, the decaying rate changes to $t^{-3/2}$.

It can be shown that the free-surface amplitude, similar to the $\gamma \simeq 1$ case, asymptotically grows linearly with time,

$$\eta_a^{\text{cl}}(kc_1 t \gg 1) = \eta_0 \left(1 - C_0 + C_t kc_1 t\right), \tag{18}$$

where coefficients C_0 and C_t are defined in (2005).

3.2   Effects of mass ablation

To carry out stability analysis of the ablation front during the shock transit time we consider a sharp-boundary model, idealizing the ablation region by a surface of discontinuity. Since the shock travels ahead of the heat front, the wave equation in region II [Eq. (16)] and the Hugoniot relations [Eqs. (13)-(15)] are still valid in this case.

The boundary conditions at the ablation front, however, must be modified to account for the thermal conduction effects mentioned in Introduction. Such modifications were first derived (for the modes localized inside the conduction zone, $kD_c \gg 1$) by (1997). In particular, it was shown that the thermal conduction in plasma corona introduces a jump in the transverse velocity $\tilde{v}_x$, $[\![\tilde{v}_x + \tilde{\rho} V/\rho - kV\eta_a]\!] = 0$, where $[\![Q]\!] = Q^{\text{bl}} - Q_1$, $V(x < x_a) = V_a$ and $V(x > x_a) = V_{\text{bl}}$ is the equilibrium fluid velocity in the ablation-front frame of reference, V_a is the ablation velocity, V_{bl} is the blowoff velocity, η_a is the ablation-front perturbation, and superscript "bl" denotes quantities in the blowoff region.

Equation (16) with the appropriate boundary conditions at the shock and ablation front was solved in (Goncharov, 1999; Goncharov et al., 2005). The solution yields the following expression for the perturbation amplitude:

$$\frac{\eta_a}{\eta_0} = \eta_a^{\text{cl}}(t) - 1 + C_{(0)} - C_t kc_1 t + e^{-2kV_a t} \left(\alpha_0 \cos \omega t + \beta_0 \sin \omega t\right) + \eta_v(t), \tag{19}$$

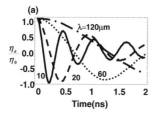
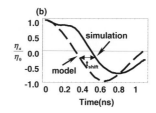

Figure 2. *(a) Ablation-front modulations calculated using Eq. (19) for different perturbation wavelengths. (b) Comparison of the ablation-front modulation.*

where $\eta_a^{\rm cl}$ is the perturbation amplitude in the absence of the mass ablation, $\eta_v(t) = (2c_1/V_{\rm bl})e^{kV_a t}\int_\infty^{kV_a t} e^{-\xi}\Omega_a(\xi)d\xi$, is the vorticity convection term, $\omega = k\sqrt{V_a V_{\rm bl}}$, Ω_a is the normalized vorticity $-i(\nabla \times \mathbf{v})_z/kc_1$ calculated at the ablation front, and the coefficients α_0 and β_0 are defined in (2005).

Next, we consider a DD foil driven by $I = 4 \times 10^{14}$ W/cm^2 laser pulse. The theoretical prediction for the ablation-front amplitude in this case is shown in Fig. 2(a).

Observe an oscillation frequency increase with the perturbation wave number. The result of the numerical simulation performed using hydrocode $ORCHID$ (McCrory and Verdon, 1991) for the same conditions is plotted in Fig. 2(b). A detailed comparison between theory and simulations shows a time shift in the ablation amplitude evolution calculated using the hydrocode results. Such a discrepancy is due to a violation of the model validity condition at the beginning of the pulse. It takes some time for the conduction zone to grow to satisfy $kD_c > 1$.

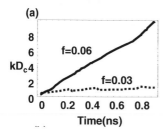
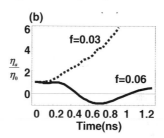

Figure 3. *Conduction zone size (a) and ablation-front amplitude (b) calculated using different flux-limiter values.*

The result presented in Fig. 2 indicates that the rate of the conduction zone growth is an important parameter characterizing the perturbation evolution at the ablation front. In the example considered in this section, the perturbation modes become localized inside the conduction zone early in the pulse (a few hundred picoseconds after the pulse start). The perturbations in this case are stable and experience oscillatory behavior as predicted by Eq. (19). In the opposite limit, when the condition $kD_c \ll 1$ is satisfied for a time much longer than a sound wave transit across the perturbation wavelength, the modes are expected to be unstable due to the LD instability.

This instability condition can be satisfied, for example, by reducing the thermal flux from the laser absorption region toward the colder shell. The flux in a hydrosimulation is controlled by a "flux limiter" coefficient f (Malone et al., 1975), which limits the Spitzer flux (Spitzer, 1953) $q_{\rm SH} = -\kappa\nabla T$ to $fq_{\rm FS}$, where $q_{\rm FS} = nTv_T$ is the free-stream limit, κ is thermal conductivity, T

is the electron temperature, n is the electron density, $v_T = \sqrt{T/m}$, and m is the electron mass.

Figure 3(a) plots the conduction zone size for the DD foil driven by $I = 2 \times 10^{14}$ W/cm² laser pulse using two flux-limiter values, $f = 0.03$ and $f = 0.06$. For the reduced flux limiter, the perturbation wavelength ($\lambda = 20\mu m$) is always larger than D_c and perturbations grow exponentially [see Fig. 2(b)]. For the larger flux, the modes become localized inside the conduction zone early in the pulse and perturbation amplitudes are stable, oscillating in time. Since the growth of D_c is very sensitive to the thermal transport, benchmarking the simulation results against the experimental measurement of the perturbation evolution during the shock propagation is a good test of thermal conduction modeling used in hydrocodes.

4   Perturbation evolution: experiment

To test the theoretical predictions for the perturbation evolution, a series of early-time perturbation measurements (Gotchev et al., 2005) was carried out on the OMEGA Laser System. The CH foil was driven by the laser pulse with the maximum intensity of $\sim 4 \times 10^{14}$ W/cm². The foil was radiographed with x rays produced by a U foil. The backlighter signal then was imaged with the high-resolution, large-angle Kirkpatrick–Baez microscope, coupled to high-current streak camera (Gotchev et al., 2004).

Comparison of the experimental results and simulations is shown in Fig. 4(a) where the

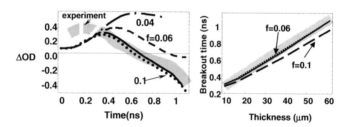

Figure 4. (a) Optical depth modulation of 40μm-thick CH foil. (b) Shock breakout time vs. foil thickness. Thick gray line shows the experimental data, thin lines correspond to hydrosimulations.

optical depth (defined as logarithm of the image intensity) modulation is plotted for the 20-μm perturbation.

Different values of the flux limiter f were used in these simulations. An increase in the flux limiter results in an increased conduction zone size D_c. The simulation with $f = 0.1$ (dotted line) gives the fastest growth in D_c and the earliest phase reversal, in good agreement with the experimental data. Such a large value of the flux limiter, however, is not consistent with the shock timing measurements presented in Fig. 4(b), where the shock breakout time is plotted for different foil thicknesses.

Thus, the RM and shock velocity measurements indicate that a single flux limiter-model cannot explain both experiments since different flux-limiter values fit different experimental data. A new transport model was proposed by (2005). The model solves a simplified Boltzmann equation to obtain an analytic expression for the heat flux. The results of the model are in good agreement with the shock velocity measurements [see Fig. 4(b), solid line].

To apply the results of the nonlocal transport model to two-dimensional hydrosimulations, an effective flux limiter was obtained by taking the maximum ratio of the heat flux in the laser-deposition region calculated using the model to the free-stream value q_{FS}. The result yields a time-dependent value for the flux limiter. The flux limiter decays from $f = 0.1$ down to $f = 0.05$ at the end of the pulse. Then, substituting these values into the hydrocode, the optical depth modulation is calculated and the results are plotted in Fig. 4(a) with a solid line. The good agreement between the simulation and experimental data for both shock timing and perturbation evolution increase our confidence in thermal transport modeling of an ICF implosion.

In conclusion, an analytical model for the perturbation evolution during the shock transit time has been developed. A comparison of the model results with the numerical simulations reveals importance of the conduction zone size D_c. The modes with $kD_c \gg 1$ are totally stabilized by the dynamic overpressure created by the ablation flow. In the opposite limit $kD_c \ll 1$, the modes experience exponential growth due to the LD instability. To match the simulation results against the experimental measurements, a new nonlocal thermal transport model was developed and applied to hydrocodes.

5   Acknowledgements

This work was supported by the U.S. Department of Energy Office (DOE) of Inertial Confinement Fusion under Cooperative Agreement No. DE-FC03-92SF19460, the University of Rochester, and the New York State Energy Research and Development Authority. The support of DOE does not constitute an endorsement by DOE of the views expressed in this article.

References

Aglitskiy Y, Velikovich A L, Karasik M et al., 2001, *Phys. Rev. Lett.* **87** 265001.

Atzeni S and Meyer-Ter-Vehn J, 2004, *The Physics of Inertial Fusion* (Clarendon press, Oxford).

Baker L M and Hollenbach R E, 1972, *J. App. Phys.* **43** 4669.

Boehly T R, Brown D L, Craxton R S et al., 1997, *Opt. Commun.* **133** 495.

Chandrasekhar S, 1961, *Hydrodynamic and Hydromagnetic Stability* (Clarendon, Oxford).

D'yakov S P, 1954, *Zh. Eksp. Teor. Fiz.* **27** 288.

Goncharov V N, Betti R, McCrory R L at al., 1996, *Phys. Plasmas* **3** 1402.

Goncharov V N, 1999, *Phys. Rev. Lett.* **82** 2091.

Goncharov V N, Gotchev O V, Vianello E et al., 2005, submitted to *Phys. Plasmas.*

Gotchev O V, Jaanimagi P A, Knauer J P et al., 2004, *Rev. Sci. Instrum.* **75** 4063.

Gotchev O V, 2005, *Ph.D. thesis* (University of Rochester, Rochester).

Ishizaki R and Nishihara H, 1997, *Phys. Rev. Lett.* **78** 1920.

Landau L D and Lifshitz E M, 1982, *Fluid Mechanics* (Pergamon, New York).

Malone R C, McCrory R L, and Morse R L, 1975, *Phys. Rev. Lett.* **34** 721.

McCrory R L and Verdon C P, 1991, in *Computer Applications in Plasma Science and Engineering*, editor Drobot A T (Springer-Verlag, New York).

McKenty P W, Goncharov V N, Town R P J et al., 2001, *Phys. Plasmas* **8** 2315.

Piriz A R, Sanz J, and Ibanez L F, 1997, *Phys. Plasmas* **4** 1117.

Piriz A R and Portugues R F, 2003, *Phys. Plasmas* **10** 2449.

Richtmyer R D, 1960, *Commun. Pure Appl. Math.* **13** 297.

Sanz J, 1994, *Phys. Rev. Lett.* **73** 2700.

Velikovich A L, Dahlburg J P, Gardner J H, and Taylor R J, 1998, *Phys. Plasmas* **5** 1491.

Spitzer L and Härm R, 1953, *Phys. Rev.* **89** 977.

Zaidel P M, 1965, *J. Appl. Math. Mech.* **24** 316.

Index

Printed and bound by CPI Group (UK) Ltd, Croydon, CR0 4YY

24/10/2024

01778719-0001